Ceramic Processing Science

Related titles published by the American Ceramic Society:

Innovative Processing and Synthesis of Ceramics, Glasses, and Composites (Ceramic Transactions Volume 85)
Edited by Narottam P. Bansal, Kathryn V. Logan, and J.P. Singh
© 1997, ISBN 1-57498-030-0

Advanced Synthesis and Processing of Composites and Advanced Ceramics II (Ceramic Transactions Volume 79)
Edited by Kathryn V. Logan, Zuhair Munir, and Richard M. Spriggs
© 1996, ISBN 1-57498-017-3

Microwaves: Theory and Application in Materials Processing III (Ceramic Transactions Volume 59)
Edited by David E. Clark, Diane C. Folz, Steven J. Oda, and Richard Silberglitt
© 1995, ISBN 1-57498-002-5

Advanced Synthesis and Processing of Composites and Advanced Ceramics (Ceramic Transactions Volume 56)
Edited by Kathryn V. Logan
© 1995, ISBN 1-57498-000-9

For information on ordering titles published by the American Ceramic Society, or to request a publications catalog, please contact our Customer Service Department at 614-794-5890 (phone), 614-794-5892 (fax), <customersrvc@acers.org> (e-mail), or write to Customer Service Department, 735 Ceramic Place, Westerville, OH 43081. Visit our on-line book catalog and web site at <www.acers.org>.

Ceramic Transactions
Volume 83

Ceramic Processing Science

Edited by

Gary L. Messing
The Pennsylvania State University

Fred F. Lange
University of California at Santa Barbara

Shin-ichi Hirano
Nagoya University

Published by

The American Ceramic Society
735 Ceramic Place
Westerville, Ohio 43081

Proceedings of Ceramic Processing '97 held in Santa Barbara, California, September 7–10, 1997.

COVER PHOTO: "SEM photograph of the obtained fibers with various seed contents at 0 wt%," is courtesy of Atsuya Towata, Hae-Jin Hwang, Masaki Yasuoka, Mutsuo Sando, and Koichi Niihara, and appears as figure 4(a) in their paper, "Synthesis of Al_2O_3/YAG Composite Fibers," which begins on page 407.

Library of Congress Cataloging-in-Publication Data
A CIP record for this book is available from the Library of Congress.

For information on ordering titles published by The American Ceramic Society, or to request a publications catalog, please call 614-794-5890.

Printed in the United States of America.

1 2 3 4–01 00 99 98

ISSN 1042-1122
ISBN 1-57498-056-4

Contents

Colloid Dispersion and Consolidation

Films, Coatings, and Fibers

Microstructures of Complex Ceramics

Preface

This volume contains papers presented at the Ceramic Processing '97 meeting that was held in Santa Barbara, CA, Sept 7–10, 1997 under the auspices of The American Ceramic Society, the Ceramic Society of Japan, the European Ceramic Society, and supported by the Materials Research Laboratory at UC Santa Barbara. This was the sixth meeting of the International Conference on Ceramic Processing Science and was hosted by Fred Lange and co-organized with Shin-ichi Hirano and Gary L. Messing.

There were nine invited papers and 117 contributed papers. About half of the papers are published here and, where complete papers were not submitted, an abstract is included to record the technical content of the meeting.

As in past volumes originating from the Ceramic Processing Science series, this book reports the most recent developments in ceramic processing. This meeting focused on three main topics: colloidal powder processing, chemical solution routes to ceramics, and processing routes for unique microstructures.

A number of people assisted with the review and editing of papers including Matt Seabaugh, Byung-Sei Jun, Mike Mandanas, Huseyin Yilmaz, Bhaskar Brahmaroutu, Ed Sabolsky, Cihangir Duran, Jeff Wei, Paul Rehrig, Koji Watari, Ender Suvaci, and Rob Carneim. We are particularly grateful for the excellent assistance of Wendy Bathgate in coordinating the review process.

We look forward to seeing you at the next meeting in Nagoya in 2000.

G.L. Messing, K.S. Mazdiyasni, J.W. McCauley, and R.A. Haber, eds., *Ceramic Powder Science*, American Ceramic Society, Columbus, OH, 825 p. (1987).

G.L. Messing, E.R. Fuller, Jr., and H. Hausner, eds., *Ceramic Powder Science II*, American Ceramic Society, Columbus, OH 1207 p. (1988).

H. Hausner, G.L. Messing, and S. Hirano, eds., *Ceramic Powder Processing Science*, Deutsche Keramische Gesellschaft, Cologne, Germany, 960 p. (1989).

G.L. Messing, S. Hirano, and H. Hausner, eds., *Ceramic Powder Science III*, American Ceramic Society, Westerville, OH, 995 p. (1990).

S. Hirano, G.L. Messing, and H. Hausner, eds., *Ceramic Powder Science IV*, American Ceramic Society, Westerville, OH, 758 p. (1991).

H. Hausner, G.L. Messing, and S. Hirano, eds., *Ceramic Processing Science and Technology*, American Ceramic Society, Westerville, OH, 820 p. (1995).

Gary L. Messing
Fred Lange
Shin-ichi Hirano

Ceramic Synthesis

SYNTHESIS OF NANO-CERAMICS WITH POLYMERS

Shin-ichi Hirano
Department of Applied Chemistry
Nagoya University
Japan

ABSTRACT

Nano-ceramics/polymer hybrids have been receiving great attention for applications in emerging areas. The conventional blending of mm-particles into oligomer or polymer generally causes the particle agglomeration, leading to the degradation of the desired properties. The newly developed processing is based on the nucleation and growth of crystalline oxide particles in organic matrix through the hydrolysis and polymerization control of metallorganic molecules with vinyl group below 100°C. Nano-sized crystalline iron oxide particles/polymer hybrids could successfully be synthesized below 100°C using a designed iron-bearing molecule precursor, which is susceptible to hydrolysis and polymerization. The hybrids exhibit the typical super paramagnetic properties as well as quantum effect. The control of the hydrolysis and polymerization reaction was found to influence strongly the microstructure of the hybrids. Nano-sized $BaTiO_3$, $PbTiO_3$ and $MgAl_2O_4$ particles were also crystallized in polymer matrix under controlled conditions The structure of metallorganic molecule was found to be an important key factor to synthesize crystalline nano-ceramics a low temperature and bind the particles chemically with polymer matrix. This novel processing is of practical use for introducing a new family.

THE STRUCTURE AND SURFACE PROPERTIES OF HYDROLYSATES DERIVED FROM CHEMICALLY-MODIFIED TITANIUM ALKOXIDES

Paul A. Venz[1], James L. Woolfrey[2], John R. Bartlett[2] and Ray L. Frost[1]

[1] Centre for Instrumental and Developmental Chemistry, Queensland University of Technology, 2 George Street, Brisbane, QLD, 4001, Australia.

[2] Materials Division, Australian Nuclear Science and Technology Organisation, Private Mail Bag 1, Menai, NSW, 2234, Australia.

ABSTRACT

The chemical modification of alkoxides can be used to control their reactivity during sol-gel processing, by affecting hydrolysis and condensation reactions. However, only limited results has been published on their effect on hydrolysate properties.

The composition, structure and surface properties of hydrolysates and sols produced from tetraisopropyltitanate modified with carboxylic acids ($C_nH_{2n+1}COOH$, where $n = 1, 2$ and 3) have been investigated by Raman spectroscopy, electrokinetic sonic analysis and titration.

The concentration of residual carboxylate species in the as-washed hydrolysates increased with increasing n. The relative abundance of terminal and bridging hydroxyls in the hydrolysates also varied with n, and the total concentration of surface hydroxyls increased with increasing n. Hydrolysates exhibited iso-electric points (IEP) at pH 5.9 to 7.3 for $n<3$. No IEP was observed when $n \geq 3$, and the corresponding hydrolysates exhibited a negative surface charge at pH > 2, attributed to sorbed carboxylate-species.

Hydrolysates initially formed at < 40 °C were amorphous oxy-hydroxides, while those initially, or subsequently, processed at higher temperatures crystallised to anatase. The rate of crystallisation at a given temperature decreased with increasing n.

Such drastic changes in surface properties would affect subsequent processing methods, including sol-gel or colloidal processing, electrophoretic deposition, *etc.*

INTRODUCTION

The conditions generally used in colloidal sol-gel processing initially produce oxide or hydroxide precipitates (hydrolysates) [1,2]. These precipitates may be used as powders, or de-aggregated by the addition of acid or base (peptisation), producing colloids (sol). The sol may be further processed to produce particles, coatings or monolithic solids. Modification of alkoxides by nucleophilic substitution of chemical species into the coordination sphere of the metal atom (*e.g.* carboxylic acids) can be used to control the reactivity of the alkoxide [3,4,5], affecting subsequent hydrolysis and condensation reactions. Only limited results have been published on their effect on hydrolysate properties.

The composition, structure and surface properties of hydrolysates and sols produced from a titanium alkoxide (tetraisopropyltitanate, TPT) modified with carboxylic acids ($C_nH_{2n+1}COOH$, where $n = 1$, 2 and 3) have been investigated by Raman spectroscopy, electrokinetic sonic analysis and titration. The effect of alkoxide modification on the composition, structure and surface properties of the hydrolysates and sols produced, and on subsequent processing methods is discussed.

EXPERIMENTAL

Pure TPT and TPT/$C_nH_{2n+1}COOH$ mixtures (1:1 mole ratio, $n = 1,2$, and 3) were added to water at 298 K (50 moles H_2O per mole of TPT). The resulting hydrolysates were washed with water, and the pH of the slurries adjusted to ~2 by addition of small quantities of 4 M HNO_3. The acidic slurries were analysed by electrokinetic sonic analysis (ESA, Matec MBS-800) and potentiometric titration (Radiometer Automatic Titrator) over the pH range from 2 to 12. The pH of the slurries was adjusted during titration by addition of 0.5 or 1.0 cm^3 aliquots of dilute NaOH, at five-minute intervals.

Raman spectra of the as-prepared and titrated hydrolysates were obtained with a Biorad FT-Raman II spectrometer, using a laser power of 500 mW, and an instrument resolution of 4 cm^{-1}. Dried hydrolysate powders were characterised by XRD using a D500 Diffraktometer (Siemens), with Co K_α radiation.

RESULTS AND DISCUSSION

Surface Effects

XRD powder patterns of the dried, as-washed hydrolysates showed only broad, diffuse peaks, suggesting that the hydrolysates were amorphous. In contrast, the corresponding Raman spectra exhibited bands at ~200, 440 and 570 cm^{-1}, which have been attributed to amorphous titania, or a highly-disordered structure with a small degree of short-range, "rutile-like" ordering [6]. In addition, bands were observed in the Raman spectra at 1416, 1455 and 1540 cm^{-1}, and assigned to residual sorbed carboxylate species with bidentate and chelating coordination geometries [6], Figure 1. The concentration of residual surface carboxylates which could not be removed by water washing increased with increasing ***n***.

The electrokinetic properties of transition metal oxides are determined largely by the speciation of surface hydroxyl sites [7]. Such hydroxyls are amphoteric, and depending on the ambient pH, can act either as bases (generating positively-charged sites):

$$\gtrdot Ti{-}OH \quad + H^+ \quad \rightleftharpoons \quad \gtrdot Ti{-}OH_2^+ \quad + H_2O$$

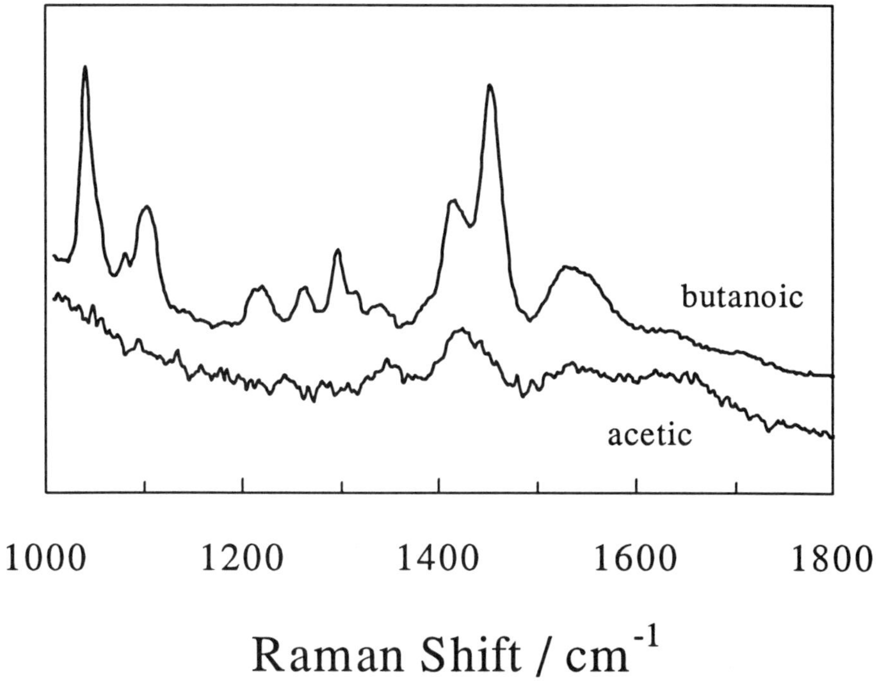

Figure 1. Raman spectra of residual carboxylate species in washed hydrolysates.

or acids (generating negatively-charged surface sites) [1,9,10,11,12]:

$$\equiv Ti{-}OH \quad + [OH]^- \rightleftharpoons \equiv Ti{-}O^- \quad + H_2O$$

Titration of the hydrolysates with dilute NaOH revealed three acidic surface species [14]:

$$\equiv Ti{-}OH_2^+ \quad + [OH]^- \rightleftharpoons \equiv Ti{-}OH \quad + H_2O \quad (pK_1)$$

$$\equiv Ti{-}O(H){-}Ti\equiv \quad + [OH]^- \rightleftharpoons \equiv Ti{-}O^-{-}Ti\equiv \quad + H_2O \quad (pK_2)$$

$$\geqslant Ti{-}OH \quad + [OH]^- \rightleftharpoons \; \geqslant Ti{-}O^- \quad + H_2O \quad (pK_3)$$

The corresponding dissociation constants are summarised in the Table I.

Table I. Surface Properties of titania hydrolysates.

Hydrolysate[1]	pK_1	pK_2	pK_3	IEP[2,3]	moles -OH per mole Ti		
					Bridging	Terminal	Total
Unmodified	2.9	6.3	8.6	5.9	0.08	0.20	0.28
Acetic	2.7	6.1	9.4	6.9	0.47	0.16	0.63
Propanoic	2.8	6.3	9.4	nd	0.43	0.11	0.54
Butanoic	2.7	6.0	9.4	nd	0.79	0.08	0.87

[1] refer to hydrolysates prepared from unmodified TPT, and from TPT modified with acetic, propanoic and butanoic acid, respectively.
[2] IEP, iso-electric point
[3] nd, no IEP was detected

The total concentration of surface hydroxyls increased with increasing *n*, while the relative abundance of bridging and terminal hydroxyls also varied with *n*.

With such profound changes in the abundance of surface hydroxyls, it was not surprising that the presence of residual carboxylate species had a significant effect on the electrokinetic properties of the hydrolysates. Typical variations in the electrophoretic mobility of the hydrolysates with pH are illustrated in Figure 2, while the IEP's of the hydrolysates are summarised in Table I. Modified hydrolysates with $n<3$ exhibited IEP's at pH 5.9 to 7.3. In contrast, no IEP was observed when $n \geq 3$; such hydrolysates exhibited a negative surface charge at pH $\geq$ 2. These effects were attributed to sorbed carboxylate species[13], which increase in abundance with increasing chain length, Figure 1.

The value of pK_3 for hydrolysates produced from pure TPT is significantly different to the corresponding values for modified precursors, Table I, suggesting that the presence of residual carboxylate species in the hydrolysate (see above) decreases the acidity of the terminal hydroxyl groups. Two explanations for this effect have been proposed [14]:

- the formation of hydrogen bonds between the terminal hydroxyl species and carboxylate species would be expected to decrease the strength of the O-H bond. However, this would be expected to increase the acidity of the hydroxyl species, leading to a corresponding decrease in pK_3;
- in contrast, steric hindrance of the terminal hydroxyl species by surface carboxylates would be expected to hinder access to the hydroxyl site, leading to a corresponding decrease in the effective acidity of the site, and an increase in pK_3.

The experimental observations favour the latter explanation.

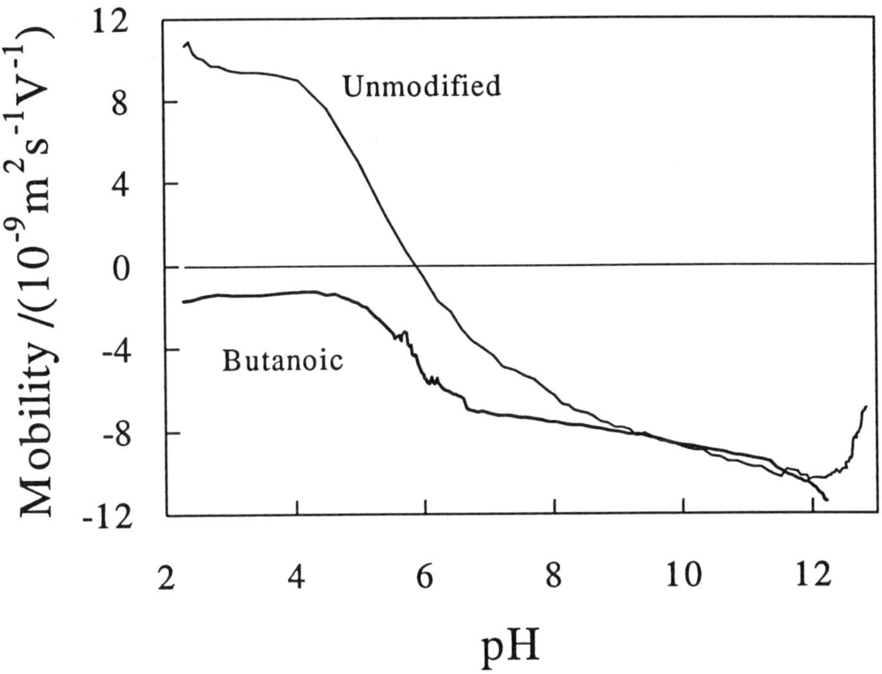

Figure 2. Electrokinetic properties of titania hydrolysates obtained from unmodified TPT and TPT modified with butanoic acid.

Crystallisation in Solution

The chain length of the carboxylic acid in the modified precursors significantly influenced the rate at which changes occurred in the anatase bands of the corresponding hydrolysate spectra during peptisation with HNO_3, or heating in water, at temperatures > 40 °C [15]. All hydrolysate systems exhibited a characteristic time, t_c, after which no further appreciable changes were observed in component band position, intensity or the full-width at half-maximum of the band, Table II.

The value of t_c increased monotonically with the length of the carboxylic acid's hydrocarbon chain, with that of the TPT hydrolysate midway between the values for the corresponding acetic and propanoic modified systems. It was concluded that unreacted carboxylate species retained within the hydrolysate structure after hydrolysis and condensation have a significant influence on the rate of crystallisation of the hydrolysates.

Table II. Effect of carboxylate chain length on the crystallisation kinetics at 60 °C

Precursor	t_c (minutes)	
	Samples peptised with HNO_3	Samples heated without HNO_3
Unmodified	180	≥360
Acetic	120	360
Propanoic	240	≥360
Butanoic	>420	>>420

t_c is the time to the end of crystallisation

Hydrolysate crystallisation can be viewed as the result of two key processes, both of which may be influenced by the presence of residual carboxylate species:

(a) A thermally-induced structural transformation; the effect of carboxylic acids on this process is shown in Table II. Crystallisation of the hydrolysates occurred very slowly at ambient temperature (requiring a period of several weeks), but at 60 °C it occurred within hours, regardless of whether or not HNO_3 was present [15].

(b) The condensation (olation, alcoxolation and/or oxolation) reactions involving hydroxyl groups in the hydrolysates [1,3].

Evidence that these reactions are involved in hydrolysate crystallisation is provided by the marked effect of acid catalysis on the rate of crystallisation during peptisation, when compared to systems heated without HNO_3, Table II. In the presence of HNO_3, hydroxyl groups on the hydrolysate surfaces are protonated by the acid, enhancing the rate of condensation [1,3] and, apparently, crystallisation. In addition, the carboxylic acid modifiers influenced the proportions in which the two predominant OH-species were present, Table I, and therefore the rate and extent of condensation during peptisation.

Effect on Sol-Gel Processing

Changes in surface properties would affect subsequent processing methods, such as sol-gel or colloidal processing, electrophoretic deposition, etc. In normal colloidal systems, the presence of an IEP means that the surface charge can be manipulated by changing the pH:

- The strength of the charge can be controlled by the deviation from the IEP and is useful in sol-gel processing to enhance the rate of peptisation of the hydrolysate to form sols, using either acidic or basic solutions.
- In addition, an approach to the IEP can be used to destabilise and coagulate the colloidal system, promoting gelation.
- Charge manipulation can also be used to promote deposition at a selected electrode during electrophoretic deposition.

In contrast, in modified hydrolysates where $n \geq 3$, no IEP is observed and the hydrolysates exhibited a negative surface charge at $pH \geq 2$, due to sorbed carboxylate species. Such behaviour would restrict the choices available during subsequent sol-gel processing.

CONCLUSIONS

The effect of alkoxide modification on the composition, structure and surface properties of the hydrolysates and sols produced from tetraisopropyltitanate (TPT) modified with carboxylic acids ($C_nH_{2n+1}COOH$, where $n = 1$, 2 and 3) has been investigated by Raman spectroscopy, electrokinetic sonic analysis and titration:

- Initial hydrolysates were amorphous but crystallise with a short-range order similar to that of anatase.
- Residual carboxylate species were retained in the as-prepared hydrolysates; the concentration of carboxylates increased with increasing n.
- The relative abundances of terminal and bridging hydroxyls varied with n, and the total concentration of surface hydroxyls increased with increasing n.
- Hydrolysates exhibited iso-electric points (IEP) at pH 5.9 to 7.3 for $n \leq 3$. In contrast, no IEP was observed when $n \geq 3$; such hydrolysates exhibited a negative surface charge at pH $\geq$ 2, due to sorbed carboxylate species.
- During heating of amorphous hydrolysates at > 40 °C, an anatase sol was formed, but crystallisation occurred at a faster rate in the presence of HNO_3.
- The carboxylic acid species also influenced the rate and extent of condensation during peptisation, and the rate at which the hydrolysates crystallised to anatase.

Such drastic changes in surface properties would affect subsequent processing methods, such as sol-gel or colloidal processing, electrophoretic deposition, etc.

ACKNOWLEDGEMENTS

The Centre for Instrumental and Developmental Chemistry, the Research Centre of the School of Chemistry, of the Queensland University of Technology is gratefully acknowledged for financial support for this project. One of the authors (PAV) extends sincere thanks to The Australian Institute for Nuclear Science and Engineering (AINSE) for both financial and in-kind support.

REFERENCES

1 C.J. Brinker and G.W. Scherer, *Sol-Gel Science*, Academic Press, New York, 1990.

2 D. Segal, *Chemical Synthesis of Advanced Ceramic Materials*, Cambridge University Press, Cambridge, 1989.

3 J. Livage, M. Henry and C. Sanchez, Prog. Solid State Chem., 18 [4] 259 (1988).

4 S. Doeuff, M. Henry, C. Sanchez and J. Livage, J. Non-Cryst. Solids, 89 206 (1987).

5 M.J. Percy, J.R. Bartlett, J.L. Woolfrey, L. Spiccia and B.O. West, "The Influence of Acetylacetone During the Hydrolysis of Zirconium (IV) Propoxide", Proceedings of the Second International Symposium on Sol-Gel Science & Technology, Cairns. July, 1996, Trans. Amer. Ceram. Soc.,81 *in press* (1998).

6 P. A. Venz, R. L. Frost, J.R. Bartlett and J.L. Woolfrey, Spectrochim. Acta Part A, 53 969 (1997).
7 R.J. Hunter, *Zeta Potential in Colloid Science* - Principles and Applications, Academic Press, New York, 1981.
8 R.J. Hunter, *Zeta Potential in Colloid Science* - Principles and Applications, Academic Press, New York, 1981.
9 C. Contescu, V.T. Popa, J.B. Miller, E.I. Ko, E.I. and J.A. Schwarz, J. Catal. 157 244 (1995).
10 N. Spanos, I. Georgiadou and A. Lycourghiotis, J. Colloid Interface Sci. 172 374 (1995).
11 A. Mpandou and B. Siffert, J. Colloid Interface Sci. 102 138 (1984).
12 C. Ludwig and P.W. Schindler, J. Colloid Interface Sci. 169 284 (1995).
13 P.A. Venz, J.R. Bartlett, R.L. Frost and J.L. Woolfrey, "Electrokinetic Studies of Titania Derived from a Chemically-Modified Titanium Alkoxide", Proceedings of the Second International Symposium on Sol-Gel Science & Technology, Cairns. July, 1996, Trans. Amer. Ceram. Soc.,81 *in press* (1998).
14 P.A. Venz, J.R. Bartlett, R.L. Frost and J.L. Woolfrey, "Electrokinetic Studies of Titania Derived from a Chemically-Modified Titanium Alkoxide", Proceedings of the Second International Symposium on Sol-Gel Science & Technology, Cairns. July, 1996, Trans. Amer. Ceram. Soc.,81 *in press* (1998).
15 P.A. Venz, J.R. Bartlett, R.L. Frost and J.L. Woolfrey, "Crystallisation of Titania Hydrolysates During Peptisation", Proceedings of the Second International Symposium on Sol-Gel Science & Technology, Cairns. July, 1996, Trans. Amer. Ceram. Soc.,81 *in press* (1998).

PREPARING SUPERFINE TiO_2 OF PTC CERAMICS BY HYDROLYSIS

Zan Ling, Zhong Jiao-shen, Luo Qi-rong, and Zao Wei
Adv. Materials Laboratory
Wuhan University
Wuhan, China 430072

ABSTRACT

In this paper, we reported a route to prepare superfine TiO_2 of PTC ceramics from the titanic ($TiO_2.nH_2O$) that is middle product of producing titanium white from ilmenite. We studied the hydrolysis processes of titanyl nitrate solutions. A new instrument that can recorded the preparing processes of superfine particles from the homogeneous solutions which lead to the solid particles is designed and used in this experiment. The influence of PH, concentration of reactant, anions present on the characteristic (size, crystalline) of final products was discussed. In this experiment, we obtained the superfine TiO_2. The particle size was about 100 nm to 200 nm measured by TEM. The rutile concentration of product was more than 90% calcining at 820°C. The impurities of Fe, Al were about 20ppm. Adding trace other anions, we can obtain the different product, the small particles (about 100 nm) tend to aggregate and format the larger spheres of uniform size containing a huge number of submits that are about 1um. The characteristic of PTC ceramic cells made by this TiO_2 and $BaCO_3$ were measured.

ENZYMATICALLY ASSISTED HOMOGENEOUS PRECIPITATION OF ALUMINA PRECURSORS

Hidero Unuma, Shinichi Kato, Toshitaka Ota, and Minoru Takahashi
Ceramic Research Laboratory
Nagoya Institute of Technology
Asahigaoka, Tajimi 507, Japan

ABSTRACT

Spherical aluminum hydroxide aggregates were precipitated at room temperature by a modified homogeneous precipitation technique involving enzymatic decomposition of urea. The rate of the nucleation of the precipitates was dependent on the urea concentration. The mean diameter of the aggregates was controllable in a range from 0.3 to 2.0 microns. In contrast to the conventional technique, the precipitation of pseudoboemite could be avoided. The precipitates transformed to gamma-and alpha-alumina after heat-treatments at 1273 and 1473K, respectively.

MORPHOLOGICAL DEVELOPMENT AND MODELING DURING AEROSOL GENERATION OF ZNO

O. Milosevic, M. Mirkovic, J. Vodnik*, N. G. Nakazey,** and T. V. Tomila**
Inst. Tech. Sci. of Servian Academy of Science and Arts
Belgrade, Yugoslavia

*Inst. Nuclear Sci., "Vinca"
Belgrade, Yugoslavia

**Inst. for Problems in Mat. Sci., NASU
Kieve, Ukraine

ABSTRACT

Spray pyrolysis, as a dispersion phase powder processing method, is applied In zinc oxide sub-micron to nanosized powder synthesis. The process involves aerosol formation from the precursor nitrate solutions and control over aerosol decomposition in a high temperature tubular flow reactor in the temperature range from 873 to 1173 K. The aerosol was generated by means of twin-fluid as well as ultrasonic mist generator with a resonant frequency of 2.5 MHz. X-ray diffraction (XRD), differential scanning calorimetry (DSC), infrared spectroscopy (IR) and scanning electron microscopy (SEM) was used in particle characterization. It was shown that twin-fluid derived ZnO particles are hollow spheres ranged from 1 to 4 μm with the crust thickness 0.2-0.3 μm. The particle crust, derived at the temperatures ranged from 873 to 1073 K, is composed of anisotropic, non-agglomerating, equiaxed ZnO mirocrystals the size of 200-500 nm. The presence of needle-like ones, which belong to intermediate $Zn_5(NO_3)_2(OH)_8\ 2H_20$ phase, is also evident. At 1173 K the particle crust is composed mostly of ZnO phase (more than 95 %) as aggregates of equiaxial microcrystals that sintered to disc-like structures after prolonged residence time. For 0.004 mol/l precursor solution, spherical, smooth, 100-200 nm diameter ZnO particles are obtained ultrasonically. Their bonding and sintering are evident after sufficient residence time. Experimental data for the ultrasonically derived oxide powders were compared with a computer simulation based on a developed mathematical model of evaporation from droplets containing dissolved solids. The model includes all stage of evaporation: initial heating and evaporation, quasi-equilibrium evaporation, crust formation and growth, boiling and porous particle drying. The model was also tested for several others' ceramics oxide precursor solutions, such as cobalt nitrate, cobalt chloride, magnesium nitrate and colloidal silicon oxide. It was shown that a good agreement between the experimental results and numerical predictions for variety of evaporation regimes and particle stoichiometry is achieved.

SYNTHESIS OF NANOPARTICLE ZnO POWDERS BY CONTROLLED PRECIPITATION

J.E. Rodríguez-Páez[#], A.C. Caballero, M. Ocaña*, C. Moure, P. Durán, J.F. Fernández.
Departamento de Electrocerámica. Instituto de Cerámica y Vidrio (CSIC)
28500 Arganda del Rey – Madrid/España.
*Instituto de Ciencia de Materiales. Centro de Investigaciones Científicas la Cartuja (CSIC) Sevilla – España.
[#]Departamento de Física – Universidad del Cauca/Colombia.

ABSTRACT.

Zinc oxide particles with different morphologies were obtained by controlling different parameters of the precipitation process: molar concentrations, pH, washing medium and pyrolysis step. In this controlled precipitation process, reaction between Zn^{2+} aqueous solutions and NH_4^+ leads to precipitate formation. In this case the formation of fine particles from solution may be considered as a three–stage process: the neutralization of the existing acid in the solution, the formation of embryos which become solid particulate nuclei and their subsequent growth. In this work both the mechanisms of precipitate formation and the dependence of the particle morphology with the chemical speciation of complexes have been studied. The formation of zinc complex was monitored by titration curves, specific conductance and thereafter XRD and FTIR of each sediment, and SEM and TEM of the obtained powders. Through this process it is possible to obtain from random to platelet structured agglomerates of zinc oxide nanoparticles.

INTRODUCTION.

Zinc oxide powders are widely used in rubber vulcanization, ceramics, textiles, protective paints, pharmaceuticals, electrographic copying, production of varistors, etc. [1, 2] Powders have been obtained by spray pyrolysis,[3] aqueous precipitation,[4] evaporative decomposition of solutions,[5] hydrolysis of zinc salt in a polyol medium[6] and calcination of basic zinc carbonate powders[7] among other procedures. For fine powder production the solution methods offer better control of particle size and shape as well chemical homogeneity. [8]

By comparasion with solid-state reactions, chemical methods have the following advantages:[9] (1) high chemical homogeneity of precipitates; (2) the particles have very active surfaces and (3) the characteristics of the powder (size, shape, structure, composition, etc.) can be controlled by tailoring the reaction parameters.

This work describes the preparation of zinc oxide nanoparticles through the formation of metastable zinc complexes by reaction between aqueous solutions containing Zn^{2+} and NH_3. These phases decompose during the washing and the calcined steps to obtain the oxide nanoparticles. The mechanisms of precipitate formation and the dependence of the particle morphology with the chemical speciation of complexes have been discussed.

EXPERIMENTAL PROCEDURE.

0.34M zinc acetate dihydrate aqueous solutions were titrated. These solutions were acidified with 0.25N HNO_3. Both sets of electrodes the pH meter (titrino DMS 716 – Metrohm) and the conductivity dip-type cell (conductivity meter WTW LF 320/SET Tetracon 325) were used. The solution was titrated with 12M NH_4OH in 0.1 ml increments by means of the teflon capillary tube submerged below the solution-gas interface while the solution was continuously stirred. The conductance and the pH values were measured simultaneously as a function of the volume of consumed base until pH ~ 9 was reached.

From the potentiometric and conductimetric titration results the precipitation process was controlled. 12M NH_4OH was slowly added while stirring on 0.34M zinc acetate solutions up to pH 7.7 (batch A) and pH 8.5 (batch B). The processes were carried out at room temperature.

The resulting white precipitate was filtered and redispersed in ethanol several times. The resultant sediment was then oven dried at 60°C for two days. Pure ZnO cannot be obtained after drying, and calcining in air at 320°C for 1 hour was required.

Both the 60°C dried precipitate and zinc oxide powder obtained after thermal treatment were characterized by X-Ray Diffraction (XRD), using CuK_α radiation, to determine the crystalline phases. Infra-Red measurements on un-washed precipitates in the 400 – 4000 cm^{-1} range were done. The particle size and morphology were examined by Scanning Electron Microscopy (SEM) and Transmission Electron Microscopy (TEM).

RESULTS AND DISCUSSION.

(A) Potentiometric and conductimetric titration.

Figure 1 shows the potentiometric and conductimetric titration curves. The potentiometric titration curve shows an initial rise in pH. In this region, the OH^- ions liberated by NH_4OH decomposition neutralize the acid. Binding of OH^- ions by zinc and other type of species, which is an OH^- consuming process, is indicated by the appearance of an extended near-horizontal platform. The range over which this plateau extends lies roughly between pH ~ 6.5 and pH ~ 7. A second rise in pH occurs after this plateau until a near constant value about pH ~ 8.5 is reached.

Small "snowflakes" were formed and dissolved within solution at the beginning of the plateau. A stable colloidal suspension was formed at pH ~ 6.7. At higher pH a viscous gel was obtained.

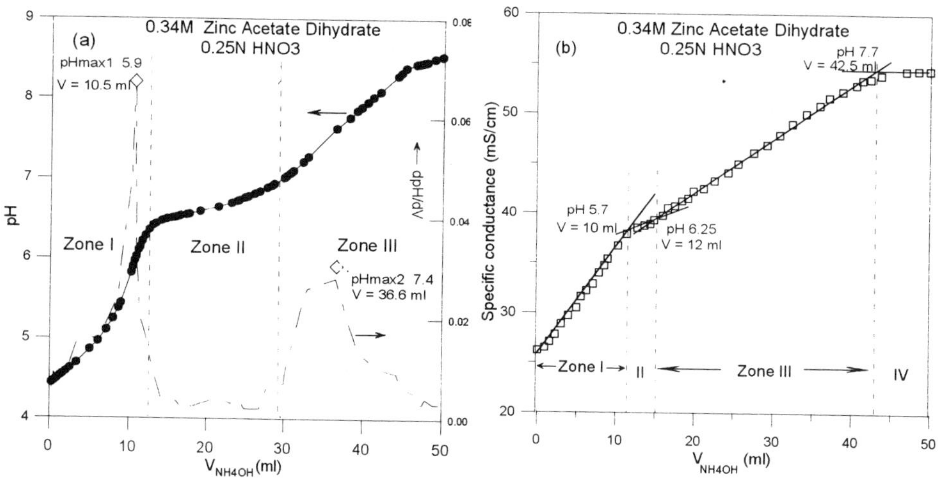

Fig. 1 . Potentiometric (a) and conductimetric (b) titration curves.

The conductivity variations may be used to follow the course of acid-base and precipitation reactions. The conductimetric titration curve shows four linear regions as defined in figure 1b. It is during the two intermediate linear regions when the colloids can be progressively formed. In the last region addition of the weakly ionized base to solution of its salt causes the conductivity to almost level off. The end-points obtained from the intersection of these linear regions are 10 ml (pH 5.7), 12 ml (pH 6.25), 36.6 ml (pH 7.7), respectively.

The phenomena that happened during the addition of NH_4OH can be interpreted in the framework of Tezak's scheme.[10] The initial zinc solution had an ideal liquidus structure and contains the constituent components: Zn_{aq}^{2+}, CH_3COO^-, NO_3^-, H^+, OH^- and because the dissociation is not complete there is an aqueous $Zn(CH_3COO)_2$ phase. In the initial rise of pH the solution is in a saturated or supersaturated state depending on the advance of the process. At zone I the neutralization of excess strong acid and weak acid complexes are the origin of both the pH and specific conductivity increase. The aggregation of constituent complexes produce the embryos at zone II of the conductivity curve.

During the plateau of potentiometric titration curve, that corresponding to zone III of the specific conductivity curve, the embryos overtake crical size and they will shape stable units, the nucleus. At that moment the colloidal suspension became stabilized.

The nuclei growth took place during the second increase of pH, zone III of the potentiometric titration curve. There are no tools to elucidate when the nucleus starts to growth. Both the nucleation and the nuclei growth processes can be happen simultaneously. It is expected that at the beginning of the zone III nuclei formation could be the main mechanism that contributes to increasing the specific conductivity in association with a nearby stable pH value meanwhile at the end of the zone III nuclei growth could be the main mechanism due the increasing pH value. Thus, the nuclei formation is the more OH^- consuming process. The interface of nucleus is very reactive and interacts with the ions that remain in solution.[11] Adsorption of metal ions and ligands can exceed a critical surface concentration and lead to the formation of new phases on the surface. This process has been named surface precipitation.[12 - 13] This may produce a complete coating of the initial surface and the nucleus growth could be inhibited. At this time the system could be represented as primary particles of colloid dimensions.

Specifically adsorbable ions originate the net surface charge on colloids. This charge induces the electrophoretic migration that is responsible of the increasing specific conductivity.

(B) Powder synthesis.

The X-Ray diffraction pattern and morphology of particles of batch A, before washing, are shown in figure 2. Two crystalline phases have been identified: nonstoichiometric amine zinc hydroxide nitrate, $Zn_5(OH)_8(NO_3)_2.(2\text{-}x)H_2O.xNH_3$ (PDF 40-1459), AZHN, and nonstoichiometric zinc acetate acetamide hydrate, $Zn(CH_3COO)_2.(2\text{-}x)H_2O.xCH_3CONH_2$, ZAAH. In addition to that, a well crystallized, unidentified phase showing low diffraction angles has been found.

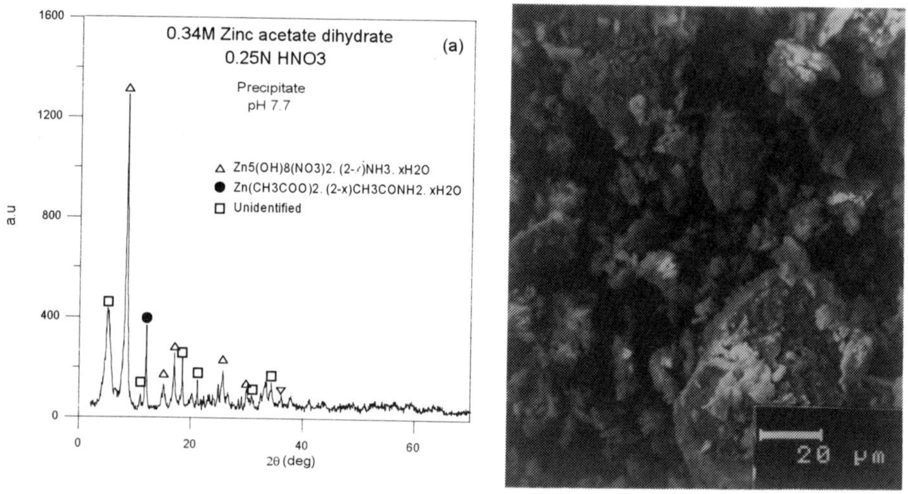

Fig. 2. The x-ray diffraction pattern (a) and SEM photographs of particles of batch A (before washing).

After washing the powder of batch A (figure 3), nonstoichiometric zinc acetate acetamide hydrate compound disappeared, the presence of nonstoichiometric amine zinc hydroxide nitrate decreased and ZnO appeared. The unassigned crystalline phase peaks were more relevant after washing the pecipitates. This sample leads to spherical shaped agglomerates if 1 – 2 μm diameter with primary particles of 50 nm (figure 3c) when the powders were calcined at 320°C.

Figures 4 and 5 show the results of XRD and SEM corresponding to batch B as precipitate and after washing, respectively. The dried precipitates presented AZHN as the main crystalline phase. After washing the main crystalline constituent of the powder was ZnO. These ZnO powders had a very special morphology that consists in platelets of 1 μm in diameter with primary particles of 50 nm size (figure 5c).

During the washing step the intermediate zinc complexes could be modified by weathering and dissolution, phase transformation and recrystallization. The chemical weathering and dissolution rates of colloids were changed with the ethanol addition that preserved the homogeneity until the critical supersaturation was reached and nucleation of new phase ocurred. The ethanol can retard the particle growth by its adsorption on the particle surface. This surface layer hinders the rearrangement of cations.[14] The continuous stirring assured the dissipation of local supersaturation regime. Chemical weathering and dissolution and other phenomenons than occur within the washing process are clearly surface controlled.[11 - 13]

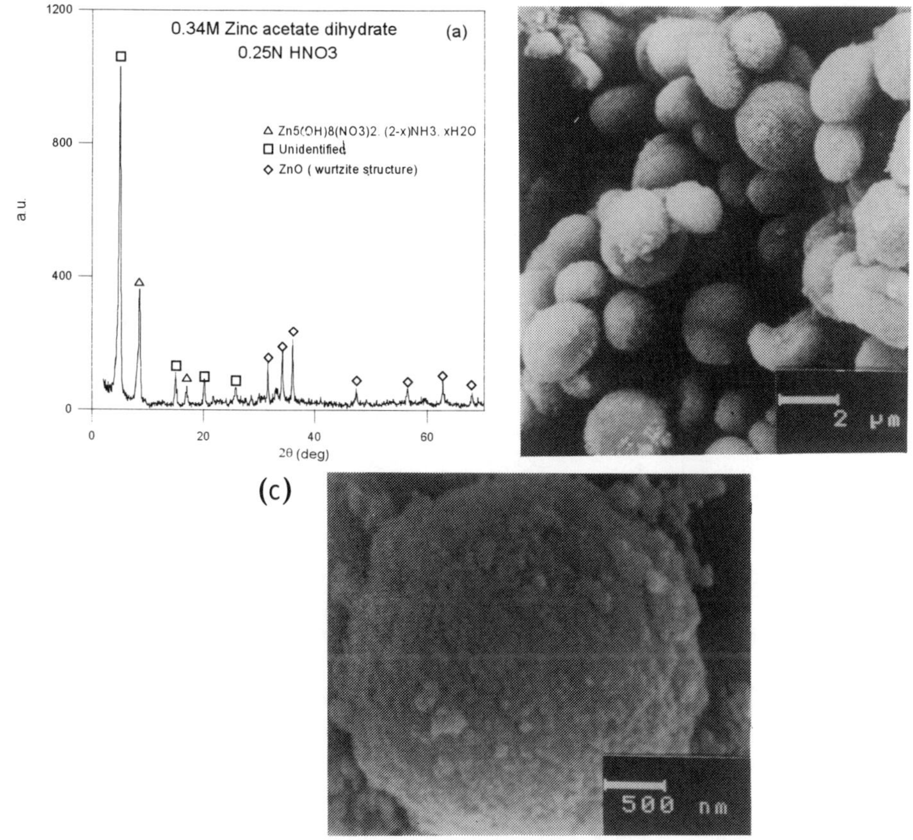

Fig . 3. XRD pattern (after washing) (a) and micrograph (after calcination at 320°C) (b and c) of particles of batch A .

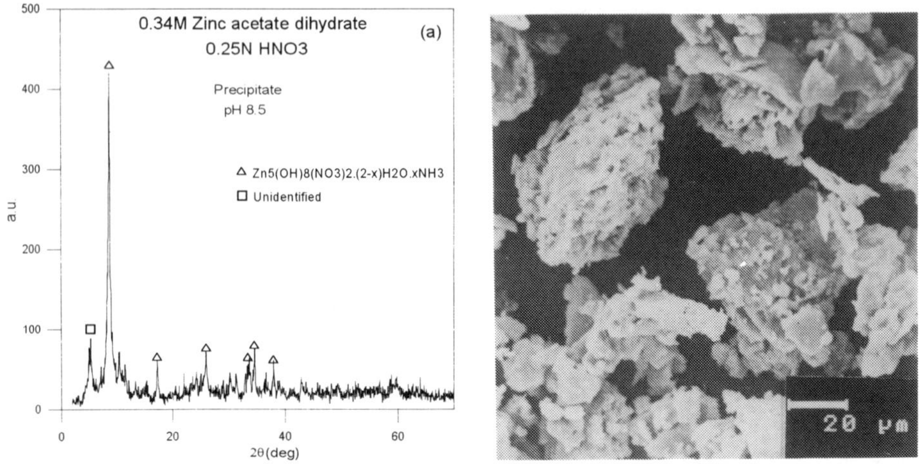

Fig. 4. XRD pattern (a) and SEM photograph (b) of particles of batch B (before washing).

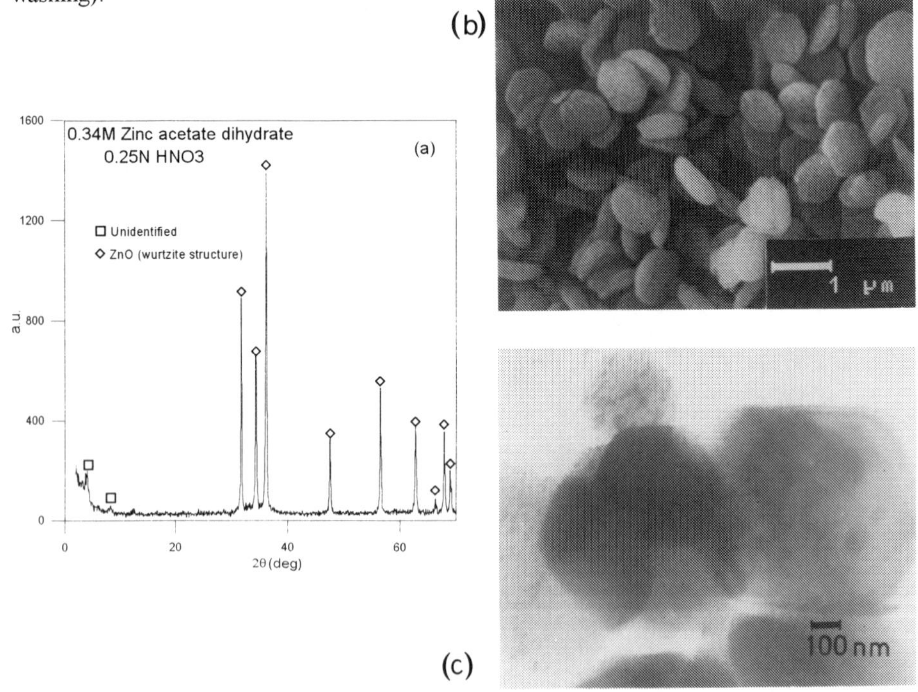

Fig. 5. XRD pattern (after washing) (a). SEM (b) and TEM (c) photograph (after calcination at 320°C) of particles of batch B.

Infra-red spectra of precipitates of batch A and B are shown in figure 6. In these spectra the characteristic IR absorption bands that correspond to amine and amide groups and carboxylic acid and ammonium salts were recorded in addition to the appearance of broad bands at ~3500 cm^{-1}. These higher frequency absorption bands indicate the existence of inter and intra-hydrogen bonds between chemical units that could be associated with dimer and polymer structures. These polymer-type structure could be clearly associated with the un-assigned low angle diffraction peak determined by XRD.

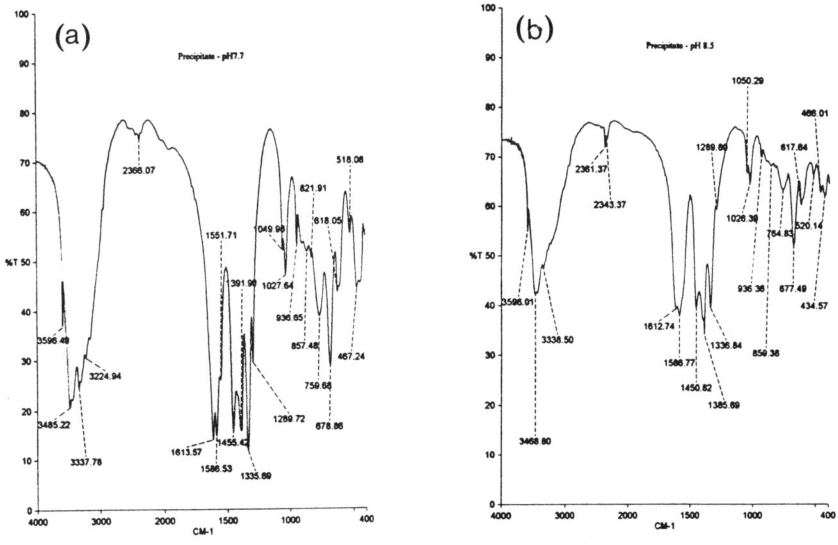

Fig. 6. Infra-red spectra of batch A (a) and B (b) samples before washing.

CONCLUSIONS

1.- The titration curves correspond to the following three process: (a) the neutralization of existing acid in the solution; (b) the embryo formation and nucleation of the solid phase and (c) the growth of the nuclei. When nuclei have been formed in the aqueous solution, unreacted free ions still remain in the solution and will adsorb onto the surface of the nuclei allowing the nuclei growth.

2.- The specific conductivity of the aqueous solution of zinc during the precipitation process was affected by the mechanisms indicated above. The surface charge that induces the electrophoretic migration in the colloidal suspension was caused by the adsorbed ions.

3.- The results of FTIR and XRD indicated that the principal zinc complexes formed during the precipitation process were: (a) zinc acetate acetamide hydrate, $(Zn(CH_3COO)_3.(2-x)H_2O.\ xCH_3CONH_2)$; (b) amine zinc hydroxide nitrate $(Zn_5(OH)_8(NO_3)_2.(2-x)H_2O.\ xNH_3$ and (c) unidentified phases having low angle diffraction peaks which could be associated with basic zinc acetates or zinc acetate - ammonia complexes.

ACKNOWLEGMENTS.

This work was supported by CICYT MAT94-807 (Spain). J.E.Rodríguez-Páez expresses his gratitude to COLCIENCIAS (Colombia) for support.

REFERENCES

[1] M. Farnsworth and C. H. Kline (Eds). " Zinc Chemicals", International Lead Zinc Research Organization, New York, 1973.

[2] L. M. Levinson (Ed.), "Advances in Varistor Technology", Ceramic Transactions Vol. 3, The American Ceramic Society, Inc., Westerville, Ohio, 1988.

[3] T. Q. Liu, O. Sakurai, N. Mizutani and M. Kato, " Preparation of Spherical Fine ZnO Particles by the Spray Pyrolysis Method Using Ultrasonic Atomization Techniques", J. Mater. Sci., **21** [10] 3698 - 3702 (1986).

[4] S. M. Haile, D. W. Johnson Jr., G. H. Wiseman and H. K. Bowen, "Aqueous Precipitation of Spherical Zinc Oxide Powders for Varistor Applications", J. Amer. Ceram. Soc., **72** [10] 2004 - 2008 (1989).

[5] D. W. Spronson, G. L. Messing and T. J. Gardner, "Powder Synthesis for Electronic Ceramics by Evaporative Decomposition of Solutions", Ceram. International, **12**, 3 – 7 (1986).

[6] D. Jézéquel, J. Guenot, N. Jouini and F. Fiévet, "Submicrometer Zinc Oxide Particles: Elaboration in Polyol Medium and Morphological Characteristics", J. Mater. Res., **10** [1] 77 - 83 (1995).

[7] M. Castellano and E. Matijevic, "Uniform Colloidal Compounds of Various Morphologies", Chem. Mater., **1** [1] 78 - 82 (1989).

[8] E. Matijevic, "Preparation and Properties of Uniform Size Colloids", Chem. Mater. **5** [4], 412 – 426 (1993).

[9] Y. Arai, "Chemistry of Powder Production", Chapman & Hall, 1996.

[10] B. Tezak, "Methorics of the Precipitation From Electrolytic Solutions", Croat. Chem. Acta, **40** [2] 63 - 77 (1986).

[11] W. Stumm, "Chemistry of the Solid - Water Interface", Wiley - Interscience, 1992.

[12] K. F. Farley, D. A. Dzombak and F. M. M. Morel, " A Surface Precipitation Model for the Sorption of Cations on Metal Oxides", J. Colloid Inter. Sci., **106** [1] 226 - 242 (1985).

[13] D. A. Dzombak and F. M. M. Morel, "Surface Complexation Modeling: Hydrous Ferric Oxide", Wiley - Interscience, New York, 1990.

[14] M. E. V. Costa and J. L. Baptista, " Characteristics of Zinc Oxide Powders Precipitated in the Presence of Alcohols and Amines", J. Eur. Ceram. Soc. **11**, 275 – 281 (1993).

Synthesis and characterization of dense, monolithic gel derived by sol-gel method

Hirofumi Matsuda, Takeshi Kobayashi, and Makoto Kuwabara
Department of Materials Science, University of Tokyo, Tokyo 113-8656, Japan

Hirokazu Shimooka and Seiji Takahashi
Department of Applied Chemistry, Kyushu Institute of Technology, Fukuoka 804-8550, Japan

Abstract

Pure, transparent and highly crystallized $BaTiO_3$ monolithic gels were synthesized by sol-gel technique at down to 90 ℃ by using a high concentration solution of Ba- and Ti-alkoxides. The gels consisted of fine particles having an average diameter of ~10 nm showed an X-ray powder diffraction pattern of a pseudo-cubic $BaTiO_3$ system at room temperature, although a tetragonal structure was more stable for bulk crystals. Dense $BaTiO_3$ ceramics which have relative density of ~95 % were obtained by sintering dried gels at 1100 ℃ for 2 hours in an oxygen atmosphere. Scattering of visible light at interfaces of fine particles were drastically reduced and optical transmission spectra of the polycrystalline gels were similar to those for single crystalline $BaTiO_3$. The estimated optical gap energies for the gels was larger by ~0.1 eV than those for single crystal by quantum size effects. Surface area and porosity analyses by nitrogen adsorption/desorption measurements showed that pore diameter distribution increased with firing temperature and the gels lost transparency at some extent due to the roughening the surfaces and growth of aggregates. Micropores below 1 nm diameter were not observed.

I Introduction

Traditional sol-gel method has been applied mostly to synthesize monodispersive powder oxides to sinter ceramics at lower temperatures or to form thin films and fibers of high purity. To obtain denser, more homogenous ceramics, much finer powders are need. However, it is difficult to prevent fine powders from aggregation. Recently, we reported an improved sol-gel process by increasing the concentration of precursor solutions, typically 1 mol/L, to obtain highly crystallized, dense monolithic gels[2,3] instead of powder xerogels. Directly firing the monolithic gels at considerably low temperatures of 1000 to 1100 ℃, dense $BaTiO_3$ ceramics were obtained, exceeding their relative densities of 95 %. This powder-less process will give a route to prepare dense and homogenous ceramic bodies with little aggregate structure because the average diameter of primary particles is 10~20 nm.

We then slowed down, homogenized hydrolysis reaction, and succeeded to synthesize polycrystalline transparent $BaTiO_3$ monolithic gels. Our process may have possibilities for polycrystalline electro-optic device applications because sol-gel process has advantages of high purity preparations and ease control of constituents and compositions.

In this article, we report details of synthesizing polycrystalline, transparent $BaTiO_3$ monolithic gels, the results of optical absorption measurements for $BaTiO_3$ fine particles, aggregated structure measured by nitrogen adsorption/desorption studies, and the structural effect on the optical gap energies and transparency is discussed.

II Experimental

Equimolar, Ba-ethoxide (99.9 %) and Ti-iso-propoxide (99.999 %) were dissolved at concentration of up to 1.2 mol/L in mixed solvent of 60 vol.% ethanol and 40 vol.% 2-

methoxyethanol in N_2 dry box. Hydrolysis of the well-stirred precursor solutions were performed at 0 ℃ by adding water vapor carefully for over 10 hours. Hydrolyzed sol solutions were held at 30 ℃ then at 50 ℃ for several days to proceed gelation and consolidation reaction (aging process). After aging process wet gels were dried in N_2 gas at 90 ℃ for several days and disk-shaped crystalline $BaTiO_3$ dried monolithic gels were obtained. Dried monolithic gels were fired at temperature up to 500 ℃ in O_2 flow, or the gels turned to be opaque after firing over the temperature. X-ray diffraction (XRD) patterns for dried and fired gels were evaluated to confirm single phase and crystallinity. Details of crystallization and sintering behaviors of $BaTiO_3$ monolithic gels were reported elsewhere.[2,3] To study the effects of residual organic ligands on optical absorption and microstructure, samples with shorter aging time or smaller amount of hydrolysis water were also prepared.

Optical transmission spectra for the gels dried at 90 ℃ and fired at 500 ℃ were measured at room temperature in the excitation wavelength range of 200～900 nm by using VT-550 UV/Visible Light Spectrometer (Nihon Bunko, Japan). The disk-shaped sample with typical thickness of 0.3 mm was immersed into liquid paraffin to reduce scattering of light at the surfaces. The transmission spectra for $BaTiO_3$ single crystal were also measured at room temperature.

Relative surface areas and porosity of gels were analyzed from isothermal adsorption/desorption curves for nitrogen using Autosorb 1 (Quanta-Chrome, USA). Gels were evacuated down to 10^{-7} Torr at 80 ℃ for 8 hours before measurements. Porosity of gels were analyzed by using Dollimore-Heal (DH) method[4] which is a model suitable for analyzing mesopores.

III Results and discussion

Figure 1 shows a top view of transparent $BaTiO_3$ monolithic gel only after drying at 90 ℃ and Fig. 2 shows the XRD patterns for the gel at room temperature. Quasi-stable, pseudo-cubic $BaTiO_3$ single phase which is crystallized already only after drying at 90 ℃. The TEM study showed that the average diameter of primary crystalline particles was ～10 nm. It has been pointed out theoretically that the instability of tetragonal phase arises from delicate cancellation between favorable long-range electrostatic force from ionicity of Ti and O ions and short-range overlap force from covalency of Ti-O bond. [5] In fine particles, long range electrostatic interaction may be suppressed and cubic phase may become stable. After firing the gels at 500 ℃, pseudo-cubic reflections of sharp peaks may result from the growth of crystallites.

Firing transparent monolithic gels in oxygen revealed good behavior of densification. The density of 5.7 g/cm^3 (relative value of ～95 %) was achieved by sintering the gels at 1100 ℃ in oxygen. This sintering temperature is much lower than that for conventional method of sintering starting from metal oxides (typically 1350～1400 ℃).

Figure 3 shows transmission spectra for dried and fired $BaTiO_3$ monolithic gels and $BaTiO_3$ single crystal. [6] Due to scattering of incident light at sample surfaces, pores, and grain interfaces, transmitted intensity of incident light with longest wavelength (900 nm) through the gels decreased, though, the overall structures and the positions of absorption edges appeared similarly.

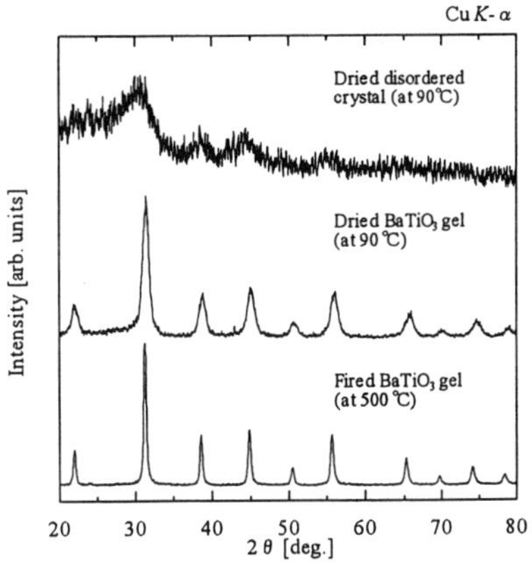

Fig. 1 Pure, transparent and highly crystallized $BaTiO_3$ monolithic gel, by hydrolyzing high concentration precursor solution (1.0 mol/L) of Ba- and Ti-alkoxide, aging the wet gel at 30 ℃ then 50 ℃, and drying at 90 ℃.

Fig. 2 X-ray diffraction patterns for dried and fired $BaTiO_3$ monolithic gels. Only the reflections of pseudo-cubic $BaTiO_3$ are observed. The gel prepared for short period of aging shows broad peaks of $BaTiO_3$ structure indicating imperfect consolidation and disordered crystal structure.

Electronic structure calculations of $BaTiO_3$ predicted that lowest excitation by light was indirect transition from the valence band edge to the conduction band edge. [7,8] Figure 4 shows excitation energy dependence of square root of absorption coefficient α to evaluate optical gap. The α for indirect transition is proportional to density of states (DOS) at top of the valence band (initial state) and DOS at bottom of the conduction band (final state). Assuming parabolic dispersion of bands,

$$\alpha(\varepsilon) \propto A(T)(\varepsilon+\varepsilon_p-\varepsilon_0)^2+B(T)(\varepsilon-\varepsilon_p-\varepsilon_0)^2 \quad (1)$$

where ε is excitation energy, ε_p is energy of phonon which is absorbed or emitted at transition, ε_0 is gap energy which is energy difference between top of valence band and bottom of conduction band, T is temperature, and A(T) and B(T) are constants independent of excitation energy. When $\alpha^{1/2}$ is plotted as a function of ε, extrapolation of tangential line from high excitation energy region intersects $\alpha=0$ at $\varepsilon=\varepsilon_0+\varepsilon_p$.

Table I shows the estimated optical gap energies $\varepsilon_0+\varepsilon_p$ for polycrystalline $BaTiO_3$ gels and $BaTiO_3$ single crystal. The optical gap energies excited by light with polarization parallel to the c-axis (E//c) and that polarization perpendicular to the c-axis ($E \perp c$) are 3.1 and 3.0, respectively, and the difference of these energies (anisotropy of optical gap energy) ~0.1 eV. Since these values are consistent of the results measured for $BaTiO_3$ single crystals reported previously, [9,10] we made similar estimations for polycrystalline $BaTiO_3$ gels (Table I).

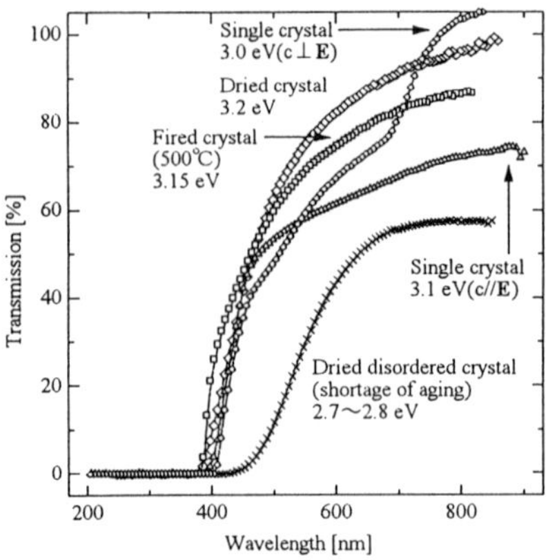

Fig. 3 Optical transmission spectra for dried and fired $BaTiO_3$ monolithic gels and single crystals at room temperature. E//c and E⊥c denote the directions of polarization vector parallel and perpendicular to the c-axis, respectively.

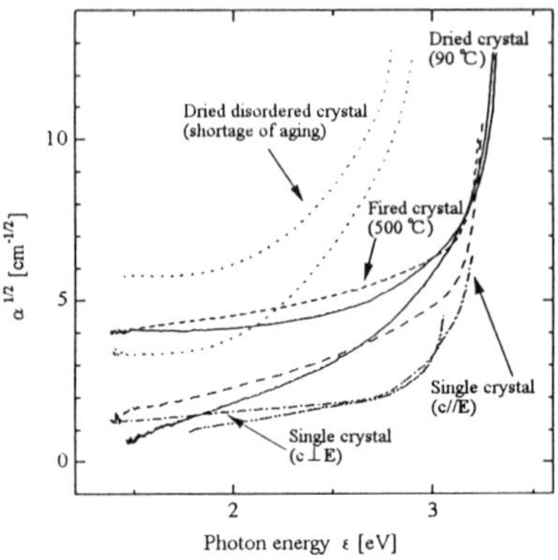

Fig. 4 Excitation energy dependence of absorption coefficient for dried and fired $BaTiO_3$ monolithic gels and single crystals. In crystalline samples, the position of absorption edge energy hardly depends on light scattering in lower energy region.

Table I Estimated optical gap energies for $BaTiO_3$ gels (in eV)

Sample	Dried 90℃	Fired 500℃	Disordered Crystal 90℃	Single crystal (E//c)	Single crystal (E⊥c)
$\varepsilon_0+\varepsilon_p$	3.20	3.15	2.7～2.8	3.10	3.00

According to literature,[9,10] the optical gap energy for cubic phase is 0.05～0.1 eV smaller than that for E⊥c. The observed optical gap energies for dried gels, however, are apparently 0.15～0.2 eV larger than that for E⊥c in $BaTiO_3$ single crystal. This energy shift may be introduced by the confinement in fine primary particles as explained for nanocrystalline silicon.[11,12] Reducing the size of the crystal, valence electrons lose kinetic energy and the valence band becomes less dispersive, the difference between the valence and conduction band may increase. In addition, the optical gap energies for fired gels are smaller than those for dried gels, while the absorption edge structure for fired gels is steeper than that for dried gels. These facts may also indicate a size effect induced blue shift. By heating at higher temperature, the crystalline primary particles may grow larger and may be more ordered.

As in Fig. 4, excitation energy dependence of α for disordered $BaTiO_3$ crystal showed a large tail absorption compared to steepler absorption for crystallized gels. Structural disorder or dangling bond states introduce localized states between the gap and this tail region absorption process may be associated with the localized states. The optical gap energies for disordered crystals ranged from 2.7～2.8 eV. Considering that samples with shorter aging time in preparation showed larger loss of weight on heating than for highly crystallized samples,[2,3] larger number of residual organic ligands may exist. Optical absorption may be associated with these organic ligand states between the gap and absorption edge energies decrease.

We have to draw attention that the optical absorption edge for fine particles (Fig. 4) are not directly comparable to theoretical results of bulk electronic structures. In addition, localized states of dangling bonds at the surface of crystallites may exist and makes it difficult to define the band gap. Thus more precise experiments using much thinner samples are needed to investigate excitation energy dependence of absorption coefficient for much higher value.

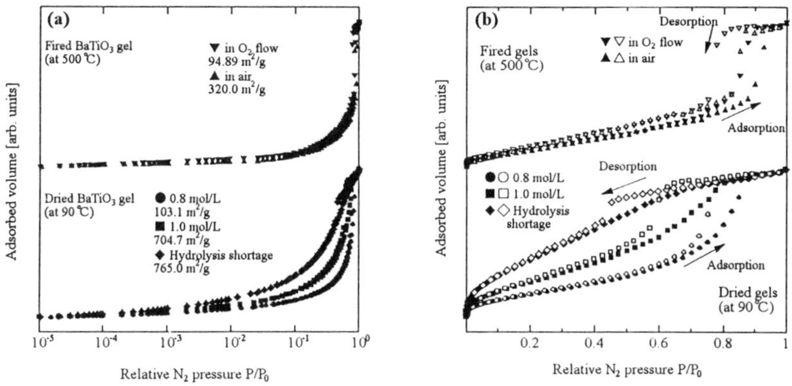

Fig. 5 Isothermal adsorption/desorption hysteresis curves of nitrogen by $BaTiO_3$ gels, **(a)** relative surface area evaluated by the BET method and low pressure adsorption/desorption by small pores (<~10 nm) and **(b)** high pressure adsorption/desorption by large pores (>~100 nm). The concentration of precursor solutions for fired gels was 1.0 mol/L.

Figures 5 (a) and (b) show isothermal nitrogen adsorption/desorption hysteresis curves for $BaTiO_3$ gels. In the region of low relative pressure in Fig. 5 (a), no evidence for adsorption by micropores (with the diameter below 1 nm)[13] was observed. The hysteresis between adsorption and desorption as observed in the high relative pressure region in Fig. 5 (b) is generally attributed to pot-shaped pores, which decrease when gels are fired. This may indicate that the difference in diameters of the bottom of the pot-shaped pores and the neck become smaller by grain growth on firing, even at 500 ℃.

Figure 6 shows distributions of pore volume analyzed from isothermal adsorption curves by DH method[4]. With firing, the gels lose transparency some extent, the shift of pore volume distribution to larger diameter indicates light scattering at the pores become more frequent due to

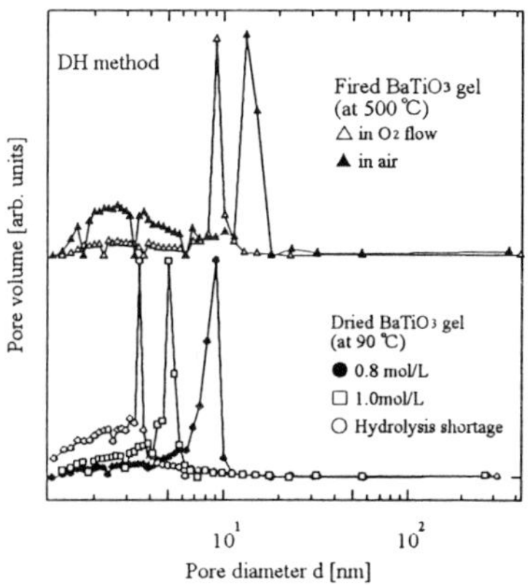

Fig. 6 Pore volume distribution for $BaTiO_3$ gels analyzed from isothermal adsorption curves by DH method.

growth of aggregates by necking.[1,13] However, assuming that the Rayleigh scattering (by which intensity ratio of transmitted light to incident light $I/I_0 \propto \exp(-d^6/\lambda^4)$ where d is diameter of body to scatter light and λ is wavelength of incident light) is dominant, pore diameter distributes 10~30 nm even for the gel fired in air and is too small to scatter visible light. As shown in Fig. 6, the peak of pore diameter distribution is somewhat asymmetric. Thus more precise adsorption/desorption curves in high-pressure region are needed to decompose the asymmetric peaks and evaluate the effect of the upper tail region of pore diameter and the structural features of aggregates on scattering of light. In addition, the gels fired in O_2 flow showed a sharper distribution with a smaller peak pore diameter than that for the gel fired in air. This fact could be interpreted as growth of aggregates occurring via oxygen vacancies and firing in O_2 flow may slow the growth of these grains, thus suppressing the scattering of light at rough surfaces and interfaces of grains.

IV Conclusion

Optical absorption measurements for fine-grained, polycrystalline $BaTiO_3$ monolithic gels were reported. The structure of absorption spectra for polycrystalline gels synthesized at only 90 ℃ are similar to those for $BaTiO_3$ single crystals grown at >1000 ℃ due to the small size of particles and aggregates prevent from scattering light at these interfaces. The optical gap energies for the gels, though, apparently become larger influenced by quantum size effect of crystalline primary particles. By firing, the aggregates of crystalline primary particles in the gels grow and more frequently scatter the visible light to lose transparency.

Acknowledgments

This research is supported by Grant-in-Aid for Science Research (Grant Nos. 08555155 and 09305043) from the Ministry of Education, Science, and Culture of Japan.

References

[1]J. R-Ortega and L. Equivias, "Structural Models of Dense Gels", J. Sol-Gel Sci. Tech., **8**, 117 (1997).
[2]H. Shimooka and M. Kuwabara, "Preparation of Dense $BaTiO_3$ Ceramics from Sol-Gel-Derived Monolithic Gels", J. Am. Ceram. Soc., **78**, 2849 (1995).
[3]H. Shimooka and M. Kuwabara, "Crystallinity and Stoichiometry of Nano-Structured Sol-Gel-Derived $BaTiO_3$ Monolithic Gels", J. Am. Ceram. Soc., **79**, 2983 (1996).
[4]D. Dollimore and G. R. Heal, J. Appl. Chem., **12**, 109 (1964).
[5]R. E. Cohen and H. Krakauer, "Lattice dynamics and origin of ferroelectricity in $BaTiO_3$; Linearized-augmented-plane-wave total-energy calculations", Phys. Rev. B**42**, 6416 (1990).
[6]H. Matsuda, M. Kuwabara, K. Yamada, H. Shimooka, and S. Takahashi, "Optical absorption in crystalline $BaTiO_3$ fine particles", submitted to Appl. Phys. Lett.
[7]Y.-N. Xu, W. Y. Ching, and R. H. French, "Selfconsistent band structures and optical calculations in cubic ferroelectric perovskites", Ferroelectrics **111**, 23 (1990).
[8]Y.-N. Xu, H. Jiang, X-F. Zhong, W. Y. Ching, "Comparison of electronic structure and optical properties of $BaTiO_3$ in the cubic and tetragonal phases", Ferroelectrics **153**, 19 (1994).
[9]S. H. Wemple, Phys. "Polarization Fluctuations and the Optical-Absorption Edge in $BaTiO_3$", Rev. B**2**, 2679 (1970).
[10]L. Hafid, G. Godefroy, A. E. Idrissi, and F. M.-Calendini, "Absorption spectrum in the near U.V. and electronic structure of pure barium titanate", Solid State Commun. **66**, 841 (1988).
[11]S. Nomura, X. Zhao, Y. Aoyagi, and T. Sugano, "Electronic structure of a model

nanocrystalline/amorphous mixed-phase silicon", Phys. Rev. B**54**, 13974 (1996).
[12]G. Allan, C. Delerue, and M. Lannoo, "Electronic Structure of Amorphous Silicon Nanoclusters", Phys. Rev. Lett. **78**, 3161 (1997).
[13]E. R. Leite, M. A. L. Nobre, M. Cerqueira, and E. Longo, "Particle Growth during Calcination of Polycation Oxides Sythesized by the Polymeric Precursors Method", J. Am. Ceram. Soc., **80**, 2649 (1997).

HYDROTHERMAL SYNTHESIS AND FORMATION MECHANISMS OF SHAPE ANISOTROPIC $BaTiO_3$ PARTICLES

Fatih Dogan
Department of Materials Science and Engineering
University of Washington, Roberts Hall, Box 352120
Seattle, WA 98195

ABSTRACT

The effect of hydrothermal processing conditions on the morphology of barium titanate particles was investigated. Precursor materials were barium hydroxide solution and titania powders with fine or coarse particle size. The mechanisms of particle formation and the reaction kinetics were studied with respect to particle morphology by microscopy techniques. When ultrafine titania powders are used, it is proposed that $BaTiO_3$ particles less than 100 nm form by crystallization of an amorphous intermediate phase. On the other hand, shape anisotropic $BaTiO_3$ particles are obtained when coarse titania powders are used as starting powders. It is suggested that the nucleation and growth of shape anisotropic particles occur in the solution supersaturated with Ti-ions by a dissolution and crystallization mechanism.

INTRODUCTION

Controlling of particle shape during powder synthesis is a key requirement towards processing of grain oriented ceramics. These ceramics exhibit improved physical properties due to the their microstructural anisotropy. Recent research activities in synthesis of ceramic powder are focused on controlling process parameters to obtain shape anisotropic particles which can be oriented by conventional ceramic processing methods, such as tape casting or extrusion.

Due to the uncontrolled particle growth and agglomeration during solid-state reactions, novel techniques (e.g., molten-salt synthesis, evaporation and condensation, controlled precipitation or hydrothermal method) have been developed to synthesize shape anisotropic powders from liquid or gas phases [1,2]. Although a variety of simple oxide powders with characteristic particle shape have been prepared by these methods, direct synthesis of multi-component oxides has been rather difficult and requires a delicate control of more complex processing parameters.

Shape anisotropic particles were synthesized by hydrothermal methods in various oxide systems, e.g., acicular particles of $PbTiO_3$ [3], hexagonal platelets of $BaFe_{12}O_{19}$ [4], cubelike particles of lead zirconium titanate (PZT) [5], octahedral crystals of Y_2O_3-stabilized ZrO_2 [6], whiskers of hydroxyapatite [7] and fibrous $BaTiO_3$ powders using fur-fibres of $2K_2O \cdot 11TiO_2 \cdot 3H_2O$ and $Ba(OH)_2$ [8,9]. In hydrothermal reactions, two possible particle formation mechanisms were proposed: (i) formation of inorganic oxide by consolidation and collapsing of the hydrous metal oxide structure and, (ii) dissolution-crystallization from a supersaturated solution [10]. According to the free energy relationships affected by the pressure and temperature of the system, metastable intermediate phases may form, redissolve and recrystallize.

Previous research conducted on hydrothermal synthesis of barium titanate has placed little emphasis on controlling particle morphology or the related reaction mechanisms and kinetics. The objectives of this work are the synthesis of hydrothermal $BaTiO_3$ particles with controlled morphology and a clear understanding of reaction mechanisms and kinetics during particle formation.

EXPERIMENTAL PROCEDURE

Starting materials were TiO_2 soot (Type P-25, Degussa Corp.) with a mean particle size of 25 nm and $Ba(OH)_2 \cdot 8H_2O$ (J. T. Baker Inc.). Coarse titania powders were obtained by heat treatment of TiO_2 soot for three hours at 800°C and 900°C which are referred as T8 and T9, respectively. Barium hydroxide was dissolved in deionized water at 80°C and hot filtered to remove insoluble $BaCO_3$ particles. $BaTiO_3$ powders were synthesized by mixing of TiO_2 powders with barium hydroxide solutions at low (0.5M) and high (1M) concentrations which are referred as LC and HC, respectively. The initial Ti:Ba ratio was 1:1.2. The suspensions were stored in sealed polyethylene bottles at around 90°C for several days without stirring. The reaction products were sampled at various time intervals, washed with CO_2-free water to remove excess barium hydroxide, and analyzed by SEM and EDS techniques.

RESULTS AND DISCUSSION

Fine $BaTiO_3$ powders with an average particle size around 100 nm were prepared at low reaction temperatures (<100°C) using titania soot. Crystalline powders were obtained as nearly spherical particles with a narrow particle size distribution (Fig. 1). Reaction mechanism of fine $BaTiO_3$ was attributed to the formation and recrystallization of a metastable intermediate phase which exhibits an

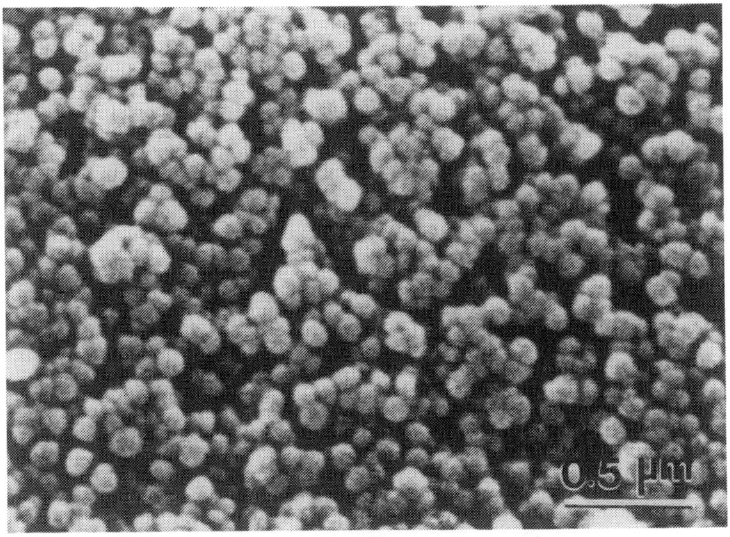

Figure 1. Fine $BaTiO_3$ particles synthesized using TiO_2-soot and 1M $Ba(OH)_2$ solution at 90°C.

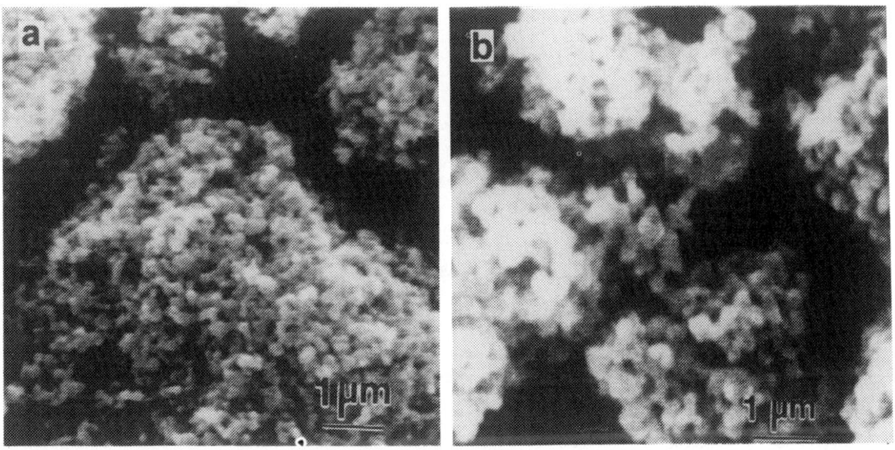

Figure 2. Coarsening of particles in titania soot after heat treatment at (a) 800°C and (b) 900°C.

amorphous structure and contains hydroxyl groups of barium and titania. It was proposed that rapid nucleation and growth kinetics are responsible for obtaining fine particles. The reaction mechanisms and kinetics during formation of nanophase barium titanate powders and thin films are described in detail elsewhere [11,12].

The rate limiting factor for nucleation and growth of $BaTiO_3$ in barium hydroxide is the dissolution rate of titania particles in the solution. If the dissolution rate of titania is low (coarse powders with low surface area), the solution becomes supersaturated with titanium ions. Hence, nucleation of $BaTiO_3$ occurs within the supersaturated solution without formation of intermediate phases. The growth rate of barium titanate particles is then controlled by the dissolution rate of titania particles.

To control the shape and particle size of $BaTiO_3$ powders, two processing parameters were varied during powder synthesis: Particle size of titania precursor powders and concentration of barium hydroxide solution. The morphology of titania powders, heat treated at 800°C and 900°C , are depicted in Fig. 2. The powders reveal an agglomerated morphology and coarsening of primary particles with increasing temperature. The nucleation and growth kinetics of barium titanate were studied with respect to the dissolution rate of TiO_2 powders which is controlled by the particle size (or surface area) and concentration of barium hydroxide solution.

Fig 3a shows the morphology of $BaTiO_3$ particles prepared using T8-powder and LC-solution after 3 days of reaction time. Nearly spherical particles with a moderately uniform size distribution were obtained. In this case, the particle size is significantly larger than that prepared from untreated titania soot. When HC-solution was used, $BaTiO_3$ particles with a similar morphology but smaller size were obtained (Fig. 3b). The decrease of the particle size with increasing concentration of barium hydroxide solution can be attributed to higher dissolution rate of titania and faster nucleation and growth rates of barium titanate particles. The results using T8-powder indicate that the reaction rates are still too fast to obtain shape anisotropic particles.

Synthesis of shape anisotropic $BaTiO_3$ particles was achieved using coarse T9-titania powder. Fig. 4a shows a typical anisotropic morphology of a barium titanate particle obtained by reaction in the LC-solution. After 3 days, the presence of undissolved titania particles indicates that the reaction is not completed. This can be attributed to a slow dissolution rate of titania which, in turn, translates into relatively slow growth rate of barium titanate particles. On the other hand, when HC-solution was used, shape anisotropic $BaTiO_3$ particles were obtained with various morphology (e.g., columnar and plate-like shape). The differences in particle morphology may be due to the local concentration differences of reaction

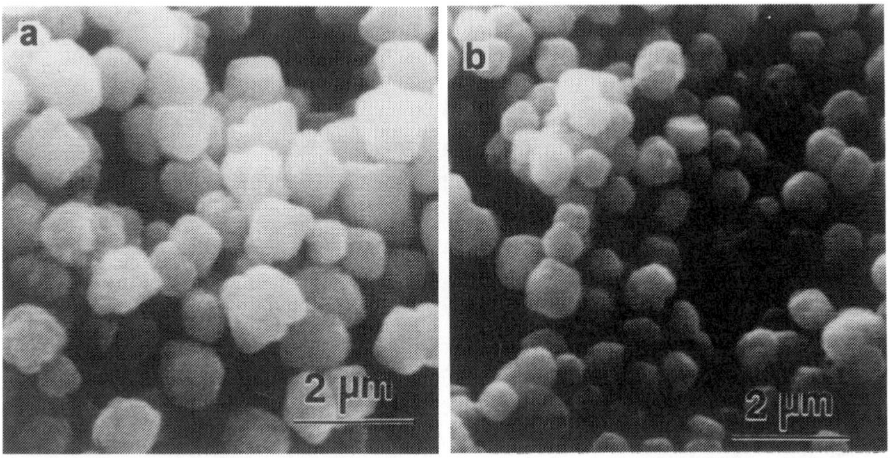

Figure 3. Coarse $BaTiO_3$ particles prepared using, (a) 0.5M $Ba(OH)_2$ solution and TiO_2 powder heat treated at 800°C for 3 hours, and (b) 1M $Ba(OH)_2$ and TiO_2 heat treated at 900°C for 3 hours.

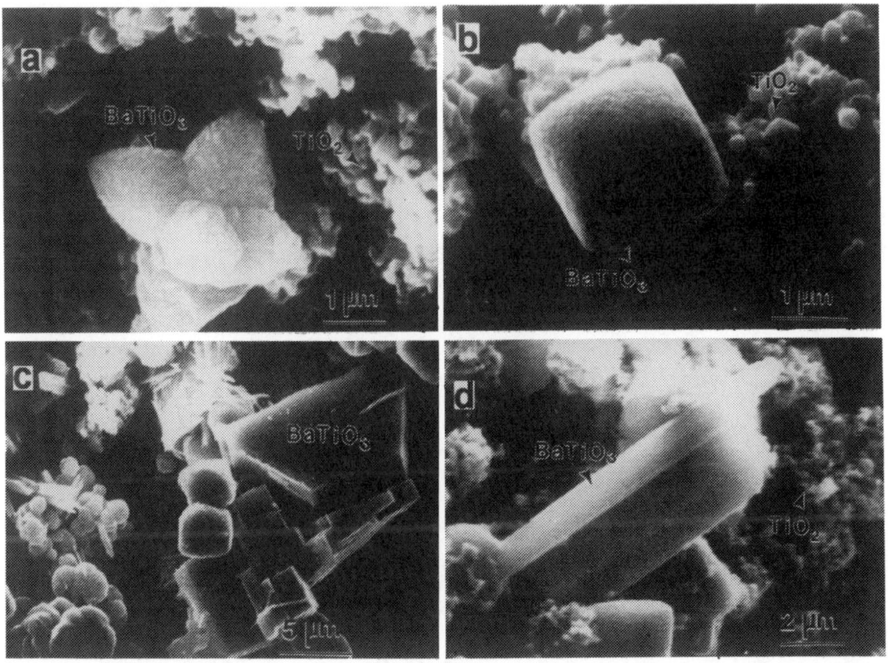

Figure 4. Shape anisotropic $BaTiO_3$ particles synthesized by reaction of TiO_2 powders, heat treated at 900°C for 3 hours, with (a) 0.5M $Ba(OH)_2$ and (b), (c), (d) 1M $Ba(OH)_2$ solution.

products in the suspension which was not stirred. Further experiments are planned to study the effect of stirring of the suspension on the growth morphology of particles.

The proposed mechanism for the formation of shape anisotropic particles is the slow dissolution of titania in barium hydroxide solution followed by nucleation and growth of crystalline barium titanate particles from the solution saturated with titanium ions. It is suggested that at the initial stage of the reaction, a limited number of $BaTiO_3$ particles nucleate which then grow with a faceted morphology to larger particles. The nucleation of second-generation $BaTiO_3$ particles at a later stage of the reaction would then depend on the balance between the dissolution rate of titania and growth rate of $BaTiO_3$. If the dissolution rate of titania would exceed the growth rate of barium titanate (e.g., increase of the reaction temperature), then, second-generation barium titanate particles would nucleate and grow. Since fluctuations in the local concentration of reaction products can lead to nonuniformity of shape anisotropic particles, a more uniform particle shape and size could be achieved, as mentioned previously, when the suspensions are stirred during hydrothermal synthesis. Future studies will address a clear understanding of the relationship between the reaction parameters and particle morphology of hydrothermally prepared powders.

SUMMARY

$BaTiO_3$ powders were synthesized under hydrothermal conditions in order to control particle morphology. Fine and nearly spherical $BaTiO_3$ particles were obtained when fine titania precursors were used. In this case, particles formed at a fast reaction rate by a mechanism proposed as recrystallization of an amorphous metastable phase. On the other hand, shape anisotropic $BaTiO_3$ powders were synthesized using coarse titania powders which dissolve in barium hydroxide solution at a relatively slow rate. It is proposed that formation of shape anisotropic particles takes place by a dissolution and recrystallization mechanism in the solution supersaturated with titanium ions.

REFERENCES

[1] D.W. Johnson Jr, "Nonconventional Powder Preparation Techniques", Am. Ceram. Soc. Bull. 60, 221-224 (1981).

[2] E. Matijevic, "Colloidal and Interfacial Aspects of Ceramic Powders", pp. 423-437 in Advances in Ceramics, Vol. 21: Ceramic Powder Science, Edited by G. L. Messing, J. W. McCauley, K. S. Mazdiyasni and R. A. Haber. The American Ceramic Society, 1987.

[3] D.J. Watson, C. A. Randal, R. E. Newnham, and J. H. Adair, "Hydrothermal Formation Diagram in the Lead Titanate System", pp. 154-162 in

Ceramic Powder Science II, edited by G. L. Messing, E. R. Fuller, Jr., H. Hausner, The American Ceramic Society, Inc., 1988.

[4] M. Kiyama, T. Takada, N. Nagai, and N. Horiishi, "Formation of Hexagonal Platelike Particles of $BaFe_{12-2x}Ti(IV)_xCo(II)_xO_{19}$ wth $0<x<1.2$", pp.51-57 in Advances in Ceramics, Vol. 15. Edited by F. F. Y. Wang. The American Ceramic Society, Inc., 1985.

[5] K.C.Beal, "Precipitation of Lead Zirconate Titanate Solid Solutions under Hydrothermal Conditions", pp. 33-41 in Advances in Ceramics, Vol. 21: Ceramic Powder Science, Edited by G. L. Messing, J. W. McCauley, K. S. Mazdiyasni and R. A. Haber. The American Ceramic Society, Inc., 1987.

[6] K. Nakamura, S. Hirano, and S. Somiya, "Hydrothermal Growth of Y_2O_3-Stabilized ZrO_2 Crystals", Am. Ceram. Soc. Bull., 56, 513-515 (1977).

[7] M. Yoshimura, H. Suda, K. Okamoto, and K. Loku, "Hydrothermal Synthesis of Biocompatible Whiskers", J. Mater. Sci. 29, 3399-402 (1994).

[8] Y. Ohara, K. Koumoto, and H. Yanagida, "Barium Titanate Ceramics with High Piezoelectricity Fabricated from Fibrous Particles", J. Am. Ceram. Soc., 68 [4] C108-C109 (1985).

[9] Y. Ohara, M. Miyayama, T. Shimizu, and H. Yanagida, "Barium Titanate Fabricated from Fur-Fibres", J. Mater. Sci., Mater. in Electronics 7, 27-33 (1996).

[10] J. H. Adair, R. P. Denkewicz, F. J. Arriagada, and K. Osseo-Asare, "Precipitation and In-situ Transformation in the Hydrothermal Synthesis of Crystalline Zirconium Dioxide", pp. 135-145 in Ceramic Powder Science II, edited by G. L. Messing, E. R. Fuller, Jr., H. Hausner, The American Ceramic Society, Inc., 1988.

[11] F. Dogan and M. Sarikaya, "Mechanisms of Hydrothermal Barium Titanate Thin Film Formation", pp. 95-99 in Polycrystalline Thin Films: Structure, Texture, Properties, and Applications II, Eds.: H. J. Frost, C. A. Ross, M. A. Parker, and E. A. Holm, Mat. Res. Soc. Symp. Proc., Vol.403, 1996.

[12] F. Dogan, S.O'Rourke, M.-X. Qian, and M. Sarikaya, "Low Temperature Hydrothermal Synthesis of Nanophase $BaTiO_3$ and $BaFe_{12}O_{19}$", pp. 69-74 in Nanophase and Nonocomposite Materials II, Eds.: S. Komarneni, J. C. Parker, H. J. Wollenberger, Mat. Res. Soc. Symp. Proc, 1997.

THEORETICALLY DENSE AND NANOSTRUCTURED CERAMICS BY LOW-TEMPERATURE PRESSURELESS SINTERING OF NANOSIZED Y-TZP POWDERS

P. Duran, M. Villegas, J. F. Fernandez, F. Capel, and C. Moure
Instituto de Cerámica y Vidrio (CSIC)
Electroceramics Department
28500 Arganda del Rey
Madrid, Spain

ABSTRACT

Nanosized yttria-doped tetragonal polycrystalline zirconia (Y-TZP) powder was prepared by the coprecipitation method using isopropanol solutions of organic and inorganic precursors. An isopropanol-washing step led to a powder mainly constituted by very soft agglomerates with 6 nm crystallite size. Such a weakly agglomerated powder compacted uniformly (average diameter pore size = 6 nm) with a homogeneous shrinkage to full density. The small particle size of the Y-TZP powder, the small pores in the green compacts, and its homogeneous pore size distribution led to theoretically dense and nanostructured (<100 nm) Y-TZP bodies by pressureless sintering at 1000°C for 8-100 hrs or at 1070°C for 2 hrs, in spite of the relative low green density (~43% theoretical). Comparatively, a commercial powder consisting of spherical weak agglomerates resulted in a fully densified ceramic but at higher sintering temperature (>1200°C) and with a grain size higher than 0.25 μm.

ENHANCED CRYSTALLISATION KINETICS OF AIR-CALCINED YTTRIUM DOPED HYDROUS ZIRCONIA BY SEEDING

J. Tartaj, P. Durán, J.F. Fernández, and C. Moure
Instituto de Cerámica y Vidrio (CSIC), Electroceramics Department
28500-Arganda del Rey, Madrid, Spain

ABSTRACT

Addition of nanometric-sized Y-TZP seed particles to yttrium-zirconium cation sols resulted in a lowering (50 to 100^{0}C) of the air-crystallisation temperature of the Y-TZP (3 mol% Y_2O_3) phase, and a strong increase in the crystallisation rate (about one order of magnitude higher) for the formation of Y-TZP phase in air-calcined coprecipitated amorphous powders relative to the unseeded Y-TZP ones. The isothermal crystallisation kinetics for Y-TZP formation in unseeded and seeded (10wt%) polycrystalline Y-TZP followed an Avrami-type kinetic relationship. Addition of nanometric-size Y-TZP seed particles to Y-TZP amorphous precursors reduced the apparent activation energy for the Y-TZP crystallisation process from 184 to 119 kJ/mol

INTRODUCTION

It is widely accepted that chemical methods of synthesis can yield powders with no compositional fluctuation and controlled particle shape and size (1). In the case of the production of ultrafine yttria-doped tetragonal zirconia polycrystalline (Y-TZP) powders, a gel-precipitation technique using metal salts as chemical precursors has been widely used (2,3). In such a chemical route, which involves the coprecipitation of metal hydroxides with ammonia, a thorough washing with an alcohol plays a crucial role in obtaining dried powders with low strength agglomerates. The final step is the calcination of dried powders at temperatures $\cong$500°C leading to crystalline Y-TZP powder with high intra-agglomerate porosity (4). Such a heat treatment inevitably leads to the formation of strong interparticle bonds, which detrimentally influence both the sinterability and the compaction of the so obtained Y-TZP powders. In order to get better powder characteristics the suggestion is to find the adequate processing which allows the use of lower crystallisation temperatures to avoid the fomation of the undesirable interparticle bonds. The claim of the present paper is the use, for the first time in the case of the Y-TZP, of the "seeding fundamental concepts", successfully applied by Messing`s group (5) in the case of Al_2O_3 to lower the crystallisation temperature and/or modify the crystallisation kinetics of the air-calcined Y-TZP precursors, as an alternative to other chemical routes.

EXPERIMENTAL PROCEDURE

The starting materials were co-precipitated gels. To do that, $Zr(C_4H_9O)_4 \cdot C_4H_9OH$ and $Y(NO_3)_3 \cdot H_2O$ were dissolved in pure isopropanol slightly acidified with HNO_3. This solution was labeled as A. On the other hand, a suspension of well crystallised Y-TZP seed particles (size $\cong$ 8 nm) in pure isopropanol was also prepared as described elsewhere (4). The amount of seed particles was in proportion of 10 wt % relative to the final Y-TZP mass. This suspension was labeled as B. Then the solution A and the suspension B were mixed and coprecipitated with a mixture of pure isopropanol-NH_4OH (1/1). The dried powders were heated from 0.1h up to 60h in an isothermal mode in the temperature range of 320 to 400°C.

After heating the amount of crystallised tetragonal phase was measured by X-ray diffraction (Siemens D5000 with CuKα) on the 111 peak area using a previously calibrated curve. A differential thermal analyzer (DTA/TG, Netzsch STA-409) was also used to study the onset, maximum and the end temperatures of the crystallisation process.

RESULTS AND DISCUSSION

Crystallisation characteristics

Figure 1 compares the results of the differential thermal analysis of the seeded Y-TZP amorphous precursors with those of the pure hydrous zirconia and unseeded Y-TZP powders. The exothermic peak represents the crystallisation temperature for pure tetragonal zirconia and yttria-doped tetragonal zirconia respectively. The crystallisation temperature of tetragonal zirconia increases with the addition of Y_2O_3 from about 420 to 455^{0}C for the unseeded Y-TZP amorphous precursors, and decreased up to about 418^{0}C for the seeded Y-TZP samples. The asymmetry of the exothermic peak increased from the pure hydrous zirconia to the seeded Y-TZP amorphous precursors, which is indicating a certain heat releases associated, probably, with the enthalpy of the tetragonal solid solution formation. Such heat release starts to be detected at about 418 and 310^{0}C for unseeded and seeded Y-TZP amorphous precursors, respectively. This means that the yttria-doped tetragonal zirconia solid solution crystallisation starts at a temperature, at least, 100^{0}C lower in the seeded Y-TZP amorphous precursors. A completely crystallised tetragonal zirconia phase could be obtained by air calcining the seeded Y-TZP amorphous precursors in the temperature range of 310 to 418^{0}C.

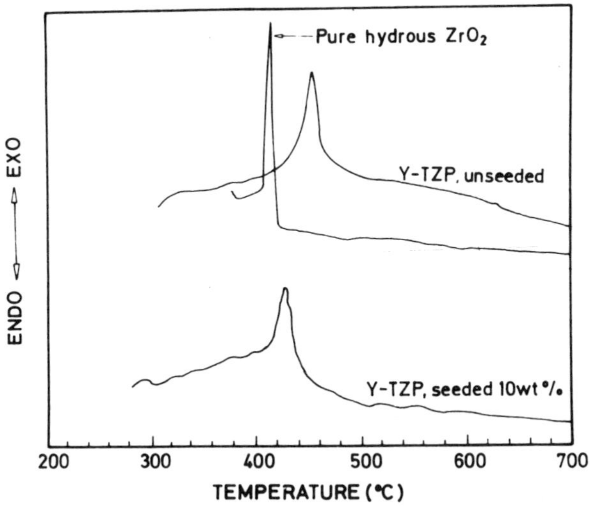

Fig.1, DTA of unseeded and 10 wt% seeded Y-TZP amorphous precursors.

Livage et al (6) established that pure zirconia starts to crystallise at 300^{0}C; and a minimum temperature of 350^{0}C would be required for complete crystallisation. Theunissen et al (7) observed the onset of crystallisation for Y-TZP at 400^{0}C and Boutz et al (8) established by DSC experiments the amorphous to crystalline transition for Y-TZP at 450^{0}C. According to our surprising results both the decrease in the onset of the crystallisation temperature and the increase of the Y-TZP crystallisation rate could be attributed, in principle, to a lower activation energy required to nucleate the Y-TZP phase from the seeds. If as suggested by Livage et al (6) the crystallisation of pure tetragonal zirconia from amorphous powders started by nucleation at a favoured points then it should be assumed that an increasing of the nucleation density sites would shift the onset of the crystallisation to lower temperatures. Figures 2 show XRD patterns for unseeded and seeded Y-TZP amorphous precursors which were calcined at temperatures in the range of 350^{0}C to 400^{0}C for 0.5 to 1 h. As it can be observed the unseeded samples remained amorphous at 350 and 375^{0}C , and no more than 30 % of crystallised tetragonal phase was obtained at 400^{0}C.

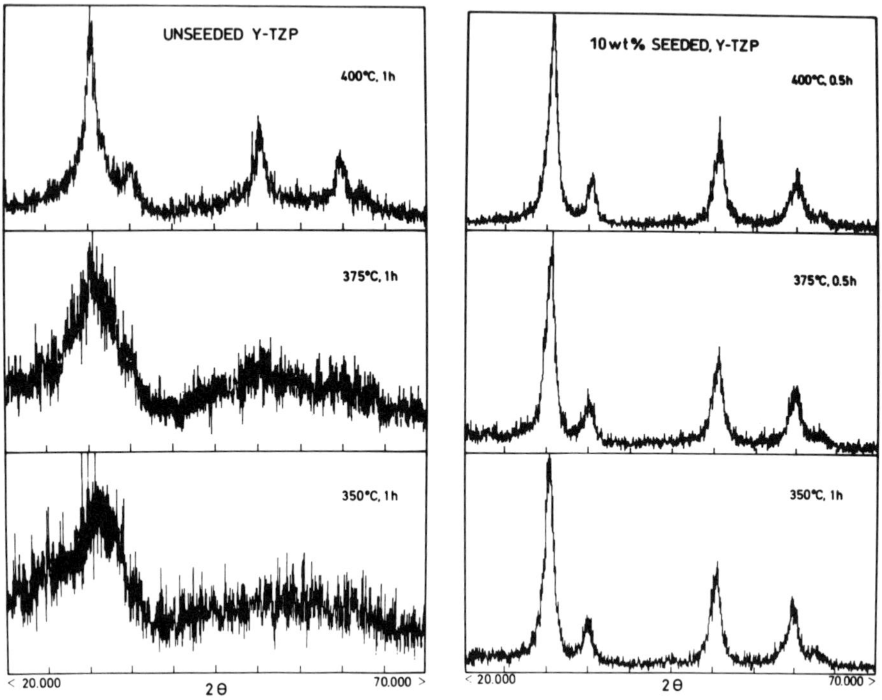

Fig.2, XRD for unseeded and 10 wt% seeded Y-TZP amorphous precursors.

The relative amount of this phase strongly increased in 10 wt% seeded Y-TZP sample. As Y-TZP seeds were homogeneously distributed, they promoted heterogeneous nucleation of tetragonal phase. The rapid crystallisation of the newly formed tetragonal phase was, probably, favoured because of epitaxial growth on the surface of the homogeneously distributed isostructural crystalline Y-TZP seed nanoparticles. The very small particle size of the Y-TZP seeds (≅8 nm) did not show detectable changes in the XRD intensities, but the epitaxial crystallisation has also been evidenced in many other seeded-gels to ceramic transformations with nanosized seed particles (9-10). Therefore, it can be assumed that the Y-TZP seeds accelerated the amount of crystalline Y-TZP formed in seeded Y-TZP amorphous precursors, by providing a great number of epitaxial nuclei with the same structure of the final Y-TZP phase.

Crystallisation kinetics

It has been suggested that the crystallisation of pure tetragonal zirconia could be considered as a nucleation and growth process (6,11). By plotting the fraction of crystallised Y-TZP phase as a function of log time, curves with a sigmoidal shape should be obtained in close agreement with Avrami model (12), which assumes that growth is controlled by nucleation. Figure 3 shows such a plotting for unseeded and seeded Y-TZP amorphous precursors at 350^0C. The curves have the same shape and slope which indicates that seeding does not influence the crystallisation mechanism but only the crystallisation rate. It is observed that the influence of seeding on the crystallisation rate of Y-TZP is so high that the seeded crystallisation is almost complete when the unseeded begins to crystallize.

From the above results it seems evident that the kinetics of the Y-TZP crystallisation were reasonably enhanced by seeding Y-TZP amorphous precursors with polycrystalline Y-TZP nanometer-size particles. As established by the Avrami-type nucleation and growth kinetics, an induction time in which no crystallisation was detected seems to exist. In our case such an induction time was obtained by extrapolation of the linear part of the sigmoidal crystallisation curves to 0% of the formed Y-TZP phase. Thus, the induction time for the seeded Y-TZP crystallisation was reduced about one order of magnitude by comparison to the unseeded Y-TZP amorphous powders at the same temperature. According to the above assumption, the Avrami transformation kinetic expression was used to evaluate the crystallisation data. The relationship between volume fraction and time is given by:

$$x = 1 - \exp(-kt^n) \qquad [1]$$

where n is a constant dependent on the nucleation and growth mechanisms, and k is the rate constant. For isothermal conditions, the experimental data were better fit according to the following linear transformation of the above equation:

$$\ln[-\ln(1-x)] = \ln k + n \ln t \qquad [2]$$

where x is the weight fraction of crystallised Y-TZP phase. Plots of ln[-ln(1-x)] as a function of ln t yields the values of n and k.

The rate constants, shown in Table I, were determined from the intercept at each isothermal temperature. As it is shown, comparable rate constants were measured at temperatures, at least, 60^0C, lower in seeded Y-TZP relative to unseeded samples. The temperature dependence of k apparently follows an Arrhenius relationship, according to:

$$k = A \exp(-E_a/RT) \qquad [3]$$

where A is the frequency factor, E_a the apparent activation energy of crystallisation, and R the gas constant.
Plots of log k vs 1/T,as shown in Fig.4, yielded straight lines with very good correlation coefficients ($\cong$0.99). The activation energies calculated were 184 $\pm$ 9.2 KJ/mol and 119 $\pm$ 6 kJ/mol for the crystallisation of the unseeded and seeded Y-TZP amorphous precursors, respectively. The activation energy of crystallisation obtained in the case of the seeded Y-TZP amorphous precursors is much lower than that reported for the crystallisation of doped hydrous zirconia (4 mol % Y_2O_3), 262 kJ/mol (13). These results support the statement that the presence of nanometer-size Y-TZP seed particles lowered the activation energy for the crystallisation process of the seeded Y-TZP amorphous precursors. It could be also assumed that the characteristics of the seed Y-TZP nanoparticles being isostructural with the final Y-TZP product and having a close lattice matching between them, will favour a decreasing of the activation energy for nucleation and reducing, thus, the overall energy requirements for epitaxial growth from the provided nuclei of the new synthesised Y-TZP phase. Such a growth process could take place by a surface diffusion mechanism, which needs a low activation energy to be carried out.

Table I. Kinetic constant k, and activation energy, E_a, for the crystallisation of unseeded and 10 wt% seeded Y-TZP amorphous precursors.

TEMPERATURE (K)	k (sec^{-1})	
	Y-TZP, unseeded	Y-TZP,seeded 10 wt %
593	-	3.20×10^{-5}
623	7.21×10^{-6}	5.75×10^{-5}
648	3.20×10^{-5}	-
673	9.98×10^{-5}	5.88×10^{-4}
E_a (kJ/mol)	184$\pm$9.2	119$\pm$6.0

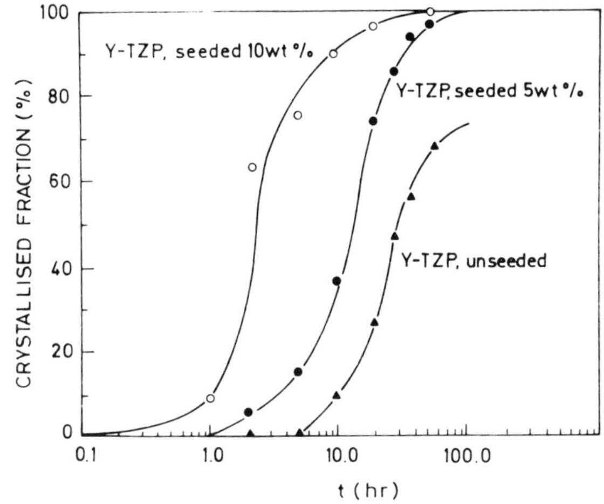

Fig.3, Isothermal kinetics for the Y-TZP crystallisation at 350^0C.

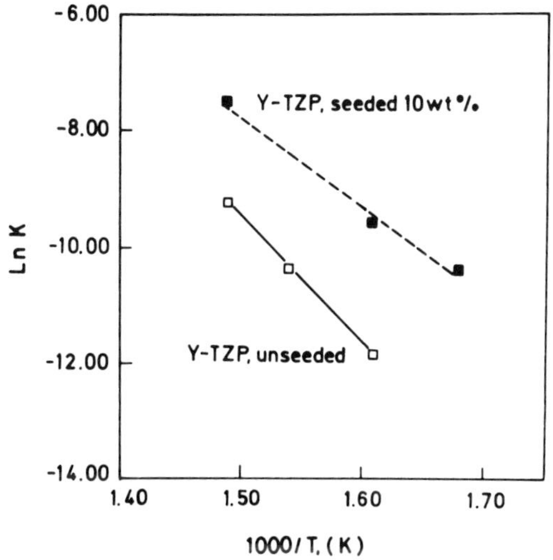

Fig.4, Arrhenius plots of cristallisation rate constant for air calcined unseeded and 10 wt% seeded Y-TZP amorphous precursors.

CONCLUSIONS

Seeding with Y-TZP nanometric particles (10 wt%) induces a lowering (about 60°C lower) in the crystallisation temperature of the Y-TZP phase and a strong increase in the crystallisation rate (about one order of magnitude higher) of Y-TZP formed in air-calcined amorphous powders relative to the unseeded Y-TZP ones. The isothermal crystallisation process is best fit by an Avramy-type kinetic relationship for a random nucleation process, with an activation energies of 184±9.2 kJ/mol (44.16±2.2 kcal/mol) and 119±6 kJ/mol (28.56±1.44 kcal/mol) for unseeded and 10 wt % seeded Y-TZP amorphous powders, respectively. Since the Avrami model assumes that growth is controlled by diffusion, then the low activation energy found for the Y-TZP crystallisation process support the contention of surface diffusion as the dominant mechanism for material transport at this thermal level.

REFERENCES

[1] R. J Brook, in Concise Encyclopedia of Advanced Ceramic Materials, ed. by R. J. Brook, p.1. Pergamon Press (1991).

[2] P. Durán, P. Recio, J.R. Jurado, C. Pascual, C. Moure, "Preparation, sintering and properties of translucent erbia-doped tetragonal zirconia",J. Am. Ceram. Soc., 72 2088-2093 (1989).

[3] W.F.M. Groot Zevert, G.S.A.M. Theunissen, A.J.A.Winnubst, A.J. Burgraaf, "Powder preparation and compaction behaviour of ultra-fine Y-TZP", J. Mat. Sci., 25 3449-3455 (1990).

[4] P.Durán, M.Villegas, F.Capel, P.Recio, C.Moure, "Low-temperature sintering and microstructural development of nanocrystalline Y-TZP powders", J. Europ. Ceram. Soc., 16 945-952 (1996).

[5] M. Kumagai, G.L. Messing, "Controlled transformation and sintering of a boehmite sol-gel by α-alumina seeding", J. Am. Ceram. Soc., 82[1-3] 277-285 (1986).

[6] J. Livage, K. Doi, C. Mazieres, "Nature and thermal evolution of amorphous hydrated zirconium oxides", J. Am. Ceram. Soc., 51 [6] 349-353 (1968).

[7] G.S.A.M. Theunissen, Ph. D. Thesis, Chapter II, University of Twente, 1991.

[8] M.M.R. Boutz, R..J..M. Olde Scholtenhuis, A..J..A. Winnubst, A..J. Burggraaf, "A hydrothermal route for production of dense nanostructured Y-TZP", Mater. Res. Bull., 29 31-40 (1994).

[9] G. Vilmin, S. Komarneni, R. Roy, "Crystallisation of $ThSiO_4$ from structurally and/or compositionally diphasic gels", J. Mater. Sci., 2 489-493 (1987).

[10] R. Roy, Y. Suwa, S. Komarneni, "Nucleation and epitaxial growth in diphasic (crystalline + amorphous) gels", pp. 247-258 in Science of Ceramic Chemical Processing. Edit. By C. C: Hench and D. Urich, Wiley, New York, (1986).

[11] R.P. Denkewicz, K.S. Tenhuisen, J.H. Adair, "Hydrothermal crystallisation kinetics of m-ZrO_2 and t-ZrO_2", J. Mater. Res, 5[5] 2698-2705 (1990).

[12] M. .Avrami, "Kinetics of phase change", J. Chem. Phys., 7[12] 1103-1112 (1939); Ibid. 9[2] 177-184 (1941).

[13] R. Srinivasan, R.V. Muraleedharan, S.K. Roy, P.K.K. Nayar, "Dehydration and crystallisation kinetics of zirconia yttria gels", J. Am. Ceram. Soc., 78[2] 429-432 (1995).

ACKNOWLEDGMENTS

This work was supported by the CICYT under the contract MAT-94-871.

PROCESSING OF $LiMn_2O_4$ POWDERS BY PRECURSOR METHOD

Yumi H. Ikuhara and Yuji Iwamoto
Fine Ceramics Research Association, Synergy Ceramics Laboratory
2-4-1, Mutsuno, Atsuta-ku, Nagoya, 456-8585, Japan
Koichi Kikuta
Department of Crystalline Materials Science, Nagoya University
1, Furo-cho, Chikusa-ku, Nagoya, 464-8503, Japan
Shin-ichi Hirano
Department of Applied Chemistry, Graduate School of Engineering,
Nagoya University
1, Furo-cho, Chikusa-ku, Nagoya, 464-8603, Japan

ABSTRACT

Single-phase $LiMn_2O_4$ fine powders could be prepared using alkoxide precursors at temperatures as low as 250°C. Stable and homogeneous precursor solutions were synthesized from lithium ethyl ethoxide, $LiOC_2H_4OEt$, and manganese ethyl ethoxide, $Mn(OC_2H_4OEt)_2$, by ligand exchange of LiO^iPr and $Mn(OEt)_2$ with 2-ethoxyethanol. Fine crystalline $LiMn_2O_4$ particles were prepared at 250°C with a particle size of around 20nm.

INTRODUCTION

Over the last ten years, lithium intercalated materials such as $LiMn_2O_4$ have been extensively studied as promising cathode materials for a secondary lithium ion battery because of their large electrochemical capacities and good cycling charge-discharge behavior [1-7]. In order to complete this cycling process, large particle surface area is required. Therefore the cathode needs to be prepared from fine powders with high crystallinity because the electrochemical behaviuor is highly affected by the microstructure of the cathode materials [8,9]. The high firing temperature usually results in extensive microstructural developement. Therefore powders of single phase of $LiMn_2O_4$ are desired to be synthesized at lower temperatures. In recent years, processing of $LiMn_2O_4$ powders at low temperature has been reported [10-13].

Chemical precursor methods through a solution route are advantageous for preparing fine powders with precise compositions at low temperatures. However, it has previously been difficult to prepare homogeneous manganese alkoxide precursor solutions because most of the manganese alkoxides cannot dissolve in organic solvents.

This paper describes details for a new route for processing lithium manganite precursor particles through modified metal alkoxides.

EXPERIMENTAL PROCEDURE

All reactions were conducted under dry nitrogen. Lithium isopropoxide, LiO^iPr and manganese ethoxide, $Mn(OEt)_2$, were used as starting reagents. Lithium ethyl ethoxide, $LiOC_2H_4OEt$, was prepared by reacting LiO^iPr with 2-ethoxyethanol, followed by refluxing for 3 hours. Manganese ethyl ethoxide, $Mn(OC_2H_4OEt)_2$, was also prepared by the addition of $Mn(OEt)_2$ to 2-ethoxyethanol, followed by a rapid stirring and refluxing for 3 hours. Ligand exchange reactions during formation of the modified alkoxides were examined by taking samples of the solutions, evaporating the solvents and analyzing the residue by FT-IR.

A homogeneous alkoxide solution was prepared by mixing the two solutions at a Li/Mn molar ratio of 0.5. A precursor powder of ultrafine particles was prepared by evaporating the solvent. Thermal decomposition mass spectroscopy (TDMS), thermal gravimetric analysis (TG), and differential thermal analysis (DTA) were employed to clarify the thermal decomposition process during preparation of the precursor powders.

The powder was heated up to 500°C in oxygen and X-ray diffraction measurements were made on prepared particles before and after heat treatment. To determine the morphology of the precursor derived particles, microstructural characterization by transmission electron microscopy (TEM) was conducted.

RESULTS AND DISCUSSION

Fig.1 shows the comparison of the FT-IR spectra of the starting lithium isopropoxide and the solution after ligand exchange reaction. Fig.1(a) exhibits a strong absorptions bands characteristic to the isopropoxy group at $1382cm^{-1}$ and $1390cm^{-1}$. These distinct absorptions ascribed to the C-H bending vibration and are due to the configurations in which two methyl groups are neighbor to the same carbon atom.

These absorptions have been diminished in the refluxed product spectrum as shown in Fig.1(b). The spectrum shows a strong absorption at $1119\ cm^{-1}$ due to

These absorptions have been diminished in the refluxed product spectrum as shown in Fig.1(b). The spectrum shows a strong absorption at 1119 cm^{-1} due to

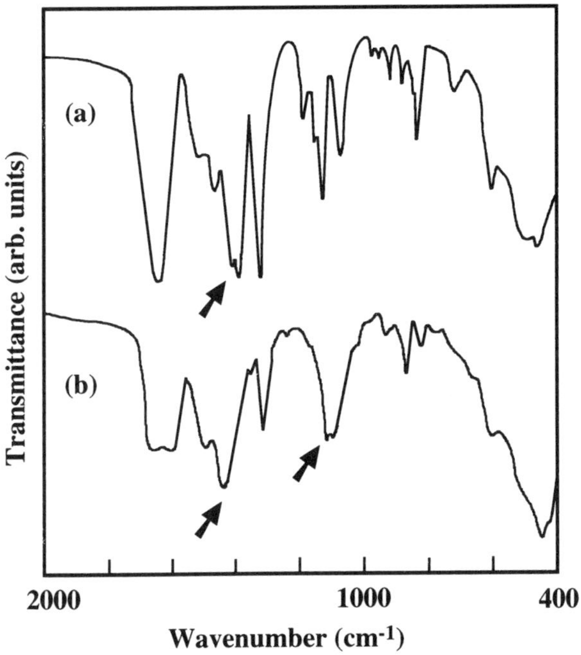

Fig.1. IR spectra of (a) starting alkoxide, LiO^iPr and (b) the reaction products after ligand exchange reaction.

asymmetrical C-O-C stretching. This is the characteristic response of ethers and is associated with the stretching of C-O-C bonds in the ethoxy ethyl group. The absorption band at 1443cm^{-1} is attributed to the asymmetrical bending of the methyl group in the chain structure of lithium ethyl ethoxide. The FT-IR analysis of the resulting reaction product after evaporating the solution suggests that the ligand exchange occurs from lithium isopropoxide to lithium ethyl ethoxide.

The IR spectra of the starting $Mn(OEt)_2$ and the solution after ligand exchange reaction are compared in Fig.2. Stretching vibration of C-O-C bond in ethoxyether is observed at 1119cm^{-1} region due to the asymmetrical ether stretching. Most characteristics of these spectra were the change in absorptions in the lower wavelength region. The broad band at 500cm^{-1} in Fig.2(a) can be attributed to the typical absorption of Mn^{II+}-O in $Mn(OEt)_2$. On the other hand, the reaction products contain triple absorptions that are typical for Mn^{III+}-O at $400cm^{-1}$, $500cm^{-1}$ and $700cm^{-1}$ as shown in Fig.2(b). Therefore, these absorption bands derived from Mn-O vibration in the refluxed products also changed due to the ligand exchange reaction of $Mn(OEt)_2$ to $Mn(OC_2H_4OEt)_x$.

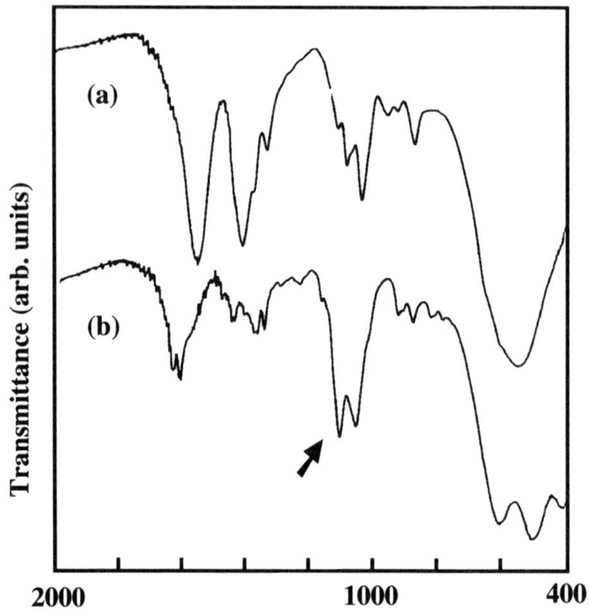

Fig.2. IR spectra of (a) starting alkoxide, $Mn(OEt)_2$ and (b) the reaction products after ligand exchange reaction

Ligand exchanges of those alkoxides are successfully controlled and a stable alkoxide precursor solution was prepared by mixing these solutions in appropriate molar ratio of metallic ions. Fine lithium manganite precursor particles could be prepared after evaporation of solvent.

Fig.3 shows DTA-TG curve for the lithium manganite precursor powder. The DTA curve shows a large exothermic reaction between 150°C and 230°C. The TG measurement indicates that the majority of the weight loss occurs in the temperature region from 100°C to 200°C. From the total weight loss of 32%, nearly 90% of the weight loss occurs below 200°C, while the remaining 10% occurs above 200°C.

The decomposion of organics for the precursor derived powders in oxygen can be more clearly observed from TDMS spectra of individual mass fragments. By

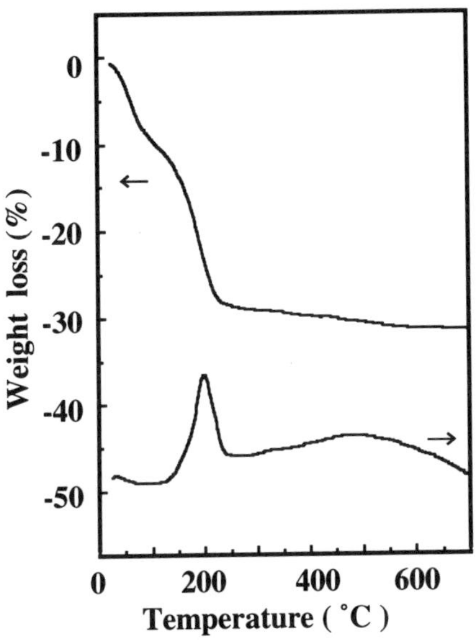

Fig.3. Thermogravimetric and differential thermal analysis of [Li-Mn-O] precursor at a rate of 5˚C/min.

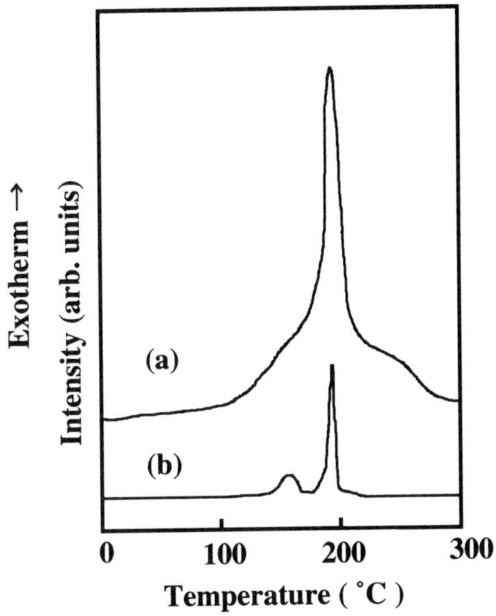

Fig.4. Thermal decomposition spectra for (a) mass 44 and (b) mass 45, extracted from the precursors in oxygen.

recording mass spectra and the temperature as a function of time, single mass species from the products can be extracted at a certain temperature. Fig.4 shows the change in mass spectroscopic ion signal for the $-CH_2OCH_2$ group, $-CH_2CH_2O$ group and CO_2 (mass 44) and for the $-OCH_2CH_3$ group (mass 45). The decomposition of the $-OCH_2CH_3$ group is observed to occurr around 150˚C to 170˚C and 180˚C to 200˚C. The relative intensity of the mass 44 species is much higher than that of mass 45. Thus, the $-CH_2CH_2OCH_2CH_3$ group in $LiOC_2H_4OEt$ and $Mn(OC_2H_4OEt)_x$ decomposed and combustion with oxygen resulted in evolution of CO_2.

The data of DTA and TDMS are in agreement since both show a sharp decomposition at low temperature, indicating that most of the organic components in the precursor decomposed continuously in the temperature region from 100˚C to 200˚C. Therefore, heat treatment was carefully conducted up to 200˚C in oxygen, followed by crystallization at various temperature up to 500˚C.

Fig.5 shows XRD data of the precursor powders before and after heat treatment at temperatures from 200°C to 500°C for 3 hours in oxygen atmosphere. The XRD profile of the precursor powder measured immediately after evaporation indicates that this [Li-Mn-O] precursor powder is X-ray amorphous. Single phase of lithium manganite spinel could be prepared at a temperature as low as 200°C, which is acceptable for the stable cathode materials. No other phases such as Mn_3O_4 were detected in powders heat-treated at temperatures below 500°C.

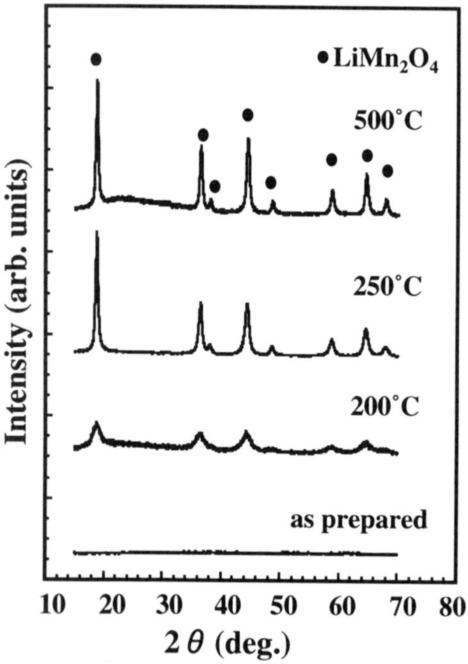

Fig.5. Powder X-ray diffraction for reaction products of [L-Mn-O] precursor as a function of the synthesis temperatures.

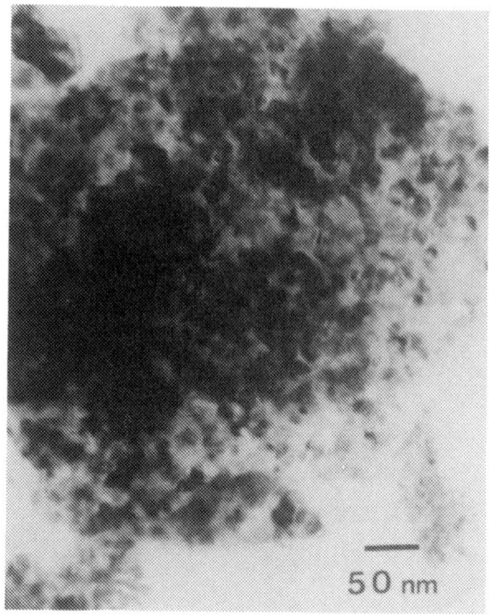

Fig.6. TEM micrograph of $LiMn_2O_4$ particles (10-20nm) with heat treatment at 200°C.

To understand the morphology and the microstructure of the powders for cathodic materials, TEM observation was conducted for the heat-treated powder. Fig.6 shows a bright-field transmission electron micrograph of an agglomerate of $LiMn_2O_4$ particles crystallized at 200°C. The agglomerate is composed of fine particles with 10 to 20 nm in size. The amorphous [Li-Mn-O] precursor has been decomposed and nucleation of the $LiMn_2O_4$ particles has occurred as shown in Fig.6. Fig.7(a) shows a bright field image of $LiMn_2O_4$ particles prepared at 250°C. The agglomerate is composed of polycrystals as identified in the diffraction patterns of the particles. By using the spots from the (400) plane of this diffraction pattern, the dark field image is obtained as shown in Fig.7(b). The particles with white contrast correspond to the (400) plane which are identified as single crystals. The particle size is measured to be around 20nm from this figure.

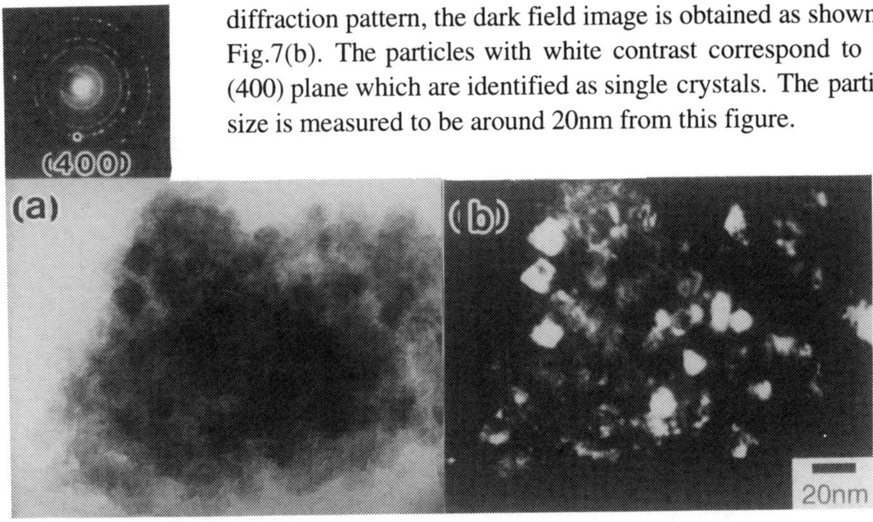

Fig.7(a) Bright field image of $LiMn_2O_4$ particles synthesized at 250°C and diffraction pattern obtained from the particles.
(b) Dark field image of selected (400) plane in the diffraction pattern, showing an agglomerate of single crystals.

The $LiMn_2O_4$ crystalline powder with a particle size of 20-30nm was also prepared with heat treatment at 500°C in oxygen. From the microstructural observations, there are no other secondary phases present to decrease the electrochemical properties, which is in agreement with the results of XRD analysis. A key factor for the preparation of such fine $LiMn_2O_4$ powder is how to the design a stable precursor by ligand exchange of metal alkoxides.

CONCLUSIONS

A stable and homogeneous alkoxide precursor solution $LiOC_2H_4OEt$ and $Mn(OC_2H_4OEt)_2$ in 2-ethoxyethanol was synthesized using ligand exchange reaction. [Li-Mn-O] precursor powder could be synthesized at temperature as low as 200°C by evaporating the solvent, which could be crystallized to fine particles of around 20nm in size at temperatures as low as 250°C. This powder should be suitable for preparing electrodes of $LiMn_2O_4$ with high surface area.

ACKNOWLEDGMENT

Work supported by NEDO as part of the Synergy Ceramics Project under the Industrial Science and Technology Frontier (ISTF) Program promoted by AIST, MITI, Japan. The authors are members of the Joint Research Consortium of Synergy Ceramics.

REFERENCES

1. M. M. Thackeray, P. J. Johnson and L. A. de Picciotto, P. G. Bruce and J. B. Goodenough, Mat. Res. Bull., 19, (1984), 179.
2. M. H. Rossouw, A. de Kock, L. A. de Picciotto and M. M. Thackeray, W. I. F. David, and R. M. Ibberson, Mat. Res. Bull., 25, (1990), 173.
3. T. Ohzuku, M. Kitagawa, and H. Taketsugu, J. Electrochem. Soc., 137, (1990), 769.
4. H. Huang and P. G. Bruce, J. Electrochem. Soc.,141, (1994), 185.
5. J. M. Tarascon, W. R. McKinnon, F. Coowar, T. N. Bowmer, G. Amatucci and D. Guyomard, J. Electrochem. Soc., 141, (1994), 1421.
6. R. J. Gummow, A. de Kock, and M. M. Thackeray, Solid State Ionics, 69, (1994), 59.
7. C. Masquelier, M. Tabuchi, K. Ado, R. Kanno, Y. Kobayashi, Y.Maki, O. Nakamura, J. B. Goodenough, J. Solid State Chem., 123, (1996), 256.
8. Z. Jiang, and K. M. Abraham, J. Electrochem. Soc., 143, (1996), 1591.
9. T. Tsumura, A. Shimizu, and M. Inagaki, Solid State Ionics, 90, (1996), 197
10. S. Bach, M.Henry, N. Baffier, and J. Livage, J. Solid State Chem. 88, (1990), 325.
11. P. Barboux, J. M. Tarascon, and F. K. Shokoohi, J. Solid State Chem. 94, (1991), 185
12. T. Tsumura, A. Shimizu, and M. Inagaki, J. Mater. Chem.,3, (1993), 995
13. W. Liu, G. C. Farringtonaa, F. Chaput and B. Dunn, J. Electrochem. Soc., 143, (1996), 879.

Non-Oxide Ceramics from Polymers and Gels

Ralf Riedel
Technische Universität Darmstadt, Fachbereich Materialwissenschaft, Fachgebiet Disperse Feststoffe, Petersenstr. 23, D-64287 Darmstadt, Germany

The paper highlights the synthesis, processing and properties of non-oxide silicon-based ceramics derived from organosilicon polymers and gels. In particular, the preparation of the first crystalline compounds in the ternary Si-C-N system with the compositions SiC_2N_4 and Si_2CN_4 is reported [1]. The crystallization behavior, the phase transformations and the thermal stability of the synthesized amorphous and polycrystalline silicon carbonitrides in different atmospheres are discussed. Furthermore, the fabrication of dense bulk ceramics in the Si-C-N system was achieved by the novel hybrid processing. The hybrid processing involves cross-linking of the polymeric ceramic precursor followed by a polymer to ceramic transformation step [2]. The cross-linked precursor can be milled, compacted and pyrolyzed to form dense, additive-free, amorphous silicon carbonitride monoliths or polycrystalline Si_3N_4/SiC/C micro/nano composites which withstand oxidation up to 1600 °C in air [3].

Polyborosilazanes synthesized from single source precursors can be transformed to silicoboron carbonitrides ($SiB_xC_yN_z$) exhibiting enhanced thermal stability with respect to crystallization and decomposition in the temperature range between 1400 and 2000 °C [4]. The ultra-high temperature stability of the silicoboron carbonitrides is discussed in terms of the Si/B- and C/N-ratios.

The sol-gel process for the synthesis of oxide glasses and ceramics has been the subject of intensive investigations in materials science and is used in a wide range of technical applications. However, it was limited hitherto to systems containing oxygen. We now report on the first anhydrous sol-gel process for the synthesis of oxygen-free silicon-based ceramics in the Si-(B)-C-N system [5]. The novel sol-gel route is studied with respect to its oxide counterpart and allows the implementation of a whole series of technologically relevant production processes, which are already known from the conventional sol-gel technique. Preliminary results of our investigations concerning sol-gel derived silicon carbonitride ceramics in form of nano-spherical powders, fibers, and membranes as well as on the coating and infiltration of porous substrates will be reported.

References:

[1] Riedel, R., Greiner, A., Miehe, G., Dreßler, W., Fueß, H., Bill, J., Aldinger, F., Angew. Chem. Int. Ed. Engl. **36** (1997) 603 - 606.

[2] Riedel, R., Passing, G., Schönfelder, H. & Brook, R.J. Nature **355**, 714 - 716 (1992).

[3] Riedel, R., Kleebe, H.-J., Schönfelder, H. & Aldinger, F. Nature **374**, 526 - 528 (1995).
[4] Riedel, R., Kienzle, A., Dreßler, W., Ruwisch, L., Bill, J. & Aldinger, F. Nature **382** (1996) 796 - 798.
[5] a) Gabriel, A.O., Riedel R., Angew. Chem. **109**, 371-373 (1997); Angew. Chem. Int. Ed. Engl. **36**, 384-386 (1997).
b) Gabriel, A.O., Riedel R., Storck S., Maier W.-F., Appl. Organomet. Chem. **11**, 833-841 (1997).

[Si-Ti-Y-O-N] MULTICOMPONENT CERAMICS DERIVED FROM POLYMERIC PRECURSORS

Yuji Iwamoto, Ko-ichi Kikuta* and Shin-ichi Hirano**
Fine Ceramics Research Association, Synergy Ceramics Laboratory
2-4-1 Mutsuno Atsuta-ku, Nagoya 456-8587, Japan
*Department of Crystalline Materials Science, Nagoya University
1 Furo-cho, Chikusa-ku, Nagoya 464-8503, Japan
**Department of Applied Chemistry, Graduate School of Engineering, Nagoya University, 1 Furo-cho, Chikusa-ku, Nagoya 464-8603, Japan

ABSTRACT

A polymeric precursor for [Si-Ti-Y-O-N] multicomponent ceramics was prepared by chemical modification of perhydropolysilazane with titanium tetrachloride and yttrium tri-methoxide. The polymeric precursor yielded TiN nano particles dispersed Si_3N_4-Y_2O_3 ceramics by pyrolysis at 1000 °C in NH_3 followed by heat treatment at 1800 °C in N_2.

INTRODUCTION

Silicon nitride (Si_3N_4) is well known for its high-strength, low thermal expansion and thermal shock resistance. Recently, ceramic particle-reinforced Si_3N_4 composite has drawn considerable attention because of substantial enhancement in the mechanical properties of Si_3N_4 ceramics, and titanium nitride (TiN) is one of the useful particle-reinforcement for Si_3N_4 ceramics which improves the fracture toughness and strength.[1, 2)]

There has been much interested in the use of silicon-based metall-organic precursors for manufacturing high-performance ceramics such as Si_3N_4.[3,4)] One of the advantages of this route for the preparation of ceramics is the ability of the synthesis of silicon-based multicomponent ceramics through chemical modification of silicon-based metall-organic precursors.[5,6)] Authors have also reported that poly-yttrosilazane derivatives derived from commercially available perhydropolysilazane (PHPS), n-alkyl alcohol and yttrium alkoxide were useful for the fabrication of [Si-Y-O-C-N] multiconponent ceramics such as nano-scale SiC particle dispersed Si_3N_4-Y_2O_3 ceramics.[7)] If molecular titanium source is introduced to PHPS by chemical modification along with yttrium alkoxide, the resulting polymer is expected to be a novel precursor for [Si-Ti-Y-O-N] multicomponent ceramics of nano-scale TiN particle dispersed Si_3N_4-Y_2O_3 ceramics. The present paper describes (1) a synthetic study of a poly-yttrosilazane derivative from a commercially available PHPS, titanium tetrachloride and yttrium alkoxide and (2) synthesis of Si_3N_4-TiN-Y_2O_3 ceramics through the polymeric precursor route.

EXPERIMENTAL PROCEDURE

Commercially available PHPS (Tonen Co. N-N610, average molecular weight of 5156, Tokyo, Japan) was used as a starting polymer. Reagents of titanium tetrachloride ($TiCl_4$, Wako Pure Chemical Industries, Ltd., Osaka, Japan) and yttrium tri-methoxide ($Y(OCH_3)_3$, High Purity Chemical Laboratory Co., Saitama, Japan) were used for the chemical modification of PHPS in this study. As-received PHPS (6.65 g: Si, 63.5; N, 24.8; O, 0.7; C, 2.4 wt.%, hence Si=150 mmol) was dissolved into dry dichloromethane (CH_2Cl_2, 200 ml) and cooled to -20 °C. $TiCl_4$ (2.85 g, 15 mmol, hence an atomic ratio of Si/Ti=10) was added slowly by syringe while stirring. After the addition was completed, the reaction mixture was warmed to room temperature, then CH_2Cl_2 was removed by distillation using a rotary evaporator to give a light green powder of titanium-modified PHPS, (Ti-PHPS, 9.1 g, 96 wt.%). Ti-PHPS was further reacted with $Y(OCH_3)_3$ (1.09 g, 6 mmol, hence an atomic ratio of Si/Y=50) in dry pyridine according to the published procedure[7] to give a pale brown powder of titanium and yttrium-modified PHPS, (Ti-Y-PHPS, 12.7 g, 125 wt.%, over yield caused by inclusions of pyridine and pyridine hydrochloride). This polymer was found to be insoluble in organic solvents such as toluene, ether and chloroform. To examine the chemical structure of Ti-modified PHPS, as-received PHPS was also reacted with $Y(OCH_3)_3$ (6 mmol, Si/Y=50) to give a white powder of yttrium-modified PHPS, (Y-PHPS, 6.9 g, 90 wt.%).

Ti-Y-PHPS was ground to fine powders using a mortar and pestle, then pyrolyzed in an quartz tube furnace (Motoyama, Osaka, Japan) under flowing ammonia (NH_3, 400 ml/min) by heating from room temperature to 1000 °C in 3h, maintaining the temperature at 1000 °C for an additional 3h, followed by furnace cooling down to room temperature to give an [Si-Ti-Y-O-C-N] amorphous, black powder, product (65 wt. % based on the total amount of PHPS and reagents used in the chemical modification reactions). As-received PHPS was also pyrolyzed under the same condition as mentioned above to give a brown [Si-N] amorphous product (75 wt.%). The pyrolyzed sample was ground to fine powders using a mortar and pestle, then sieved through a 250 μm screen before uniaxial pressing at 150 MPa (15 mm diameter, 1.5g). The green compacts were placed on a BN plate within a Si_3N_4 crucible and heat-treated in a graphite resistance-heated furnace (Model High Multi 10000, Fujidempa Kogyo, Osaka, Japan) at temperatures of 1200 °C, 1400 °C, 1600 °C and 1800 °C. N_2 pressures applied in this heat treatment procedure were 196 kPa at 1200 °C, 392 kPa at 1400 °C and 980 kPa above 1600 °C.

Infrared (IR) spectra were recorded on KBr pellets containing polymer samples (Model System 2000, Perkin Elmer, Connecticut, USA). Elemental analyses for Si,Ti, Y, O, C and N were carried out on the pyrolyzed and heat treated samples. X-ray diffraction (XRD) measurements were performed on the pyrolyzed and heat treated samples with CuKα radiation using an automated powder diffractometer equipped with a monochrometer (Model RINT 2500, Rigaku, Tokyo, Japan).

Apparent density of the heat-treated specimens was measured by the Archimedes method. Relative density (RD) was evaluated using the theoretical density (3.365 g/cm^3) calculated from the composition of the Ti-Y-PHPS-derived amorphous powder.

Microstructure of the heat-treated specimens was studied by scanning electron microscopy (SEM, Model S-800, Hitachi, Tokyo, Japan) and transmission electron microscopy (TEM, Model JEM-100CX, JEOL, Tokyo, Japan).

RESULTS AND DISCUSSION

The IR spectrum analysis of the as-received PHPS resulted in absorption peaks at 3400 cm^{-1} (N-H), 2950-2900 cm^{-1} (C-H), 2150 cm^{-1} (Si-H), 1250 cm^{-1} ($Si\text{-}CH_3$), 1180 cm^{-1} (N-H) and 840-1020 cm^{-1} (Si-N-Si). Figure 1 shows a typical change in IR spectra of chemically modified samples. Compared with the as-received PHPS (Fig. 1(a)), a remarkable decrease in absorption intensity at 1180 cm^{-1} (N-H) is observed in the IR spectra of Ti-PHPS (Fig. 1(b)) and Y-PHPS (Fig. 1(c)). In our previous study using IR and ^{1}H-NMR analysis, reaction of PHPS with $Y(OCH_3)_3$ formed N-Y bonds.[7)] Narula et al. also reported that hexamethyldisilazane ($HN[Si(CH_3)_3]_2$) reacted with $TiCl_4$ to form N-Ti bond.[8)] According to our IR spectroscopic analysis and the published result, PHPS with $TiCl_4$ seemed to form N-Ti bonds.

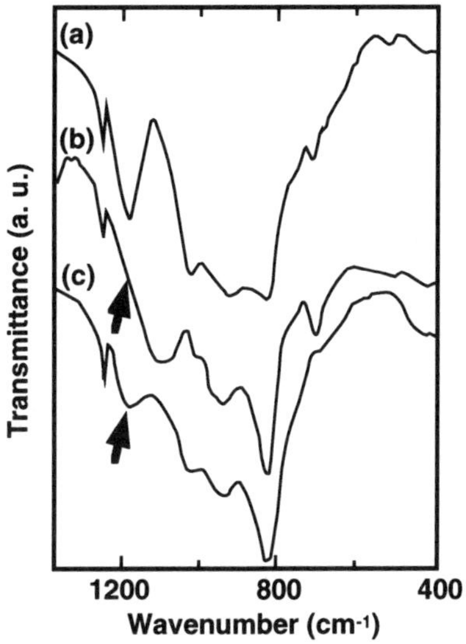

FIG. 1. IR spectra of (a) as-received PHPS, (b)Ti-PHPS and (c) Y-PHPS.

After chemical modification of Ti-PHPS with $Y(OCH_3)_3$, the polymer sample (Ti-Y-PHPS) was pyrolyzed. The chemical composition of the pyrolyzed samples is shown in Table I. The Si/Ti and Si/Y atomic ratios of the pyrolyzed sample derived from Ti-Y-PHPS are 13 and 62, respectively, which are slightly larger than the calculated values of 10 and 50, and carbon content is also increased in comparison with that of the pyrolyzed sample derived from as-received PHPS.

Table I. Chemical composition of pyrolyzed samples.

Polymeric precursor	Composition (wt.%)						Atomic ratio	
	Si	Ti	Y	O	C	N	Si/Ti	Si/Y
Ti-Y-PHPS	60.5	8.1	3.1	5.6	4.2	21.0	13	62
As-received PHPS	71.8	0.0	0.0	0.4	1.8	25.0	—	—

One reason for the increase in carbon content is considered to be caused by an inclusion of pyridine hydrochloride ($HCl \cdot NC_5H_5$) in Ti-Y-PHPS according to the following reaction scheme.

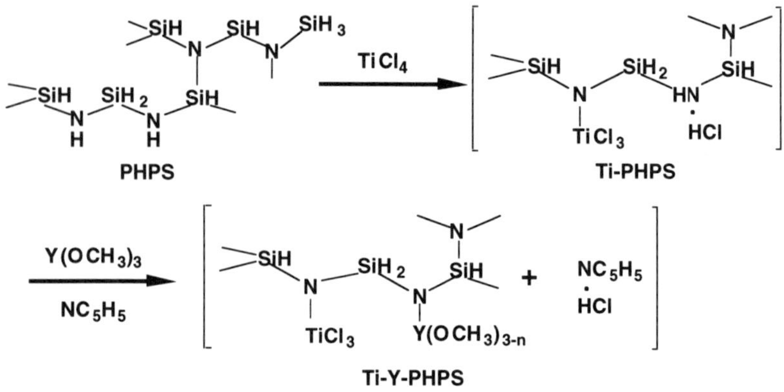

However, in comparison with the as-received PHPS, titanium and yttrium were successfully doped in the pyrolyzed sample by a two-step chemical modification reaction of PHPS.

The XRD patterns of the pyrolyzed samples are shown in Fig. 2. The sample derived from the as-received PHPS exhibits a broad amorphous diffraction line (Fig. 2(a)), while the sample derived from Ti-Y-PHPS shows distinct peaks at around 2Θ (degrees) of 36.5, 42.5 and 61.7, which indicates that the crystallization of TiN or Ti(C,N) started during pyrolysis.

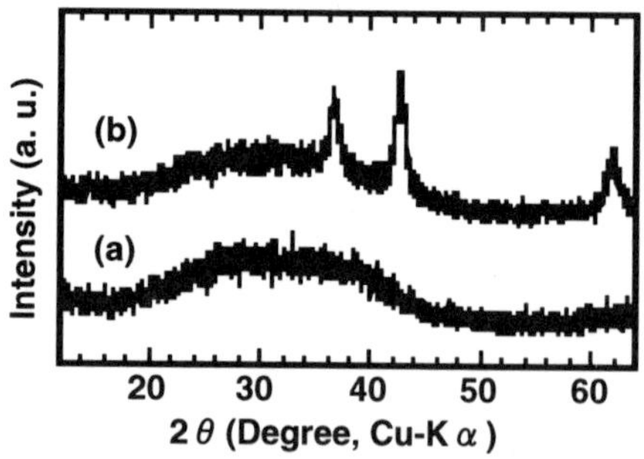

FIG. 2. XRD patterns of pyrolyzed samples, derived from (a) as-received PHPS and (b) Ti-Y-PHPS.

Further crystallization behavior of Ti-Y-PHPS-derived powder compact is shown in Fig. 3. At 1200 °C, α-Si_3N_4 crystallization is observed. At 1400 °C, both α- and β-Si_3N_4, and Si_2N_2O peaks begin to appear. The Si_2N_2O peaks disappear and the α-/β-Si_3N_4 transformation is completed at 1800 °C. On heating up to 1800 °C, the distinct peaks observed in the pyrolyzed sample are sharp. However, at 1200 °C, the peaks are slightly shifted to lower angles, which indicates the formation of $Ti(C_{0.3}, N_{0.7})$ (based on JCPDS Card No. 42-1488). Shift to higher angles with increasing temperature were observed and, at 1800 °C, the XRD pattern was identical to TiN (based on JCPDS Card No. 38-1420).

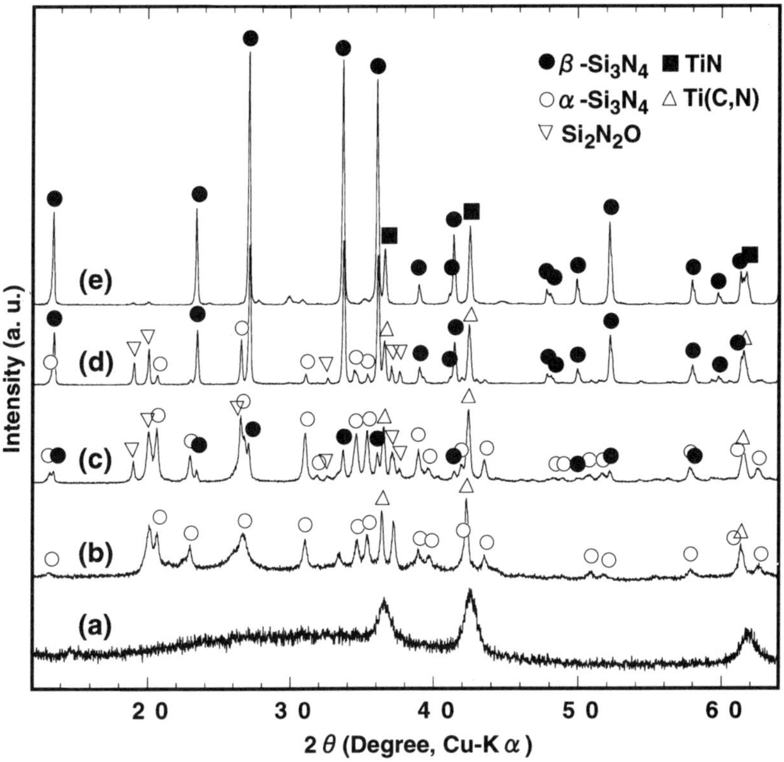

FIG. 3. Crystallization behavior of Ti-Y-PHPS-derived powder compact. (a) 1000 °C-pyrolyzed in NH_3 flow, (b) 1200 °C in N_2 (196 kPa), (c) 1400 °C in N_2 (392 kPa), (d) 1600 °C in N_2 (980 kPa) and (e) 1800 °C in N_2 (980 kPa).

Figure 4 shows carbon, oxygen and nitrogen content, and weight change of the Ti-Y-PHPS-derived powder compact during heat treatment up to 1800 °C. Both carbon and oxygen contents decrease with increasing temperature, especially remarkable decreases in carbon and oxygen contents are observed during heating process from 1200 °C to 1600 °C, while nitrogen content increases with increasing temperature. During these compositional changes, weight loss of the compact increases with increasing temperature.

The crystallization behavior observed below 1600 °C indicates that [Si-Y-O-N] liquid phase promoted both α-Si_3N_4 crystallization and α-/β-Si_3N_4 transformation below 1400 °C. The formation of Si_2N_2O at 1400 °C is due to the excess oxygen introduced from $Y(OCH_3)_3$ (O/Y=3) and the Ti-Y-PHPS-derived powders may be highly reactive towards ambient moisture which leads to an increase of total oxygen content (Table I). During the heating process from 1400 °C to 1800 °C, Si_2N_2O decomposed to Si_3N_4, with SiO and O_2 as volatile species.

Figure 5 shows the ellingham diagram for Si-Ti-C-N_2 system. N_2 pressures applied in this study are also plotted in Fig. 5. Values of the thermodynamic equations were taken from literature.[9] As shown in Fig. 5, N_2 pressure applied below 1600 °C are higher than the equilibrium pressure for the equation of 3/2SiC +N_2 = Si_3N_4 + 3/2C, and the amount of TiC in the solid solution of Ti(C_{1-x}, N_x) can be determined by its own stability.[10] However, as shown in Table I, Ti-Y-PHPS-derived powders were [Si-Ti-Y-O-C-N] multicomponent amorphous phase with excess oxygen, and carbon in the amorphous phase was spent for reactions such as carbothermal nitridation of SiO_2. The Ti-Y-PHPS-derived [Si-Ti-Y-O-C-N] multicomponent amorphous phase was thermodynamically metastable and was able to give off gases such as CO and SiO during heat treatment up to 1800 °C. This lead to the reduction of the amount of TiC in Ti(C_{1-x}, N_x) and weight loss of the powder compact. Finally, at 1800 °C, carbon in the amorphous phase was spent out and thermodynamically stable β-Si_3N_4 and TiN were formed.

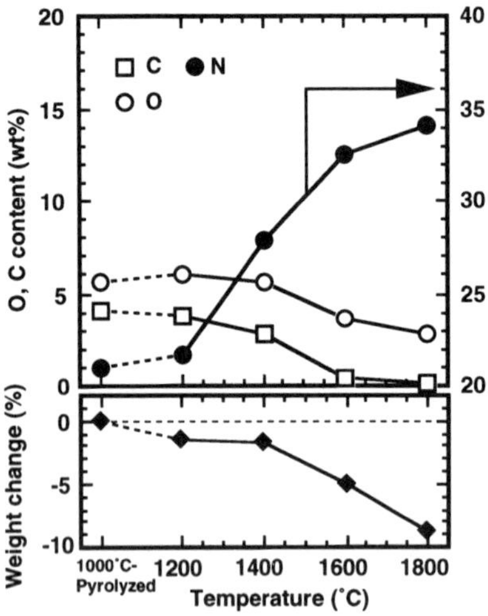

FIG. 4. Carbon, oxygen and nitrogen content, and weight change of Ti-Y-PHPS-derived powder compact during heat treatment.

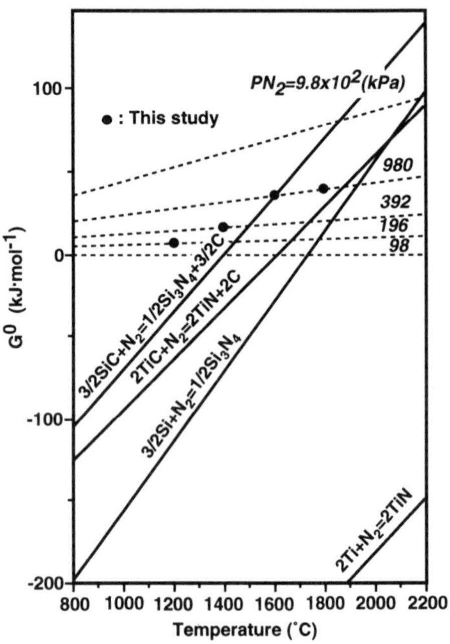

FIG. 5. The Ellingham diagram for Si-Ti-C-N_2 system.

Figure 6 shows microstructures of the powder compact heat-treated at 1800 °C. The density of the heat-treated compact was low (75% RD), and the Ti-Y-PHPS-derived powder yielded a uniformly fine-grained microstructure (Fig. 6(a)). Some nanometer particles are finely dispersed within β-Si_3N_4 grains (Fig. 6(b)), which are considered to be TiN detected by XRD as a minor phase (Fig. 3(e)). These observations suggest that the polymeric precursor route investigated in this study has a potential to synthesize TiN nano particle dispersed Si_3N_4-based ceramics.

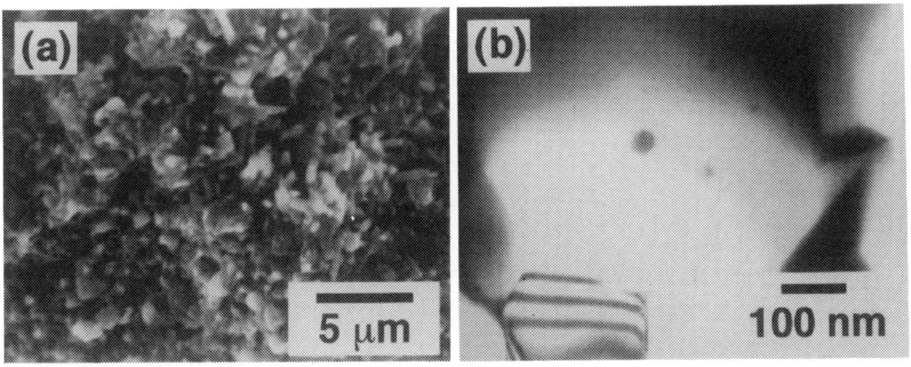

FIG. 6. Microstructure of Ti-Y-PHPS-derived powder compact heat-treated at 1800 °C for 1h in N_2 (980 kPa). (a) SEM micrograph of fracture surface and (b) TEM micrograph observed within the matrix β-Si_3N_4 grains.

CONCLUSIONS

[Si-Ti-Y-O-C-N] multiconponent amorphous powders were synthesized by the pyrolysis at 1000 °C, in NH_3, of chemically modified PHPS using $TiCl_4$ and $Y(OCH_3)_3$. The [Si-Ti-Y-O-C-N] amorphous powders yielded uniformly fine-grained Si_3N_4-TiN-Y_2O_3 ceramics by heat treatment at 1800 °C in N_2. The resulting ceramics revealed that nano-scale TiN particles were finely dispersed within the matrix β-Si_3N_4 grains and the polymeric precursor route investigated in this study has a potential to synthesize TiN nano particle dispersed Si_3N_4-based ceramics.

ACKNOWLEDGMENT

Work supported by NEDO as part of the Synergy Ceramics Project under the Industrial Science and Technology Frontier (ISTF) Program promoted by AIST, MITI, Japan. The authors are members of the Joint Research Consortium of Synergy Ceramics.

REFERENCES

1) A. Bellosi, S. Guicciardi and A. Tampieri, "Development and Characterization of Electroconductive Si_3N_4-TiN Composites," J. Europ. Ceram. Soc., **9**, 83-97(1992).
2) M. Miyata, Y. Yasutomi, Y. Sawai and T. Kanai, "Microstructure and Mechanical Properties of Self-Composed Si_3N_4-TiN Ceramics Prepared from Si, TiC and TiO_2 Mixed Powder," J. Ceram. Soc. Jpn., **105**[9], 761-77(1997).
3) D.Seyferth and G.H.Wiseman, "Silazane Precursors to Silicon Nitride," pp.2656-71 in Ultrastructure Processing of Ceramics, Glasses and Composites, edited by L. L. Hench and D. R. Ulrich, Willey-Interscience, New York, 1984.
4) O. Funayama, M. Arai, Y. Tashiro, H. Aoki, T. Suzuki, K. Tamura, H. Kaya, H. Nishii and T. Isoda, "Tensile Strength of Silicon Nitride Fibers Produced from Perhydropolysilazane," J. Ceram. Soc. Jpn., **98**[1], 104-107(1990).
5) O.Funayama, T. Kato, Y.Tashiro and T.Isoda, "Synthesis of Polyborosilazane and Its Conversion into Inorganic Compounds," J. Am. Ceram. Soc., **76**[3], 717-723(1993).
6) R. Reidel and W. Dressler, "Chemical formation of Ceramics," Ceram.Int., **22**, 233-239(1996).
7) Y. Iwamoto. K. Kikuta and S. Hirano,"Microstructural development of Si_3N_4-SiC-Y_2O_3 ceramics derived from polymeric precursors," J. Mater. Res., **13**[2], 353-361(1998).
8) C. K. Narula, B. G. Demczyk, P. Czubarow and D. Seyferth, "Preparation of Silicon Nitride-Titanium Nitride and Titanium-Titanium Nitride Composites from $(CH_3)_3SiNHTiCl_3$-Coated Si_3N_4 and Ti Particles," J. Am. Ceram. Soc., **78**[5], 1247-51(1995).
9) JANAF Thermodynamical Table, 3rd Edition, US National Bureau of Standards, Washington D. C., 1986.
10) M. Herrmann, B. Balzer, Chr. Schbert and W. Hermel, "Densification, Microstructure and Properties of Si_3N_4-Ti(C,N) Composites," J. Europ. Ceram. Soc., **12**, 287-296(1993).

SYNTHESIS OF INORGANIC-ORGANIC HYBRIDS FROM METAL ALKOXIDES AND PHENYLTRIETHOXYSILANE

Ikuko Yoshinaga, Noriko Yamada, and Shingo Katayama
Advanced Technology Research Labs
Nippon Steel Corporation
3-35-1 Ida, Nakahara-Ku
Kawasaki 211, Japan

ABSTRACT

Inorganic-organic hybrids represent a new class of materials with varying properties depending on constituent inorganic and organic species. Most of the hybrids have been synthesized by utilizing the hydrolysis of organoalkoxysilane. Hybrids synthesized from phenyltriethoxysilane have higher thermal stability and mechanical strength. TEOS is often used as an inorganic source with which phenytriethoxysilane can be copolymerized. The properties of the hybrids are anticipated to be widely altered by the incorporation of other inorganic components. In this work, inorganic-organic hybrids have been synthesized from phenyltriethoxysilane and alkoxide of Al or Ti. The formation of Ti-O-Si bonds in the hybrids synthesized with Ti ethoxide was confirmed by an infrared absorption peak at 922 cm-1 attributed to the n(Ti-O-Si) vibration. Dynamic mechanical measurements of the hybrids synthesized with Al ethoxide were carried out in the temperature range of -150° to 300°C. The hybrids exhibited a storage modulus of 3 Pa below -130°C. As the temperature increased, a drop in the storage modulus was observed around -70° and 150°C. The tand exhibited a sharp peak at -70°C and a broad peak at 150°C corresponding to the drops in the storage modulus. This indicates that the hybrid has two constituents showing different glass transition temperature. Research supported by NEDO, under the Synergy Ceramics Project of the ISTF program promoted by AIST, MITI, Japan.

COMPOSITES IN THE Ti-Si-C SYSTEM

Chang-Sun Park, Feng Zheng,* Sam Salamone,* and Rajendra Bordia*
Department of Metallurgical Engineering
Chosun University, Korea

*Department of Materials Science and Engineering
University of Washington
Seattle, WA

ABSTRACT

The ternary compound Ti_3SiC_2, in the M-SI-C system, has received increased attention due to its unique combination of properties. These include machinability. retaining ductility at elevated temperatures, refractory behavior, and oxidation resistance. Previous synthesis routes included CVD, Self-propagating high temperature synthesis (SHS), Arc malting, and solid state reactions combining elemental species and carbides (SiC or TiC). Our goal is to make in-situ composites of Ti_3SiC_2 with TiC or SiC by solid state reaction during sintering with elemental Si, Ti, and C powders. This process allows the in-situ production of composites and strict control over the volume percent of the phases. The thermal and kinetic aspects of the phase formation were examined using standard dry powder processing techniques and pressureless sintering in an inert atmosphere. The ratio of the volume fraction of Ti_3SiC_2 to the volume fraction of SiC or TiC has been varied over a wide range by control of the powder stoichiometry and heat treatment. The composites were characterized by semi-quantitative XRD to determine the phase composition and the relative percentage of each phase present. The kinetic response was investigated using two different particle sizes of Ti powder. The focus of this presentation will be the effect of stoichiometry and heat treatment on phase evolution. It is shown that composites with varying amounts of Ti_3SiC_2 with TiC can be produced by controlling the heat treatment at a fixed stoichiometry (under certain conditions minor amounts of Ti_5Si_3 are also produced). However, in order to make composites of Ti_3SiC_2 and SiC, the stoichiometry has to be changed. The results are rationalized using existing phase diagrams for this system.

ALUMINIUM MODIFIED POLYSILAZANE PRECURSORS FOR SI-C-N CERAMICS

B. Bauer and B. Derby
Department of Materials
Oxford University
Parks Road, OX1 3PH, UK

ABSTRACT

Ceramics fabricated from Si containing polymeric precursors can only be sintered to full density with difficulty when in bulk form. This is because of the covalent character of the Si-C and Si-N bonds. In order to achieve full density without the application of pressure, sintering aids are commonly added. These must be added after polymer processing and pyrolysis and reduce the effectiveness of the pre-ceramic polymer fabrication route. Here we describe a novel route by which metal oxide species believed to aid sintering are added as side chains to a polysilazane precursor polymer. Methylaluminiumdiethoxide and diethylaluminiumethoxide reagents are reacted with a commercial polysilazane (NCP 200 Nihimen, Japan). The chemical structure of the resulting modified polymer has been characterized by nuclear magnetic resonance and fourier transform infra-red spectroscopy which show the formation of Al-N bonds. The pyrolysis product has been characterized by porosimetry and microanalysis to determine its structure and chemical composition and show an amorphous Siliconoxycarbide with a uniform pore size distribution. X-ray diffraction and transmission electron microscopy have been used to determine the final product after a high temperature crystallization and densification heat treatment.

PROCESSING OF SILICON CARBIDE CERAMICS USING CHEMICALLY MODIFIED POLYCARBOSILANE CONTAINING FLUOROALKOXY GROUPS AS BINDER

Yuji Iwamoto, Sachiko Okuzaki, Ko-ichi Kikuta* and Shin-ichi Hirano**
Fine Ceramics Research Association, Synergy Ceramics Laboratory
2-4-1 Mutsuno Atsuta-ku, Nagoya 456-8587, Japan
*Department of Crystalline Materials Science, Nagoya University
1 Furo-cho, Chikusa-ku, Nagoya 464-8503, Japan
**Department of Applied Chemistry, Graduate School of Engineering, Nagoya University, 1 Furo-cho, Chikusa-ku, Nagoya 464-8603, Japan

ABSTRACT

Chemically modified polycarbosilane (PC) containing fluoroalkoxy groups as side chains (PCOCF) could be synthesized from a commercially available PC and fluoroalkyl alcohol. PCOCF was found to be act as an efficient binder for SiC powder compaction in forming process and PCOCF also yielded β-SiC microcrystallites on SiC powder surfaces after pyrolysis at 1000 °C.

INTRODUCTION

Silicon carbide (SiC) is well known for its high thermal and chemical stability, high strength, hardness and thermal conductivity. Recently, increasing attention has been directed to the use of silicon-based polymeric precursors for fabrication of SiC ceramics. One of the advantages of this route is that the solubility and fluidity of the polymeric precursors afford potential routes for the preparation of SiC fibers[1)] and coatings[2)], and the use of the polymeric precursors as a binder in the conventional powder processing.[3, 4)]

Polycarbosilane (PC) is the most typical polymeric precursor for SiC base-ceramics, and PC can be chemically modified to yield binary systems such as SiC-TiC[5)] and SiC-ZrC.[6)] The authors have also synthesized polyaluminocarbosilane (PCOAl), and SiC-based ceramics with submicron sized platelet grains were synthesized from PCOAl and SiC powders.[7)]

In this study, chemically modified PC containing fluoroalkoxy groups (PCOCF) was synthesized to improve the fluidity of PC. This is expected to be essential for improving green density of SiC powder compacts when combined with SiC powders. This paper describes (1) a synthetic study of PCOCF, (2) effect of PCOCF coating on green density of SiC powder compacts and (3) pyrolysis behavior of PCOCF on SiC powder surfaces.

EXPERIMENTAL PROCEDURE

Figure 1 shows a synthetic route for PCOCF. As reactive functional groups, Si-Cl bonds were introduced to a commercially available PC (Nippon Carbon Co. Ltd., Type L, mean molecular weight of 800) by refluxing with carbon tetrachloride (CCl_4, Nakarai chemical, reagent grade) in the presence of catalytic amount of benzoyl peroxide (BPO, Nakarai chemical, reagent grade) for 24 h. Sodium fluoroalkyl alkoxide ($NaOCH_2(CF_2)_3CF_2H$) was in-situ prepared from 1H,1H,5H-octafluoro-1-pentanol ($CF_2H(CF_2)_3CH_2OH$, PCR Incorporated, reagent grade) and sodium hydride (NaH, Nakarai chemical, reagent grade). $CF_2H(CF_2)_3CH_2OH$ was added dropwise to a suspension of NaH in dry tetrahydrofuran (THF, Nakarai chemical, reagent grade) under nitrogen at 0 °C, then the mixture was stirred for 30 min at room temperature to give a suspension of $NaOCH_2(CF_2)_3CF_2H$ in THF. To this suspension, a solution of the chlorinated PC (PC-Cl) in dry THF was added dropwise at 0 °C, then the reaction mixture was refluxed for 3 h. After addition of ice-water, the mixture was stirred at room temperature for 1h in order to neutralize excess reagent (NaH) and hydrolyze Si-Cl bonds which didn't react with $NaOCH_2(CF_2)_3CF_2H$, followed by concetrating in a rotary evaporator. The residue was suspended in water and extracted with toluene. The organic extracts were washed with water, saturated aqueous sodium chloride, dried over anhydrous magnesium sulfate, and concentrated in a rotary evaporator to give PCOCF.

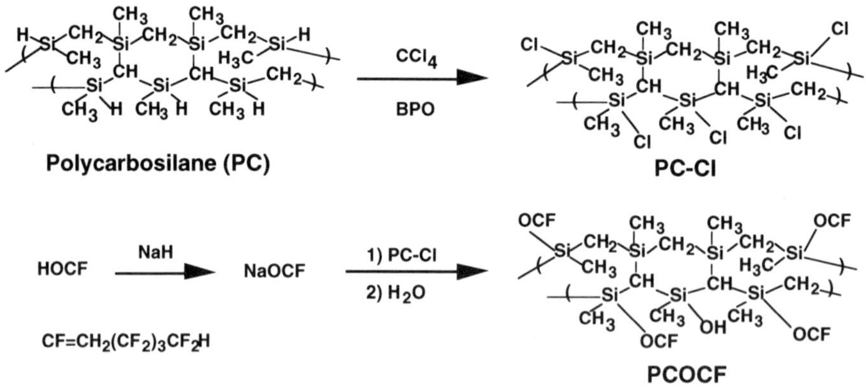

FIG. 1 Synthetic route for chemically modified PC with $CHF_2(CF_2)_3CH_2OH$ (PCOCF).

The synthesized PCOCF was dissolved in acetone, and α-SiC powders (Showa Denko, DU A-1, d50= 0.45 μm) were soaked in the solution for 24 h at room temperature. The amount of PCOCF was 5 wt.% of the total weight of the mixture with SiC powders. The suspension was dried, sieved and uniaxially pressed at 650 kg/cm^2 into pellets of 14 mm diameter.

Infrared (IR) spectra were recorded on KBr pellets containing samples (Model System 2000, Perkin Elmer, Connecticut, USA). Thermogravimetric (TG) analysis was performed on PCOCF-coated SiC powders up to 1000 °C at a heating rate of 10°C /min under helium (He) flow using Rigaku Thermoflex of TAS 300 system. Mass analysis of PCOCF during pyrolysis was recorded using a Shimadzu gas chromatograph mass spectrometer of PARVUM (QP-500). Elemental analyses of oxygen and fluorine were also obtained for PCOCF-coated powders. After pyrolysis at 1000 °C, the crystallization behavior of PCOCF was observed by transmission electron microscopy (TEM, Model JEM-100CX, JEOL, Tokyo, Japan). The sample was prepared from fine powders mixed with acetone on a micromesh screen. The green densities were calculated by measuring the weight and volume of SiC based pellets. Relative density (RD) was calculated using the theoretical density of SiC (3.21 g/cm^3) and that of as-received PC or PCOCF (1.1 g/cm^3).

RESULTS AND DISCUSSION

Figure 2 shows FT-IR spectra of polymer samples and $CF_2H(CF_2)_3$-CH_2OH. In the IR spectrum of as-received PC (Fig. 2(a)), absorptions are observed at 2900, 2950 cm^{-1} (C-H), 2100 cm^{-1} (Si-H), 1350-1450 cm^{-1} (C-H), 1250 cm^{-1} (Si-CH_3), 1020 cm^{-1} (Si-CH_2-Si) and 840 cm^{-1} (Si-C). The spectrum of PC-Cl shows a remarkable decrease in absorption intensity at 2100 cm^{-1} (Si-H) and a new absorption band at 490 cm^{-1} assigned to Si-Cl bonds (Fig.2 (b)). $CHF_2(CF_2)_3CH_2OH$ has C-F bonds which show strong absorption bands at 1143 cm^{-1} and 1180 cm^{-1} (Fig. 2 (c)). Compared with Fig. 2(b), in the IR spectrum of PCOCF (Fig. 2(d)), the absorption band at 490 cm^{-1} disappears and new absorption bands are observed at 1143, 1180 cm^{-1} (C-F) and 1100 cm^{-1} (Si-O-C). These changes in Fig. 2(d) reveals that Si-O-C bonds have been successfully formed by the reaction of Si-Cl bonds in PC-Cl with $NaOCH_2(CF_2)_3CHF_2$.

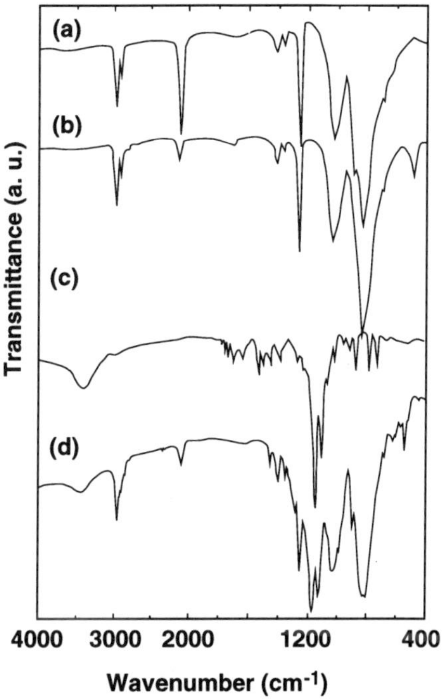

FIG. 2 FT-IR spectra of (a) as-received PC, (b) PC-Cl, (c) $CHF_2(CF_2)_3CH_2OH$ and (d) PCOCF.

Another weak and broad absorption band at 3350 cm^{-1} in Fig. 2(d) is assigned as Si-OH groups, which is considered to be formed by hydrolization of unreacted Si-Cl bonds in PCOCF during work-up process as mentioned in the experimental procedures. According to these IR spectroscopic analyses, PCOCF is a chemically modified PC which has $OCH_2(CF_2)_3CHF_2$ groups as side chains.

Table I shows green density of SiC powder compacts. In the case of as-received SiC powders, the green density is 1.88 g/cm^3 (58.6% RD), which can be improved to 1.94 g/cm^3 (62.6% RD) by coating with 5 wt.% as-received PC, and further improved to 2.03 g/cm^3 (65.5% RD) by coating with 5 wt.% PCOCF. It is apparent that the green density is remarkably improved by coating with PCOCF, which is considered to be result in the increased fluidity given to PC by the fluoroalkoxy groups.

Table I Green density of SiC powder compact.

SiC powder	Density (g/cm^3)	Relative density (%)
5 wt.% PCOCF coated	2.03	65.5
5 wt.% PC coated	1.94	62.6
As-received	1.88	58.6

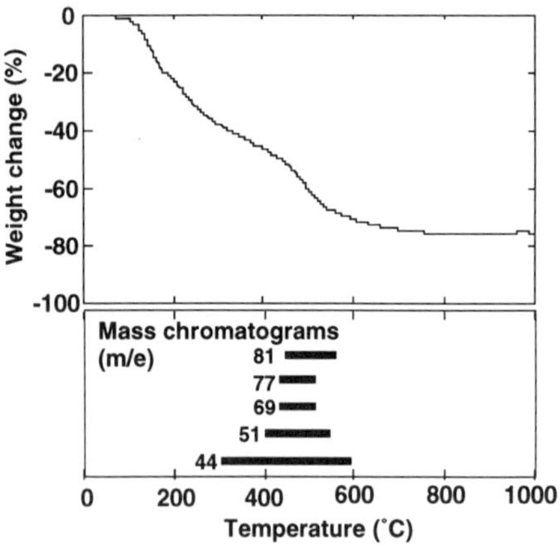

FIG. 3 TG curve of PCOCF coated on SiC powders and some gases detected by chromarograms with various m/e. (Heating rate :10 ˚C /min in He flow)

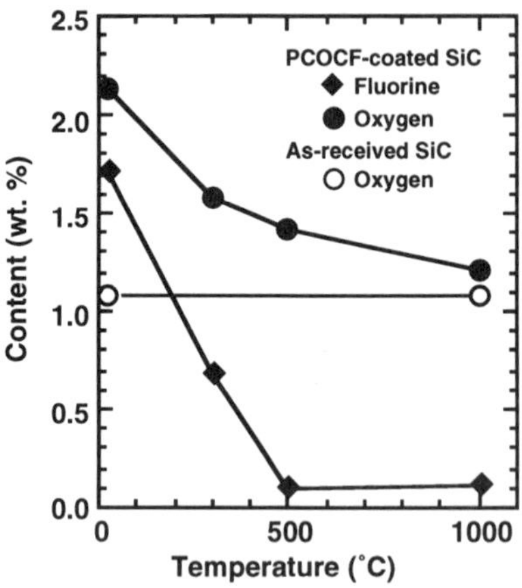

FIG. 4 Oxygen and fluorine content in PCOCF-coated SiC powders, and oxygen content in as-received SiC powders during pyrolysis.

The pyrolysis behavior of PCOCF on SiC powders was further studied. The TG curve of PCOCF-coated SiC powders is shown in Fig. 3. Some gases measured by chromatograms with various mass numbers are also shown in Fig. 3. The TG curve shows that the weight loss starts at around 100 ˚C and is almost completed at 800 ˚C. At 1000 ˚C, the ceramic yield derived from PCOCF is 24 wt. %. It is slightly larger than that of as-received PC itself (19 wt. %). During pyrolysis from 300 °C to 600 °C, some gases measured by chromatograms with mass numbers (m/e) of 44, 51, 69, 77, and 81 were detected. Figure 4 shows oxygen and fluorine content of PCOCF-coated SiC powders, and oxygen content of as-received SiC powders. Both oxygen and fluorine in PCOCF-coated SiC powders decrease with increasing temperature, especially, a remarkable decrease in fluorine content is observed up to 500 °C and the fluorine content is less than 0.1 wt.% above this temperature. The oxygen in as-received SiC powders is 1.1 wt.% and constant up to 1000 °C. The increase in the initial oxygen content of PCOCF-coated SiC powders was originated from PCOCF, and the oxygen and fluorine originated from PCOCF were almost removed during pyrolysis up to 1000 °C.

These chemical composition changes suggest that the side chains of $OCH_2(CF_2)_3CHF_2$ groups in PCOCF were removed during pyrolysis from 300 °C to 500 °C. Some gases detected from 300 °C to 600 °C were considered as decomposition species of Si-$OCH_2(CF_2)_3CHF_2$ moieties in PCOCF and their chromatograms with m/e of 44, 51, 69, 77, and 81 were considered as CO_2, CF_2H, CF_3, $Si(CH_3)_2F$ and $Si(CH_3)F_2$, respectively.

The weight loss observed below 300 °C is considered as volatilization of oligomer in PCOCF, while that observed above 600 °C is due to the decomposition of side chains such as Si-CH_3 groups, which are often observed during pyrolysis of various kinds of PC.[8, 9] The slight increase in ceramic yield after pyrolysis at 1000 °C is probably explained by interaction between PCOCF and SiC powder surfaces, for example, Si-O-Si bond formation between Si-OH groups in PCOCF and Si-OH groups existed on SiC powder surface and PCOCF yielded microcrystallite β-SiC on SiC powder surfaces after pyrolysis at 1000 °C as shown in Fig. 5. These results suggest that PCOCF is a useful precursor for fabrication of SiC-based ceramics and the microstructure control of SiC matrix.

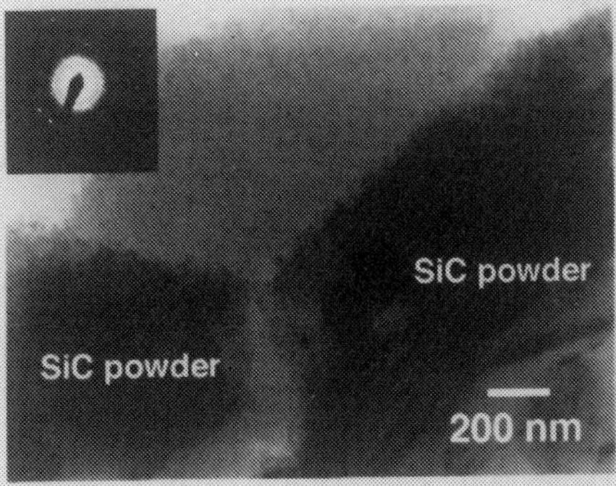

FIG. 5 TEM micrograph and selected-area diffraction pattern of PCOCF on SiC powder surfaces after pyrolysis at 1000°C.

CONCLUSIONS

Chemically modified polycarbosilane (PC) containing fluoroalkoxy groups as side chains (PCOCF) could be synthesized from a commercially available PC and fluoroalkyl alcohol. Coating of PCOCF on SiC powders was found to be effective in increasing the green density of uniaxially pressed SiC powder compact, which was considered to be result in the increased fluidity given to PC by the fluoroalkoxy groups. The fluoroalkoxy groups in PCOCF were removed during pyrolysis from

300 °C to 500 °C and PCOCF yielded β-SiC microcrystallites at 1000 °C. These results suggest that PCOCF is a useful precursor for fabrication of SiC-based ceramics with controlled microstructure.

ACKNOWLEDGMENT

Work supported by NEDO as part of the Synergy Ceramics Project under the Industrial Science and Technology Frontier (ISTF) Program promoted by AIST, MITI, Japan. The authors are members of the Joint Research Consortium of Synergy Ceramics.

REFERENCES

1) S. Yajima, J. Hayashi and M. Omori, "Continuous Silicon Carbide Fibers of high Tensile Strength," Chem.Lett., [9], 931-34(1975).
2) P. Colmbo, T. E. Paulson and C. G. Pantano, "Synthesis of Silicon Carbide Thin Films with Polycarbosilanes (PCS)," J. Am. Ceram. Soc., **80**[9], 2333-40(1997).
3) K. J. Wynne and R. W. Rice, "Ceramics via Polymer Pyrolysis," Annu. Rev. Mater. Sci., **14**, 297-334(1984).
4) P. Czubarow and D. Seyferth, "Application of poly (methylsilane) and Nicalon polycarbosilane precursors as binders for metal/ceramic powders in preparation of functionally graded materials," J. Mater. Sci., **32**, 2121-30(1997).
5) S. Yajima, T. Iwai, T. Yamamura, K. Okamura and Y. Hasegawa, "Synthesis of Polytitanocarborosilane and Its Conversion into Inorganic Compounds," J. Mater. Sci., **16**, 1349-55(1981) .
6) G. D. Soraru, A. Ravagni and R. Campostrini, "Synthesis and Characterization of Si-Zr-C-O Ceramics from Polymer Precursors", J. Europ. Ceram. Soc., **8**, 29-34(1991).
7) S. Okuzaki, Y. Iwamoto, S. Kondoh, K. Kikuta and S. Hirano, J. Mater. Res., submitted.
8) Y. Hasegawa, "Synthesis of continuous silicon carbide fiber, Part 6: Pyrolysis process of cured polycarbosilane fiber and structure of SiC fiber," J. Mater. Sci., **24**, 1177-90(1989).
9) E. Bouillon, F. Langlais, R. Paller R. Naslain, F. Cruege, P. V. Huong, J. C. Sarthou, A. Delpuech, C. Laffon, P. Lagarde, M. Monthioux and A. Oberlin, "Conversion mechanisms of a polycarbosilane precursor into an SiC-based ceramic material," J. Mater. Sci., **26**, 1333-45(1991).

STRUCTURAL CHARACTERIZATION OF NEW TERNARY TRANSITION METAL NITRIDES SYNTHESIZED BY THE AMMONOLYSIS OF COMPLEXED METALLOORGANIC PRECURSORS

K.Scott Weil and Prashant N. Kumta
Department of Materials Science and Engineering
Carnegie Mellon University
Pittsburgh, PA 15213

Jekabs Grins
Department of Inorganic Chemistry
Arrhenius Laboratory, Stockholm University
Stockholm, Sweden S-106 91

ABSTRACT

A number of new ternary transition metal nitrides have been synthesized using a complexed metalloorganic precursor approach. In this technique, two metal species are simultaneously dissolved as chlorides in a common solvent such as acetonitrile and then reacted with triethylamine to form a semi-viscous air stable solution which separates from the acetonitrile phase. Removal of the solvent yields a solid precursor. When this precursor is heat treated in ammonia under the appropriate conditions, a ternary single phase nitride is formed. The approach has been used to synthesize two new series of ternary transition metal nitride compounds, MTN_2 and M_3T_3N (where M = Fe, Co, or Ni and T = Mo, W, or Zr), whose crystal structures have been determined by Rietveld analyses of the respective powder x-ray diffraction data. Although the ratio of the early and late transition metals is the same in both types of compounds and the final soak temperature required to form each during ammonolysis often only varies by 100 - 150°C, the crystal structure and nitrogen composition of each nitride type differs dramatically. The compositional and structural differences between these two series of compounds will be illustrated in this paper by examining of the latest ternary nitrides we have synthesized, $NiWN_2$ and Ni_3W_3N, in further detail.

INTRODUCTION

Over the past decade or so, a growing list of new ternary and higher order nitride compounds has been discovered. In a vast majority of these new nitride systems, one of the elements involved is an alkali or alkaline-earth metal, giving rise to an inductive effect whereby the electropositive element stabilizes the formation of a ternary transition metal or post-transition metal nitride.[1-3] While the application of the inductive effect has proven quite successful in preparing ternary nitrides, the number and type of potential new multicomponent nitrides synthesized by this method is obviously limited to those containing a highly electropositive element. As a result, only a few of the ternary nitrides found to date do not contain an alkali or alkaline-earth element.

In the last few years, however, several research groups have focused their attention on the synthesis of ternary transition metal - transition metal nitrides[4,5]. Two of the motivating factors responsible for the interest in these materials include: 1) the higher order nitride compounds are expected to display different, more complex crystal structures and more exotic magnetic, optical, and electrical properties in comparison to their binary transition metal counterparts and 2) to date, the preparation of this type of ternary nitride using traditional approaches has proven to be quite challenging and may require the development of new chemical synthesis techniques, which could eventually be generally applicable for discovering other new compounds. In the absence of the inductive effect, the difficulty in synthesizing multicomponent transition metal nitrides lies in the fact that most of these compounds have small free energies of formation due in part to the strong triple bond in one of the potential decomposition products, N_2. Thus the use of standard solid state approaches to form these phases at the high temperatures necessary to obtain reasonable diffusion and reaction rates is often precluded because the phases tend to be stable only at lower temperatures.

In an attempt to identify new low-temperature, molecular-based routes for synthesizing multicomponent transition metal nitrides, we have developed an alternative mixed metalloorganic synthesis technique and have used it to prepare over ten new ternary nitride compositions. In particular, due to the general lack of research on multicomponent transition metal nitrides, we have focused our efforts on exploring new and potentially exciting combinations of the early and late transition metals with nitrogen. In this paper, we briefly describe our synthesis strategy and report on the details of the crystallographic structures of $NiWN_2$ and Ni_3W_3N, contrasting the different types of metal-to-metal and metal-to-anion bonding found in each compound.

EXPERIMENTAL PROCEDURE

Materials Synthesis

Both of the nitride compounds discussed here were synthesized by the ammonolysis of a complexed mixed-metal precursor, using an approach similar to that previously reported by us.[6] The anhydrous chlorides, $NiCl_2$ and WCl_6, were dissolved in anhydrous HPLC grade acetonitrile in a 1:1 molar ratio. These two chloride species were then simultaneously complexed with triethylamine, forming a viscous liquid. This liquid was heated to ~90°C to evaporate off the acetonitrile and leave behind a black, waxy solid precursor. In preparing both $NiWN_2$ and Ni_3W_3N, the precursor was heat treated in electronic grade ammonia using a horizontal flow-through tube furnace. The initial steps of the heat treatment schedule were the same to form both nitrides. Prior to heating, the furnace tube was purged for ten minutes with pre-pure nitrogen, then for an additional minutes with ammonia. Under 130ml/min flowing ammonia, the furnace was heated as follows: 1) ramp from room temperature to 120°C at 5°C/min and hold at 120°C for one hour; 2) ramp from 120°C to 250°C at 5°C/min and hold at 250°C for four hours; 3) ramp from 250°C to the final soak temperature at 2°C/min and hold at this temperature for eight hours. The soak temperature used to form $NiWN_2$ was 575°C and that used to form Ni_3W_3N was 700°C. Although Ni_3W_3N crystallizes at 700°C within eight hours, a more crystalline material and therefore sharper diffraction pattern could be obtained using a twenty hour hold at this soak temperature.

Materials Characterization

Quantitative chemical assays on two samples of each ammonolyzed powder product were conducted by Galbraith Laboratories (Knoxville, Tennessee). They employed a wet chemical analysis technique to determine the concentration of the metal species and combustion analyses to evaluate the nitrogen and carbon contents. In preparing samples for x-ray analysis, each nitride powder was slurried in acetone and ground with an agate mortar and pestle for five to ten minutes. Two sets of x - ray diffraction patterns for each powder sample were used in the structural analysis work. One pattern, used for indexing and lattice parameter measurements, was generated by the Guinier-Hägg technique using $CuK\alpha_1$ radiation over a 2θ range of 0 - 90°. An internal silicon standard was incorporated with each powder sample prior to conducting the diffraction experiment in order to prepare very accurate lattice parameter measurements. A series of computer programs, TREOR,[7] DICVOL,[8] and PIRUM,[9] were then used to determine the Bravais lattice and indexing sequence and to refine the lattice parameters for each compound.

A second set of x-ray diffraction data, which was used for structural refinement, was collected with a Rigaku theta-theta diffractometer employing graphite monochromated $CuK\alpha$ radiation under the following step scan settings: 10 - 110° range, 0.02° step size, and 15 second collection time. Prior to the diffraction experiments, the diffractometer was aligned using a (100) oriented silicon single crystal for accurate peak positions. The Rietveld refinement program, RIETAN-94,[10] was applied to refine the crystal structure of each nitride, including the following variables: space group, atomic parameters, and nitrogen stoichiometry. In each refinement, the diffraction peak profile shape was represented by the modified pseudo-Voigt function, with profile asymmetry introduced by employing a multi-term Simpson's rule integration. Atomic thermal displacements were assumed to be isotropic. Crystal structure graphics and geometric calculations were prepared using Ca.R.Ine software.[11]

RESULTS

Results from the chemical analyses, shown in Table I, verify the target

Table I. Chemical Composition

Compound	at% Ni	at% W	at% N	at% C
$NiWN_2$ (experimental)	25.08	25.10	49.62	0.08
$NiWN_2$ (theoretical)	25.00	25.00	50.00	-
Ni_3W_3N (experimental)	43.11	43.08	13.69	0.06
Ni_3W_3N (theoretical)	42.86	42.86	14.28	-

stoichiometries of the nickel-tungsten precursor ammonolyzed under two different heat treatment conditions as $NiWN_2$ and Ni_3W_3N. Carbon impurities were found to be present in relatively insignificant concentrations in both compounds, with the remaining compositional balance likely due to small amounts of oxygen and/or chlorine impurities. As seen in other ternary transition metal-transition metal nitrides,[4,5] the nitrogen contents in these compounds are slightly substoichiometric.

Based on the observed d spacing and intensities from the powder x-ray diffraction pattern for $NiWN_2$, shown in Fig. 1(a), the compound was indexed to a hexagonal unit cell with a = 2.8614(2) Å and c = 15.631(6) Å and is similar to that reported for $LiMoN_2$,[3] with systematic absences in the (10*l*) and the (00*l*) lines. This corresponds to a choice of two space groups for the structure: $R\bar{3}m$ (No.166) and R3 (No.146). The only difference between these two space groups is the nitrogen coordination about the tungsten atoms; $R\bar{3}m$ places the tungsten in octahedral holes, while R3 places it in trigonal prismatic sites.[3] A series of preliminary refinements comparing each space group with the experimental data indicated that the best fit was achieved with the R3 space group. Thus, further structural analysis and refinement was limited to this space group only.

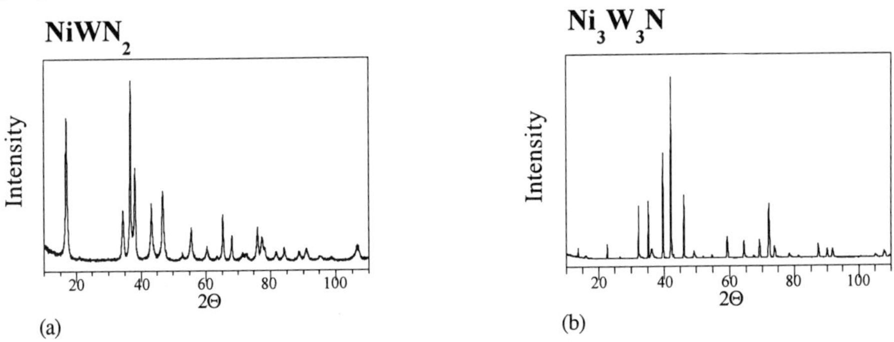

FIG. 1. Powder X-Ray Diffraction Patterns for (a) $NiWN_2$ and (b) Ni_3W_3N

The powder x-ray diffraction pattern for Ni_3W_3N, shown in Fig. 1(b), was indexed as a cubic Bravais lattice with a lattice parameter of a = 11.0205(8) Å. Systematic absences in the indexing, along with Rietveld refinement of the PXRD pattern indicate that the compound crystallizes in the space group $Fd\bar{3}m$ (227). Given below in Table II are the Wyckoff positions of the constituent atoms in both of the new nitrides.

Table II. Atomic Parameters for $NiWN_2$ and Ni_3W_3N

$NiWN_2$
a = 2.8614(2) Å
c = 15.631(6) Å

Atom	x	y	z	$100U_{iso}$(Å^2)
Ni	0	0	0.8199(8)	1.2(2)
W	0	0	0	0.3(2)
N(1)	0	0	0.2641(6)	1.6(6)
N(2)	0	0	0.4056(10)	1.4(8)

Ni_3W_3N
a = 11.0205(8) Å

Atom	x	y	z	$100U_{iso}$(Å^2)
Ni(1)	0.4097(6)	0.4097(6)	0.4097(6)	0.3(4)
Ni(2)	$^5/_8$	$^5/_8$	$^5/_8$	0.2(4)
W	0.1926(12)	0	0	0.1(8)
N	$^1/_8$	$^1/_8$	$^1/_8$	0.7(6)

DISCUSSION

The crystal structures of nearly all ternary transition metal nitrides can be classified as one of three types, either: 1) layered hexagonal or rhombohedral structures such as $(Fe_{0.8}Mo_{0.2})MoN_2$[4] and $MnWN_2$;[5] 2) cubic, η-nitride structures such as Fe_3W_3N;[12] or 3) cubic, anti-perovskite structures, such as Fe_3NiN and Mn_3CuN.[13] As expected, each of the two new nitrides, $NiWN_2$ and Ni_3W_3N, can be placed into one of these categories. What is interesting about the structures of these two compounds is that they both display a form of segregated metallic clustering; that is, the two transition metal species, nickel and tungsten, in each compound arrange into their own separate polyhedral metal clusters linked together in a periodic fashion. This is most clearly seen in the layered structure, that of $NiWN_2$, in which the nickel and tungsten atoms segregate onto two different sets of hexagonal layers in the compound.

Herle et al.[14] were the first to synthesize $NiWN_2$ and indexed the compound with a hexagonal lattice, choosing the hexagonal group $P6_3/mmc$ (194) as the space group. However, upon reviewing the published pattern and the systematic absences of the $P6_3/mmc$ space group, we found an incongruity in the indexing and suspected that this compound is isostructural with $LiMoN_2$, exhibiting a rhombohedral unit cell. Thus, we decided to synthesize and characterize $NiWN_2$ in our laboratory to examine the structure of this compound in further detail.

As determined from indexing and preliminary refinement of the Guinier-Hägg diffraction pattern, $NiWN_2$ crystallizes in the rhombohedral space group R3. Using hexagonal notation, the unit cell has lattice parameters of a = 2.8614(2) Å and c = 15.631(6) Å. As shown in Fig. 2(a) and (b), the ideal structure of $NiWN_2$ consists of alternating hexagonal close-packed sheets of nickel, nitrogen, tungsten, nitrogen, etc. with the nickel atoms sandwiched within the octahedral interstices of the two adjacent nitrogen layers and the tungsten atoms sitting in trigonal prismatic coordination between the two neighboring nitrogen layers.

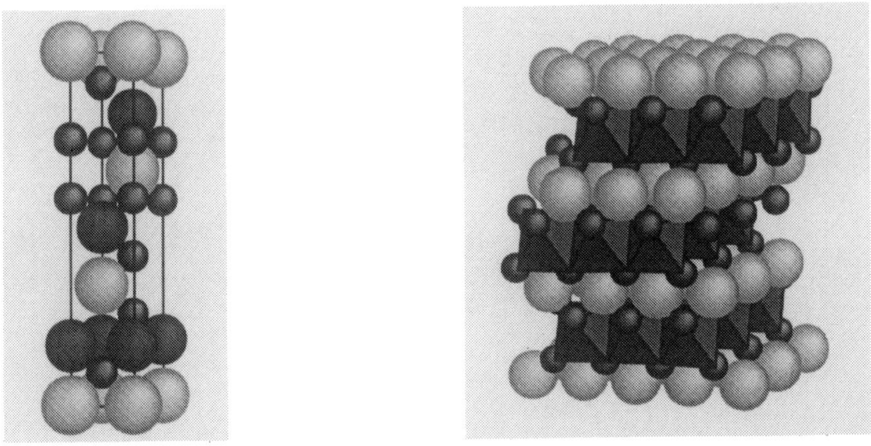

(a) (b)

FIG. 2. Proposed structure for $NiWN_2$: (a) unit cell and (b) extended polyhedra model. In Fig. (a), the dark gray, large spheres represent the nickel atoms, the light gray, large spheres represent the tungsten atoms, and the small, dark gray spheres represent the nitrogen atoms. In Fig. (b), the octahedral nickel positions are shown as filled polyhedra. The tungsten and nitrogen positions are indicated again by the large and small spheres, respectively.

The stacking arrangement of these layers along the c-axis can be represented as AbCaCaBcBcAbA, where the upper case letters, A, B, and C, represent the hcp metal sheets and the close-packed nitrogen sheets are equated with the lower case letters, a,b, and c. With regard to metallic clustering, this compound can be described as a two-dimensional cluster structure in which the nickel and tungsten atoms congregate on separate layers each forming a close-packed array of flat hexagonal cells which share common edges.

The second new nitride, Ni_3W_3N, also displays metallic clustering, in this case with three-dimensional connectivity. Indexing and structural analysis of the powder diffraction pattern of Ni_3W_3N revealed that this compound crystallizes in the cubic $Fd\bar{3}m$ space group and has lattice [a = 11.0205(8) Å] and atomic parameters consistent with those of the η-carbide structure. A schematic of the unit cell of Ni_3W_3N is shown in Fig. 3(a) and polyhedral representations of the atomic coordination of this metal rich nitride are displayed in Fig. 3(b) through (e). As seen in Fig. 3(b) and (c), the nitrogen atoms are octahedrally coordinated by tungsten, forming NW_6 octahedra. On the hand, the nickel atoms cluster into groups of eight, arranged into four tetrahedra each of which share a face with a fifth central tetrahedron, thus forming a stella quadrangula. The arrangement of these Ni_8 stellae quadrangulae are displayed in Fig. 3(d) and (e). These two polyhedral arrangements are related; one is a central tetrahedron capped on all four faces by additional tetrahedra, while the other is a central octahedron capped on four (of its eight) faces by additional octahedra. The NW_6 octahedra corner share in a manner which is similar to that seen in the pyrochlore structure. The Ni_8 stellae quadrangulae also corner share on a structural lattice separate from that formed by the NW_6 octahedra. Because these two lattices interpenetrate, each occupying the interstices of the other, their unit cell edges are identical in size. Calculations of the W-W and Ni-Ni bond distances indicate that metallic bonding exists between the metal atoms in each cluster type.

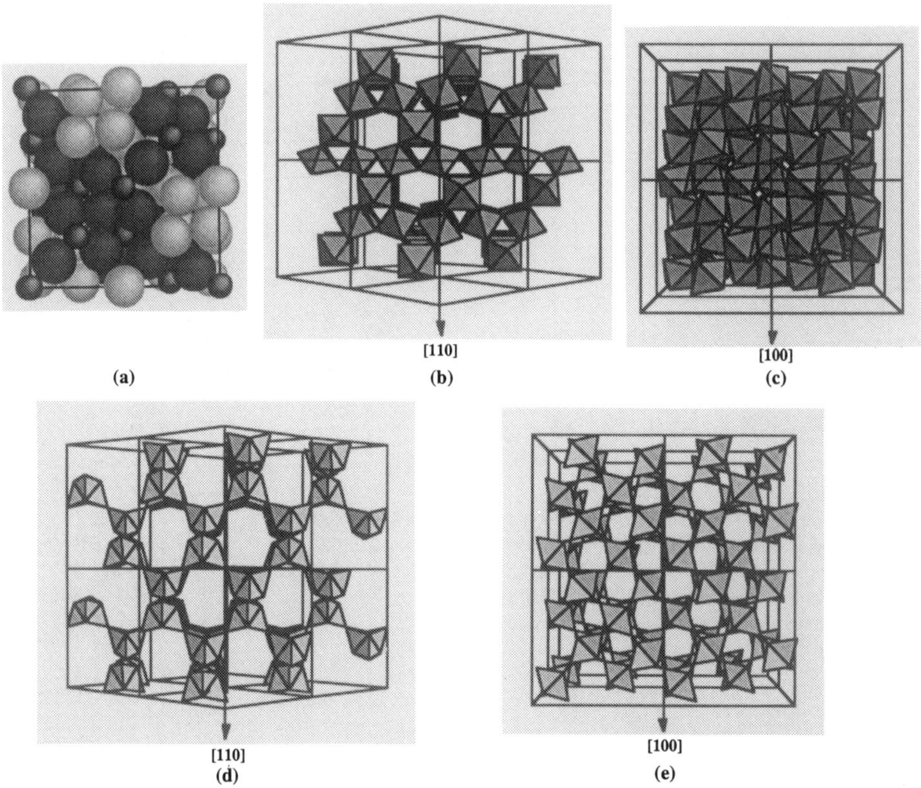

FIG. 3. The proposed structure of Ni_3W_3N: (a) unit cell and (b) - (e) polyhedral models. In Fig. (a), the dark gray, large spheres represent the tungsten atoms, the light gray, medium size spheres represent the nickel atoms, and the small, dark gray spheres represent the nitrogen atoms. In Fig. (b) and (c), each nitrogen atom is coordinated by six tungsten atoms, forming slightly distorted NW_6 octahedra. The octahedra share corners in a manner similar to that seen in the pyrochlore structure. Fig. (b) shows the view of these octahedra along the [1 1 0] direction, while Fig. (c) displays the view along the [1 0 0] direction. In Fig. (d) and (e), clusters of eight nickel atoms are arranged into four tetrahedra each which share a face with a fifth central tetrahedron, forming a stella quadrangula. These composite polyhedra corner-share in this structure. Fig. (d) shows the view of these stellae quadragulae along the [1 1 0] direction, while Fig. (e) displays the view along the [1 0 0] direction.

Interatomic Bonding

Bond distances between the constituent atoms in both of these structures were calculated from the refined unit cell data. The metal-metal and metal-nitrogen bond distances found in $NiWN_2$ are similar to those found in other ternary transition metal nitrides in this family. The Ni-N bond distances in $NiWN_2$ of 2.321(10) Å and 2.118(4) Å compare well with that for the Mn-N bond in $FeWN_2$, 2.145(5) Å[15]. Likewise, the W-N bonds in $NiWN_2$ of 1.997(4) Å and 2.011(8) Å are similar to that found in β - $MnWN_2$ of 2.258(2) Å[5] and in $FeWN_2$, 2.156(5) Å.[15] As noted by Bem et al. for $FeWN_2$, the metal-nitrogen bond lengths in these ternary transition metal nitrides are significantly longer in comparison to bond lengths existing in the alkali- and alkaline-earth-transition metal nitrides, which implies that the bonding in $NiWN_2$ is at least partially covalent. Similarly, the separation between opposing sheets of nitrogen atoms is 2.246(12) Å which is shorter than the in-plane N-N separation of 2.861(2) Å. This observation also suggests that the metal-nitrogen bond is substantially covalent in nature and implies that there may be direct bonding between opposing nitrogen layers. Based on this data and on electronic structure calculations by Singh[16] for the prototype structure $LiMoN_2$, it is suspected that $NiWN_2$ consists of strongly covalent WN_2 sheets with unusually pronounced direct N-N bonding across the trigonal prismatic coordinated tungsten atoms and that the nickel layers may interact ionicly with the WN_2 sheets to build the crystal structure along the c-axis.

The metal-metal and metal-nitrogen bond distances found in Ni_3W_3N are similar to those found in other ternary transition metal nitrides and carbides in this family. The W-N bond distance in Ni_3W_3N of 2.122(6) Å compares well with that for the W-N bond in Fe_3W_3N, 2.119(5) Å.[12] Likewise, the W-W bond distances in Ni_3W_3N of 2.902(6) Å and 3.098(2) Å are similar to that found in Fe_3W_3N of 2.901(2) Å and 3.090(5) Å[12] and in Co_3W_3C, 2.881(2) Å and 3.110(5) Å.[17] Similarly, the Ni-Ni bond distances of Ni_3W_3N are very nearly equal to the comparable metal bonds in Fe_3W_3N[12] and Co_3W_3C.[17] The metal-nitrogen bond lengths in these ternary transition metal nitrides are significantly longer in comparison to bond lengths existing in the alkali- and alkaline-earth-transition metal nitrides. For example, the W-N bond distances in Na_3WN_3[1] and Ba_3WN_4[2] range from 1.78(1) - 1.93(1) Å, nearly 10 - 20 % shorter than the W-N bonds in Ni_3W_3N. The longer metal-nitrogen bond lengths imply that the bonding in Ni_3W_3N is at least partially covalent. Similarly, comparisons of the metallic bond lengths in the NW_6 octahedra with those in metallic tungsten suggest that the longer metallic bonds in Ni_3W_3N are due to covalent bonding within the octahedra. On the other hand, the metal to metal bond lengths within the Ni_8 stellae quadrangulae clusters are nearly equivalent to those found in metallic nickel, suggesting that within these clusters, the nickel atoms form metallic bonds.

CONCLUSIONS

Using a complexed mixed-metal precursor synthesis approach, we have synthesized a number of new ternary transition metal nitride compounds with crystal structures in the MTN_2 and M_3T_3N series. Although the ratio of the early and late transition metals is the same in both types of compounds and the final soak temperature required to form each during ammonolysis often varies only by 100 - 150°C, the crystal structure and nitrogen composition of each nitride type differs dramatically. Of the two compounds discussed here in detail, $NiWN_2$ was found to be isostructural with the layered nitride, $LiMoN_2$, while Ni_3W_3N was found to crystallize in the face-centered cubic η-carbide structure. With regard to metallic clustering, $NiWN_2$ can be described as a two-dimensional cluster structure in which the nickel and tungsten atoms congregate on separate layers each forming a close-packed array of flat hexagonal cells which share common edges. Bond length calculations imply that the bonding in this compound is substantially covalent, at least across the separate metal and nitrogen layers. On the other hand, the second new nitride, Ni_3W_3N, displays three-dimensional metallic clustering. Again, bond length calculations imply some covalency within the corner-sharing NW_6 octahedra. However, the bonding within and across the corner-sharing Ni_8 stellae quadrangulae clusters appears to be metallic. While both $NiWN_2$ and Ni_3W_3N have the same metals ratio, differences in the nitrogen composition predicate significant structural differences between each compound in terms of the spatial arrangement and type of near neighbor bonding of the metal species.

ACKNOWLEDGMENTS

The authors would like to acknowledge the assistance of Lars Göethe at Stockholm University for preparing the Guinier Hägg data and to Jason Wolf at Carnegie Mellon University for his help in calibrating the Rigaku x-ray diffractometer. This work is supported by the National Science Foundation under Grant No. CTS 9700343. One of the authors (KSW) would also like to thank the Electrochemical Society for the F. M. Beckett Award which provided the opportunity to work abroad.

REFERENCES

[1]D. Ostermann, U. Zachwieja, and H. Jacobs, "Natriumnitridometallate, Na_3MN_3, von Molybdän (VI) und Wolfram (VI) mit $[CrO_2O_{2/2}]$-isosteren $[MoN_2N_{2/2}]^{3-}$-Ketten," Journal of Alloys and Compounds, 190 [1] 137-40 (1992).

[2]A. Gudat, P. Höhn, R. Kniep, and A. Rabenau, "Ternäre Nitride des Bariums mit den Elementen Mo und W," Zeitschrift fur Naturforschung, 46b [3] 566-74 (1992).

[3]S. H. Elder, L. H. Doerrer, F. J. DiSalvo, J. B. Parise, D. Guyomard, and J. M. Tarascon, "$LiMoN_2$: The First Metallic Layered Nitride," Chemistry of Materials, 4 [4] 928-37 (1992).

[4]D. S. Bem, H. P. Olsen, and H-C zur Loye, "Synthesis and Electronic and Magnetic Characterization of the Ternary Nitride $(Fe_{0.8}Mo_{0.2})MoN_2$," Chemistry of Materials, 7 [10] 1824-28 (1995).

[5]J. Grins, P-O Kåll, and G.J. Svensson, "Synthesis and Structural Investigation of the Ternary Nitride, $MnWN_2$," Journal of Materials Chemistry 5 [7] 571-9 (1995).

[6]K. S. Weil and P. N. Kumta, "Synthesis of Ternary Transition Metal Nitrides Using Chemically Complexed Precursors," Materials Science and Engineering B, 38 [1-2] 109-17 (1996).

[7]P. -E. Werner, L. Eriksson, and M. Westdahl, "TREOR, A Semi-Exhaustive Trial-and-Error Powder Indexing Program for All Symmetries," Journal of Applied Crystallography, 18 [5] 367-70 (1985).

[8]A. Boultif and D. Louer, "Indexing of Powder Diffraction Patterns for Low Symmetry Lattices by the Successive Dichotomy Method," Journal of Applied Crystallography, 24 [6] 987-93 (1991).

[9]P. -E. Werner, "Refinement of the Unit Cell," Arkiv for Kemi, 31 [2] 513-516 (1969).

[10]F. Izumi, "Rietveld Analysis Programs RIETAN and PREMOS and Special Applications," Chapter 13 in The Rietveld Method, 1st ed. Edited by R. A. Young. Oxford University Press, Oxford, 1993.

[11]C. Boudias and D. Monceau, Ca.R.Ine Cristallographie, version 3.0 (1989).

[12]K.S. Weil and P.N. Kumta, "Fe_3W_3N: A Rietveld Refinement Using X-ray Powder Diffraction," Acta Crystallography C, 53[12] 1745-48 (1997).

[13]H. H. Stadelmaier, "Metal-Rich Metal-Metalloid Phases," p. 141-180 in Developments in the Structural Chemistry of Alloy Phases, 1st ed. Edited by B. C. Giessen. Plenum Press, New York, NY 1969.

[14]P. S. Herle, N. Y. Vasanthacharya, M. S. Hegde, and J. Gopalakrishnan, "Synthesis of New Transition Metal Nitrides, MWN_2 (M = Mn, Co, Ni)," Journal of Alloys and Compounds, 217 [1] 22- 24 (1995).

[15]D. S. Bem, C. M. Lampe-Önnerud, H. P. Olsen, and H-C zur Loye, "Synthesis and Structure of Two New Ternary Nitrides: $FeWN_2$ and $MnMoN_2$," Inorganic Chemistry, 35 [3] 581-5 (1996).

[16]D. J. Singh, "Electronic Structure of the Layered Nitride $LiMoN_2$," Physical Review B, 46 [15] 9332-5 (1992).

[17]Z. Bojarski and J. Leciejewicz, "Carbon Cobalt Tungsten $[Co_3W_3N]$," Structure Reports, 32A, 45 (1967).

PHASE SELECTION IN INDUCTION PLASMA TREATMENT OF MOLYBDENUM DISILICIDE POWDERS

Xiaobao Fan, and Takamasa Ishigaki

National Institute for Research in Inorganic Materials
1-1 Namiki, Tsukuba-shi, Ibaraki 305, Japan

Abstract In this investigation, the in-flight modification of $MoSi_2$ powders has been carried out in the Ar-H_2 and Ar-N_2 induction plasma process respectively. Attention was principally paid to the modification of the crystal phase and the chemical composition in the induction plasma treated $MoSi_2$ powders. Under the processing conditions used in this study, all the compounds and the polymorphic phase in the Mo-Si system (excepting Mo) have been observed in the plasma treated $MoSi_2$ powder particles. The reason of the formation of metastable β–$MoSi_2$ phase in the plasma process has been explained by comparison of the critical free energy for nucleation for α-$MoSi_2$ and β–$MoSi_2$. The formation of low silicides, Mo_5Si_3 and Mo_3Si, and free Si was explained in terms of metastable eutectic reaction during rapid solidification processing. The discrepancy in the phase composition observed in Ar-N_2 induction plasma treated $MoSi_2$ powders with those treated by Ar-H_2 induction plasma has been discussed. The lower thermal conductivity of plasma gas N_2, the heterogeneous nucleation mechanism during the solidification, and the chemical reaction between N_2 and $MoSi_2$ are the major reasons leading to the different result in the $MoSi_2$ powders treatment.

1. INTRODUCTION

A simple principle governs the energy efficiency of heat engine: the higher is the operation temperature, the higher the efficiency. Considerable effort has been directed at increasing the working temperature in propulsion system in civil and military aircraft. In this sense, $MoSi_2$ becomes one kind of scientific and technologically strategic materials for the 21 century, because of its significant potential to meet the demands of advanced high-temperature structural applications in the range 1500~2000 K, in both oxidizing and aggressive environment (1).

The last several years saw the development of plasma spraying process as a viable technique for producing dense monolithic and composite forms of $MoSi_2$ (2~10). Plasma spraying process has proved to be able to reduce segregation and residual stresses in $MoSi_2$ that typically limit the formability of such a brittle material. The ductility and tensile strength have been improved (11).

Many researchers have reported the formation of the metastable β–$MoSi_2$ phase, and low silicides, Mo_5Si_3 and Mo_3Si, in the plasma sprayed $MoSi_2$ coatings and deposits, while the initial feedstock put into the plasma plume is the α-$MoSi_2$ powders. This has been, in general terms, attributed to the feature of rapid quenching and the silicon depletion occurring during plasma spraying process (5~8).

In this work, we present a detailed explanation on the phase selection mechanism of $MoSi_2$ during plasma spraying process. The experimental work of this investigation included the $MoSi_2$ powders' modification by Ar-H_2 and Ar-N_2 induction plasma respectively. Reactor pressure, and plate power level. were the experimental parameters used to alter the thermal history of the injected powder particles. Attention was principally given to the modifications of the crystalline phase and of the chemical composition in the induction plasma treated $MoSi_2$ powders. The theoretical explanation included (i) the calculation of critical free energy for nucleation from the melt of $MoSi_2$, which explained the phase competition mechanism in the polymorphic $MoSi_2$; (ii) the construction of metastable equilibrium phase diagram, which illustrated how the low silicide phases and silicon may be produced during the plasma spraying of $MoSi_2$; and (iii) the complicated chemical equilibrium calculations, which revealed the chemical reactions in the plasma process may influence the eventual phase content. On completion of this investigation, the phase selection mechanism of $MoSi_2$ in plasma spraying process and other rapid solidification process has been explained.

2. EXPERIMENTAL PROCEDURE

The experimental installment has been described elsewhere (12). The induction plasma torch (Model PL50, TEKNA Plasma Systems Inc., Canada) uses a water-cooled ceramic plasma confinement tube, in which a 3-turn induction coil is incorporated. A water cooled stainless steel probe, the tip of which is located at the center of the plasma, penetrates axially through the torch head to inject the $MoSi_2$ powders into the plasma. An airtight cylindrical reactor vessel (250 mm inner diameter and 1000 mm long) is made as a double-wall, water cooled stainless steel structure.

The initial $MoSi_2$ powders used in this investigation are of the "fused and crushed" type (Japan New Metals Co.), containing exclusively tetragonal phase α-$MoSi_2$ and traces of free residual silicon. The chemical composition of starting materials was measured of Si 36.83, C 0.02, O_2 0.26, Fe 0.09 and Mo 62.67 in weight percentage. The screen size gradation is -38+20 μm.

The experiment conditions are summarized in Table I. As is well known, plasma-particle interaction is a complex phenomenon and while a large number of factors can influence this interaction, the variable parameters selected in this work are those expected to have significant impacts on the particles' "in plasma" residence time and thus their heat exchange behavior with the plasma. They are (i) the reactor pressure, and (ii) the plasma power level.

$MoSi_2$ powders after plasma treatment were collected on the reactor wall and base. The collected powders were examined by the scanning electron microscope (SEM, Akashi, ISI-DS130) to observe their surface morphologies and cross-section microstructures, and by the X-ray diffractometry (XRD, Philips PW unit) to obtain the XRD patterns. The crystalline phases were also examined by the electron diffraction pattern (EDP) from the transmission electron microscope (TEM, Hitachi-500).

Table I. Summary of the experiment conditions

Fixed Parameters	Plasma gas flow rates (slpm) Plasma gas: Ar 30 Sheath gas: Ar 85 H_2 (or N_2) 5 Carrier gas: Ar 5 Powder feed rate: 4 g/min
Variable Parameters	Plate power: 20 ~ 40 kW Pressure of reactor: 40.0 ~ 80.0 kPa

3. RESULTS

Figure 1 shows the X-ray diffraction patterns of the $MoSi_2$ powders treated in Ar-H_2 and Ar-N_2 plasma, respectively. The common crystalline phases identified in both cases consist of tetragonal α-$MoSi_2$, hexagonal β-$MoSi_2$, Mo_5Si_3 and Si. The α-phase is the predominant phase in the Ar-N_2 induction plasma treated $MoSi_2$ powders,

Table II. Comparison of the phase content (wt.%)

	Ar-H_2 plasma 120/5 l/min 35 kW, 40 kPa	Ar-N_2 plasma 120/5 l/min 35 kW, 40 kPa
α-$MoSi_2$	37.6	46.2
β-$MoSi_2$	45.7	24.6
Mo_5Si_3	14.7	18.1
Mo_3Si	no detectable	8.4
Si	2.0	2.8

while β-phase is the major phase in the Ar-H_2 induction plasma treated $MoSi_2$ powders. Mo_3Si formed only in the Ar-N_2 induction plasma treated powders. The plasma treated powders were further crashed in an agate mortar and were examined by TEM. Fig. 2 shows EDP of β-phase observed in the $MoSi_2$ powder particles. A comparison of phase contents of $MoSi_2$ powders treated in Ar-H_2 plasma and Ar-N_2 plasma is given in Table II. The phase contents were determined from the XRD patterns using "adiabatic principle of X-ray diffraction analysis of a mixture" (12).

Figure 3(a) displays the typical surface morphology of faceted spheres that is commonly found in the induction plasma treated $MoSi_2$ powder particles. The fraction of the faceted spheres, by visual inspection, in the powder particles treated by Ar-N_2 induction plasma is less than those treated by Ar-H_2 induction plasma. This observation is related to another experimental result, that the weight fraction of β-$MoSi_2$ in the powders treated by Ar-N_2 induction plasma is lower than those treated by Ar-H_2 induction plasma. Another typical particle morphology, "polygonal granule", is discernible in Ar-N_2 induction plasma treated $MoSi_2$ powders, see Fig. 3(b). In this kind of particle, the melting was so insufficient that the particles retain their polygonal shapes, rather than spherical form. However, crystal growth did

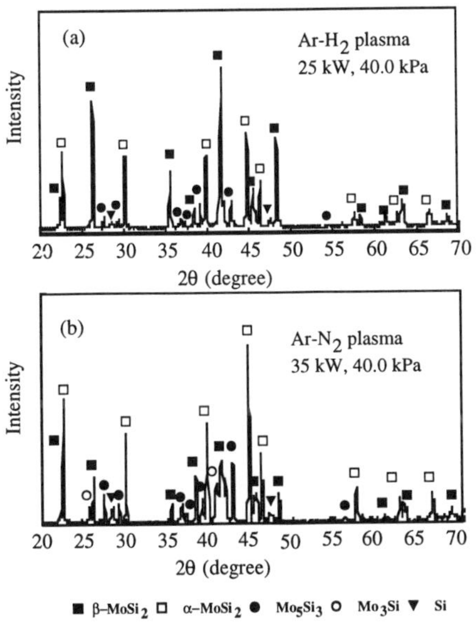

Fig. 1 XRD patterns of the $MoSi_2$ powders treated in Ar-H_2 plasma and Ar-N_2 plasma respectively.

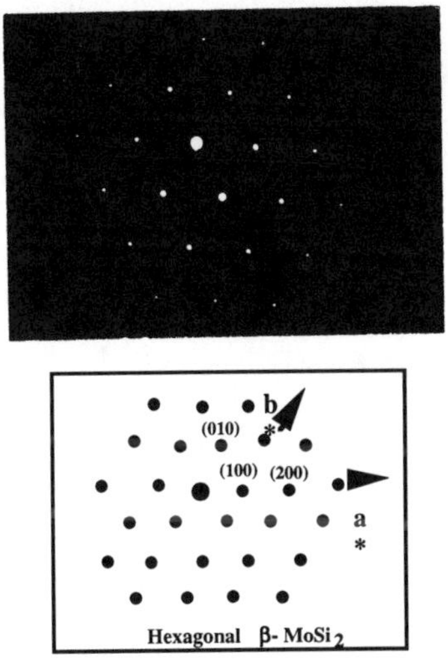

Fig. 2 EDP of β-phase formed in the plasma treated $MoSi_2$ powder particles.

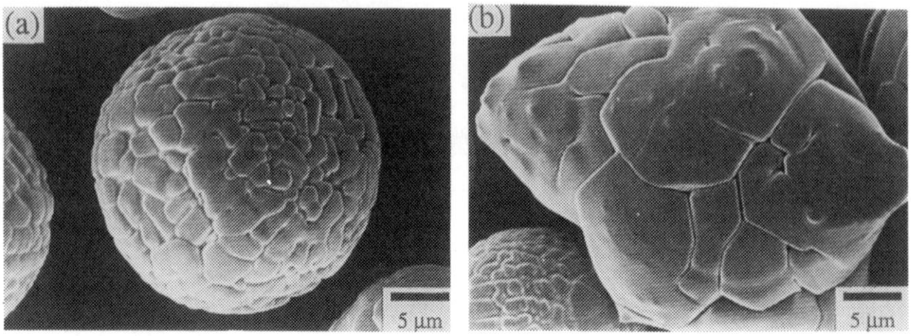

Fig. 3 The morphology of (a) faceted sphere; and (b) polygon granule, uniquely found in Ar-N_2 plasma treated $MoSi_2$ powders.

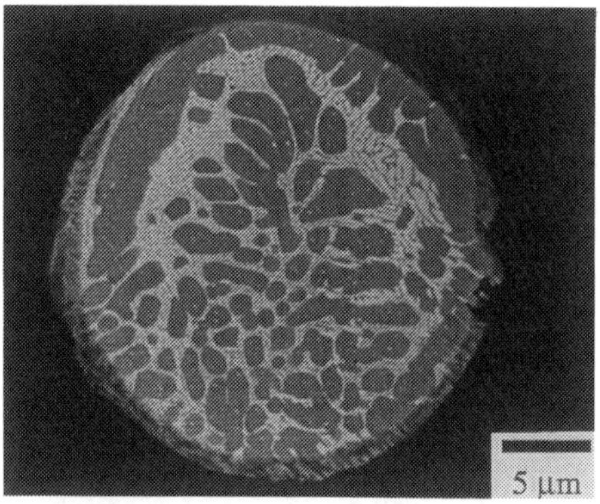

Fig. 4 The cross-sectional microstructure of $MoSi_2$ powders treated in induction plasma.

occur to form the crystalline facets on the non-spheroidized particles. The heterogeneous nucleation mechanism may play a governing role during the solidification of these powder particles.

Figure 4 is the micrographs of cross-sections made from back-scattered electrons image (BEI). The dark areas, examined by energy dispersion spectrum (EDS), correspond to $MoSi_2$ phase. The white or gray areas have been determined to be Mo-rich and are believed to be eutectic mixtures of Mo_5Si_3 /Si and/or Mo_3Si. These micrographs provided evidence for eutectic structure in the induction plasma treated $MoSi_2$ powders.

4. DISCUSSION

4.1 The metastable phase formation

The hexagonal structure of β-$MoSi_2$ is a high temperature metastable phase (13), but it forms predominately in Ar-H_2 plasma modified $MoSi_2$ powders, as well as in those treated by the Ar-N_2 induction plasma. The evolution of metastable structures is closely tied to the materials' thermal history, especially the achievement of high undercooling prior to nucleation (14). Turnbull (15) put forward the concept "energize and quenching" as the principle for the production of metastable phases. Shingu (16) has further estimated the attainable excess energy by different methods of metastable phases formation.

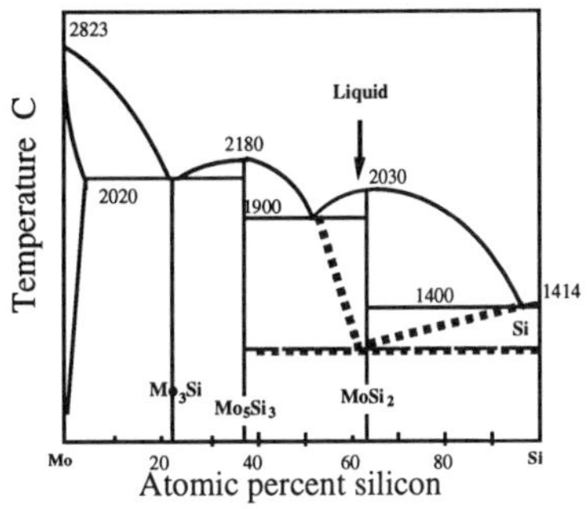

Fig. 5 Metastable eutectic reactions.

The present authors has explained the metastable β-$MoSi_2$ phase formation in the plasma process by referenceto a comparison of the critical free energies of nucleation for α-$MoSi_2$ and β-$MoSi_2$ (17). It is concluded that the critical free energy for nucleation of β-$MoSi_2$ is lower than that of α-$MoSi_2$ at solidification temperature and thus β-$MoSi_2$ formed by a homogeneous nucleation mechanism will be the prevailing phase in the droplets. The tetragonal structure of α-$MoSi_2$ can also form by a homogeneous nucleation mechanism as the undercooling is considerable in the plasma spraying process and the ratio of critical Gibbs free energy of both α-$MoSi_2$ and β-$MoSi_2$ approaches unity when the undercooling is <0.8 T_m. Another important source of the α-phase $MoSi_2$ in plasma treated $MoSi_2$ material is its formation via solid-state transformation. The hexagonal-to-tetragonal phase transformation occurs even at the very rapid solidification rates existing in the plasma spray processing and is characterized by the formation of twins in the crystal grain (5).

The high percentage of α-phase in the Ar-N_2 plasma treated $MoSi_2$ powders may be related to the heterogeneous nucleation mechanism during the rapid solidification. As well known, the "heat conduction potential" *I(T)* of N_2 is almost about one order of magnitude lower than that of H_2 in the range 500~10000 K. If the particles have the same residence time in plasma, the particles treated in the Ar-N_2 plasma may experience incomplete melting (18), and heterogeneous nucleation is likely to occur. When there is a unmelted core (tetragonal α-phase of the initial materials) inside the liquid droplet, the formation of α-$MoSi_2$ is much easier than the situation of homogeneous nucleation. The "polygonal granules" shown in Fig. 3(b) may give evidence of the heterogeneous nucleation.

4.2 The metastable eutectic reactions

The formation of low silicides, Mo_5Si_3 and Mo_3Si, and a free Si phase can be explained in terms of metastable eutectic reaction during rapid solidification processing (19). Fig. 5 provides a schematic Mo-Si binary phase diagram. The extension (dash lines in Fig. 5) of the equilibrium liquidus phase boundaries to an intersection yields the location of each metastable eutectic point.

During the liquid-solid phase transformation in the in-flight $MoSi_2$ liquid droplet, the first event should be the congruent solidification to produce $MoSi_2$ phase, when the undercooling takes the melt below the metastable melting point. The metastable melting point of β-$MoSi_2$ may be only a few degrees Celsius below the equilibrium melting point (2303 K), according to W.J. Boettinger *et al.* (13). Following the competition of nucleation rates, α-phase $MoSi_2$ can also form directly from the melt, as stated above, and be transformed from β-$MoSi_2$. The partial vapor pressure of silicon of $MoSi_2$ is high, silicon volatilization during particle flight brings about the Si-deficient liquid droplets. Once the composition of liquid droplets is displaced towards the liquidus that bounds the metastable L+Mo_5Si_3 two-phase field, the Mo_5Si_3 phase will segregate. Finally, following temperature decline, metastable eutectic reaction occurs, as marked in Fig. 5:

$$Liquid \rightarrow Mo_5Si_3 + Si. \qquad (1)$$

This reaction mechanism explains why Mo_5Si_3 phase always exists in various rapid solidification processes of $MoSi_2$ and the silicon formation in the induction plasma treated $MoSi_2$ powders.

The existence of Mo_3Si may be explained by the similar construction of metastable equilibrium region in the phase diagram. The course of the solidification of in-flight $MoSi_2$ melt droplet proposed above is confirmed by the evident eutectic microstructure seen in the cross-sections of induction plasma treated $MoSi_2$ powders, shown in Fig. 4.

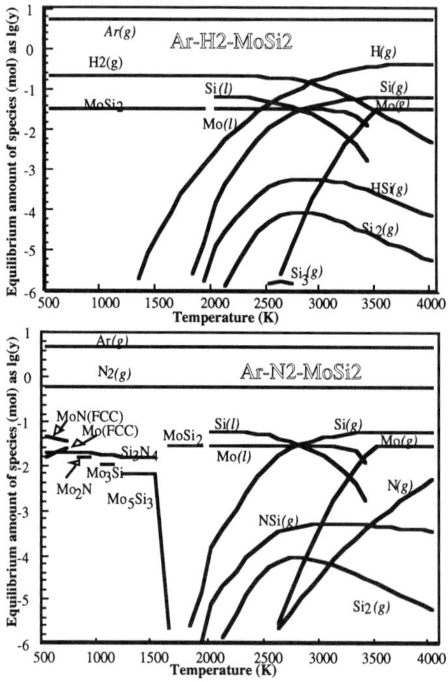

Fig. 6 Thermodynamic equilibrium for $MoSi_2$ in Ar-H_2 and Ar-N_2 systems.

4.3 The chemical reactions

The thermodynamic equilibrium compositions were calculated for the (Ar-H_2-$MoSi_2$) and (Ar-N_2-$MoSi_2$) systems respectively. Computations were carried out for simulating the real conditions of the induction plasma process. Typical results are shown in Fig. 6. The difference of the thermodynamic equilibrium compositions in the two systems is obvious. In the (Ar-H_2-$MoSi_2$) system, Fig. 6(a), $MoSi_2$ is stable until its melting point, then simply experiences "melting", from the solid phase into the liquid phase. However, in the (Ar-N_2-$MoSi_2$) system, many chemical reactions between N_2 gas and $MoSi_2$ may occur. In the temperature range from 500 to 1700 K, nitride Si_3N_4, and Mo, MoN, Mo_2N, Mo_3Si and Mo_5Si_3 are all the possible chemical products in succession. Only after 1700 K, $MoSi_2$ can be thermodynamically stable until the melting point. Since there is the chemical reaction between plasma gas and the injected material, the $MoSi_2$ powders treated in the Ar-N_2 plasma are expected to be different from those treated in the Ar-H_2 plasma. This has been evidenced by checking the phase quantity of various phase produced in two different plasma processes (see Table II).

5. CONCLUSION

$MoSi_2$ powders, treated by the induction plasma process, are composed of β-$MoSi_2$, α-$MoSi_2$, Mo_5Si_3, Mo_3Si and Si phase. The classical nucleation theory by comparison of the critical free energy for nucleation for α-$MoSi_2$ and β-MoSi2 has explained the formation of metastable β-$MoSi_2$ phase in the plasma process. The formation of low silicides, Mo_5Si_3 and Mo_3Si, and free Si was explained in terms of metastable eutectic reaction during rapid solidification processing. The discrepancy in the phase composition observed in Ar-N_2 induction plasma treated $MoSi_2$ powders with those treated by Ar-H_2 induction plasma is due to the lower thermal conductivity of plasma gas N_2, the heterogeneous nucleation mechanism during the solidification, and the chemical reaction between N_2 and $MoSi_2$.

REFERENCES

[1]A.K. Vasudévan, and, J.J. Petrovic, "A Comparative Overview of Molybdenum Disilicide Composites", Materials Science and Engineering, **A155**, 1-17(1992).

[2]H. Kung, R.G. Gastro, A.H. Barlett, and J.J. Petrovic, "The Structure of Plasma Sprayed $MoSi_2$-Al_2O_3 Microlaminate Tubes", Scripta Metall. Mater., **32**, 179-83(1995).

[3]R.G. Gastro, H. Kung, and P.W. Stanek, "Reactive Plasma Spraying of $MoSi_2$ Using An Ar-10% CH_4 Powder Carrier Gas", Mater. Sci. Eng., **A185**, 65-70(1992).

[4]R. Tiwari, H. Herman, and S. Sampath, "Vacuum Plasma Spraying of $MoSi_2$ and Its Composites", Mater. Sci. Eng., **A155**, 95-100(1992).

[5]T.E. Michell, R.G. Castro, and M.M. Chadwick, "{112}<111> Twins in Tetragonal $MoSi_2$", Philos. Mag. **A65**, 1339-51(1992).

[6]M. Fukumoto, T. Ueda, and I. Okane, "Protecting Properties of Oxidation Resistant Plasma Sprayed $MoSi_2$ Coating", Yosha, **30**, 1-6(1993) (in Japanese)

[7]A. Ohomori, and M. Fukuoda, "Characterization of Plasma-Sprayed Silicide Intermetallic Compound Coatings"; pp.433-37 in *NTSC Proceedings*, Anaheim, CA, 7-11 June 1993.

[8]R.G. Castro, R.W. Smith, A.D. Rollett, and P.W. Stanek, "Ductile Phase Toughening of Molybdenum Disilicide by Low Pressure Plasma Spraying", Mater. Sci. Eng., **A155**, 101-107(1992).

[9]Y. Jeng, J. Wolfstine, E.J. Lavernia, D.E. Bailey, and A. Sickinger, "Low-Pressure Plasma Deposition of SiC-Reinforced $MoSi_2$", Scripta Metall. Mater., **28**, 453-458(1993).

[10]D.E. Lawrynowicz, J. Wolfenstine, E.J. Lavernia, S. Nutt, D.E. Bailey, A. Sickinger, and A.M. Hirt, "Reactive Synthesis and Characterization of $MoSi_2$/SiC Using Low-Pressure Plasma Deposition and 100% Methane", Scripta Metall. Mater., **32**[5], 689-93(1995).

[11]R.W. Smith, and R. Knight, "Thermal Spraying I: Powder Consolidation - From Coating to Forming", JOM, **8**, 32-39(1995).

[12]X. Fan, T. Ishigaki, and Y. Sato, "Phase Formation in Molybdenum Disilicide Powders During in-flight Induction Plasma Treatment", J. Mater. Res, **12**, 1315-1326(1997).

[13]W.J. Boettinger, J.H. Perepezko, and P.S. Frankwicz, "Application of Ternary Phase Diagrams to the Development of $MoSi_2$-Base Materials", Mater. Sci. Eng., **A155**, 33-44(1992).

[14]C.G. Levi, V. Jayaram, J.J. Valencia, and R. Mehrabian, "Phase Selection in Electrohydrodynamic Atomization of Alumina", J. Mater. Res., **3**, 969-83(1988).

[15]D. Turnbull, "Metastable Structure in Metallurgy", Metall. Trans., **12A**, 695-708(1981).

[16]P.H. Shingu, "Thermodynamic Principles of Metastable Phase Formation"; pp. 1275-80 in First International Conference on Processing Materials for Properties, Ed. by H. Henein and T. Oki, TMS, Warrendale, PA, 1993.

[17]X. Fan, and T. Ishigaki, "Critical Free Energy for Nucleation from The Congruent Melt of $MoSi_2$", J. Cryst. Growth, **171**, 166-73(1997).

[18]E. Bourdin, P. Fauchais, and M. Boulos, "Transient Heat Conduction under Plasma Conditions", Int. J. Heat Mass Transfer., **26**, 567-82(1983).

[19]J.H. Perepezko, and W.J. Boettinger, "Use of Metastable Phase Diagrams in Rapid Solidification", pp. 223-40 in Alloy Phase Diagrams. Ed. by L.H. Bennett, T.B. Massalski, and B.C. Giessen, Mat. Res. Soc. Symp. Proc., Vol.19, Elsevier Science Publishing Co. Inc., New York 1983.

SYNTHESIS OF Al_2O_3/SiC POWDERS FROM ALUMINOSILICATE PRECURSORS

Carlos E. Borsa, Fabiola Spiandorello, and Ruth H. G. A. Kiminami
Departamento de Engenharia de Materiais
Universidade Federal de Sao Carlos, Brazil

ABSTRACT

Carbothermal reduction of natural aluminosilicates by carbon black was used to obtain reactive Al_2O_3/SiC powders. Three different precursors, namely, Kaolin, Kyanite and Pyrophyllite were investigated. The powders were mixed in water, in stoichiometric proportion. Additions of alpha-alumina were carried out to correct the ratio of the resultant material to 5wt%SiC. Carbothermal reduction of synthetic alumina and silica was used as control material. The dried blend was broken in mortar and heated at 1500°C for 2h in reduced atmosphere for the reaction to proceed. The resultant powder was characterized by x-ray diffraction, helium picnometry and scanning and transmission electron microscopy. The results showed that only alpha-alumina and alpha-SiC were present as crystalline phases and that the powders had similar densities. The size, morphology and the level of aggregation of the alumina and of the SiC phases obtained, depend on the type of the precursors used, i.e., particle size and purity level. SiC whiskers were observed for all compositions when metallic impurities, such as iron and titanium, were present. The whiskers were 100nm to 500nm in diameter and up to 100 microns in length and presented different morphologies. Densification results of pressureless sintered powders are also reported.

Al_2O_3/SiC NANOCOMPOSITES FROM ALTERNATIVE PROCESSING ROUTES

Carlos E. Borsa
Departamento de Engenharia de Materiais, Universidade Federal de São Carlos
CP 676- CEP 13565-905- Brazil
Richard I. Todd
Manchester Materials Science Centre, University of Manchester and UMIST, Grosvenor Street, Manchester, M1 7HS, UK
Richard J. Brook
Department of Materials, Oxford University, Parks Road, OX1 3PH, UK

ABSTRACT

This work investigated two alternative processing routes for the preparation Al_2O_3/SiC nanocomposites. One used a polymeric precursor for the SiC phase (polysilastyrene) and commercial alumina powder. The materials were mixed in toluene, dried and heated to 1500°C to crystallize the SiC phase. The resultant powder consisted of SiC particles (50-300nm in size) distributed at the surface of the alumina particles. The second processing route utilized an inorganic precursor for alumina (aluminum nitrate). This precursor was dissolved in water in the presence of nano-sized commercial SiC and urea. Controlled heat treatments, at low and high temperatures, promoted the precipitation of an alumina phase onto the SiC particles. In both processing routes, the material was milled and hot pressed at 1700°C to achieved full densification. TEM and XRD analysis were used to characterize the powders and the sintered pieces. Samples prepared from conventional mixture of commercial powders were used as the control. It is shown that the use of chemical precursors produced distinctive microstructures, with different matrix grain sizes, SiC particle sizes and location in the matrix as compared to the control material.

INTRODUCTION

The incorporation of small amounts of nano-SiC particles into a Al_2O_3 matrix has been shown to produce significant improvements in mechanical properties compared to monolithic Al_2O_3, especially at high temperatures [1]. The fabrication of the composite material is usually carried out by the mechanical mixture of commercial powders [1-3]. This method has proved to be fast and reliable for the attainment of the desired microstructure. However, the necessity of improving over the existing material has driven the research for alternative processing routes, such as those using chemical precursors. They can produce more homogeneous and finer powders [4,5].

This work investigated this possibility by processing Al_2O_3/SiC nanocomposite powders obtained from two alternatives routes. One used a polymeric precursor for the SiC phase (polysilastyrene) and commercial alumina powder. The other one utilized an inorganic precursor for the alumina (aluminum nitrate), dissolved in water in the presence of nano-sized commercial SiC and urea. The results showed that compared with materials produced by the mechanical mixture of commercial powders, the materials derived from the SiC-precursor route had smaller particle size but a poorer distribution of the SiC particles, whereas materials derived from the Al_2O_3-precursor route had smaller matrix grain size and better homogeneity of the SiC.

EXPERIMENTAL

Two alternative processing routes used to prepare the nanocomposite powder: (1) SiC-precursor route and (2) Al_2O_3-precursor route. A conventional processing route was used as control. The SiC content used in this investigation was 10wt%.

Preparation of the Nanocomposite Powder

Al_2O_3-Precursor route: In this case, SiC powder was coated with a porous layer of an Al_2O_3 precursor formed by a heterogeneous precipitation reaction. The method was used on a route for coating SiC whiskers [4]. Urea ($CO(NH_2)_2$ - Aldrich) and hydrated aluminum nitrate (AN) ($Al_2(NO_3)_3$ $9H_2O$- Aldrich) were used as the reagents in the reaction. An aqueous solution was prepared containing 12g/ liter of urea and 0.3M of AN. This was slowly heated in a spherical flask connected with a condenser, to reduce the loss of water, over a period of 2h to the process temperature 85°C. The solution was magnetically stirred and the changes in the pH were continuously measured by a pH-meter. This condition was maintained until the pH increased from 1.5 to 4. At this time a pre-heated aqueous dispersion of 5g/litre of SiC was added to the solution (using a 1:1 ratio) and aged for 24h under vigorous stirring. Under this condition, the precipitation of a complex of aluminum hydroxide (CAH) occurred, coating the SiC particles. Following the coating, the suspension was cooled to room temperature and allowed to settle, after which the supernatant was discarded and the remaining slurry washed with deionised water and dried at 80°C for 24h.

The resulting powder was calcined in Ar for 2h at temperatures varying from 600. During this treatment, CAH was converted to γ-alumina.

SiC-Precursor Route: For this route a polysilastyrene polymer (PSS 400-Nippon Soda) was used as the precursor for the SiC. Initially the polymer was dissolved in toluene (10g:100ml) and then mixed with the alumina powder. The dispersion was then ball milled for 3h using alumina balls and dried under low vacuum. The blend was broken up using a pestle and mortar and crosslinked at 400°C for 3h, then pyrolysed at 1000-1500°C for 2h. All heat treatments were done in an Ar/1%H_2 atmosphere using 5°C/h for the heating rate in the crosslinking step and 5°C/min for the pyrolysis step. A boat with graphite powder was placed inside the furnace during the pyrolysis stage in an attempt to reduce the oxidation reaction of the polymer. The powder was ball milled in propan-2-ol for 3h using alumina balls and dried at 80°C for 24h.

Control Route: As control route, nanocomposite materials were prepared by the mechanical commercial nano-sized Al_2O_3 (APK53, Sumitomo- mean particle size 0.2μm) and SiC (UF45, Lonza- mean particle size 0.2μm) powders. They were attrition milled for 1.5h (weight ratio of water to powder =1) using magnesia stabilized zirconia media, flocculated by adding HCl and finally dried for 24h at 80°C in a plastic tray. The flocculation of the matrix powder was found to prevent the segregation and agglomeration of the SiC nano-particles.

Densification of the Materials

The dried nanocomposite powder produced by the different preparation routes was broken up using a pestle and mortar and passed through a 180μm sieve. Calcined powders were uniaxially hot-pressed in a graphite die (1h/25MPa/Ar) at 1700°C.

Characterization Techniques

Sintered densities, D, were measured using the Archimedes method, X-ray diffraction was used to check for the presence of extra phases arising from oxidation or reaction. TEM (Jeol 200 CX) was used to check the particle size and the morphology of the powders used, and the distribution of nano-particles inside the grains. Thin samples were prepared by dimpling the samples down to 60μm and by ion beam thinning until a hole was produced in the material.

RESULTS AND DISCUSSION

Powder Characterization

Figure 1 shows a TEM micrograph of the powder obtained by the coating reaction (Al_2O_3-precursor route) after calcination at 600°C. The commercial SiC (dark particles) was coated by a porous phase formed by the precipitation reaction of the aluminum nitrate precursor. The coated phase, was relatively homogeneous and was identified by XRD as $\gamma-Al_2O_3$. Some SiC particles were not individually coated, which was probably due to the difficulty of dispersing all the agglomerates during the precipitation reaction. This would be easier for lower concentration of SiC, but not feasible for fabricating large amount of powder. Calcination at 1200°C transformed the coating phase from γ to α-Al_2O_3, but extensive coarsening of the powder resulted from this operation. Ball milling was used to reduce this problem.

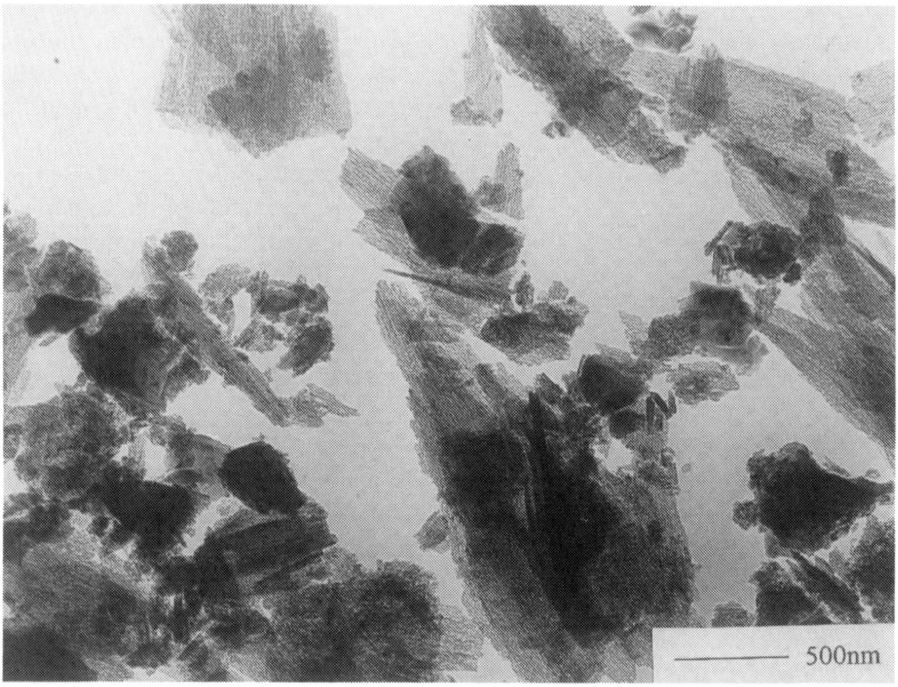

Fig. 1- TEM micrograph of the coated SiC particles obtained by the precipitation reaction in the Al_2O_3-precursor route.

Figure 2 shows a TEM micrograph of the alumina powder coated by nano-sized SiC particles prepared by the SiC-precursor route. The nano-size SiC was obtained by the controlled thermal decomposition of a polymer at the temperature of 1500°C. This temperature is necessary to crystallize the SiC particles, which can, however, lead to the formation of sintered structures as shown in Figure 2. An efficient milling stage is then needed to break down these agglomerates.

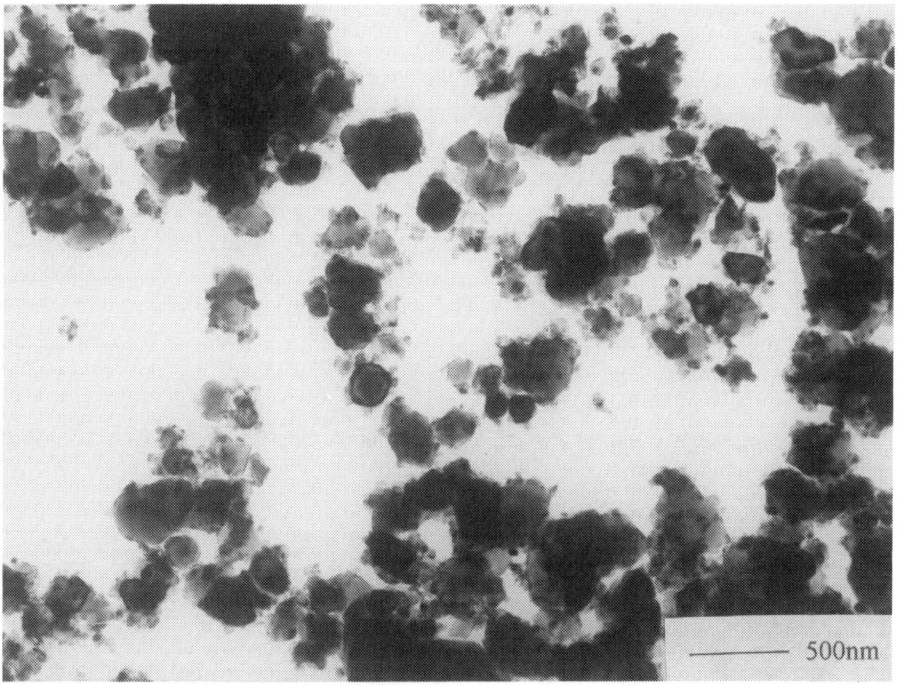

Fig. 2- TEM micrograph of alumina powder coated by 10-20nm sized SiC particles obtained by the controlled thermal decomposition of the PSS polymer at 1500°C.

Microstructure analysis

Figure 3 shows TEM micrographs of the materials prepared by the alternative processing routes. Comparing to the control route (Figure 4), significant differences in their microstructures can be observed, with variations in matrix grain size and morphology, and particle size and distribution.

Nanocomposite materials processed from powders derived from the SiC-precursor route (Fig. 3b) presented small SiC particles, as expected from the in situ chemical formation of the particles during the decomposition of the polymer [5]. The powder, however, had a heterogeneous dispersion of the SiC particles with matrix grain size varying from 0.5μm to 10μm. The heterogeneity of this material can be explained by the difficulties in dispersing this polymers during the homogenization step. PSS, a neutral polymer, does not provide steric defloculation for the alumina powder in organic media. This results in a heterogeneous mixture of the ingredients, which remains throughout the process leading to abnormal grain growth of the matrix in regions where there are no particles, and growth of the SiC particles where their concentration is higher.

For nanocomposites prepared by the Al_2O_3-precursor route (Figure 3a), the most significant feature is the small grain size of the alumina matrix and its elongated morphology. The SiC particles are well dispersed and are situated both on the alumina grain boundaries and within the grains. The reason for this distinct behavior on morphology of the alumina grains is not known in detail, but since the alumina had a different origin from the powders used in the other specimens there is no reason to expect them to give the same final microstructure.

In summary, the use of chemical precursors allow to achieve phases which were finer than it can be obtained by commercial powders. This is important when small grain size materials are required. However, the advantage of producing more homogeneous dispersion of the second phase, was not observed, in the case of the SiC-precursor route. The control route was less sensitive to the processing variables

Table I presents a summary of the microstructural characteristics for each material.

Table I- Summary of the microstructural characteristics for each material

Preparation Route	Range of SiC particle Size	Average grain size and morphology of the Al_2O_3 matrix	Predominant location of SiC particles
Al_2O_3-precursor	50-200nm	1.1μm- Elongated	both inside and at the grain boundaries
SiC-precursor	30-200nm	1.5μm- Equiaxial with bimodal grain size distribution	inside the grains
Control	50-200nm	2.5μm- Equiaxial	inside the grains

CONCLUSIONS

Alumina/SiC nanocomposites were prepared by two alternative processing routes, namely, SiC-precursor and Al_2O_3-precursor.

Compared to a control route, using the mechanical mixture of commercial powders, materials derived from the SiC-precursor route had smaller particle size but a poorer distribution of the SiC particles, whereas materials derived from the Al_2O_3-precursor route had smaller matrix grain size and better homogeneity of the SiC particles.

a)

b)

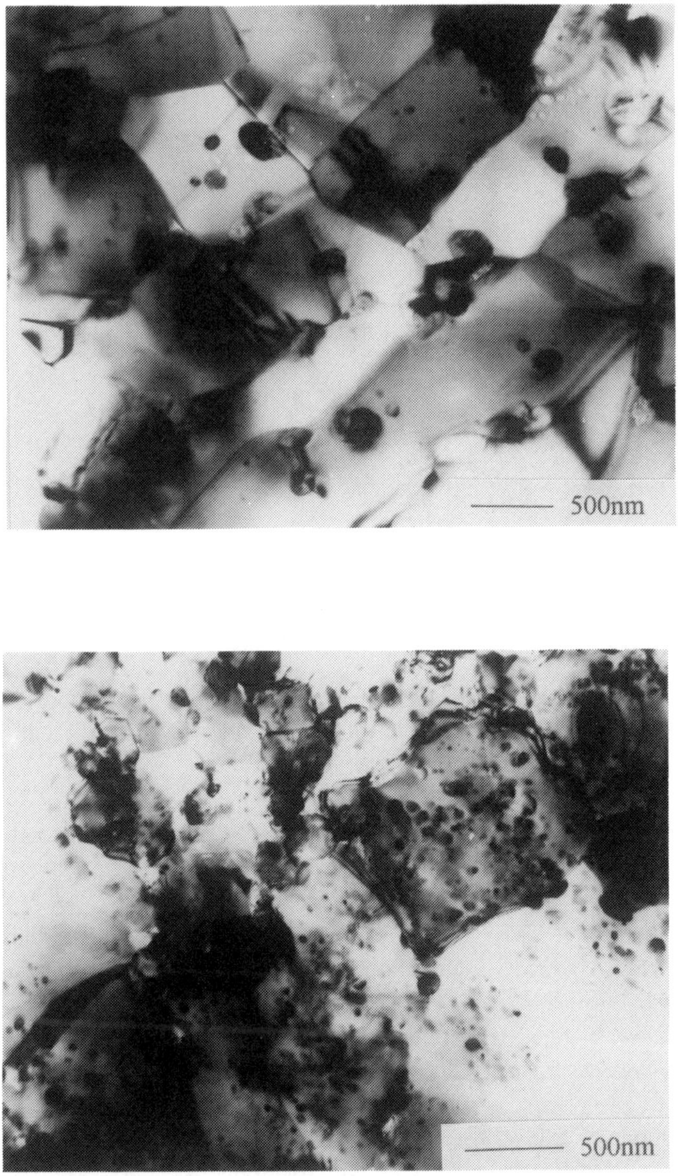

Fig. 3- TEM micrograph of dense nanocomposite materials prepared by different processing routes: a) Al_2O_3-precursor and b) SiC-precursor.

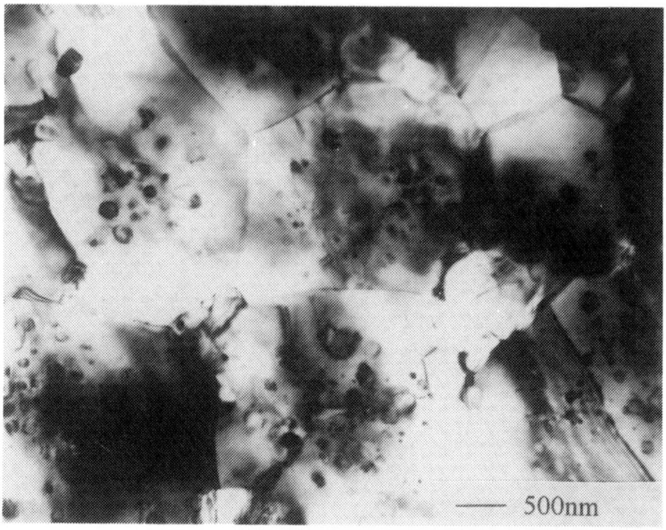

Fig. 4- TEM micrograph of dense nanocomposite materials prepared by the mechanical mixture of commercial powders (control route).

ACKNOWLEDGMENTS

The authors thank CNPq, Brazil, for the financial support. We also acknowledged the CEC and to the SERC, UK for the provision of facilities.

REFERENCES

[1]K. Niihara and A. Nakahira, "Structural Ceramic Nanocomposites by Sintering Method Roles of Nano-size Particles", Ceramics: Toward the 21st Century, Ceramic Society of Japan., 404-17 (1991).

[2] C.L. Stearns, J. Zhao and M.P. Harmer, "Processing and Microstructure Development in Al_2O_3/SiC Nanocomposites", Jounal of the European Ceramic Society, 10, 473-77(1992).

[3] C.E. Borsa, S. Jiao, R.I. Todd and R.J. Brook, "Processing and Properties of Al_2O_3/SiC Nanocomposites", Journal of Microscopy, 177, pt. 3 pp. 305-312 (1995).

[4]D. Kalponek, L.C. De Jonghe, "Particulate Composites From Coated Powder", Journal of the European Ceramic Society, 7345-51 (1991).

[5]R. Riedel, K. Strecker and G. Petzow, "In Situ Polysilane-Derived Silicon Carbide Particulates Dispersed in Silicon Nitride Composites", Journal of the American Ceramic Society, 72 [11] 2071-77 (1989).

Colloid Dispersion and Consolidation

PROCESSING CERAMIC DISPERSIONS

C. F. Zukoski, Glenn Channell and Jennifer Wall
Department of Chemical Engineering
University of Illinois
Urbana, IL 61801

ABSTRACT

Fired ceramic properties depend heavily on green body microstructures as these are often frozen into place during sintering. As a consequence, for dense ceramics, green bodies with the highest possible particle packing density are desirable. This can be accomplished in ways that limit fired flaw density by working with dispersions where the particle interaction potential is controlled to yield cohesive compacts by altering suspension conditions to the point where the particles are attractive. This process produces suspensions with yield stresses and which deform in a plastic manner under external stress. Key questions revolve around how to manipulate particle interaction potentials to produce desired slurry and green body properties. In this presentation, we discuss the role of particle interactions on shear and compressive rheology of alumina suspensions. Attractions are manipulated in such a manner that the pair potential is driven systematically from attractive to repulsive and the changes mechanical properties are investigated. For these suspensions a key variable is the gel volume fraction where the strength of attractions and the particle number density reach a level that aggregated structures span space and produce a space filling network. By normalizing volume fraction on this parameter, compressive yield stresses at modest volume fractions and parameters characterizing the shear rheology are condensed onto master curves. That a variety of particle interaction potentials produce the same flow curves when normalized to the gel volume fraction suggests that the gel volume fraction is a useful tool in deconvoluting the balance between strength of attraction and density of interparticle bonds. This scaling fails as maximum packing fraction is approached. This limiting volume fraction depends on poorly understood details of the microstructure. In this discussion we present a model for aggregated suspension microstructure where density fluctuations established during aggregation and subsequent shearing which controls ultimate control green body density.

DISPERSION AND STABILISATION OF VERY FINE PARTICLE SUSPENSIONS

Johnny Widegren and Lennart Bergström*
Department of Materials Science and Engineering
Royal Institute of Technology
SE-100 44 Stockholm, Sweden

*Institute for Surface Chemistry
SE-114 86 Stockholm, Sweden

ABSTRACT

Successful application of Electrophoretic Deposition (EPD) using very fine powders requires well-stabilized suspensions. In order to fabricate composites by EPD, it is imperative to control the zeta-potential to ensure uniform and dense packing of the particles. We have investigated the colloidal stability and electrokinetic properties of three different, commercially available powders; Al_2O_3, TiO_2 and SiC, dispersed in ethanol. The pH of the ethanolic suspensions was controlled by different types of acids and bases, and the effect on the suspension properties was evaluated. In addition, we have also studied the effect of several surface active additives.

BROWNIAN DYNAMICS VIEW ON COAGULATION KINETICS IN COLLOIDAL SUSPENSIONS

M. Huetter, L. J. Gauckler and H. C. Oettinger*
Colloidal Chemistry–Ceramic Processing Group at the ETH
*Institute of Polymer Physics at the ETH
Zürich, Switzerland

ABSTRACT

An interpretation of macroscopic material properties of colloidal suspensions in terms of particle shape, interaction potentials, the solvent viscosity and temperature can be given by Brownian Dynamics simulations. The latter can be used to study both the rheology of the stabilized suspension and the coagulation behavior. We study spherical particles interacting through DLVO-forces and establish a connection between the interaction potential parameters and diffusion- and reaction-limited aggregation behavior, respectively. The destabilization of the suspension by ionic strength and by changing the surface potential of the particles, i.e. changing the pH, are compared.

AGGREGATION EFFECTS ON PROPERTIES AND PROCESSING OF COLLOIDAL SiO_2 SUSPENSIONS

J. A. Lewis and J. Guo
Materials Science and Engineering Department
The Beckman Institute
University of Illinois
Urbana, IL 61801

ABSTRACT

We have studied the influence of colloidal stability on the properties and processing of aqueous and non-aqueous SiO_2 (monodispersed spheres) suspensions. Rheological (shear and compressional) measurements were carried out to study suspension properties as a function of solids loading, solvent composition, and salt additions. In addition, sedimentation and drying studies were carried out to explore the effects of suspension stability on structure evolution during these critical processes. Through these studies, we aim to delineate the role of interparticle, gravitational, and capillary forces on structure evolution and stress development during constrained drying of particulate films.

RELATIONSHIP BETWEEN RHEOLOGY AND INTERPARTICLE FORCES IN AQUEOUS $BaTiO_3$ SYSTEMS

Jürgen Sindel, W. M. Sigmund, B. Baretzky, and F. Aldinger
Max-Planck-Institut f. Metallforschung and Universität Stuttgart
Institute f. Nichtmetallische, Anorganische Materialien
Pulvermetallurgisches Laboratorium
Stuttgart, Germany

ABSTRACT

The rheology of aqueous $BaTiO_3$ systems is important for the production of multilayer capacitors and actuators by tape casting. We report about investigation son the pH dependence of the rheology of concentrated aqueous slurries using acrylic polymers and di-block copolymers as dispersants. Surface force distance curves were measured in - situ for the same systems with a specially derived AFM-setup. Changes in interparticle forces that are observed by the AFM technique are correlated with changes in rheological behavior. Ellipsometry indicates that these changes are due to a different kind of adsorption of the dispersants at the interface. AFM and ellipsometry are done on diluted model systems whereas concentrated slurries are studied by rheology and DRIFT. A detailed comparison for the results by these different methods will be given.

HETEROGENEITY OF ADSORPTION AND ITS IMPACT ON STABILITY OF ALUMINA SUSPENSION WITH POLYELECTROLYTE

Lu-Cun Guo, Yao Zhang, Nozomu Uchida, and Keizo Uematsu
Department of Chemistry
Nagaoka University of Technology
Kamitomioka, 1603-1
Nagaoka, Niigata, Japan

ABSTRACT

The evidence for the existence of heterogeneity of the adsorption of polyelectrolyte on alumina surfaces was shown. The impact of the heterogeneous adsorption on the stability of rheological properties was studied. It was found that the heterogeneity of adsorption occurs when the adsorption is of high affinity. At low concentration or low coverage of the polyelectrolyte, this heterogeneous adsorption can cause flocculation, even the suspension was a high zeta potential. A model describing this phenomenon was provided graphically and in formula. The formation of the concentrated or continuous "bare" surfaces was shown to be responsible for the flocculation, because much stronger attraction would rise between the partly covered particles. The implication of the discovery for ceramic colloidal processing was also demonstrated.

THE RHEOLOGICAL BEHAVIOR OF SLURRIES AND CONSOLIDATED BODIES CONTAINING INTERPENETRATING NETWORKS

Matthew L. Fisher and Fred F. Lange
University of California, Santa Barbara
Santa Barbara, CA 93106

ABSTRACT

The rheological behavior of non-aqueous slurries containing interpenetrating silicon nitride networks has been investigated. The two networks are a strongly attractive (flocculated) network consisting of silicon nitride suspended in dodecane, and a weakly attractive network exhibiting short range repulsion which is formed by suspending silicon nitride that has been reacted (alkylated) with a short chain alcohol. Dynamic, oscillatory measurements have been used to study the shear modulus and yield stress of the mixtures as a function of the volume fraction of particle sin the strongly attractive (non-alkylated) network. It was observed that the strong network overwhelmingly controls the behavior of the mixtures at concentrations above approximately 20% of the uncoated powder. Bodies consolidated by pressure filtration were tested in uniaxial compression to determine the stress/strain behavior of the saturated compacts and to ascertain the ratio of strong to weak network necessary to produce clay-like rheology.

INFLUENCE ON THE STABILIZATION MECHANISM ON THE RHEOLOGICAL CHARACTERISTICS AND SLIP CASTING OF ZIRCONIA SUSPENSIONS

J. Lara-Aguilar and H. Balmori-Ramirez
National Polytechnic Institute
Department of Metallurgical Engineering
A.P. 75-872
07300 Mexico City, Mexico

ABSTRACT

The rheological and slip casting characteristics of aqueous colloidal zirconia aqueous suspensions stabilized electrostatically or electrosterically was studied in order to select the best one in terms of their consolidation characteristics. The suspension preparation parameters were selected after stability map for zirconia suspensions in which stable and unstable regions were identified as a function of pH and ammonium polyacrylate (NH4PAA) content. This diagram was constructed with results of zeta potential, agglomerate size measurements and sedimentation tests. The influence of the stabilization mechanism on slip casting was studied by means of viscosity measurements, thixotropic behavior, characterization, cake growth rate and density variation as a function of cake height for suspensions with different solids contents (20 to 45 volume %). The electrosterically stabilized suspensions were superior in the sense that they allowed a higher solids charge and the cast cakes were denser and more homogeneous than for the electrostatically stabilized suspensions. The NH4PAA adsorption mechanism on zirconia was studied by means of thermogravimetric and FTIR analysis.

RHEOLOGY OF AQUEOUS 3Y-ZrO_2 POWDER SLURRIES: EFFECT OF POLYELECTROLYTE TYPE AND MOLECULAR WEIGHT

F. Shojai, T. Mäntylä, A. B. A. Pettersson*, and J. B. Rosenholm*
Institute of Materials Sciience
Tampere University of Technology
P. O. Box 589
FIN-33101, Tampere, Finland

*Department of Physical Chemistry
Abo Akademi University
Porthansgantan 3-5
FIN-20500 Abo, Finland

ABSTRACT

Preparation of well dispersed slips with desired rheological properties is a critical step in all ceramic processing based on slip casting. Aqueous slip casting increasingly is used in ceramic processing due to its environmentally friendly nature, as well as due to its simplicity and capability of producing complex shapes. The rheology of diluted and concentrated slurries is studied in order to evaluate the dispersibility with different stabilizing agents. This information is used to find the optimum chemical and mechanical processing of the slip. In this study, the effect of the type and molecular weight of polyelectrolyte on the rheology of aqueous slips of 3Y-ZrO_2 powders has been investigated. Physical and chemical composition of 3Y-ZrO_2 powders such as shape, specific surface area, particle size and size distributions, phase composition, bulk chemical composition and surface analysis were determined by scanning electron microscope (SEM), N2 adsorption, laser scattering, X-ray fluorescent (XRF) and X-ray photoelectron spectroscopy (XPS). The efficiency of electrostatic stabilization was discussed based on the results of zeta potential as a function of pH. The viscosity of aqueous slurries with increasing solid content was measured and it showed higher dispersibility of the powder which was pre-processed for slip casting. Ammonium polyelectrolytes and sodium-salt polyelectrolytes with the same head-group and different chain length or molecular weight were used as dispersant. Rheology of the aqueous slips were studied as a function of electrolyte concentration for each polyelectrolyte. The effect of milling time on the viscosity of the slips with and without the dispersing agent was investigated. It was found that ammonium polyelectrolytes were more efficient than sodium polyelectrolytes. The change in viscosity of slips dispersed by high molecular weight polyelectrolyte is more significant in continuous addition of polyelectrolyte. Full dispersion of the slip was accomplished with lower concentration of the polyelectrolyte of higher chain length. The differences between the dispersibility of these 3Y-ZrO_2 powders were discussed.

ELECTROKINETIC BEHAVIOR AND STABILITY OF SiC NANOPARTICLE DISPERSIONS

Pedro Tartaj and Jose S. Moya
Instituto de Ciencia de Materiales de Madrid (ICMM) CSIC
Cantoblanco 28049
Madrid, Spain

ABSTRACT

The stability of aqueous suspension containing SiC nanoparticles prepared by inductive coupled plasma has been studied. The physico-chemical analyses indicated that the as-prepared samples were constituted by monocrystalline spherical particles of the 3C-SiC polytype with a mean diameter of about 40nm. HRTEM clearly showed that the particles were covered by an amorphous surface layer which according to XPS was mainly composed by carbon. DTA and TG analyses showed that the free carbon was eliminated by heating the sample at 650°C, but the resulting powder presented an excess of amorphous SiO2 on particle surface. This silica coating was leached out by using a 5% HF solution. It has been determined, by evaluating an effective Hamaker constant, that the most stable aqueous suspensions are obtained for the sample containing the amorphous SiO2 surface layer. Finally, evidences of the presence of a strong hydrophobic force have been detected in the original sample and in the one prepared by heating at 650°C and further leaching with HF.

DEVELOPMENT OF SHORT-RANGE REPULSIVE INTERPARTICLE FORCES IN AQUEOUS Si_3N_4 SLURRIES WITH CHEM-ADSORBED SILANES

Miroslav Colic, George Franks, Matthew Fisher, and Fred F. Lange
Materials Department
College of Engineering
University of California, Santa Barbara
Santa Barbara, CA 93106

ABSTRACT

Different chem-adsorbed silane molecules have been used to produce weakly attractive silicon nitride particle networks for aqueous colloidal processing. Silanes with diamino and polyethyleneglycol hydrophilic heads yielded slurries with the lowest viscosity, longest sedimentation stability and highest packing density. Chem-adsorbed silane molecules protected silicon nitride and yttrium oxide, a common processing aid, from hydrolysis at pHs between 5.5 and 11. A novel approach was used to produce short range repulsive potentials necessary to yield the weakly attractive networks. Addition of salt to dispersed silicon nitride slurries with particles coated with polyethyleneglycol-silane, caused the collapse of the 22 atoms long chains and residual electric double layer. This produced a weakly attractive network which persisted during consolidation to yield a plastic body with a flow stress that was dependent on the counterion size. When 0.5 M tetramethylammonium chloride was used at pH 10, plastic bodies had a flow stress similar to clay, whereas lithium counterions produced bodies with a much higher flow stress.

SILANOL ADSORPTION AND PROPERTIES OF Si_3N_4

Joseph A. Yanez and Wolfgang Sigmund
Max Planck Institute, PML
Heisenbergstr. 5
Stuttgart 70569, Germany

ABSTRACT

Colloidal processing has been found to be an effective way to increase the mechanical strength of ceramic materials. Colloidal ceramic processing's success stems from its ability to remove strength limiting heterogeneities by filtration. The technique requires the introduction of a long range repulsive force prior to filtration and a short range repulsive force after consolidation. Other researchers have shown that the absorption of silanols on the surface of Si_3N_4 creates the desired behavior. An in-depth study has been made to gain a better understanding of the mechanism by which the silanols introduce these properties to a particle suspension. Sputtered Si3N4 thin films were made without sintering aids. The properties of these films were studied by ESCA and Auger spectroscopy. The adsorption kinetics and layer thickness were determined by ellipsometry and NMR. The morphology of the substrate and the adsorbed layer and the interparticle potential was determined by AFM techniques in water, vacuum, and air. The silanol adsorption behavior on the sputtered films has been compared to the adsorption behavior on the powder through adsorption isotherm curves determined by UV spectroscopy.

ALIGNMENT OF NON-SPHERICAL PARTICLES DURING CERAMIC PROCESSING

Terry A. Ring
Department of Materials Science and Engineering
University of Utah
Salt Lake City, UT 84112

ABSTRACT

Ceramic suspensions exhibit shear thinning rheological properties due to the rotation of particles during shear. Prediction of these shear thinning rheologies are given. During extrusion and coating processes a ceramic paste undergoes shear. Shear, above a threshold value, preferentially orients non-spherical particles in suspension. After the shear is removed these particles maintain their preferential orientation leading of non-uniform shrinkage during drying and sintering, as well as non-uniform mechanical (and other) properties. This poster identifies the processing conditions in tape casting and extrusion where the shear threshold is exceeded and particles are preferentially oriented. Electron micrographs of cast ceramics are shown where these orientations are observed. Examples of non-uniform shrinkage during sintering are explained by the preferential orientation of the constituent particles.

RHEOLOGICAL PROPERTIES OF SEPIOLITE WITH NANOSIZE, FIBROUS MORPHOLOGY

Esin Günay, Tarik Baykara, and T. Osman Oezkan*
Tübitak-MRC
M.C.T.R.I.
Gebze-Kocaeli, Turkey

*I.U. faculty of Engineering
Metallurgical Engineering Department
Istanbul, Turkey

ABSTRACT

Sepiolite with a general formula of $Si_{12}Mg_8O_{30}(OH)_4(H_2O)_4.8H_2O$ is a clay mineral which has a needle-like strength and high surface area and found wide applications for its absorbing, theological and catalytic properties. The structure of sepiolite consists bundles of many small fibers and does not show swelling in water because of its low cation exchange capacity and chain-like structure. However, if the bundles of sepiolite fibers are dispersed using mechanical mixing technique, many of the fibers are dragged from the bundles and entangle with other individual fibers. They form a randomly intermeshed network which entraps liquid and increases the viscosity. Anisometry of sepiolite particles along with the surface characteristics are responsible for the theological properties of this mineral. In this work, the effect of shear rate and time on the separation of fibers from the bundles of sepiolite from Sivrihisar region (Eskisehir) were studied. As a result, it was observed that the viscosity increased with increasing shear rate and time. Results were supported by TEM analysis of the microstructure.

Interaction Between Binder and Dispersant in Aqueous Slurry of Alumina

T. Nakashima*, H. Mizutani, Y.Zhang, N. Uchida and K. Uematsu
Department of Chemistry, Nagaoka University of Technology
Kamitomioka 1603-1, Nagaoka, Niigata 940-2188, Japan

Abstract

Organic polymers are widely used in ceramic processing as binders, dispersants, etc. Their adsorption on the surface of ceramic particles as well as their effect on the rheology of the ceramic slurry has been an important subject in studies on ceramic processing. In a practical system containing two or more kinds of polymers, the adsorption as well as the function of a specific kind of polymer can be affected by other coexisting polymers. In the present study, the competitive adsorption of PVA as binder and NH_4PAA as dispersant was examined in an alumina slurry. Various slurries were prepared by changing the addition order of organic additives. The adsorption of PVA on alumina surface is weaker than that of NH_4PAA. Adsorption of PVA on the alumina surface was found to be inhibited by the presence of NH_4PAA, but NH_4PAA was not affected by the presence of PVA.

Introduction

In many industrial processing of ceramics, kinds of organic polymers are added to slurry for various purposes[1-5]. The role for each of them, such as a dispersant and a binder has been examined in systems containing only one kind of additive[6-11]. Polymers for the former application adsorb on powder particles to generate a repulsive force between them, reducing the viscosity of the slurry. Those for binder applications form bridges between particles in a dried green body to provide strength. However, there has been little study reported for systems containing kinds of polymers. In an extreme case, an interaction may occur between the coexisting polymers, changing their original function drastically.

This paper examines the interaction between the dispersant and binder in an aqueous alumina slurry. The interaction is evaluated through measurements of adsorption and flow characteristics of the slurry. Polyvinyl-alcohol, PVA and ammonium salt of polyacrylic acid, NH_4PAA were used as binder and dispersant, respectively. The system chosen is one of the most common in ceramic industry. This provides a general guide line for the production of ceramics.

Experimental

(1) *materials*

Alumina powder (AA-05, Sumitomokagaku, Japan) with a mean particle size of 0.46 μm and a BET surface area of $3.27 m^2/g$, was used in this study. The purity of the powder was >99.99% α-Al_2O_3, containing (in ppm) <20 Si, <5 Na, <1 Mg, <1 Cu, and <1 Fe. Commercial PVA and NH_4PAA solution were used ; (PVA105, average molecular weight 22,000 and saponification value 98.5%, Kurare, Japan) and (D-305, 10 wt% solution, average molecular weight 150,000, Cyukyoyushi, Japan). Note that the concentration of NH_4PAA was always expressed by the solution base in the text and figures.

(2) *method*

Raw alumina slurry was adjusted to a solid loading of 30 vol%. PAA and/or PVA were added in this slurry with the following four methods : (1) single addition of either PVA or NH_4PAA, (2) simultaneous addition of PVA and NH_4PAA , (3) PVA addition after equilibium adsorption of NH_4PAA, and (4) NH_4PAA addition after equilibium adsorption of PVA. The adsorption temperature was 20℃ and the adsorption time was set from 20 min to 24 hr. The viscosity of the slurries was measured with a rotational viscometer (VT550, Haarke, Swiss). A part of theslurry was centrifuged (6300g for 5 min) and the concentration of the organic additives in the supernatant was measured after dilution using liquid chromatography (C-R7A, Shimadzu, Japan) to determine the amount of polymers adsorbed.

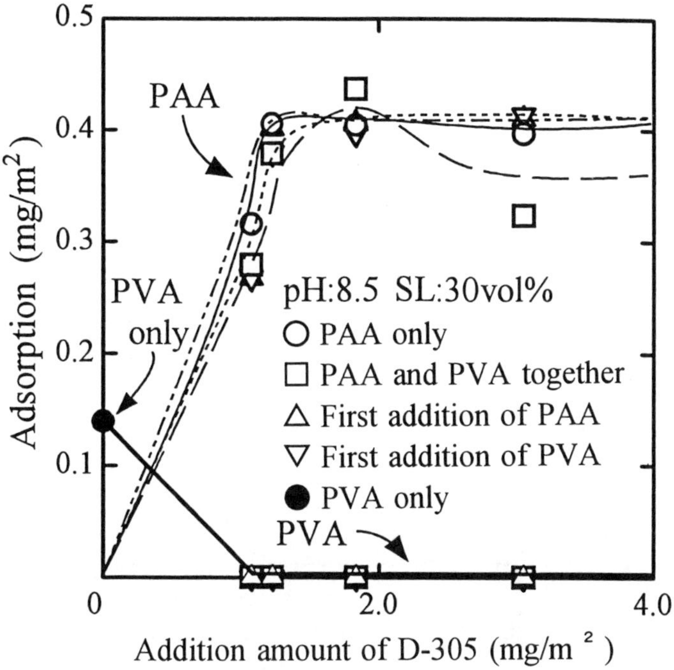

Figure 1 Relationship between the amount of PAA (10% solution base) addition and the amount of adsorption with various addition order

Result

Figure 1 shows the relationship between the amount of NH_4PAA added to the system and the amount of organic polymers adsorbed on alumina surface for slurries which were prepared with various orders of polymer addition. The amount of NH_4PAA adsorbed on alumina surface increased with increasing amount of NH_4PAA added to the slurry and became almost constant for the additions over 1.5mg/m^2 alumina. The adding orders of polymer had no marked effect on either the amount of PVA or NH_4PAA adsorbed. No PVA can adsorb on alumina surfaces under the presence of NH_4PAA over 1.5mg/m^2. Whereas the adsorption of NH_4PAA was not affected by the presence of PVA. NH_4PAA adsorbs more strongly on alumina surfaces than PVA at this condition.

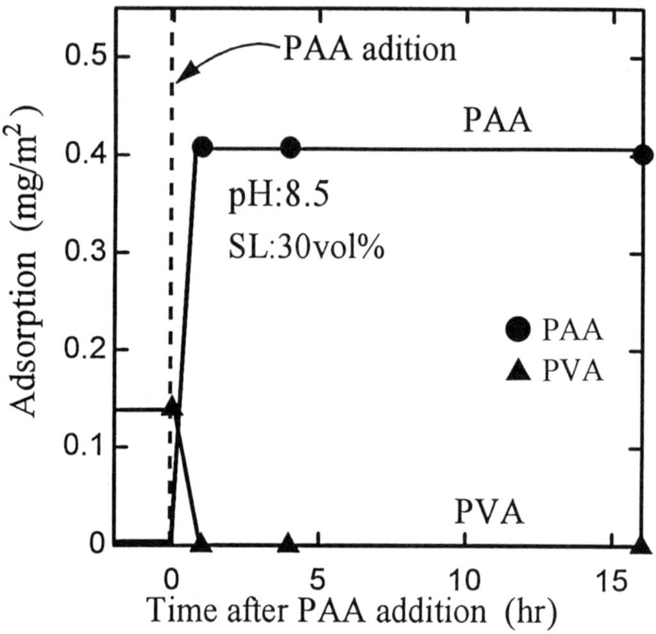

Figure 2 Adsorption of PAA(10% solution base) and desorption of PVA as a function of time

Figure 2 shows the changes in the amounts of PVA and NH_4PAA adsorbed on the alumina surface after the addition of NH_4PAA with time. In this experiment, the slurry containing only the powder and PVA was left at least for 24h to establish the equilibrium condition. Then NH_4PAA was added to the slurry at the time shown by the broken line, and the amounts of these polymers on alumina surface were measured at various times. Clearly, the adsorbed PVA is replaced rapidly by NH_4PAA. Adsorption of PVA on alumina is not possible in the presence of NH_4PAA over 1.5mg/m^2 at the pH 8.5. In contrast, the adsorption of NH_4PAA proceeded rapidly on the alumina surface and reached equilibrium in a short time.

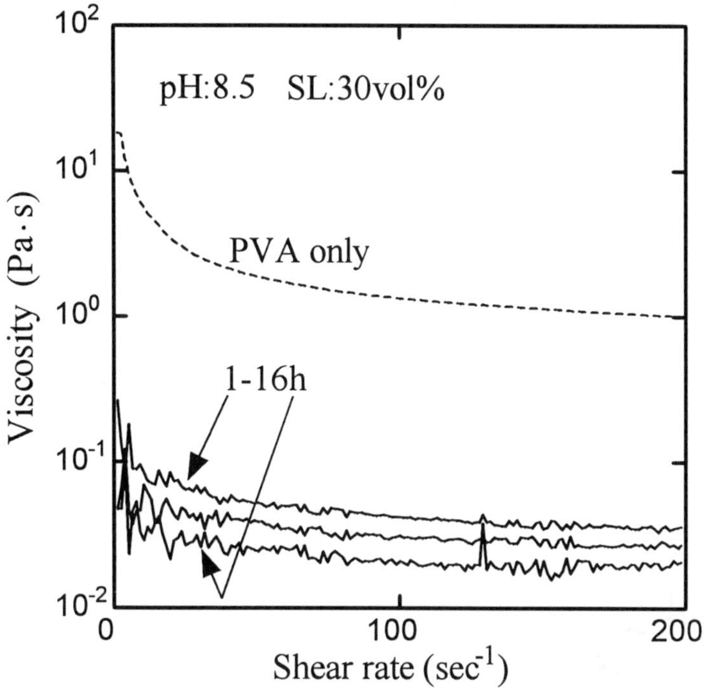

Figure 3 Effect of PAA (10% solution base) addition on the steady shear viscosity

Figure 3 shows the change of viscosity after addition of NH_4PAA (1.5mg/m^2) in the slurry for which PVA was adsorbed beforehand on the alumina surface to equilibration. The viscosity of slurry was measured to investigate the effect of NH_4PAA addition on the dispersion of the alumina powders in the slurry. The broken line shows the viscosity of slurry containing PVA only. A high viscosity was noted for this slurry, showing that the powder particles were not dispersed well without dispersant. After NH_4PAA is added to the slurry, the viscosity decreased rapidly first and then decreased gradually with time. Clearly the dispersion of particles are improved by the addition of NH_4PAA.

Figure 4 shows the change of the viscosity after addition of PVA in the slurry for which NH_4PAA (1.5mg/m^2) was adsorbed on the alumina surface beforehand to equilibration. The broken line shows the viscosity of slurry containing NH_4PAA only. The slurry has low viscosity with NH_4PAA only. The low viscosity was retained after the addition of PVA. It clearly has no influence on the viscosity of slurries for all times. The dispersant NH_4PAA governs the slurry viscosity.

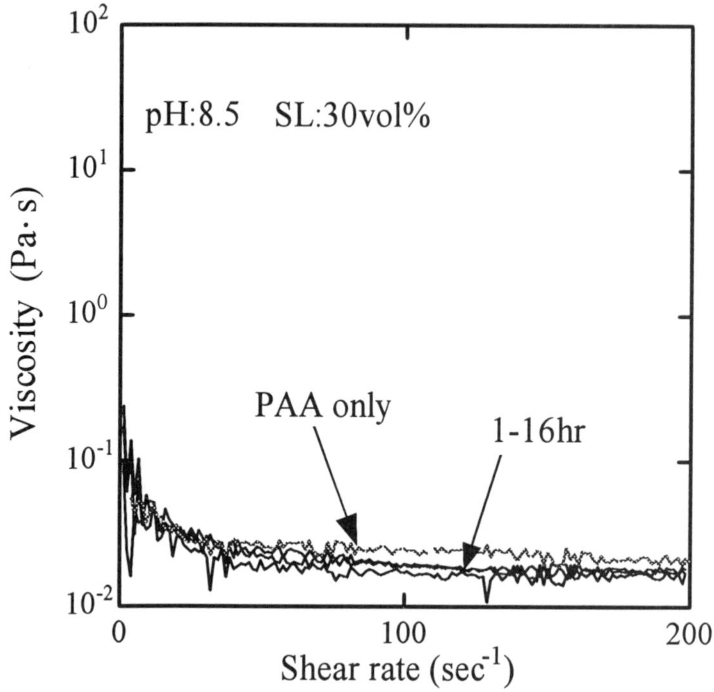

Figure 4 Effect of PVA addition on steady shear viscosity

Discussion

The adsorption of NH_4PAA was found to be much stronger than PVA on alumina particles. It can replace the PVA previously adsorbed on the surface. In contrast, PVA cannot adsorb on alumina surfaces occupied by NH_4PAA. The differences in adsorption behavior can be explained by the relevant chemical bonding. The adsorption of NH_4PAA involves a strong chemical interaction between a dissociated carboxyl group and a positive aluminum ion in the alumina surface. On the other hand, only weak hydrogen bonding occurs in the adsorption of PVA between hydroxyl group and proton at alumina surface.[13-14] These differences of chemical interactions is responsible for the stronger adsorption of NH_4PAA than PVA.

The viscosity of slurry is governed primarily by NH_4PAA. This is explained by strong adsorption characteristics and dispersing capability of the negatively charged NH_4PAA molecules. Once covered with the polymers, high negative charges are established on the particle surface, generating repulsive electrical interaction between particles . Presence of PVA in the liquid medium should neither significantly affect shielding the electrical interaction nor interfering with the surface adsorption of NH_4PAA. In contrast, the weak adsorbing capability of PVA allows adsorption of only small amounts on the surface. Only weak steric interaction is generated between particles even with large amount of adsorption. Even with sole presence, the dispersing capability of PVA should be low.

Above discussions also explain that the order of addition has no influence on the viscosity and the amount of NH_4PAA adsorbed on the alumina surface. The adsorption of NH_4PAA is independent of the presence of PVA and governs the viscosity. The same equilibrium is established regardless of the presence of PVA.

Conclusion

1. NH_4PAA is more strongly adsorbed on alumina surfaces than PVA. The order of addition does not have any influence on the final characteristics of the slurry. NH_4PAA can even adsorbed on surface covered with PVA molecules by replacing them rapidly.

2. The viscosity of the slurry depended only on NH_4PAA addition. It decreased rapidly after NH_4PAA additions to the system. PVA does not have a marked influence on the viscosity of slurry in the system. These results of adsorption and viscosity are consistent.

Reference

〔1〕 G.Y. Onoda, Ceramic Processing before Firing (1978), P.227.

〔2〕 R.J. Pugh, Surface and Colloidal Chemistry in Advanced Ceramics Processing (1994), vol.51; P.175.

〔3〕 D.W. Fuerstenau, Ceramic Transactions (1988), vol.1; P.323-51.

〔4〕 R. Moreno, Am. Ceram. Soc. Bull (1992), 71(10); P.1521-31.

〔5〕 R. Moreno, Am. Ceram. Soc. Bull (1992), 71(11); P.1647-57.

〔6〕 G.Y. Onoda, Ceramic Processing before Firing (1978), P.235-51.

〔7〕 C.S Khadilkar, Ceramic Transactions (1988), vol.1; P.397-409.

〔8〕 R.G. Frey, J.Am. Ceram. Soc (1984), 67(3); P.199-203.

〔9〕 C.W. Nies, Advances in Ceramics (1984), vol.9; P58-66.

〔10〕 R.J. Lukasiewicz, J.Am. Ceram. Soc (1989), 72(4); P.617-24.

〔11〕 K.Kendall, British Ceramic Society Proceeding (1986), vol.37; P.255-65.

〔12〕 P.C. Hidber, J.Am. Ceram. Soc. (1995), 78(7); P.1775-80.

〔13〕 C.C. Treischel, J.Am. Ceram. Soc (1946), 29(5); P.129-32.

〔14〕 M.M. Zwick, J.Appl. Polym. Sci (1965), (9); P.2393-424.

THERMOCHEMISTRY OF PAA ADSORPTION ON ALUMINA SURFACE

Kumiko Yamashita, Wataru Kabutoya,
Yao Zhang, Nozomu Uchida, Keizo Uematsu
Department of Chemistry, Nagaoka University of Technology
Kamitomioka 1603-1, Nagaoka, 940-2188, Japan

ABSTRACT

The heat of adsorption of PAA on alumina powder surface was measured directly by calorimetry in aqueous phase. Calvent type micro calorimeter was used for the measurement. The amount of adsorption was 0.59 mg/m^2. The heat of adsorption was 0.027 J/m^2. The adsorption process is endothermic.

1. INTRODUCTION

Several kinds of organic polymer additives are used to control the characteristics of slurry in ceramic forming process[1]. Such additives are thought to adsorb on the surface of the powder and develop their functions. However, their basic science has not been elucidated enough. Guo and his coworkers have studied the adsorption behavior of poly-acrylic acid (PAA), which is a typical dispersant in ceramic industry, on alumina powder surface in aqueous phase[2-4]. They found the amount of adsorption increased with the increase of temperature. It suggests that the heat of adsorption is endothermic on the basis of the Le Chatelier's principle. This phenomenon seems to be contrary to the usual polymer adsorption in which the heat of adsorption is negative i.e. exothermic.

To develop the knowledge about the unique chemistry of PAA-alumina powder-water system, another approach is required. The aim of this study is to verify the endothermic heat of adsorption directly by means of calorimetric measurement.

2. MATERIALS AND METHODS

2.1. Materials

The polymer additive used in this study is poly-acrylic acid Na salt (PAA-Na; Aldrich Chemical Company, Inc. :average molecular weight 30000). The segment formula is shown in fig 1. Ammonium salt of this polymer is used as a typical dispersant for slurry in ceramic industry. The ceramic powder is high purity commercial alumina (Advanced Alumina AA-05; Sumitomo Chemical Industry Inc. : specific surface area is 3.50 m^2/g).

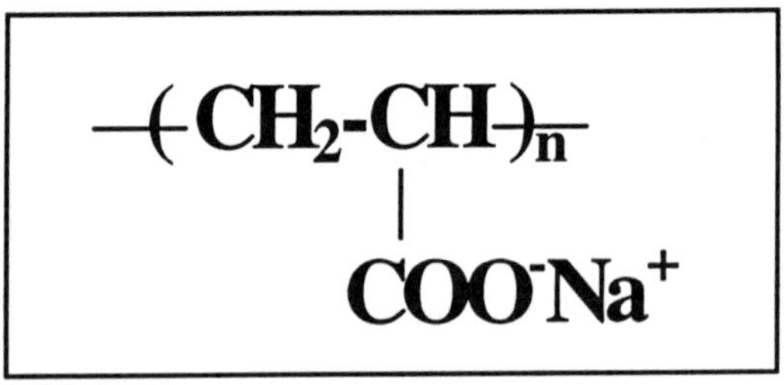

Fig. 1. Segment Formula of PAA-Na

2.2. Method

(1)Calorimetry of heat of the adsorption

Calvent type micro calorimeter (Tokyo Riko Co., Ltd., model TCC-21) was used to measure the heat of adsorption of PAA on Al_2O_3 powder surface[5]. Schematic diagram of Calvent type micro calorimeter is shown in Fig 2. There are sample and reference cells. The temperature difference (ΔT) between them is detected by a thermocouple. To establish a calibration curve, a known heat (Q) was generated in the sample cell by Joule heating, and the ΔT was plotted against time (t). The area bounded by the plot and the base line was found to be proportional to Q. PAA solution is set in each cell. A glass ampul which contains alumina slurry is set in the sample cell and an ampul which contains only distilled water is set in the reference cell. The reaction was started by breaking the glass ampuls to mix the slurry and water with the PAA solution. The heat of reaction generated in the sample cell is determined by integrating the ΔT-time diagram.

Measurement temperature is 40℃. The amount of alumina powder is 2.0 g, concentration of PAA solution is 0.5 mass %, the amount of the solution is 25 ml, and pH of slurry mixed alumina powder, distilled water and PAA solution is 8.55.

(2) Measurement of the amount of adsorption

After the calorimetry measurement, the solid and liquid phases of slurry were separated by centrifuging. The concentration of free polymer i.e. unadsorbed PAA in the liquid phase was determined by measuring the refractive index to obtain the amount of adsorbed PAA.

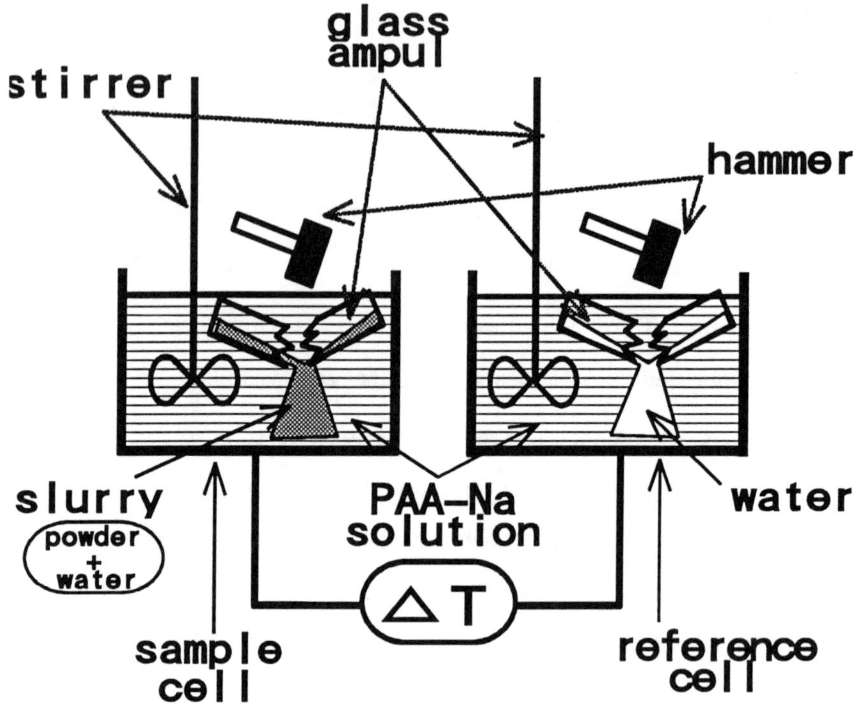

Fig. 2. Schematic diagram of Calvet type micro-calorimeter

(3)Cancellation of undesirable heat component

There are several undesirable heat components generated during the measurement. They were canceled by following manners.

a) Heat of wetting : When dry ceramic powder is soaked into water, heat of wetting is generated. To remove this undesirable heat, the alumina and water were mixed first and then the mixture was added into PAA aqueous solution. Alumina-water slurry which contained 50 mass % of Al_2O_3 was sealed in a glass ampul and set in the measurement cell .

b) Heat of dilution: When alumina slurry was mixed with PAA solution, there must be heat of dilution of PAA by water contained in slurry. To cancel this heat, the glass ampul in the reference cell contains an equal amount of water to that contained in the slurry in sample cell.

c) Heat of neutralization : PAA has -COOH groups, so heat of neutralization must be considered when it is mixed with other basic reagent. To avoid this complexity, Na salt of PAA (-COONa) was used.

d) Heat of breaking of glass ampul: Final source of undesirable heat is the difference of heat of breaking between individual glass ampuls. The difference value was estimated at 10^{-3}J by breaking several empty ampuls.

e) Drift of room temperature : Although the calorimeter itself carries a temperature controller, large drift of room temperature ($>\pm 5$℃) can introduce a serious bias to the out put signal. To avoid such bias, calorimeter was placed in air conditioned chamber.

So the value, 10^{-3}J resulted from the scatter of quality of the glass ampuls is the inevitable absolute error in this experiment.

3. RESULTS AND DISCUSSION

Figure 3 shows the result of this experiment. Temperature difference between sample cell and reference cell is plotted as a function of time. The negative ΔT shows the heat of adsorption being endothermic. It agrees with the result of the change of the amount of adsorption with temperature. The absolute value of the heat of adsorption was 0.19J from the calibration curve. So the heat of adsorption per unit surface area is 0.027 J/m^2. The amount of adsorbed PAA was 0.59 mg/m^2 Al_2O_3 powder surface area.

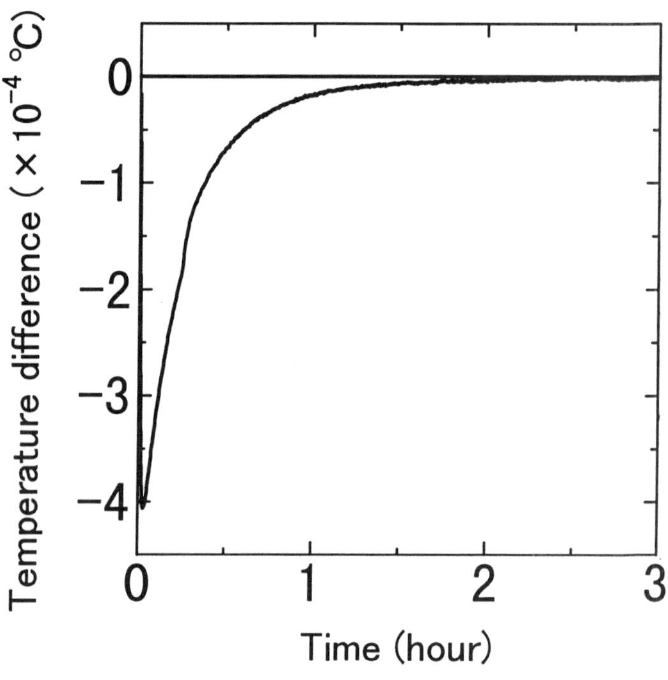

Fig. 3. The example of experimental result

Generally, adsorption of polymer on some surface is exothermic[6,7], since the polymer loses its degree of freedom. However, for PAA, a few endothermic cases have been reported[8]. When chemical reaction process proceeds, Gibbs free energy change (ΔG) of the process must be negative.

$$\Delta G = \Delta H - T\Delta S < 0$$

In this experiment, ΔH was found to be positive. So ΔS must have a large positive value to make ΔG negative. It means this adsorption reaction is entropy controlled.

Quantum chemical calculation in our laboratory showed that the heat of adsorption of PAA on alumina surface is negative in vacuum state and it turns positive in the presence of water molecules[9].

4. SUMMARY

Contrary to the usual cases of polymer adsorption, the heat of adsorption of PAA on Al_2O_3 powder surface in the presence of water molecules is endothermic. It was concluded that this adsorption process is entropy controlled.

REFFERENCES

[1]Y. Kankawa, "Organic Materials Used in New Ceramic Production", Bulletin of The Ceramic Society of Japan, 29[7]585-589(1994)

[2]Y. Tomita, L. Guo, Y. Zhang, N. Uchida and K. Uematsu, "Effect of Temperature on the Slurry Characteristics and Green Body of Alumina", J. Am. Ceram. Soc., 78[8]2153-56(1995)

[3]L.Guo, Y. Zhang, N. Uchida and K. Uematsu, "Influence of Temperature on Stability of Aqueous Alumina Slurry Containing Polyelectrolyte", J. European Ceram. Soc., 17NOS 2-3 (1997)354-50

[4]L. Guo, Y. Zhang, N. Uchida and K. Uematsu, "Adsorption Effects on Rheological Properties of Alumina Suspensions with Polyelectrolyte", J. Am. Ceram. Soc., 81[4](1998) in press.

[5]W. P. White, "The Modern Calorimetry", 194 Chemical Catalog Co., New York(1928)

[6]M.L.Machesky and P. F. Jacobs, "Titanium Calorimetry of Aqueous Alumina Suspensions Part I .Result and Comparison with Similar Studies", Colloids and Surface, 53(1991)297-328

[7]E.Killmann "Conformation and Thermodynamics of Adsorbed Macro-Molecules at the Liquid/Solid Interface", Polymer, 1976, Vol17, 864-868

[8]M.R.Böhmer, Y. E. A. Sofi and A. Foissy, "Calorimetry of Poly-(Acrylic Acid) Adsorption on TiO2" J. of Colloids and Interface Science, 164,(1994) 126-135

[9]M.Obata,Y.Zhang, N. Uchida, K. Uematsu, "Quantum Chemical Study of Interactions between Ceramic Powder Surface and Polymer Additives", in this proceeding

Quantum Chemical Study of Interactions between Ceramic Powder Surface and Polymer Additives

Mitsuyoshi Obata , Yao Zhang , Nozomu Uchida , Keizo Uematsu
Department of Chemistry , Nagaoka University of Technology
Kamitomioka 1603-1 , Nagaoka , 940-2188 , Japan

ABSTRACT

Adsorption behavior of polymer on alumina surface is studied by means of quantum chemical calculation. Semiemprical Hamiltonian PM3 on MOPAC93 was used for calculation.

The PAA-alumina system is chosen as a model. A molecule of the model polymer was placed close to an alumina cluster and the geometry of polymer was optimized to obtain the energy minimum structure. Subjects studied include adsorption geometry of polymers, adsorbing site on alumina and heat of adsorption.

1. INTRODUCTION

In the ceramic processing, several kinds of organic polymer additives[1] are used to control the slurry characteristics. The polymer additives function as dispersion or binder by adsorbing on ceramic powder surface. To control the slurry characteristics precisely, it is important to understand the interaction between organic polymer additives and ceramic powder surface. However, it is difficult to examine this adsorption behavior directly. The aim of this study is to investigate the adsorption behavior of polymer on ceramic powder surface in molecular level by means of quantum chemical calculation.

The PAA-Al_2O_3 system was chosen as a model to study the following three important subjects:

1) The effect of pH on adsorption. The pH governs both the surface charge of alumina and the degree of dissociation of PAA and thus the adsorption behavior. The effect of pH is investigated by adding or removing a proton to/from the system.

2) The adsorption morphology, since it affects the slurry characteristics. Generally, the morphology of polymer adsorption is categorized into three types. They are single point, loop-train-tail and prostrated types i.e. multi-points adsorption. It is of interest to clarify which type is preferable in this system.

3) The effect of water. In an ordinal molecular orbital calculation, the objective molecule is placed in a vacuum. However, in any real slurry system, the polymer

and alumina powder are surrounded by water molecules. The effect of solvent must be properly considered by including several water molecules into the system.

2. CALCULATION

This study applies the semiemprical hamiltonian PM3[2] on MOPAC 93[3] installed in EWS Fujitsu S-4/10. The results of calculation were displayed by using Molecule Editor on SONY Tektronix Calculation Chemical Support System CAChe. Alumina clusters used in this study are shown in Fig.1 and 2. The structure of model PAA is shown in Fig.3. Trimer and pentamer of acrylic acid were used as model PAA. The heat (enthalpy) of adsorption (ΔH_{adso}) is defined as the difference between heat of formations of isolated geometry and combined one. The bonding strength is identified by inter atomic distance and two-center energy (TCE)[4]. Although this parameter is not equal to the "bond energy" by definition, it can be a convenient measure of bond strength.

In the calculation, it is necessary to consider the drastic changes in the states of alumina surface and PAA with pH. At low pH and high pH, alumina surface are positively and negatively charged, respectively by adsorbed protons and O^- group i.e. dissociated -OH group. At intermediate pH, i.e. alumina ZPC pH[5], alumina surface is neutral. PAA starts to dissociate at pH about 4, since the pKa of carboxyl group is 4.72.[6] Cases must be divided into following three;

1. low pH where anionic PAA and cationic alumina coexist in the slurry,
2. alumina ZPC pH where anionic PAA and neutral alumina coexist,
3. high pH where anionic PAA and anionic alumina coexist. In this study, the models are constructed for the cases of low pH and alumina ZPC pH. To express the low pH state of PAA and alumina, a proton is added to the O(3)-H(5) group on the alumina cluster (Fig.1; alumina cluster has positive charge) and carboxyl groups in PAA are dissociated to COO^- anion (Fig.3). Alumina ZPC pH state is represented by the coexisting neutral charge alumina cluster and model PAA having dissociated $-COO^-$ anion.

To study the adsorption morphology, the geometries of isolated PAA and alumina cluster were optimized to obtain the each energy minimum independently as the first step. Then the structure constructed by the carboxylate anion of PAA placed close to the aluminum atom (1) in alumina cluster II (Fig.2) was optimized. The geometry of alumina is fixed to the initial one to save calculation time.

In the calculation, basically solvent was not considered i.e. it is vacuum state. However, in the real system, there are many water molecules around the PAA and alumina. Water molecules may play a important role in the adsorption. In this study, a model considering the existence of water molecule was constructed.[7)]

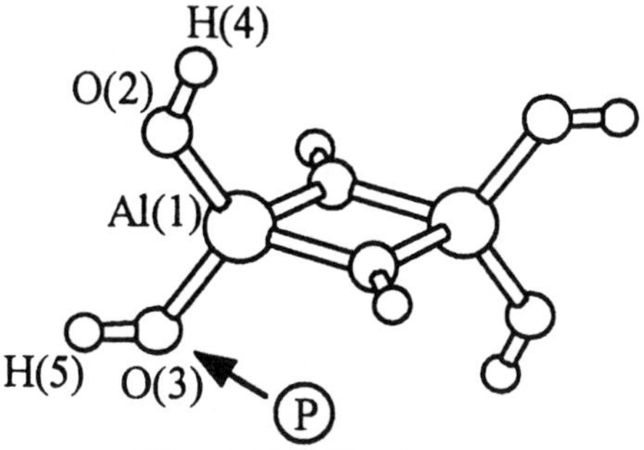

Fig.1. Alumina cluster I
(Acidic Proton added to Oxygen atom(3))

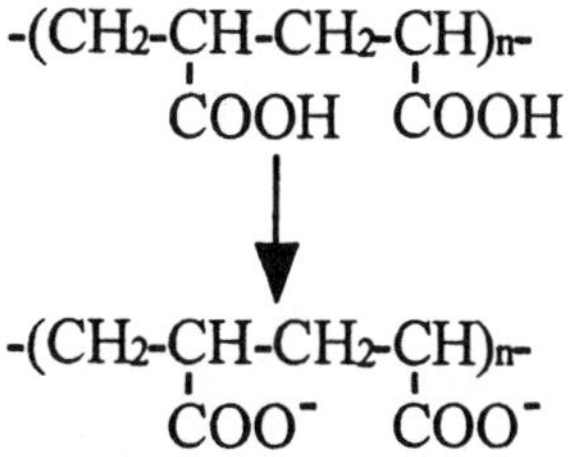

Fig.3. PAA(Poly Acrylic Acid)

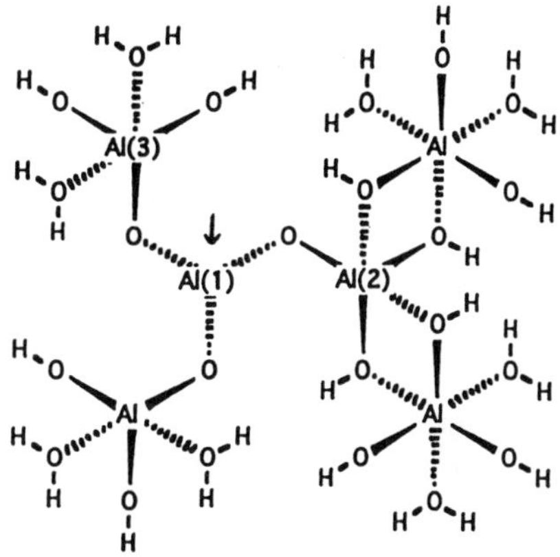

Fig.2. Alumina cluster II

3. RESULTS AND DISCUSSION

3-1. Effect of pH

Fig.4 and 5 shows the low pH adsorption structure and alumina ZPC pH one, respectively. In fig.4, the upper part of the structure is PAA and the lower one is alumina cluster which represents low pH state by adding a proton (P) to the O (3) atom in O (3) -H (5) group. A strong chemical bond is formed between the aluminum atom (1) and the oxygen atom (6). The interatomic distance between the Al atom (1) and the O atom (6) is 0.178 nm. This value is comparable with the general Al-O distance 0.172 nm. The Two-Center Energy (TCE) is -9.2 eV. This value is 61 % of usual Al-O bonding (-15 eV). The enthalpy of adsorption (ΔH_{adso}) is -485 kJ/mol. Charge transfer from PAA to the alumina cluster is 0.80. It means this interaction has a high covalency. In contrast, only weak interaction was found in the case of alumina ZPC pH adsorption structure (Fig.5). The inter atomic distance between Al atom (7) and O atom (8) is 0.250 nm. TCE is -4.6 eV. This value is only 30 % of usual Al-O bonding. The enthalpy of adsorption is 25 kJ/mol. Charge transfer is 0.19. Therefore, a strong chemical bonding is formed in the low pH zone and only weak interaction is found at the alumina ZPC pH. These results agree well with experimental result in which PAA adsorbed well in low pH, but the adsorption become poor in higher pH.

3-2. Adsorption morphology

Fig.6 and 7 show the results of calculations for multi-point adsorption and single point adsorption, respectively. In these models, first (I), center (II) and end (III) carboxyl groups of PAA were dissociated to $-COO^-$. The upper part of these structures is PAA and the lower part is the alumina cluster. In the multi-point adsorption case, PAA was initially placed in a horizontal position for alumina cluster and then the geometry of PAA was optimized. In the optimized structure (Fig.6), the shape of PAA is rather straight and the O atom (4) of the central-carboxylate anion bonded strongly to Al atom (1) in alumina cluster. Inter atomic distance between Al atom (1) and O atom (4) is 0.184 nm and TCE is -10.5 eV. The charge transfer is 0.68 electron from the PAA to the alumina cluster. The oxygen atoms of the first (I) and the end (III) carboxylate anion also interact with the alumina cluster. The interaction of the first (I) and the end (III) carboxylate anion is 0.408 nm, 0.684 nm in interatomic distance and -5.66 eV, -3.39 eV in TCE. The enthalpy of adsorption is -462 kJ/mol.

On the other hand, in fig.7, the PAA was initially placed in vertical position for alumina cluster and optimized. Oxygen atom (5) in the terminal carboxylate anion makes a strong bond with the Al atom (1) in the alumina cluster (Fig.7). The interatomic distance between the Al atom (1) and the O atom (5) is 0.183 nm and the TCE is -10.4 eV. The charge transfer is 0.57 from the PAA to the alumina cluster. These values are comparable with the main interaction in fig.6. However, other carboxylate anions have no interaction with the alumina cluster and the enthalpy of adsorption is -415 kJ/mol. Although the main interactions are comparable to each other in multi-point adsorption and single point adsorption, the multi-point adsorption becomes more stable than single point adsorption in total interaction, i.e. ΔH_{adso}. Thus, prostrated adsorption (multi-point adsorption) is favorable in this system.

3-3. Effect of solvent water

Hydrated PAA (PAA + three H_2O molecules) and hydrated alumina cluster (alumina cluster + three H_2O molecules) were calculated independently (Fig.8) and then adsorbed structure with H_2O molecules was calculated (Fig.9). Values of the enthalpy of formation are -8058 kJ/mol for the hydrated alumina cluster and -2050 kJ/mol for the hydrated PAA. The total enthalpy of formation of isolated systems is -10108 kJ/mol. In the adsorbed structure (Fig.9), the upper part of the structure is PAA and the lower part is alumina cluster. Water molecules around the PAA were displaced and a chemical bond was formed between the O atom (4) in the PAA and the Al atom (1) in the cluster. This chemical bond has the interatomic distance between Al (1) and O (4) of 0.182 nm and a TCE -9.94 eV. The enthalpy of formation is -9681 kJ/mol for adsorbed structure. The enthalpy of adsorption is 427 kJ/mol >0. It means the process is to be endothermic. Although this result appears contrary to common polymer adsorption, it is consistent with the results of calorimetry experiments[8]. In the case of adsorption without H_2O molecules (in the other words, in vacuum state), the enthalpy of adsorption is -431 kJ/mol, i.e. the adsorption is exothermic. Considering the existence of water molecule, this adsorption process was suggested to be endothermic.

4. SUMMARY

In this study, we have investigated the effect of pH, the adsorption morphology and the effect of water molecule for PAA adsorption onto alumina surface by means of quantum chemical calculation. The results are summarized as follows;

(1) PAA have strong interaction with alumina surface in low pH.
(2) Prostrated adsorption (multi-points adsorption) is preferable in this system.
(3) This adsorption process is endothermic under the existence of water molecule.

REFERENCE

1) Yosimitsu Samukawa, Ceramics 29 (7), P585-589 (1994)
2) J. J. P. Stewart, J. Comput. Chem., 10, 209, 221 (1989)
3) J. J. P. Stewart, Fujitsu Limited, Tokyo, Japan (1993)
4) M. J. S. Dewar, W. Thiel, J. Am. Chem. Soc., 99, 4899 (1977)
5) Robert J. Pugh, Lennart Bergstrom, Surfactant Sci. Series, vol.51, 136-148 (1993)
6) John Mcmurry, "Fundamentals of Organic Chemistry", (1990)
7) Karl N. Kirschner, Edward C. Sherer, George C. Shields, "Use of the Supermolecule Approach To Model the Syn and Anti Conformations of Solvated Cyclic 3',5'-Adenosine Monophosphate", J. Phys. Chem., 100, P3293-3298 (1996)
8) Kumiko Yamashita, Wataru Kabutoya, Yao Zhang, Nozomu Uchida, Keizo

Uematsu, "Thermochemistry of PAA Adsorption on Alumina surface", in this proceedings

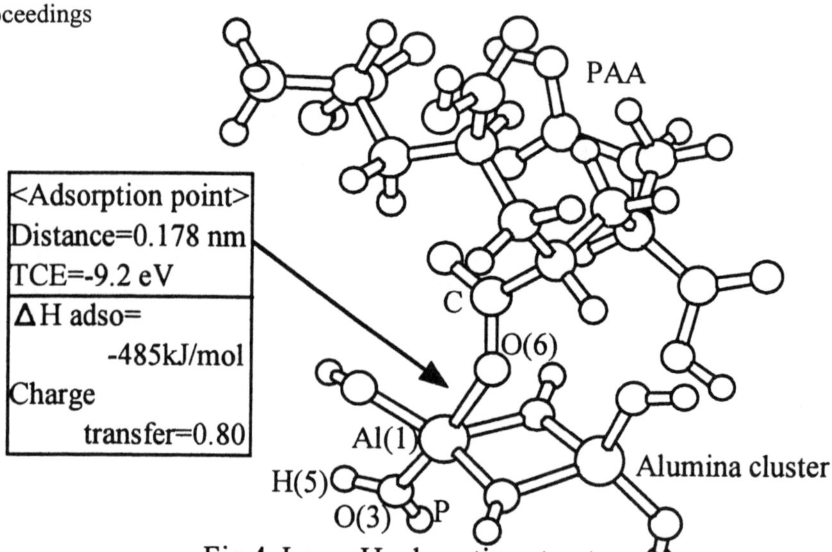

Fig.4. Low pH adsorption structure

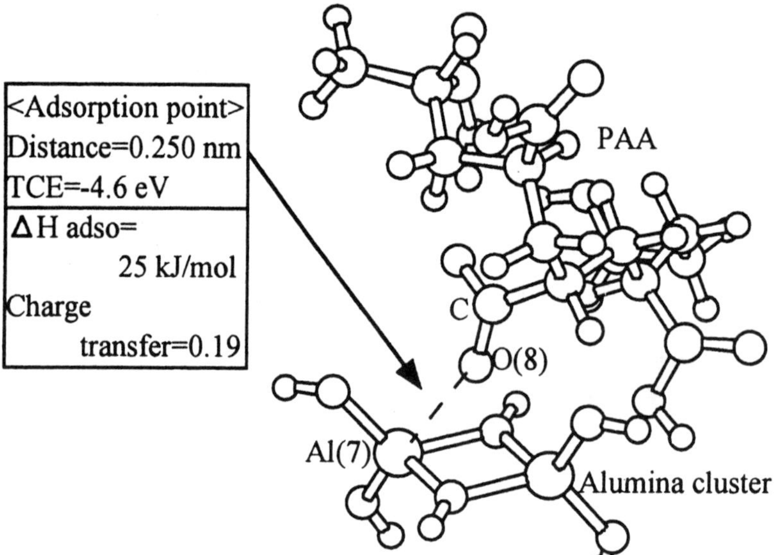

Fig.5. Alumina ZPC pH adsorption structure

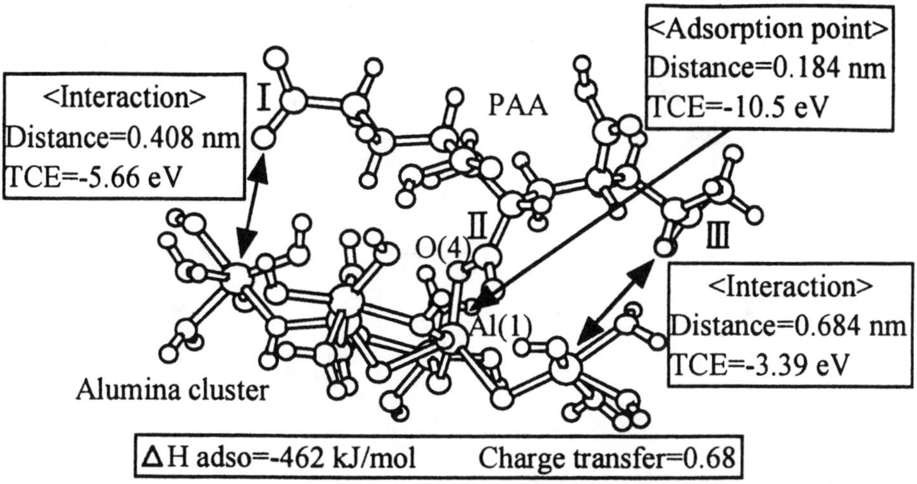

Fig.6. Central-carboxylate anion adsorption structure

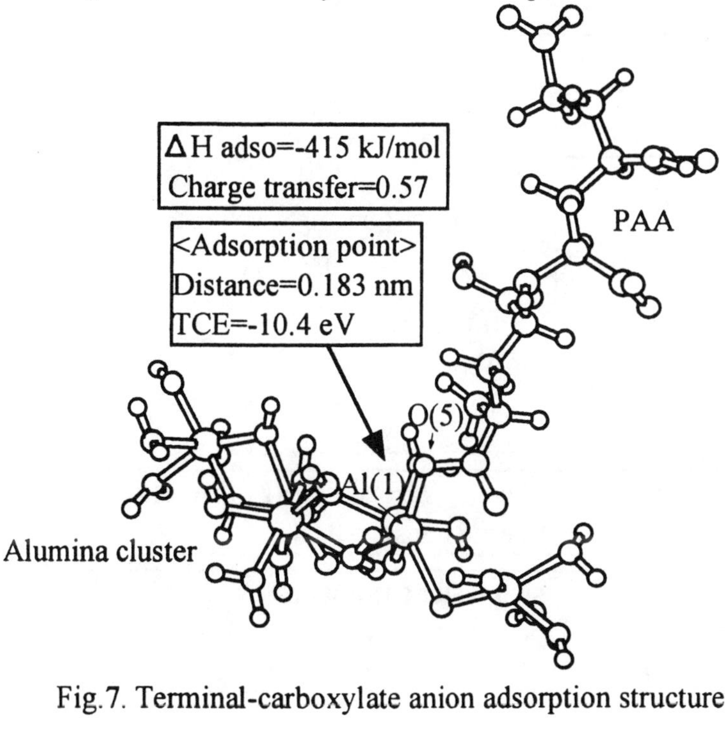

Fig.7. Terminal-carboxylate anion adsorption structure

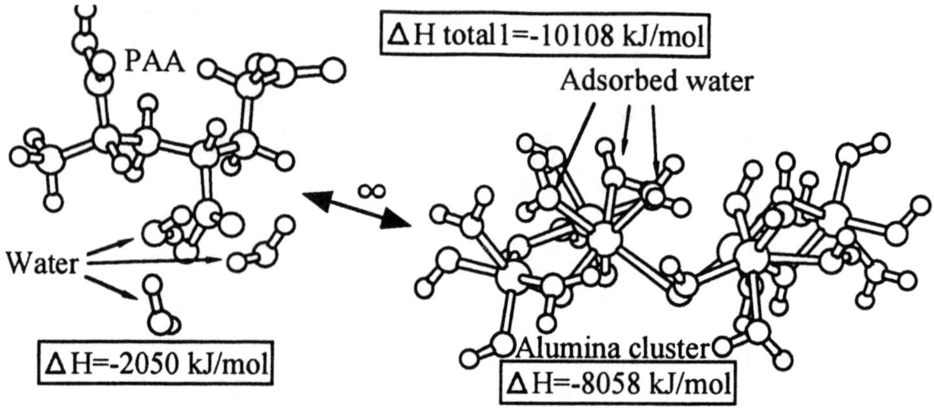

Fig.8. Hydrated PAA and hydrated alumina cluster

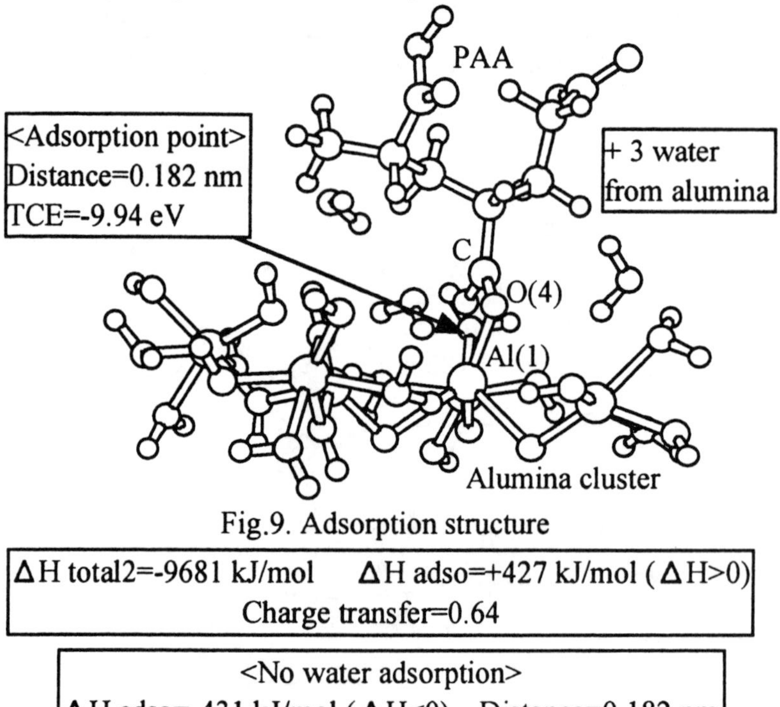

Fig.9. Adsorption structure

ΔH total2=-9681 kJ/mol ΔH adso=+427 kJ/mol (ΔH>0)
Charge transfer=0.64

<No water adsorption>
ΔH adso=-431 kJ/mol (ΔH<0) Distance=0.182 nm
TCE=-10.24eV Charge transfer=0.57

WETTING BEHAVIOR OF TAPE CASTING SLURRIES ON POLYMERIC TAPE CARRIERS

Christiane Lutz and Andreas Roosen
University of Erlangen-Nuremberg
Department of Materials Science, Glass and Ceramics
91058 Erlangen, Germany

ABSTRACT

The aim in tape casting is the production of a planar, homogeneous, defect-less green tape. An important processing step is the casting of a slurry on a tape carrier. The casting behavior and the quality of the final product is strongly influenced by the wetting behavior of the slurry on the tape carrier. Slurries of stabilized zirconia were cast on various polymeric tape carriers using doctor blade tape casting technology in order to investigate the wetting behavior. Commercial surfactants were added to the aqueous slurry system to improve the wetting behavior, the detachment ability of ceramic green tapes, and the final tape quality. The wetting behavior was characterized by measuring contact angles of surfactants and binder solutions on tape carriers by the Sessile Drop Method. The influence of these surfactants on the viscosity of aqueous slurries was measured by cone-plate-rheometry. The removal of the green tape from the polymeric carrier and the green tape quality were described by optical methods. This study reflects the importance of controlling the wetting behavior of aqueous slurries used in tape casting. It also provides possibilities to adjust and measure the wetting behavior.

INTRODUCTION

The doctor blade tape casting technology is used for manufacturing ceramic green sheets for different applications such as capacitors, sensors, substrates and piezoelectric devices. The process is well described in the literature (1-3). Currently, the substitution of organic solvents by water in tape casting slurries gains increasing importance because of environmental reasons.

In order to achieve a planar, homogeneous and defect-less green tape of constant thickness, the ceramic slurry has to wet the polymeric carrier sufficiently, avoiding inhomogeneities due to phase separation of the slurry. Of equal importance is the easy removal of the dried ceramic green tape from the polymer carrier under a small angle without residues on the carrier. This also depends on the wetting behavior. Water as a solvent, with its higher surface tension, strongly influences the wetting behavior and can result in severe problems concerning green tape homogeneity and green tape removal.

The aim of this study was to observe the influence of tape casting parameters such as solvent, binders, surfactants, and type of polymeric carrier on surface tension, contact angle, green tape removal, and green tape quality. The results should allow to define processing parameters for water based slurries to avoid phase separation during casting and to guarantee an easy green tape removal with a minimum of stretching.

MATERIALS AND METHODS

Starting Materials

Two slurries containing 8 mol% Y_2O_3 stabilized zirconia powders, (FYT13.0-002H, Unitec Ceramics Comp., UK) were used. This powder is commonly used to manufacture electrolytes for the Solid Oxide Fuel Cell. The mean particle size d_{50} was 0,84 µm (Microtrac UPA, Leeds & Northrup Comp., USA), and the specific surface area was 5,84 m²/g (ASAP2000 Nitrogen, Micromeritics, Norcross, USA).

For the aqueous slurries doubly deionized water was used as a solvent. A polyelectrolyte (Duramax, Rohm & Haas Comp., USA) was added as a dispersant. Different binders like acrylic binder 1, acrylic binder 2 (both Duramax, Rohm & Haas Comp., USA) and carboxymethylcellulose (Tylose, Hoechst AG, GER) were tested. Due to the strong foaming during processing a defoamer (Agitan, Münzing GmbH, GER) had to be used. Because organic solvents have many advantages in tape casting technology, e.g. low surface tension and good spreading on the tape carrier, a slurry based on an ethanol/toluene (25:75) mixture was also prepared to compare its properties with the water based slurries. The composition of the slurries are shown in Tab. I.

Surfactants are classified as anionic, cationic, non-ionic, or ampholytic according to the charge carried by the surface-active part of the molecule (4). The surfactants in Tab. III used for this study are nonionic wetting agents.

Table I. Slurry System Components

Components	Aqueous System	Organic System
Powder	ZrO_2	ZrO_2
Solvent	H_2O	EtOH/Toluene
Dispersant	polyelectrolyte	fatty acid
Binder	different binders	PVB
Plasticizer	different plasticizers	BBP
Defoamer	copolymer	-
Others	-	resin

For the preparation of the aqueous slurry, the powder was deagglomerated in a solvent-dispersant solution by attrition milling for 2 hrs. After the addition of the binder the slurry was mixed in an attrition mill for 2 hrs. The surfactants (< 0,1 mass%), listed in Tab. II, were added to the homogenous slurry, which was mixed for one hour. After degassing the slurries were cast on different tape carriers of coated and uncoated PET- and PP tapes (Tab. III). The organic slurry was prepared by ball milling for 12 hrs in order to deagglomerate the powder; binders and plasticizer were then added and the slurry was homogenized for 24 hrs.

Table II. Surfactants

Surfactants	Trade Name and Supplier	
EO - PO - block – polymers	Pluronic	(BASF AG, GER)
Copolymers: polysiloxane + inert filler silicic acid	Agitan	(Münzing GmbH, GER)
Silicon based surfactant	Trukalon	(Trukem GmbH, GER)
Polyglycolether	Genapol	(Hoechst AG, GER)

Table III. Polymer Tape Carriers

Tape Carrier	Trade Name and Supplier
PET (Polyethylentherephthalat)	Hostaphan RN 100.0 (Hoechst AG)
PET + silicon coating	Hostaphan RN 100.0 + Silicon
PP (Polypropylene	Trestaphan NNA (Hoechst AG, GER)
PP + polyolefine coating	Trestaphan GND (Hoechst AG, GER)
PET Mylar Type A	Mylar Type A (DuPont, USA)

Wetting Behavior of Slurries

The influence of the surfactants on the wetting behavior of the slurry and consequently on the quality of the final green tape were studied. The Young Equation (Eq. 1) describes the relation between free surface energies and the contact angle Θ (4-6), which was taken as a measure of the wetting behavior. A static measurement of the contact angles, the Sessile Drop Method (100-00-230, Rame-Hart, USA), was used. The contact angle between the pure components or the slurry and the polymeric carrier, was measured ten times for each sample (Fig. I).

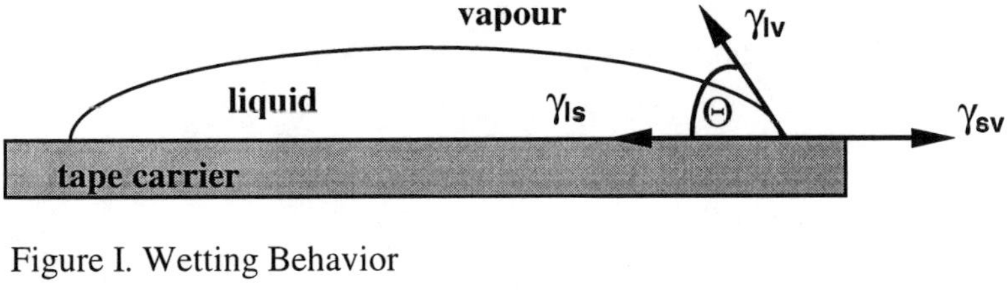

Figure I. Wetting Behavior

$$\gamma_{lv} \cos\Theta = \gamma_{sv} - \gamma_{ls} \quad \text{(Eq.1)}$$

γ_{lv}, γ_{ls} : surface energy between phase l and v, phase l and s respectively
γ_{sv} : surface free energy between phase s and v

Viscosity of Slurries and Glass Transformation Temperature of Binders

The viscosity of the slurries was detected by cone-plate rheometry (Rheometer CS, Bohlin Instr., UK) at room temperature. The stress sweep was 100 Pa for 30 sec before the measurement of viscosity under controlled stress was performed.

The glass transformation temperature of the binders was determined by dynamic differential scanning calorimetry DSC (General V 4.1C DuPont 2920, DuPont, USA). The heating and cooling rate was 5 K/min respectively. (7)

Characterization and Release Performance of Green Tapes

Concerning surface defects, cracks, and phase separations the dried green tapes were optically characterized. The contact angle at the edge between green tape and polymeric carrier was measured using micrographs. The removal of the dried tape from the carrier (Fig. II) was determined and classified into 3 groups.

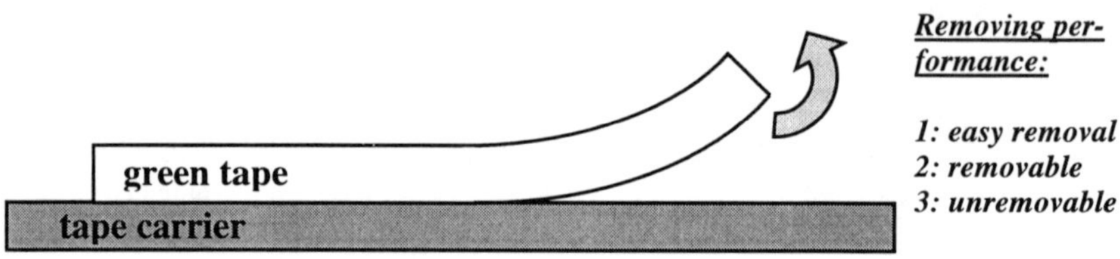

Figure II. Release Performance

RESULTS

Selection of binders for aqueous slurries

Different binders for aqueous slurries were tested in their ability to form a flexible green tape. A suitable binder for tape casting slurries must have the following properties: i) low T_g for good flexibility, ii) reducing the surface tension of water on the polymeric tape carrier and iii) easy removal from the polymeric tape carrier.

Binders with a T_g around room temperature like acrylic binder 2, carboxymethylcellulose, and polyvinylacetate show poor flexibility if no plasticizer was added (Fig. III). On the contrary the dried acrylic binder 1 has a good flexibility without using plasticizers due to the low T_g point. Furthermore, the binder allows to obtain a slurry with a high powder content in order to achieve a high green density of the tape.

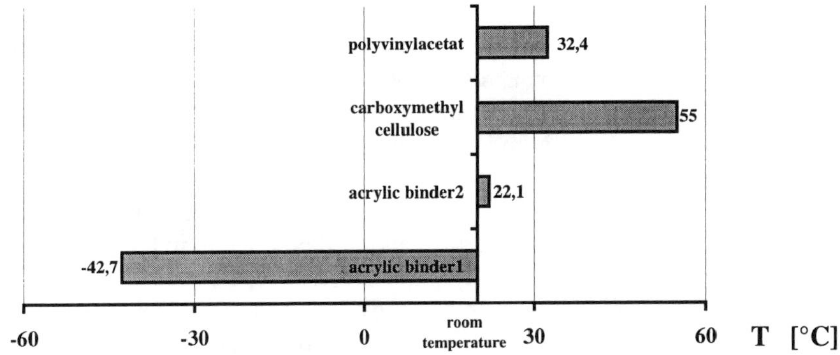

Figure III. Glass transformation temperature T_g of binders

Solved binders effect the contact angle between water and polymeric carrier. Fig. IV shows the contact angles of pure water and of binder solutions on different tape carriers. Carboxymethylcellulose does not clearly lower the surface tension of water. The acrylic binder 1 and 2 reduce the contact angle much stronger.

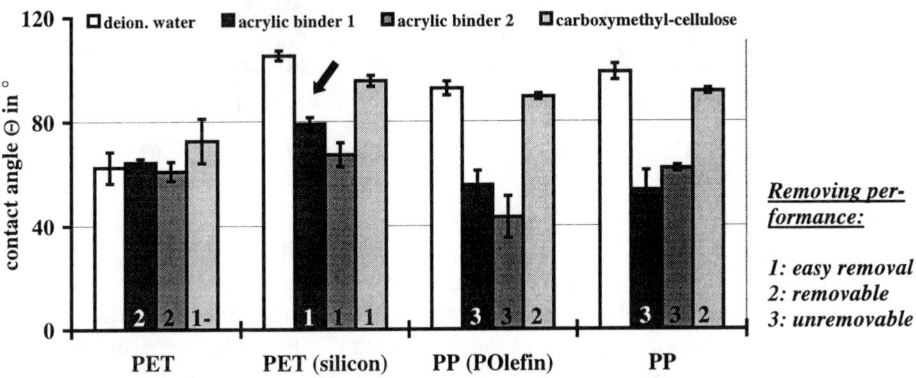

Figure IV. Contact angle and removal of binders on different tape carriers

The dried films of both polymers can be removed very easily from the silicon coated PET polymer carrier. Because the acrylic binder 1 allows to obtain the highest density of the green tapes, this binder was used for further optimization: an optimization was necessary, because the green tape of acrylic binder 1 showed the best tape removal, if silicon coated PET is used as a carrier. However, the wetting behavior of the slurry on this carrier was not sufficient for tape casting. Tape inhomogeneities and crack formation was observed. The following results show how surfactants can improve the wetting behavior.

Wetting behavior of surfactants

It is well known that short-range forces exist between molecules which are responsible for surface tension and wetting behavior. This short–range forces can be influenced by surfactants. Water has a high surface tension (72,8 mN/m) compared to ethanol (22,3 mN/m) and toluene (27,9 mN/m). In water based tape casting slurries it is necessary to lower the surface tension by adding surfactants to improve the wetting behavior of the slurries on polymer tape carriers. Fig. V shows the strong influence of surfactants on the contact angles tested on silicon coated PET, uncoated PET (Mylar) and polyolefine coated PP. The EO-PO-block-polymer and a polyglycolether do not show as good results as the copolymers 1 and 2. The silicon based surfactant reduces the contact angle drastically.

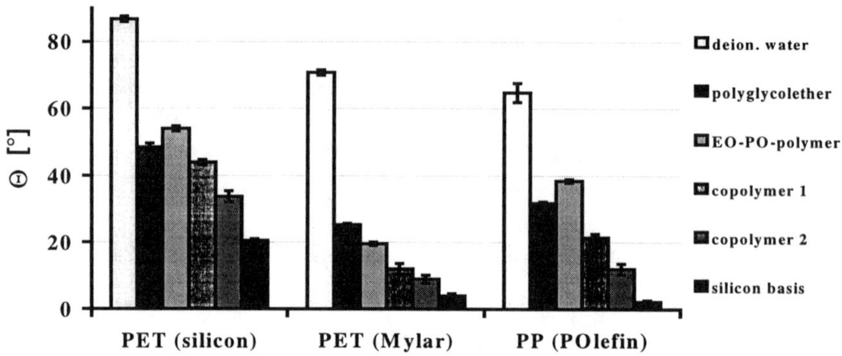

Figure V. Contact angle of water and surfactants on tape carriers

Viscosity and wetting behavior of aqueous slurries

Adding surfactants to the basic slurry does not only influence the contact angle, but also causes different changes in viscosity (Fig. VII). Polyglycolether and silicic acid filled copolymer 2 lower the viscosity of the tape casting slurries, while the other surfactants increase the viscosity. The silicon based surfactant resulted in phase separated slurries, which are of coarse not good for tape casting (Fig. VI). Phase separations are also observable in the edges of the green tapes by an optical microscope. Green tapes cast on silicon coated PET show low wetting angles, if polyglycolether and copolymer 2 are added to the tape casting slurry. The other components for example copolymer 1, EO-PO-block-polymer, and silicon based surfactant have agglomerates in the cross-section of the green tapes and a rough surface due to phase separations. For this reason the silicon based surfactant is not suitable for the aqueous tape casting slurries, even though this surfactant reduces the contact angle to a great extent (Fig. V). Using polyglycolether and copolymer 2, stable slurries of low viscosity with good wetting behavior on silicon coated PET could be obtained.

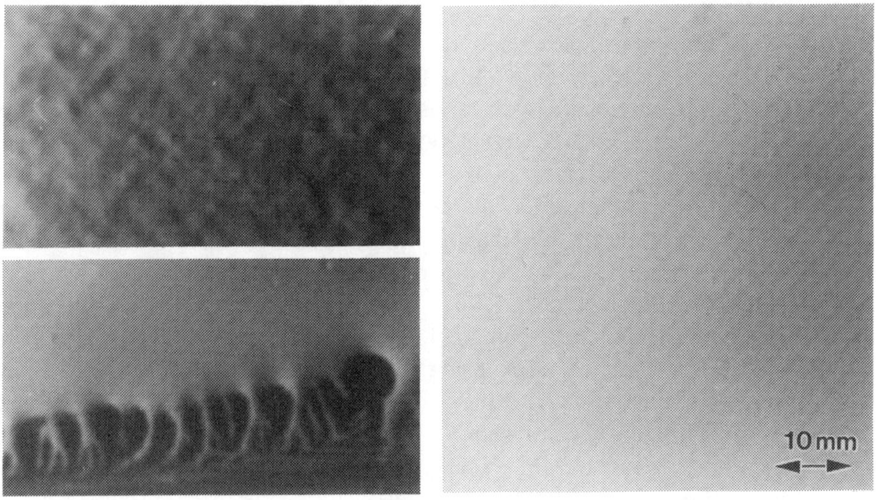

Figure VI. left: phase separation (top: slurry with acrylic binder and silicon based surfactant; bottom: slurry with carboxymethylcellulose as a binder) right: homogeneous green tape (slurry with acrylic binder and copolymer containing silicic acid)

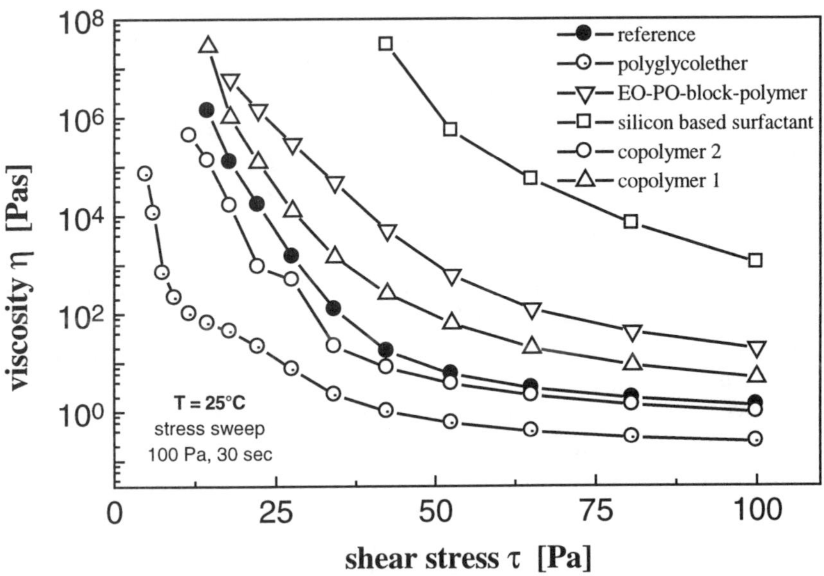

Figure VII. Influence of surfactants on viscosity

CONCLUSIONS

When studying the complex relations between wetting angle, slurry viscosity, and tape removal it is clearly shown that a small contact angle does not guarantee to meet the other requirements, such as viscosity or tape removal. Tape casting of the organic solvent based slurry with a PVB-BBP-binder-system results in a flexible green tape. This slurry shows an excellent wetting behavior on a silicon coated PET carrier and in addition no chemical reactions between additives and carrier occur. The contact angle of the ethanol/toluene slurry on silicon coated PET is small: $\Theta \sim 20°$. From this knowledge and from the results mentioned above, it becomes clear that silicon coated PET can only be used as a tape carrier for aqueous tape casting slurries, if the contact angle of water on silicon coated PET is minimized to $\Theta \sim 30\text{-}45°$ by adding nonionic surfactants. Lower contact angles result in rough surfaces and phase separations. Suitable surfactants for the aqueous slurries of stabilized zirconia were identified. By adding polyglycolether and copolymer containing silicic acid, excellent wetting of the slurry, no phase separation between slurry and carrier, and a sufficient slurry viscosity was achieved.

The current work showed that sessile drop investigations in conjunction with viscosity measurements is a simple and excellent method to study phase separations due to insufficient wetting behavior of aqueous slurries. If the contact angle of the slurry is smaller than 50°, a green tape of good quality can be attained.

REFERENCES

[1]R.E. Mistler, D,J. Shanefield, R.B. Runk, "Tape casting of ceramics"; pp. 411-448 in Ceramic Processing before Firing. Edited by G. Onoda Jr., L.L. Hench. John Wiley & Sons Ltd., New York, 1978

[2]A. Roosen, "Basic Requirements for Tape Casting of Ceramic Powders"; pp. 675-629 in Ceramic Transactions, Vol. 1, Ceramic Powder Science II. Edited by G.L. Messing, E.R. Fuller and H. Hausner. The American Ceramic Society, Westerville, Ohio, 1988.

[3]H. Hellebrand, "Tape Casting"; pp. 190-260 in Materials Science and Technology, Part 1, Processing of Ceramics. Edited by R.W. Cahn, P. Haasen and E.J. Kramer. VCH Verlagsgesellschaft mbH, Weinheim, 1996.

[4]D.J. Shaw, "Introduction to Colloid and Surface Chemistry," Butterworth-Heinemann Ltd. Oxford, 1993

[5]R.G. Horn, "Surface Forces and their Action in Ceramic Materials," Journal of the American Ceramic Society, 73 [5] 1117-1135 (1990).

[6]G. Ström, M. Frederikson, P. Stenius, "Contact Angles, Work of Adhesion, and Interfacial Tensions at a Dissolving Hydrocarbon Surface," Journal of Colloid and Interface Science, 119 [2] 352–361 (1987).

[7]L. Braun, J.R. Morris Jr, W.R. Cannon, "Viscosity of Tape Casting Slips," Bulletin of the American Ceramic Society, 65 [5] 727–729 (1985).

MICRO-FOCUS X-RAY IMAGES OF AGGREGATES IN AQUEOUS ZIRCONIA SUSPENSIONS

J C Austin*, W Liang and K Kendall
Birchall Centre for Inorganic Chemistry and Materials Science, Department of Chemistry, Keele University, Keele, Staffordshire ST5 5BG, UK. *Tel +44 (0)1782 584277, Fax +44 (0)1782 715944, *email cha28@keele.ac.uk

ABSTRACT

Aggregates have been observed in aqueous zirconium dioxide suspensions using micro-focus X-ray imaging. In a suspension of volume fraction ϕ=0.1, aggregates in the size range 100-200 μm were seen to develop over a 2 hour period. In a more dilute suspension (ϕ=0.036) de-stabilised by the addition of PVP, aggregates in the size range 20-50 μm were observed.

INTRODUCTION

Direct observation of structure in concentrated suspensions has been difficult for a long time[1]. Methods employing rheometry, light scattering, Coulter counting etc. are indirect or model dependent and therefore may not be precise. Many of these techniques also require dilute conditions which may not reflect the true nature of concentrated suspensions. Neutron scattering does provide information about particle packing[2] but is too expensive and impractical to be considered for wide industrial use. In recent years an advanced micro-focus X-ray system (HMX-160, X-Tek Systems Ltd.) has become commercially available and is capable of resolving structure down to 5 μm. In this work we use the micro-focus X-ray system to observe both the small and large scale structure of aqueous suspensions of zirconia at volume fractions (ϕ) of 0.036 and 0.1. We have observed structure on a scale of 100-200 μm forming over a 2 hour period after de-stabilisation of a sample at ϕ=0.1. Also structure at a scale of 20-50 μm has been seen in a sample with a concentration of ϕ=0.036. The scale of these structures is broadly consistent with Coulter counter readings[3].

EXPERIMENTAL METHOD

The micro-focus X-ray system used for this work consists of a maintainable electron gun assembly as the X-ray source, an X-ray sensitive image intensifier with an attached CCD camera as a detector, and two sample manipulators (5-axis and 2-axis) in between. This equipment is housed in a lead shielded enclosure.

The electron gun operates in the voltage range 25-160 kV in 5 kV steps and a beam current of 5-1000 μA. At optimum focus the electron beam strikes an area of a tungsten target no larger than 5 μm in diameter. This is effectively the resolution of the instrument. X-rays emerge from a window approximately 3.5 mm in diameter which can be either aluminium or beryllium. The spectrum of the X-rays produced has a minimum wavelength λ_{min} which corresponds to the acceleration voltage setting, a peak intensity at $3\lambda_{min}/2$ and decreasing intensities at longer wavelengths[4]. In addition a cluster of narrow high intensity peaks appear between 52 kV and 69 kV for appropriate settings. These correspond to the K lines of tungsten.

The detector is mounted on rails as are the manipulators enabling its distance from the source to be varied between 0.135-1.065 m. The X-ray sensitive area is circular with a diameter of 0.15 m. The attached CCD camera is directly connected to a monitor via a 75 Ω coaxial cable providing real time imaging. The brightness of a point in the image is proportional to the X-ray power intensity incident at the corresponding point on the detector.

The image signal from the camera is re-transmitted to a real time digital image processor (Argos 20, Hamamatsu Photonics K.K.) which is in turn connected to a bench top PC (OptiPlex GXMT 5133, Dell Computer Corp.) where images are stored and processed employing a commercial image analysis package (PC_Image ver. 2.2, Foster Findlay Associates Ltd.).

Initially a sample of an aqueous zirconia (particle size=0.76 μm) suspension was prepared at a volume fraction of ϕ=0.28 and stabilised by the addition of an ammonium-polyelectrolyte (Darvan 821A). The sample was de-stabilised by dilution to ϕ=0.1 and poured into a test cell for X-ray imaging. The test cell consisted of two 6 mm thick perspex (PMMA) plates separated by copper wire 0.55 mm in diameter and clamped to prevent leakage. X-ray images (30 kV, 985 μA) of the sample were recorded at 25 minute intervals up to 2 hours after the dilution (figures 1-3). In each case the image was accumulated 512 fold to reduce image noise.

A further sample of lower volume fraction was prepared and de-stabilised by the addition of PVP (Polyvinylpyrrolidinone). In this case the test cell consisted of two tufnol plates 1 mm thick separated by a 0.5 mm thick sheet of silicone rubber greased and clamped to prevent leakage. The sample formed a loose sediment in the cell with concentration ϕ=0.036. The use of thinner test cell plates and lower concentration samples allowed a lower beam current to be used for recording the X-ray image (35 kV, 130 μA) leading to a higher quality image (figure 4).

RESULTS AND DISCUSSION

Figures 1-3 show X-ray images of an unstable aqueous zirconia sample taken in chronological order. Figure 1 is a high magnification image taken 25 minutes after de-stabilisation by dilution to ϕ=0.1. There is very little evidence of instability at this stage. Apart from a small dark region near the top/centre of the image about 70 μm across, the grey level in the image is smooth and lacks structure. It is believed that the dark region near the top of the image is an aggregate in the process of formation and as such is the beginning of de-stabilisation.

Figure 2 is a low magnification image taken 100 minutes after dilution. At this stage de-stabilisation is evident throughout the visual field. In certain areas dark high density regions appear immediately below bright low density regions. This is evidence of local sedimentation where aggregates form and subsequently descend at higher velocity than the surrounding solid phase by virtue of Stokes' law. The brighter areas above are regions which have been vacated by rapidly descending aggregates.

Figure 3 is a close up of a typical region of the sample taken 2 hours after dilution. Micro-structure is observed at a scale of 100-200 μm which was not evident in figure 1 indicating that the observed structure was formed in the intervening period.

Figure 4 is an X-ray image of a lower concentration aqueous zirconia suspension supported in a thinner test cell. This suspension was de-stabilised by the presence of PVP, although the original concentration was ϕ=0.0216 the X-ray path was through a sediment which had settled. By measuring the total areas of the sample and the sediment, the sediment area/sample area ratio was found to be 3/5, thus providing a sediment concentration of the order of ϕ=0.036. This concentration was still lower than that of the previous sample and therefore allowed a much lower beam current setting to be used. Because it is known that high beam currents do not focus well (spot size = 20-30 μm at 985 μA), then this goes some way to explaining the superior quality and resolution of the image in figure 4 compared with that in figure 3. In this case micro-structure of a much smaller scale (20-50 μm) was observed which would allow earlier stages of flocculation to be observed.

Figure 1 Close up image of an aqueous zirconia suspension 25 minutes after destabilisation (ϕ=0.1, sample thickness=0.55 mm, acceleration voltage=30 kV, beam current=985 μA).

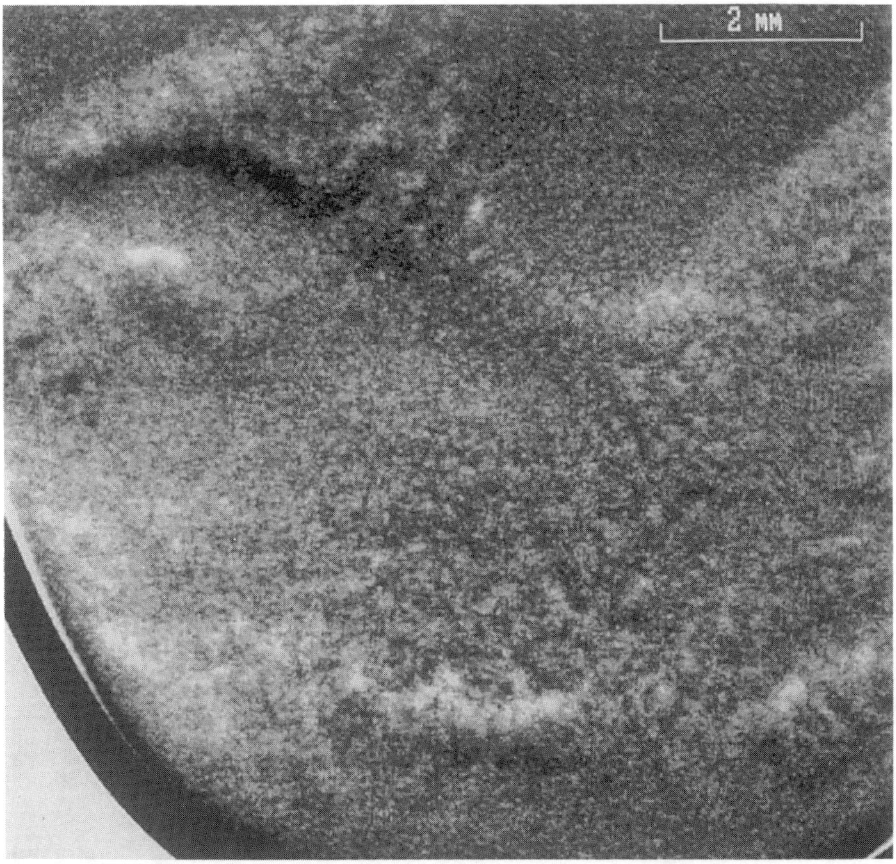

Figure 2 Large scale view of the same suspension 100 minutes after destabilisation (ϕ=0.1, sample thickness=0.55 mm, acceleration voltage=30 kV, beam current=985 μA).

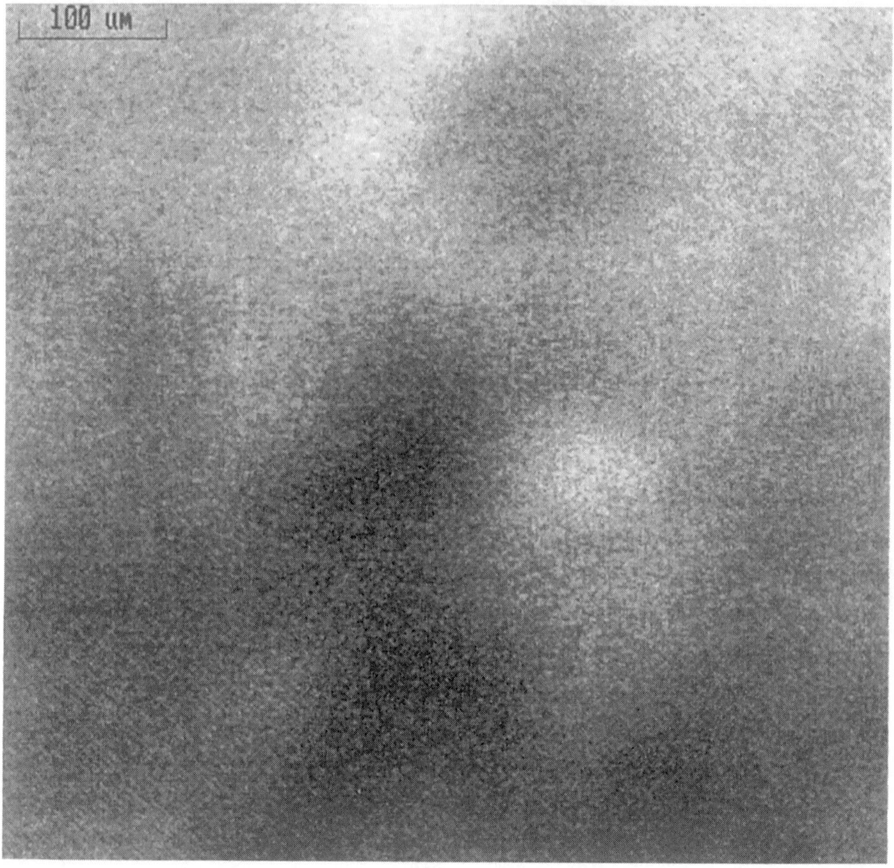

Figure 3 Close up view of the same suspension 2 hours after de-stabilisation (ϕ=0.1, sample thickness=0.55 mm, acceleration voltage=30 kV, beam current=985 μA).

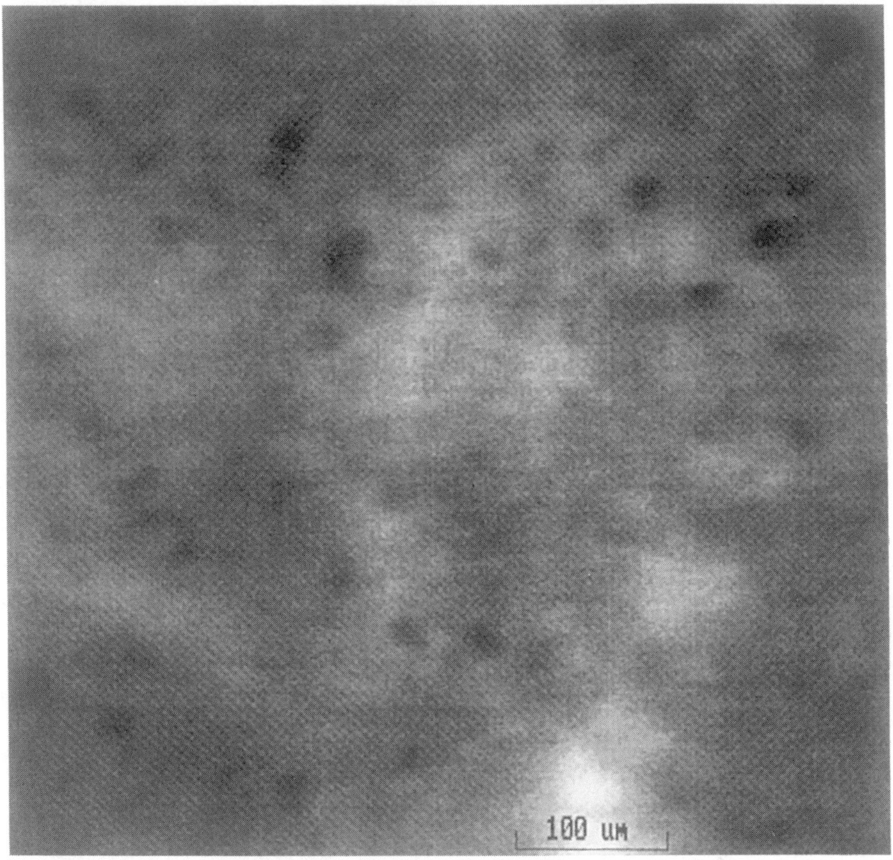

Figure 4 Close up of an aqueous zirconia suspension (sediment) de-stabilised by adding PVP (ϕ=0.036, sample thickness=0.5 mm, acceleration voltage=35 kV, beam current=130 μA).

CONCLUSION

It has been demonstrated that micro-focus X-ray imaging can be used to study the structure and density profile of any suspension with a large X-ray opacity contrast between the phases. For unstable zirconia suspensions, aggregates were easily detected and measured. Dilute suspensions were seen to develop smaller aggregates than more concentrated suspensions. Such aggregates could not be detected by light scattering because they disappeared during the necessary dilution process.

ACKNOWLEDGEMENT

The authors gratefully acknowledge EPSRC (grant No. GR/K 82352) for financial support of the work which is presented in this paper.

REFERENCES

[1]J.W. Goodwin, pp 209, in "Scientific Methods for the Study of Polymer Colloids and their Applications", Edited by F. Candau and R.H. Ottewill, Kluwer Academic Publishers, Dordrecht, 1990.

[2]R.H. Ottewill, pp 183, in "Solid/Liquid Dispersions", Edited by Th.F. Tadros, Academic Press, London, 1987.

[3]W. Liang, J.C. Austin and K. Kendall, "X-ray Imaging of Aggregates in Concentrated Zirconia Suspensions", Journal of Materials Science Letters, in press.

[4]R. Jenkins and J.L. de Vries, "Practical X-ray Spectrometry" pp. 3, 2nd ed., MacMillan and Co. Ltd., 1970.

MECHANICAL PROPERTIES OF CONSOLIDATED BODIES VS INTERPARTICLE PAIR POTENTIAL

Fred Lange
University of California
Santa Barbara, CA

ABSTRACT

Colloidal powder processing can improve the reliability and strength of ceramics by reducing the size of strength degrading heterogeneities by filtering the powder prior to consolidation. Removing heterogeneities greater than a given size is equivalent to a proof test, which can truncate the strength distribution. Although signification property improvements can be made with this approach, forming methods, consistent with the removal of flaw populations are still under development. This development requires knowledge relating the mechanical properties of saturated powder compacts to the potential between particles within the body, just as the properties of crystalline materials are related to the interatomic potential. The background to this program resides with the discovery that short-range repulsive potentials can be developed that, when combined with the pervasive attractive van der Waals potential, produce an interparticle pair potential characterized by a potential well. Three different chemical routes that shroud the particles with a barrier layer have been identified to produce the short-range repulsive potential. The repulsive pair potential initiates when their barrier layers begin to interact. This interaction produces an attractive, but non-touching particle network in which particles "sit" within a potential well at an equilibrium separation distance. The depth of the potential well and the number of particles per unit volume (controlled by the volume fraction and particle size) governs the strength of the network. The strength of the network can be characterized by measuring elastic modulus and yield stress (at a given stressing rate) of the slurry, and the response of the saturated powder body, formed by consolidating the same slurry, to stress. Brittle behavior (crack extension before flow) occurs when either the potential well is too deep, or the particles are forced together (into a deeper potential well) during pressure consolidation. Plastic behavior, required for near net-shape forming, can be achieved for other bodies. Bodies can have a flow stress that is identical to the flow stress of commercial clay. Plastic bodies can also be achieved by mixing together weak and strong particle networks in the slurry state, and then consolidating. With increasing volume fraction, the interpenetrating strong network gradually increases the flow stress of the weak network. The "workability" of these plastic consolidated bodies, i.e., their ability to be formed into complex shapes without either tearing or cracking, can be assessed experimentally for simple forming operations, and then modeled via simulations for more complex shape forming requirements.

PLASTIC OR BRITTLE? INFLUENCE OF COAGULATION MECHANISM ON CRACKING BEHAVIOR OF WET ALUMINA GREEN BODIES

B. Balzer, M. Widmer, M. K. M. Hruschka, and L. J. Gauckler
Colloidal Chemistry-Ceramic Processing Group at the ETH
Zürich, Switzerland

ABSTRACT

Alumina green bodies were produced by direct coagulation casting (DCC). Starting from electrostatically stabilized colloidal suspensions the coagulation was performed either by changing the pH towards the isoelectric point or by increasing the ionic strength in order to compress the double layer thickness. The former way lead from pH 4 to 9 (IEP of alumina) the latter was conducted at pH 9 in a suspension stabilized with diammonium citrate. Coagulation behavior and mechanical properties of the wet green bodies were investigated by impedance spectroscopy and found to depend on the coagulation mechanism. Thus, bodies made via ionic strength increase exhibit a considerably higher strength than those coagulated by pH increase. On the other hand, they can be liquefied again, if vibration is applied.

EXPERIMENTS ON SLURRY CONSOLIDATION MECHANICS

D.N.Ravishankar and B.J.Kellett
Department of Material Science and Engineering
University of Cincinnati
Cincinnati, OH 45221-0012

1. INTRODUCTION

Uniform consolidation is a prerequisite for powder processing uniform, high density ceramics as non-uniform green body density almost invariably leads to cracks during sintering.[1] Achieving uniform consolidation, reliably and consistently, will require a higher understanding of the mechanics of powder consolidation than currently available. The non-steady state conditions present during consolidation in slurry based processes such as slip casting, tape casting, etc. make it difficult to study the mechanics involved experimentally. Conventional characterizing methods used such as particle size distribution, pH value, density, viscosity, ink blot spreading, etc., are important from the standpoint of quantifying processing behavior on a strictly empirical basis. New characterization methods are needed based on the mechanics of cake formation which measure the underlying properties of the cake.[2] This article explains the application of granular mechanics experiments to determine the two fundamental properties of cake formation: permeability and density (solid volume fraction) behavior. A number of different experiments to measure these cake properties are discussed.

The consolidation of slurries into powder compacts involves both the time dependent removal of liquid, and the consolidation behavior of the powder particle skeletal network. The theory of flow through porous media is based on the phenomenological Darcy equation[3]:

$$q = -\frac{K}{\eta}\nabla u_e \qquad (1)$$

The volume flow rate of liquid divided by the cross sectional area (the liquid flux q) is proportional to an "excess" liquid pressure gradient (∇u_e). The liquid flow is with the gradient in excess liquid pressure. The liquid pressure is in excess of the absolute liquid pressure which includes the hydrostatic component produced by the self weight of the permeant and suspended particles. The proportionality constant of equation 1 defines the permeability (K) divided by the viscosity of the permeant (η).

Theoretical expressions for permeability fall within two models: capillary models and drag models. The Kozeny-Carman model is derived from capillary flow models and is defined by the specific surface area per unit volume of solid (S_V) and the solid volume faction of the cake (ϕ_c):

$$K = \frac{1}{\zeta S_v^2}\frac{(1-\phi)^3}{\phi^2} \qquad (2)$$

ξ is the tortuosity constant (typical values between 3 and 6) which is a measure of the increased distance liquid must travel to pass through the cake.

Permeability equations, like Kozeny-Carman, are functions of particle packing density. Darcy's law defines permeability (K), which is a central issue in the study of flow through porous media. The central issue for ceramic processing is powder packing uniformity. The yield curve defines the relationship between particle packing density and effective stress and is central to understanding ceramic process behavior. The effective stress is related to the liquid pressure by mechanical equilibrium[4]:

$$\nabla\sigma' = \frac{\eta}{K}\vec{q} + \phi(\rho_s - \rho_l)\vec{g}_o \qquad (3)$$

The effective stress on the powder particle skeleton network increases in the direction of fluid flow. Also included in equation 3 is the buoyancy weight of the cake where ρ_s is the density of the solid particles, and ρ_l is the liquid density. This increases in the direction of gravity. The excess liquid pressure is defined by Darcy's law and permeability. Finally the permeability is a function of the particle packing density. The yield curve is a central connecting element to determining processing behavior and the ultimate uniformity of a ceramic compact formed from slurries.

The yield curve concept defines particle packing density as a function of the effective stress. As a powder compact is forced together, the interparticle forces increase causing irreversible particle packing changes. The yield curve assumes that the particle packing density is only a function of effective stress and thus the local effective stress defines an irreversible consolidation curve. Rates of consolidation, while perhaps important, are not considered within the yield curve concept. The time dependence is solely attributed to Darcian flow of liquid through the pore volume.

The consolidation behavior of powders is highly nonlinear and frequently found to increase to the log of the effective stress:

$$\phi = \phi_o + \phi_\sigma \log\left(\frac{\sigma'_z}{\sigma_o}\right) \tag{4}$$

ϕ is the solid volume fraction, and σ'_z is the uniaxial stress.

The Darcy's law require a liquid pressure gradient across the cake. This liquid pressure gradient is supported by mechanical stresses in the cake skeleton network. For low density, weak cakes, this mechanical stress gradient leads to non uniform consolidation. The cake will be under the greatest compressive stress near the filter. Further, in first applying a liquid flux to a cake, the consolidation of the cake is brought about by the increasing liquid pressure drop across the cake. These supporting stresses in the cake drive consolidation. The degree of deformation is described by the yield curve. These issues complicate the accurate determination of the yield curve and permeability characteristics of cakes under just the conditions which are needed to determine processing behavior. It will be shown that these characteristics can be determined with a consolidation cell shown in Figure 1. A number of different types of experiments with the consolidometer will be demonstrated.

2. EXPERIMENTAL PROCEDURE

The equipment used for this study consists of a consolidometer (Figure 1) capable of operating at stress ranges ranging from 0.6 kPa to 200 kPa. It is made up of a plastic tube in which the ceramic slurry is contained between two porous filter plates. The bottom plate is fixed and the top plate is closely machined and floats on top of the slurry.

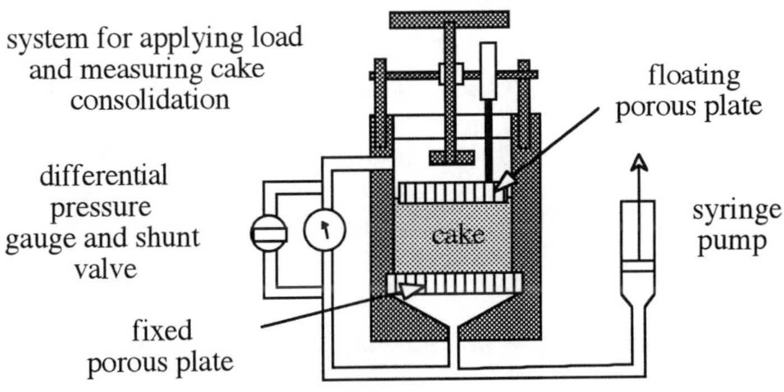

Figure 1 Instrumented consolidometer for measuring densification and permeability of cakes under constant flow rate conditions.

Fine filter paper of pore size 0.22 μm are fixed on the plates to contain even the most dispersed slurries. These effectively separate the slurry from water that is filled on both sides of the slurry and ensure that, when the syringe pump is on, water alone is transferred while the ceramic solid content is confined within. The syringe pump creates the liquid flux for consolidation by operating at flow rates varying from 0.3 to 30 μm/s. The suction pressure of gypsum mold is of the order of 100 kPa which is within the pressure range of this set-up.

Consolidometer experiments were undertaken with commercially available alumina powder (ALCOA A16SG). Aqueous slurries were prepared in plastic bottles and agitated with high purity alumina media by a SWECO vibratory mixer for about 1 hour to produce compressible slips of 20 vol. % solids. These experiments were conducted with straight DI water, flocculated slurries of viscosity ~1500 mPa. sec. The liquid pressure difference between the top and bottom of the cake was measured using a variable reluctance differential pressure transducer (DP15, Validyne Engineering Corp., Northridge CA) and the displacement of the top layer of the cake was continuously measured using a linear variable differential transformer(B50, Solartron Metrology, Buffalo, NY). The pressure transducer was connected in parallel with a shunt valve.

An Ammonium Polymethacrylate deflocculant (Tradename Darvan C, supplied by R.T.Vanderbilt Co., Inc.) was used to study the effect of slurry dispersion on the yield curve and permeability characteristics. Information provided by the manufacturer indicated a polymer loading of 25 % (dry mass basis) and molecular weights between 12000 and 14000 amu and density of 1.11 gcm^{-3}. Suspensions were ball milled on a horizontal roller for different amounts of time ranging from less than half hour to more than 24 hours, in order to facilitate dispersion of the alumina particles. The pH value was always near the isoelectric point of alumina (between 8.5 and 9.5). Slurry viscosity was measured in a Brookfield Viscometer prior to each experiment.

Flocculated slurries (those with Brookfield viscosity greater than 0.7 Pa-s) are easy to test, as their higher viscosity are easier to contain within the cell. Dispersed slurries require tighter seals and finer grades of filter paper. Whatman #4 filter paper was used for slurries with viscosities greater than 0.7 Pa-s, 0.22μm Nylon filters from Micron Separations Inc. for lower viscosity, dispersed slurries.

3. RESULTS

In the first experiment, slurry is poured into the consolidometer to a height of 1.93 cm. The floating top plate was applied to the surface of the slurry. Shown in figure 2 is a typical pressure/displacement -time response for the self consolidation of a fresh slurry (20 vol.% solids, η=1.5 Pa-s, 1.9cm initial height). The weight of the floating top plate applies an equivalent stress of 0.52kPa to the top of the slurry. The weight of the powder in the slurry produces an additional stress of 0.056 kPa. A negative pressure implies that the pressure beneath the cake is greater than above. As this is a stable flocculated slurry, the weight of the powder particles and the top plate is already partially self supported by the cake by the start of the experiment (about 1 minute after pouring in the slurry).

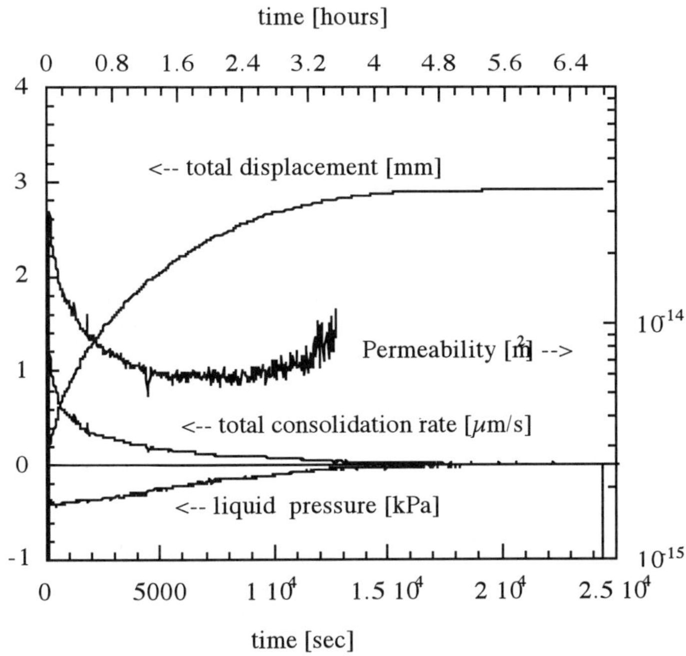

Figure 2 Liquid Pressure and Displacement response of a flocculated slurry with no applied load and an imposed liquid flux of 1 μm/sec.

Consolidation slows and stops, reaching an equilibrium value as the liquid pressure difference across the cake approaches zero. One way of looking at this condition is that the effective stress profile across the cake has reached the yield curve values. The displacement rate, measured by the LVDT, is equal to the liquid flux rising through the cake and passing through the top filter plate. The variation in the liquid flux through the cake will depend on the consolidation characteristics of the cake. The simplest assumption is uniform consolidation of the cake. For uniform consolidation the liquid flux, relative to the solid particles in the cake, will vary linearly from zero at the bottom of the cake to the velocity of the top plate, (initially about 1.5 μm/s).

Steady state is reached in about 4 hours at a displacement of 3 mm. For uniform consolidation, the permeability is calculated by integrating Darcy's law:

$$P = \int_{o}^{z_c} q \frac{\eta}{K} dz \qquad (5)$$

where z_c is the total thickness of the cake and $q = q_{max.} z/z_c$ where $q_{max.}$ is the velocity of the top plate. For uniform cake consolidation the permeability is accordingly:

$$K = \frac{q_{max.} z_c \eta}{2P} \qquad (6)$$

The decrease in permeability is associated with the increase in solid fraction from 20 vol.% to about 24 vol.%. The decreasing displacement rate and liquid pressure gradient contribute to the increase noise in the permeability measurement.

The experiment shown in figure 2 is very similar to the soil mechanics oedometer experiment.[5] In the limit of small deformations, solid mechanics theory leads to the simplified Terzaghi equation [6]:

$$\frac{\partial u_e}{\partial t} = C_v \frac{\partial^2 u_e}{\partial z^2} \qquad (7)$$

Where u_e is the excess liquid pressure. The consolidometer constant C_v is a common design parameter in soil mechanics:

$$C_v = \frac{K}{\eta} \frac{1}{m_v} \qquad (8)$$

where m_V is the derivative of the yield curve:

$$m_v = \frac{1}{\phi} \frac{d\phi}{d\sigma} \qquad (9)$$

This approach can also be used to determine the permeability and compressibility of cakes. The use of this method is demonstrated elsewhere.[7]

Shown in figure 3 is the effect of imposing a liquid flux on a flocculated slurry (20 vol.% solids, 1.4 Pa-s, 19.3 mm initial height). As in figure 2, before the syringe pump is activated, the liquid pressure first decreases to about -0.4kPa due to the weight of the top filter plate. This slight negative pressure (about -0.4kPa) is quickly overwhelmed by the syringe pump pulling liquid downward through the cake (syringe pump applied flow rate 0.5 μm/s). As shown in Figure 3, the liquid pressure and cake consolidate in a linear fashion, much like a constant piston velocity filter pressing experiment.[8] The liquid pressure increases as the cake consolidates and the average permeability of the cake decreases. After about 1 hour, the consolidation of the cake slows and both the cake displacement and liquid pressure reach steady state values at 2 hours. An equilibrium is reached when the effective stress variation through the cake reaches the cakes yield curve. At steady state the liquid flux is constant through the cake and equal to that produced by the syringe pump. The calculated average permeability is 1.15E-15 m^2 and the average solid volume fraction ϕ_f is 27.3 %. The average effective stress on the cake is 3.8 kPa.

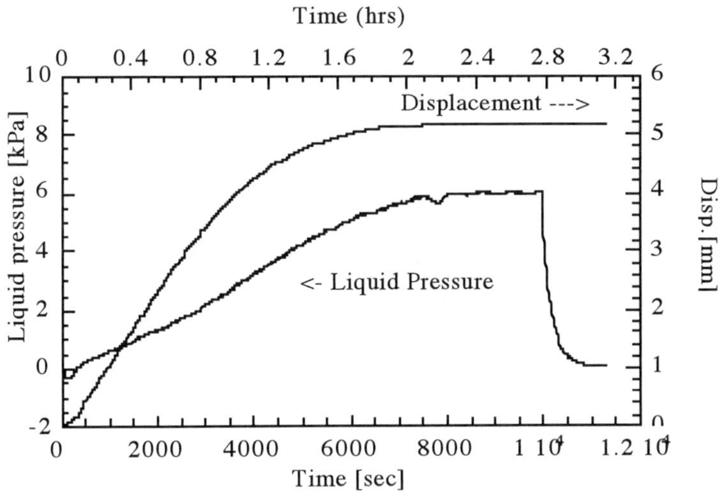

Figure 3 Liquid Pressure and Displacement response of a flocculated slurry with no applied load and an imposed liquid flux of 0.5 μm/sec.

A similar experiment with a constant top load producing an effective stress 36 kPa is shown in figure 4. This applied stress initially produced a liquid pressure of 25kPa, suggesting that the cake is already supporting 11kPa of the applied stress. This liquid pressure dissipated to zero in one hour from the Darcian flow of liquid up through the cake. The syringe pump was then turned on to produce a flux downward through the cake of 0.5 μm/s and the liquid pressure increased to a steady state pressure of 19.4 kPa in about one hour. Most of the consolidation occurs with the application of the load (4.5mm). The application of the liquid flux produced a consolidation of 0.45 mm. The average solid volume fraction at the end of the experiment is 41.4 % solids at an average effective stress of 47kPa. The average permeability is 2.74E-16 m^2.

It is important to note the ramifications of the large liquid pressures produced in these experiments. This liquid pressure gradient implies an equivalent effective stress gradient. The applied load produces an effective stress which is uniform through the cake. In figure 4, the effective stress gradient is about one half that produced by the applied load. The degree to which these effective stress gradients lead to variations in particle packing density is addressed elsewhere.[7,8]

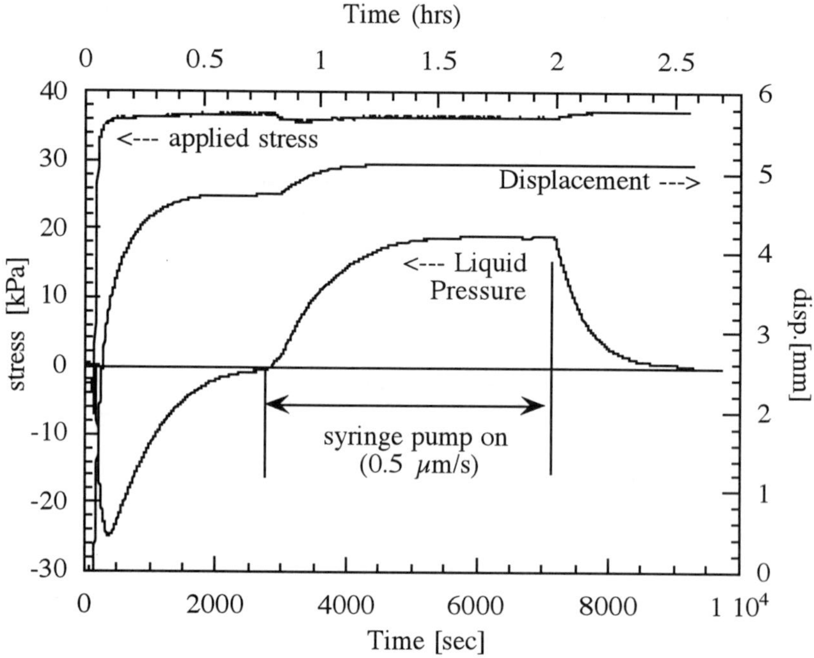

Figure 4 Liquid pressure and displacement response of the system to an applied stress of 35kPa, and then to an applied flow rate produced by the syringe pump. Each of these responses can be used to determine cake properties.

The average values of permeability, effective stress and solid volume fraction from a series of experiments are shown in Fig. 5 and 6. Shown are flocculated slurries and slurries dispersed with 0.033 wt.% Ammonium Polymethacrylate (Tradename Darvan C, supplied by R.T.Vanderbilt Co., Inc). The permeability curves on the left half of figure 5 are those determined via the technique demonstrated in figure 2. The markers on the right side of figure 5 are average permeability values. The Kozeny-Carman function shown is for S_v = 33 1/μm (from data provided by the manufacturer) and a tortuosity ζ=4. The Kozeny-Carman equation appears to be a reasonable fit to the average permeability data. The cakes formed from flocculated slurries are more permeable than those formed from dispersed slurries. Shown in Figure 6 is a log normal yield curve fitted to one set of data. The cakes formed from flocculated slurries tend to be less dense than those formed from the flocculated slurries.

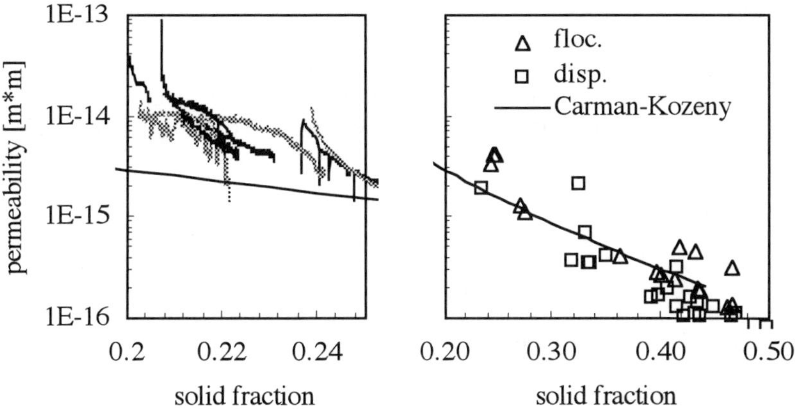

Figure 5 Permeability of flocculated and dispersed alumina slurries. Carman-Kozeny behavior is shown.

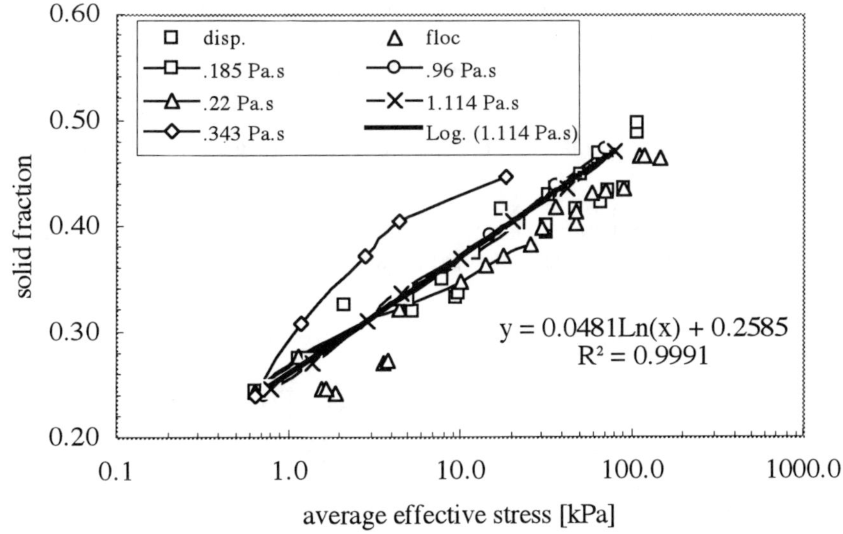

Figure 6 Yield curves for flocculated and dispersed slurries. Log-normal best fit is shown for one constant load experiment. The viscosity of the slurries measured before consolidation are shown in the legend. These are average values of effective stress and solid fraction.

4 CONCLUSIONS

A range of techniques using a cell outfitted with displacement and pressure gauges have been undertaken to measure the permeability and yield curve of alumina slurries. We observe that the permeability of dispersed slurries are somewhat lower then the flocculated slurries. The solid volume fraction of flocculated and dispersed slurries are surprisingly similar. The Kozeny-Carman equation for permeability, and the log-normal yield curve appear to reasonably fit the data presented in this article. The data shown in figures 5 and 6 are based on average values of density and permeability. The range of measured values suggest that these methods may only be accurate to a factor of 2. Better techniques which take into account the non uniformity of the cake should improve the accuracy of these measurements. This is the focus of current research activities.

REFERENCES

1 James S. Reed, Introduction to the Principles of Ceramic Processing, John Wiley & Sons (1988).

2 B.J.Kellett, C.Y.Lin and D.N.Ravishankar, “Consolidation at Constant Flow Rate and Load to determine Permeability and Yield Curve Behavior”, Ceramic Transactions, 79, Advanced Synthesis and Processing of Composites and Advanced Ceramics, ED. K.V.Logan, Z.A.Munir and R.M.Spriggs (Amer. Ceram. Soc., Westerville, Ohio)

3 Darcy, H., Les Fontainer Publiques de la Ville de Dijon, *Dalmont*, Paris (1856)

4 B.J.Kellett and Ching Yao Lin, "The Mechanics of Constant Rate Filter Pressing of Highly Flocculated Slurries," *J.Amer.Ceram.Soc.* **80** [2] 381-93 (1997).

5 R.F.Craig, Soil Mechanics, Sixith Edition, E&FN Spon (1997).

6 R.F.Craig, Soil Mechanics, Sixith Edition, E&FN Spon (1997).

7 B.J.Kellett and D.N.Ravishankar, “Permeability, Yield Curve and Compressibility Measurements on Alumina Slurries,” *to appear in Ceramic Transactions*

8 Ching Yao Lin and B.J.Kellett, "General Observations of Constant Flow Rate Filter Pressing," *to appear in the J.Amer.Ceram.Soc.* (1998).

COMPUTER SIMULATION OF MICROSTRUCTURE AND STRENGTH OF SETTLED COLLOIDAL POWDERS

Jong-Cheol Kim and David M. Martin
Department of Materials Science and Engineering,
Iowa State University, Ames, Iowa, 50011

ABSTRACT

Particle settling was studied by computer simulation using simultaneous particle condensation and rearrangement. This three-dimensional settling algorithm includes estimation of sediment strength. Volume distribution through the sediment was checked and compared with experimental findings. Packing volume depended strongly on the degree of particle rearrangement. Sediment strength and isotropy also depended on the degree of particle rearrangement. Severe particle bridging was found near sharp corners.

INTRODUCTION

Settling affects the microstructure of powder compacts. A dense and uniform powder compact is important because of its resultant high sintering density and low sintering temperatures. Most of the research on particle settling has been concentrated on continuum approaches where it is impossible to check and visualize the behavior of individual particles. There are so many complex assumptions in such modeling because it is hard to understand and visualize individual particle effects. Molecular dynamic or Monte Carlo computational approaches[1, 2] to this problem have been tried. In this research we increased the efficiency of modeling using two simple concepts; condensation and rearrangement.[3-6] We developed a program on a PC using FORTRAN 90. The settling process was visually displayed as the computation proceeded to help understand the basic mechanisms of colloidal powder sedimentation.

MODELING

All particles were incrementally settled a unit distance repeatedly until they touched other particles. When particles touched, they rearranged by upper particles rotating around lower particles. Fig. 1 shows a general flow chart of the total settling algorithm.

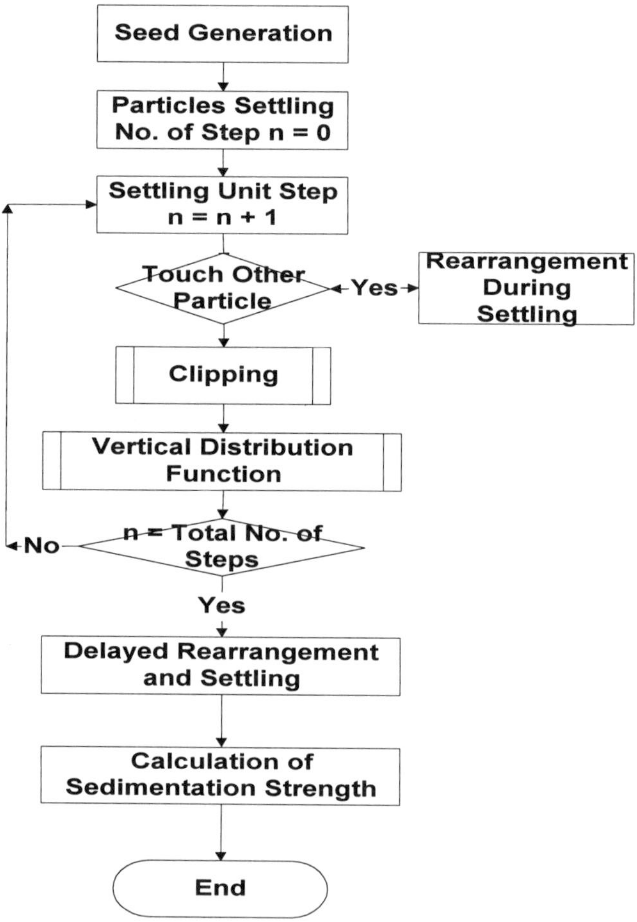

Fig. 1. Flow chart of total settling algorithm.

The rearrangement process starts to run by checking the coordination number(CN) of each particle. If CN is 1, the particle rotates over its neighboring particle in an arbitrary downward direction and if CN is 2, the particle rotates around the line joining the centers of the particles it touches, when CN is 3, the particle is fixed. Fig. 2 shows how rearrangement varies with CN.

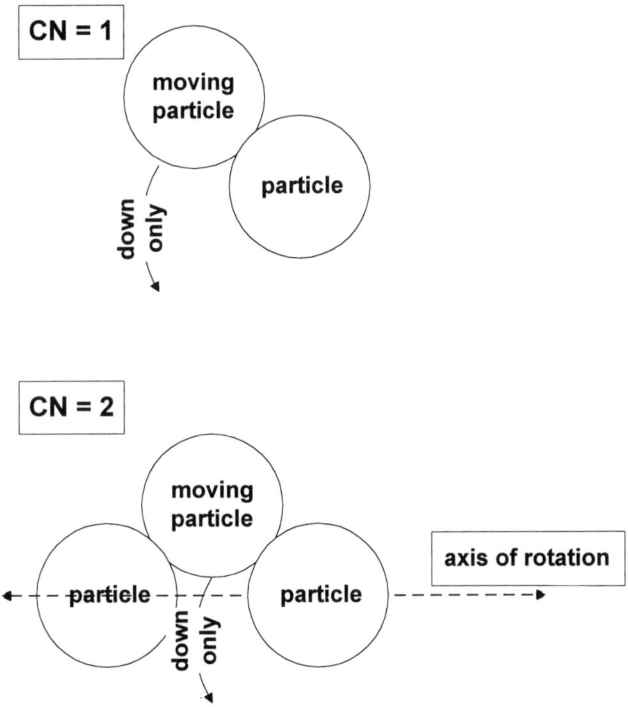

* If CN = 3, Fixed

Fig. 2. Rearrangement processes with different coordination number.

Sediment strength and/or fracture energy is important to subsequent processing.[6-10] Strength was estimated using the following formula.

$$\text{Sediment Strength} = \frac{\text{Bonding Energy per Contact} \times \text{Bonds Crossing Plane}}{\text{Area of Plane} \times 2}$$

where bonding energy is the interparticle potential energy between two particles.

This bonding energy was fixed at 100 kT for convenience in this paper.

RESULTS AND DISCUSSIONS

For comparison with experimental results the sedimentation study done by Schilling and Aksay[11] was used. Our simulation conditions follow:
Particle radius = 10.0μ
Volume concentration of agglomerates = 10.8%
Default values of the unit settling distance and the total number of movements were 2.5μm and 100, respectively.

In the case of particle settling with rearrangement, particles were allowed to rearrange 5 times by 5 degrees in a random direction. Rearrangement by more than this amount did not show much change in sediment density. Simple settling of particles without rearrangement showed a low density sediment occupying 4/5 of the original suspension volume as shown in Fig. 3. Settling of particles with rearrangement resulted in a sediment twice as dense. This was in good agreement with the corresponding experimental results.[11] The settled layer can be divided into four regions after settling. The upper region is empty of particles(region 1 in Fig. 3). The second region is a transition zone (region 2 in Fig. 3) which shows a steep increase in volume concentration of particles with depth. This surface layer is rough and has many vacant sites available for additional particles. The third is a region of constant density (region 3 in Fig. 3) where particles are compact. The last is an interfacial region between the packed particles and the bottom of the box(region 4 in Fig. 3) where settling density was quite a bit higher than inside the region of constant density due to the nature of the 2-dimensional packing(the particles were settled onto a geometrically flat plane.). The volume concentrations in Fig. 3 show the difference between layers settled with and without rearrangement; these correspond to Schilling and Aksay's results[11] for layers freshly settled(no rearrangement) and aged(rearrangement). This suggests that the transition from an "attached-particle network" to "compressed network"(in their terminology) depends on the rearrangement.

Sediment fracture energy was estimated on planes with various orientations as shown in Fig. 4. Fracture energy depends on the direction at the bottom of the box which is shown in Fig. 5. The fracture energy is a maximum in the vertical direction at the bottom of the sediment. This was caused by high packing density on the geometrically flat bottom plane. Rearranged particles showed a higher and more isotropic fracture energy than those of the not rearranged as shown in Fig. 5 and Fig. 6.

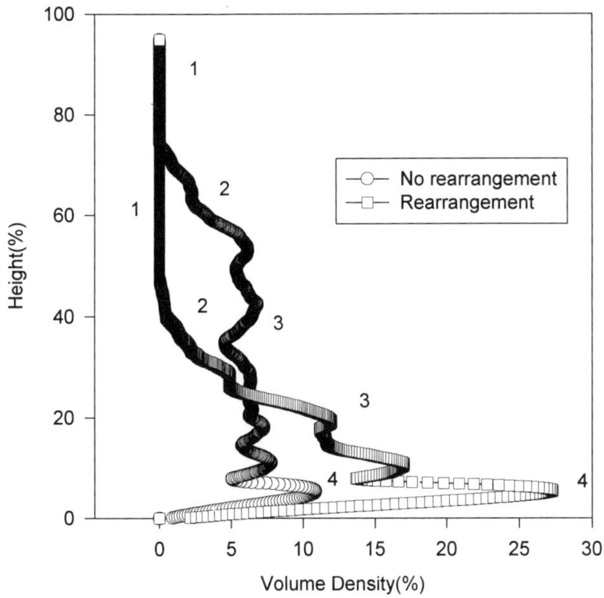

Fig. 3. Vertical volume distribution of sediments rearranged and not rearranged during settling in a compartment.

This was because particles formed more bonds with neighboring particles. Rearrangement caused strength to increase and to become isotropic.

Particles were settled in different shape boxes in order to see how angular strength depended on the settling direction and box geometry. A tilted box is shown in Fig. 7. Radial anisotropic distributions of fracture energy were measured at three different locations as shown in Fig. 7. Fracture energies at side location B and C clearly showed maxima in the settling direction in the case of unrearranged particles as shown in Fig. 8 a) . The maxima shifted towards the normal to the box sides after rearrangement as shown in Fig. 8 b).

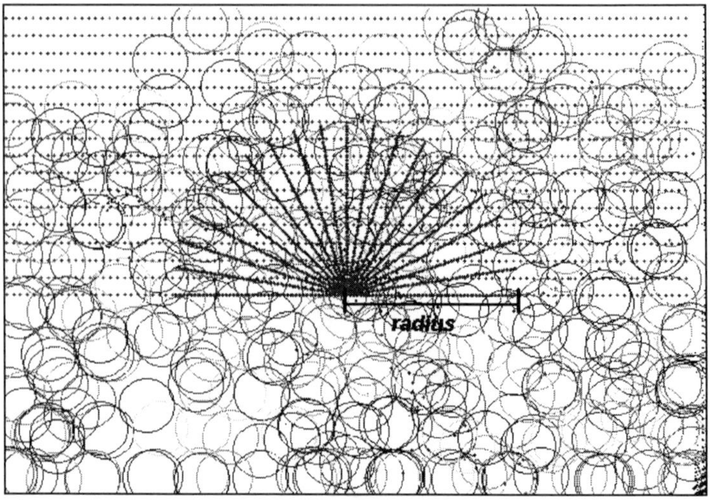

Fig. 4. Projection of settled particles for fracture energy calculation.

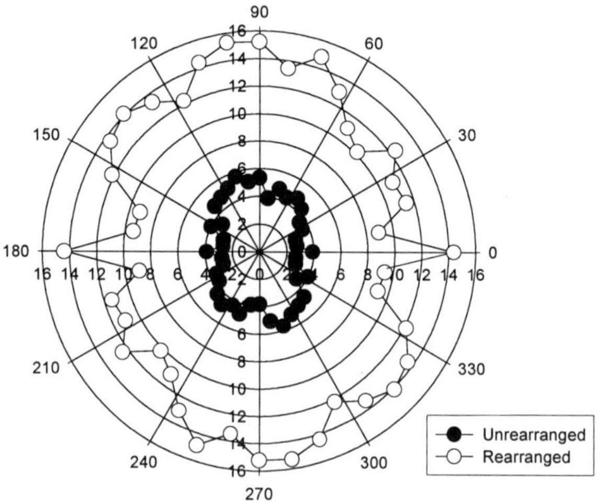

Fig. 5. Radial change in fracture energy of rearranged and unrearranged particles measured at the bottom of the settled bed (radius = 2.5 and energy unit = $J/m^2 \times 10^{-10}$)

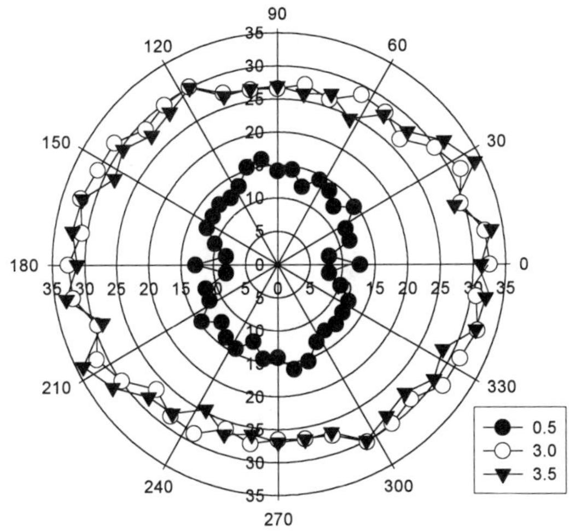

Fig. 6. Radial change in fracture energy of rearranged particles at different heights (radius = 2.5 and energy unit = $J/m^2 \times 10^{-10}$)

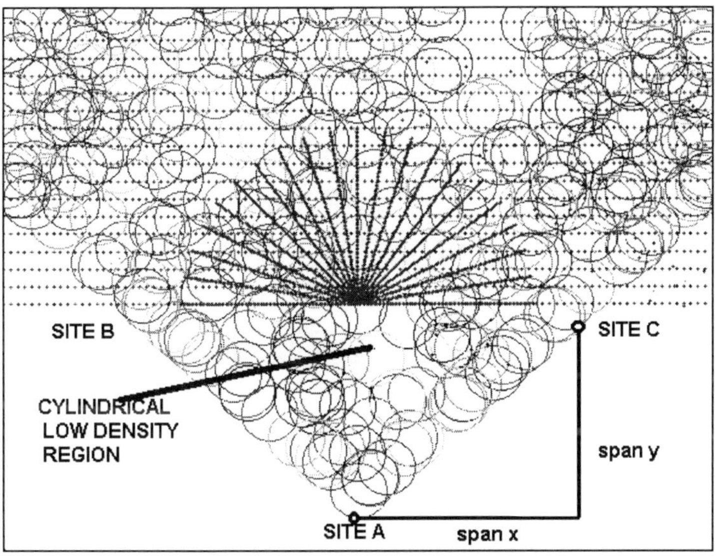

Fig. 7. Projection of settled particles with vertical and horizontal planes for fracture energy calculation in a tilted box.

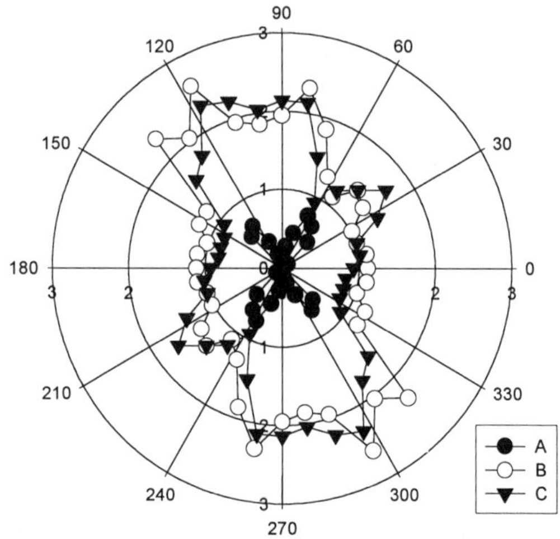

a) unrearranged at location A, B and C with radius 2.5(span x = 2.5, span y = 2.5)

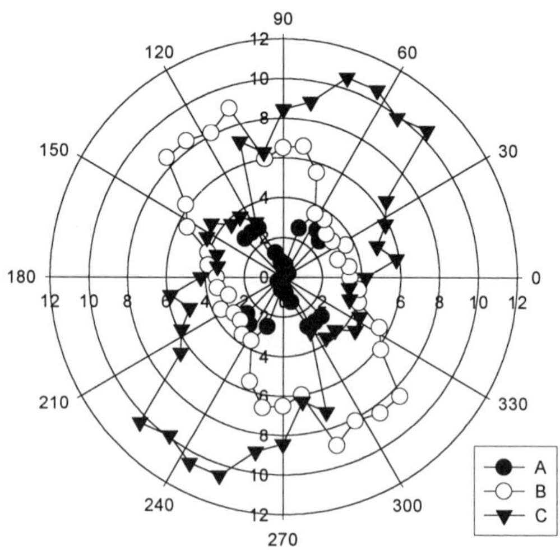

b) rearranged at location A, B and C with radius 2.0 (span x = 2.0, span y = 2.0)

Fig. 8. Radial change in fracture energy of rearranged and unrearranged particles measured at the different locations in different box shapes.

A minimum at location C appeared in the settling direction in both rearranged and unrearranged cases; this was mainly due to the low density region at the corner of the box. There were several cylindrically shaped low density regions in addition to the one at the corner in the particle compact. The settling of particles in a corner caused bridge formation and resultant low density regions in the particle compact. The same measurements were done in the case of lower degree of tilt; the results were similar as shown in Fig. 8 except that fracture energy at location A was larger. This was because there is less bridging between particles during settling.

CONCLUSIONS

A new three-dimensional settling algorithm was developed which incorporated metastable states and delayed rearrangement. Full rearrangement decreased packing volume by about 50% and tripled the strength compared to packing without rearrangement, which was in quite good agreement with experimental results. There was a radial anisotropy of fracture energy most severe for unrearranged layers. Fracture energy became more isotropic with rearrangement. The changes in orientation of fracture energy maxima in a tilted box are explained by the change in particle network structure. Particles settled in the corner of a tilted box exhibit bridging and low density regions in the particle compact due to geometrical constraints caused by the corner. Such low density regions are probably unavoidable if the radius of curvature of an acute corner is comparable to particle radius.

REFERENCES

[1]S. Warr and L. V. Woodcock, "Solid/Fluid Interfaces in Non-linear Steady States," Faraday Discuss., 95 329-46 (1993).

[2]M. J. Vold, "Computer Simulation of Floc Formation in a Colloidal Suspension," J. of Colloid Science, 18 684-95 (1963).

[3]J. Kim and D. M. Martin, "Computer Simulation of Three Dimensional Particle Packing," Submitted to J. of Amer. Cer. Soc., (1996).

[4]R. M. German, "Structure and Properties of Agglomerates"; pp.55-88 in Particle Packing Characteristics, Metal Powder Ind. Fed., Princeton, New Jersey, 1989.

[5]N. D. Aparicio and A. C. F. Cocks, "On the Representation of Random Packings of Spheres for Sintering Simulations," Acta. Metall. Mater. 43[10] 3873-84 (1995).

[6]K. E. Davis and W. B. Russel, "A Model of Crystal Growth in the Sedimentation and Ultrafiltration of Colloidal Hard Spheres," Adv. Ceram., 21 573-585 (1987).

[7]R. Buscall, "The Sedimentation of Concentrated Colloidal Suspensions," J. Colloids and Surfaces, 43 33-53 (1990).

[8]A. S. Michaels and J. C. Bolger, "Settling Rates and Sediment Volumes of Flocculated Kaolin Suspensions," Ind. Eng. Chem. Fund., 1 24-33 (1962).

[9]F. M. Tiller and Z. Khatib, "The Theory of Sediment Volumes of Compressible, Particulate Structures," J. Colloid and Interface Science, 100[1] 55-67 (1984).

[10]R. Buscall, P. D. A. Mills, J. W. Goodwin, and D. W. Lawson, "Scaling Behavior of the Rheology of Aggregate Networks formed from Colloidal Particles," J. Chem. Soc. Faraday Trans. 1, 84[12] 4249-60 (1988).

[11]C. H. Schilling and I. A. Aksay, "Gamma-Ray Attenuation Analysis of Packing Structure Evolution During Powder Consolidation"; pp.800-808 in Ceramic Transactions, Vol. 1, Ceramic Powder Science II. Ed. by G. L. Messing, E. R. Fuller, Jr., and H. Hausner. American Ceramic Society, Westerville, OH, 1988.

NUMERICAL SIMULATION OF TWO-DIMENSIONAL CAKE GROWTH IN A RECTANGULAR SLIP CASTING SYSTEM

M. Takahashi*, J. Shimizu*, H. Unuma* and T. Aragaki**

* Ceramics Research Laboratory, Nagoya Institute of Technology
Asahigaoka 10-6, Tajimi 507, Japan

** Department of Applied Chemistry, Nagoya Institute of Technology
Gokiso, Nagoya 466, Japan

ABSTRACT

On the assumption that a filter cake is incompressible and uniform, two-dimensional cake growth during slip casting of a rectangular body was simulated by FEM (Finite-Element Method). The simulation result showed that the profile of cake front changed with time from a rectangle to a parabola. The parabolic profile will appear owing to the simultaneous deposits of solid particles on the horizontal and vertical surfaces of a rectangular porous mold.

INTRODUCTION

Slip casting using a plaster mold is an attractive method to fabricate a large-sized or complex-shaped ceramic green body. The serious problems are cracking during drying and/or sintering, and low dimensional accuracy of sintered products. Both the disadvantages essentially arise from the nonuniform density distribution of a cast body.

One way to overcome these problems is to improve the uniformity of a green body as much as possible. Preparation of a highly concentrated and well-dispersed ceramic slurry is usually very effective. Another way is the application of *in situ* solidification of ceramic slurry such as gel casting[1)] and direct coagulation casting[2)]. An alternative route is to predict the density distribution. If the density distribution can be estimated by numerical simulation, it can be used to control nonuniform shrinkage of a cast body during drying and sintering. For this aim, it is necessary to understand the cake growth which is intimately related with the development of density distribution.

Adcock & McDowall[3] mathematically analyzed the slip casting rate on a flat surface, and Tiller & Hsyung[4] on a radial surface. These two approaches treat the one-dimensional filtration problem, *i. e.*, the cake surfaces always keep the similar configuration as the molds. It is interesting to examine the two- and three-dimensional cake growths when the filtration occurs simultaneously onto the bottom and side-wall surfaces in a rectangular or cylindrical system. Both the systems are quite important for industrial ceramics. It was shown in our previous work that FEM analysis is valuable to predict the three-dimensional cake development in the cylindrical system[5]. The numerical results were almost in coincidence with the experimental ones. In the present work, the two-dimensional cake growth during slip casting of a rectangular body will be analyzed.

EXPERIMENT

Slurry composed of alumina powder (AL160SG4, Showa Denko K. K.), dispersant (D-305, Chukyo Yushi Corp.), and distilled water was prepared by ball milling for 48 h. The solid concentration was 58.8 vol%, and the mass ratio of the polymer dispersant to the powder was 0.36%. The preliminary test showed that a minimum viscosity was obtained at this ratio.

The slurry was cast into a rectangular plaster mold (5x5x2 cm) sandwiched by two acrylic plates. Unfortunately, it was impossible to take a clear picture of the cake profile through the transparent plates. The boundary between the cake and slurry was vague because the solid concentration of the slurry was quite high. It was also attempted to obtain a cake by draining of the residual slurry in the mold at the specified time. However, the cake got out of shape easily. For reference, the change of cake profile in a cylindrical mold was shown in Fig. 1. It should be noted that the parabolic profile could be observed by naked eyes at the later stage of casting even in the two-dimensional rectangular system.

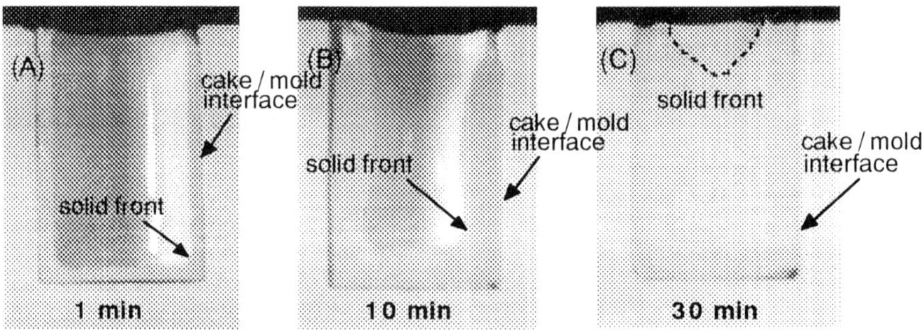

Figure 1. Cake development observed during slip casting in a cylindrical mold. Samples were cut along the symmetrical plane after drying of the mold/cake system.

PROCEDURE FOR FEM ANALYSIS

Fig. 2 shows a generalized slip casting model. Cake is formed by a filtration mechanism where the major driving force is the suction pressure P_{cap} of the plaster mold. The fluid, usually water, flows through the cake and plaster connected in series which have different permeabilities. Fluid velocity v for such a permeant flow is given by the Darcy's law.

Based on the assumption that the cake is incompressible and uniform, the differential equation for fluid pressure P is derived by substitution of the Darcy's equation into the continuity equation as shown in Fig. 3. This pressure equation, the so called Laplace's equation, can be solved easily by FEM under proper boundary conditions. Once the fluid pressure is obtained, the cake growth H_c per unit time will be calculated by combination of filtrate volume V and mass balance equation. Liquid penetration H_m into the dry mold can be calculated quite similarly. Thus, the developments of cake front (the interface between cake and slurry) and liquid front (the front penetrated into the plaster mold) with time can be simulated by repeating this step. Details of derivation of fundamental equations and FEM procedure are described elsewhere[5].

The FEM software named ICGSR (Incompressible Cake Growth Simulation in Rectangular System) was developed by our group according to the concept above mentioned. It can be easily obtained by modification of the ICGSC (Incompressible Cake Growth Simulation in Cylindrical System)[5] from a cylindrical coordinate to a two-dimensional Cartesian one. The in-house software consists of several subroutines which are the input of slurry, mold, and cake parameters, the setting of initial and boundary conditions, the division of analysis region, the solutions of liquid pressure and flow vector, the calculations of cake growth and liquid penetration on the basis of the mass balance equations, the smoothing of the new solid and liquid fronts. The calculation time for one cycle is only 0.1 s on a work station (Model 712/60, Hewlett-Packard Co.). The total calculation time was 5 min. in this work although it depends on the time interval for the next cake growth.

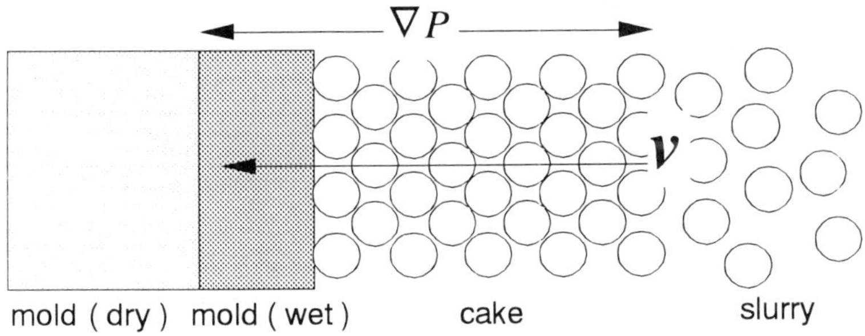

Figure 2. Generalized model for slip casting.

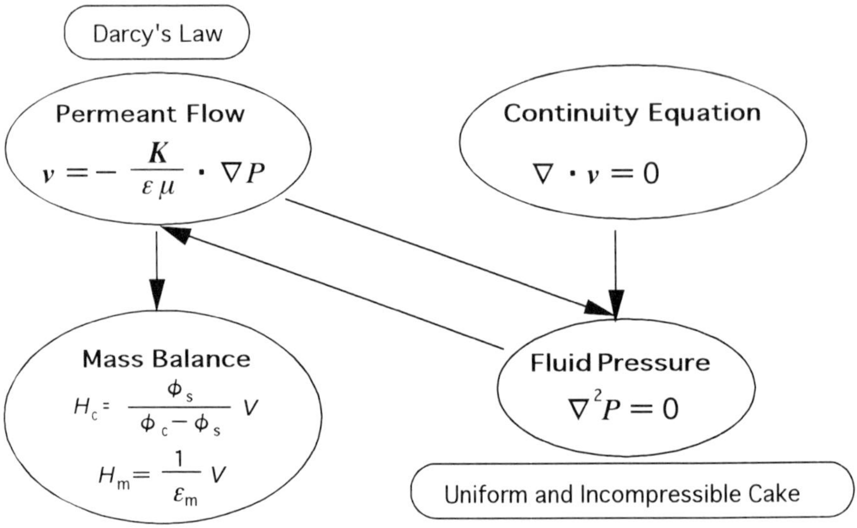

Figure 3. Fundamental equations and procedure for FEM analysis.

RECTANGULAR SLIP CASTING SYSTEM

Fig.4 shows a schematics of two-dimensional slip casting system for a rectangular shape. The boundary conditions are denoted in Fig.5. The pressure at the liquid front is given by the suction pressure of the mold, and the pressure at the cake front is expressed by a gravity force of slurry over it. Another important boundary condition is that the apparent flow vector does not change at the interface between cake and mold. The analysis region is closely divided at the corner as shown in Fig.6. The major parameters[6)] for calculation are summarized in Table 1. The values of parameters are determined by experimental measurements except for the cake permeability which is estimated using the Kozeny-Carman equation.

RESULTS AND DISCUSSION

Fig.7 shows the simulation result for cake and liquid fronts. Both the fronts grew not in one dimension but in two dimensions. The profiles of the cake front changed with time from rectangular to parabolic owing to interaction of filtration into the bottom and side-wall directions. Only numerical method can predict such unique change of cake profile. It should be noted that similar profiles for cake front were obtained experimentally[7)] and numerically[5)] even in a cylindrical system.

Fig.8 shows the filtration rates obtained by numerical and mathematical approaches. The mathematical equation by Adcock & McDowall[3)] can be applied to filtration only when it occurs onto the horizontal or vertical mold surface

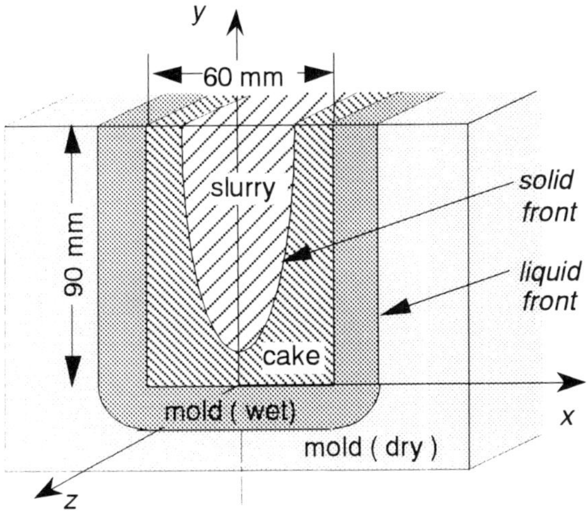

Figure 4. Slip casting in a two-dimensional rectangular system.

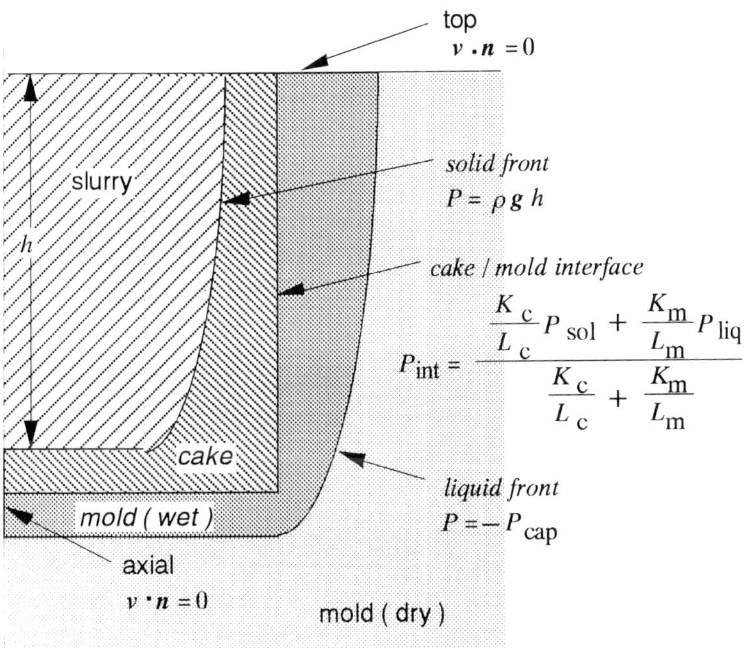

Figure 5. Boundary conditions for FEM analysis.

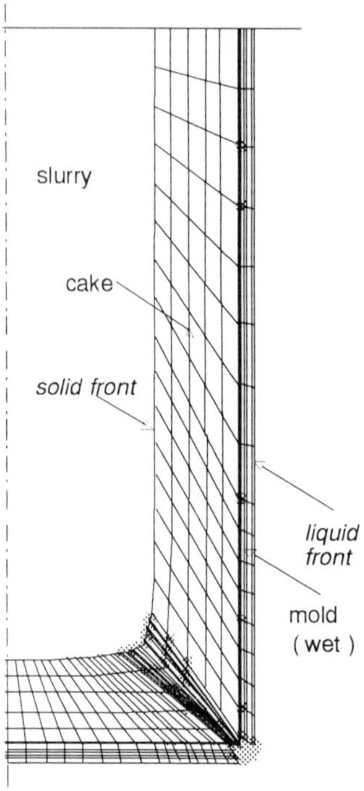

Figure 6. Analyizing region with 480-finite elements.

Table 1. Values of major parameters used for FEM analysis.

Alumina Powder	
Density	3.93×10^3 (kg/m^3)
Surface Area	5.30×10^3 (m^2/kg)
Slurry	
Volume Fraction , f_s	0.588 (-)
Density	2.72×10^3 (kg/m^3)
Water Viscosity, m	1×10^{-3} (Pa • s)
Cake	
Volume Fraction, f_c	0.640 (-)
Permeability , K_c	5.25×10^{-17} (m^2)
Plater Mold	
Suction Pressure, P_{cap}	46.4×10^3 (kPa)
Permeability, K_m	2.39×10^{-14} (m^2)
Porosity, ε_m	0.499 (-)

independently. Simple addition of the two filtrate volumes is greater than the numerically obtained one. Compared to the one-dimensional rectangular system neglecting the interaction of perpendicular flows, the filtration rate becomes gradually slower in the two-dimensional system because the effective surface area for filtration actually decreases with time as shown in Fig. 7.

CONCLUSION

Cake growth and filtration rate during slip casting of a rectangular body was analyzed by FEM. The results showed unique behaviors when the deposit occurs onto both the vertical and horizontal mold surfaces. In this work, the most important assumption is the incompressibility and uniformity of the cake. In order to simulate not the cake profile but also the cake density distribution, further investigation is required for the compressible cake[8].

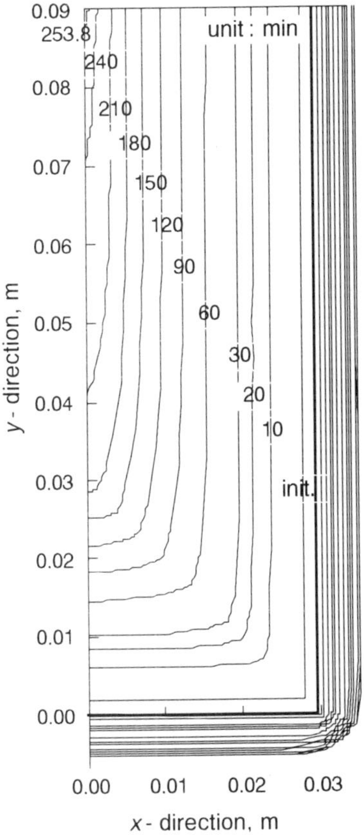

Figure 7. Numerical result for development of solid and liquid fronts.

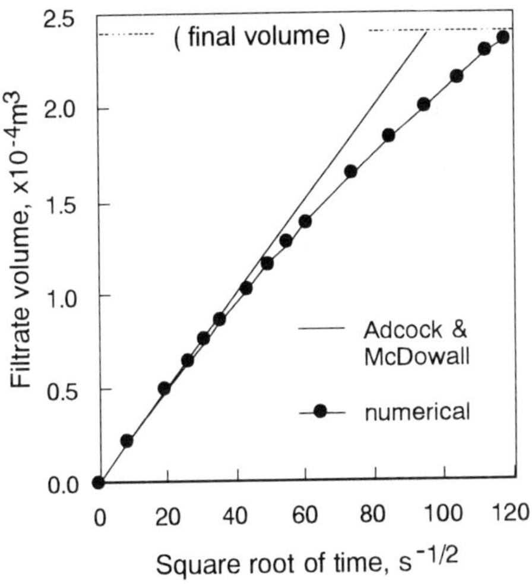

Figure 8. Relation between filtrate volume and square root of time.

REFERENCES

[1]A. C. Young, O. O. Omatete, M. A. Janney, and P. A. Menchhofer, "Gelcasting of Alumina," J. Am. Ceram. Soc., **74**, 612-618 (1991).

[2]T. J. Graule, W. Si, F. H. Baader, and L. J. Gauckler, "Direct Coagulation Casting (DCC): Fundamentals and Forming Process for Ceramics," Ceram. Trans., **51**, 457-461 (1995).

[3]D. S. Adcock and I. C. McDowall, "The Mechanism of Filter Pressing and Slip Casting," J. Am. Ceram. Soc., **40**, 355-362 (1957).

[4]F. M. Tiller and N. B. Hsyung, "Theory of Filtration of Ceramics: II, Slip Casting on Radial Surfaces," J. Am. Cearm., Soc., **74**, 210-218 (1991).

[5]J. Shimizu, M. Takahashi, T. Aragaki, H. Unuma, and M. Ueki, "Finite Element Method Analysis of Three-Dimensional Cake Growth During Slip Casting," J. Ceram. Soc. Japan, **105**, 891-896 (1997).

[6]M. Takahashi, J. Shimizu, H. Unuma, S. Matsubayashi, M. Ueki, and H. Suzuki, "Effect of Characteristics of Plaster Mold on the Slip Casting of High Concentrated Alumina Slurry," J. Ceram. Soc. Japan, **104**, 510-516 (1996).

[7]M. Takahashi, J. Shimizu, H. Unuma, S. Matsubayashi, and M. Ueki, "Experimental Analysis of Three Dimensional Consolidation Behavior in Slip Casting," J. Ceram. Soc. Japan, **103**, 1160-1166 (1995).

[8]M. Takahashi, T. Aragaki, T. Ohashi, and H. Unuma, "Computer Simulation of Cake Growth and Particle Consolidation during Slip Casting," will be published in the Proceedings of the 9th CIMTEC - Word Ceramics Congress & Forum on New Materials, Florence, Italy (1998).

PROCESSING OF NANOSIZED CERAMIC POWDERS - A BIMODAL SLIP CASTING APPROACH

P. Bowen*, O. Charvin, and H. Hofmann
Powder Technology Laboratory, EPFL, MXD, CH 1015 Lausanne, Switzerland.

C. Carry
U.R.A. n°446, ISMa, Bâtiment 410-415, F91405 Orsay cedex, France.

C. Hérard
Maret SA, Rue des Croix 43, 2104 Bôle, Switzerland.

ABSTRACT

The slip casting of nanosized γ-Al_2O_3 powders (primary particle sizes 10-30nm) has been investigated with particular emphasis on the particle size distributions of the various powders. To improve packing and consequently sintering of these aluminas, monomodal and bimodal distributions have been investigated. Both monomodal and bimodal powders produced by attrition milling gave improved sintered densities (97%) when compared with simple isostatic pressing of the as received powders (80% density). Final grain sizes were similar in the sintered samples at around 1μm. The influence of the processing parameters, powder characteristics and green body structure are discussed.

INTRODUCTION

The promise of nanograined materials with new mechanical, optical, magnetic and electrical properties has recently led to a great deal of research activity.[1] One of the main obstacles in realizing this promise has been the difficulty of processing nano-sized powders and producing fully dense components. When trying to achieve full density, grain growth out of the nano-regime (100nm) is often difficult to avoid.[2] Frequently the starting nanopowder is not the thermodynamically stable phase at the processing temperature and there are phase transformations during the densification process which also complicate the issue - a typical example is the gamma alumina investigated in the present study.[3] Nanopowders are invariably made up of agglomerates 10-20 times larger than the primary nanoparticle and one major factor in the densification of nanopowder compacts is to stimulate inter-agglomerate densification and not have a predominance of intra-agglomerate densification leaving large stable pores in the final piece.

Use of bimodal particle size distributions (PSD) to improve packing has been thoroughly studied over many years.[4] The main limitations in achieving better properties through the use of bimodal powders has been in the difficulty of producing a homogeneous green body[5] and that the final properties are linked to the larger rather than the finer particles used.[4] In the current study an attempt has been made to use aggregates of around 1.0 and 0.1 microns. The main difference when compared to previous approaches is that all primary nanoparticles have the same size (10-20 nm) and similar intra-aggregate pore size distributions. By the use of a bimodal particle distribution the with the 0.1μm aggregates filling the large inter-aggregate pores a more homogeneous pore size distribution and particle coordination numbers throughout the whole green body may be created leading to a more homogeneous sintering step.

Slurry processing has shown lots of promise for the production of homogeneous green bodies.[6] Recent work using a bimodal PSD and optimization of slip solid content show that achieving a homogeneous green body using a bi-modal PSD is possible.[7]

This study presents the slip casting of gamma alumina using monomodal and bimodal distributions. The resulting green body densities measured and their pore size distributions characterised with nitrogen adsorption. The samples were then sintered (pressureless) and the microstructures of fracture surfaces compared.

BIMODAL PACKING

The sintering of a powder compact is closely linked to the packing of the powder particles in the green body. One way of enhancing the packing density of randomly packed spheres is to use a bimodal distribution where the small fraction can fit into the interstices of a matrix of the large fraction. For the random packing of bimodal spheres (monodispersed) with a coarse particle diameter (Dc) 10 times that of the fine particle diameter (Df); the packing fraction can be theoretically increased from 0.64 for coarse particles only to 0.82 for a bimodal distribution, with the volume of the coarse particles being 73.5%.[8] Experimentally Taruta et al[7] achieved a packing fraction of 0.78 for a Dc/Df of 10 and coarse fraction of 70% volume. The small discrepancy between these theoretical and experimental packing fractions can attributed to two factors i) the coarse and fine powders are not monodispersed and have size distributions ii) the packing of the bimodal mixture is not perfectly homogeneous. In the present study the use of coarse and fine aggregates of nanoparticles will suffer from both these limitations and a 2D schematic representation of the model spherical bimodal mixture and the nanoaggregates is presented in Fig.1.

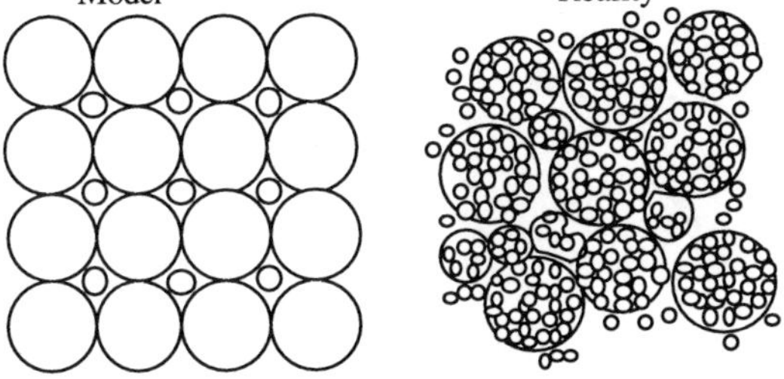

Fig. 1. Schematic representation of bimodal particle packing of spheres and nanoaggregates

EXPERIMENTAL

Samples were slip cast in cylindrical (10mm dia.) rubber moulds. The suspending medium for the slip preparation was a 6wt% polyacrylic acid (PAA mol. wt. 2000) aqueous solution (pH, 6). Modification of the powder particle size distributions was carried by attrition milling for 1 or 3 hrs (PE075 Netzsch, Germany). The milling media were 1.25 mm zirconia beads. After slip casting, the samples were dried in a desiccator with silica-gel for 7 days. The green density was calculated from the weight and cylindrical volume. The PAA was removed by heating in air for 1 hr at 600°C (heating rate 1°/min). The pore size distributions were measured using both nitrogen isotherms (ASAP 2010, Micromeretics,USA, desorption curve - BJH cylindrical model[9]) and mercury porosimetry (Carlo Erba Porosimeter 2000, cylindrical model). Samples were sintered without applied pressure in air at 1600°C (no isotherm hold, heating rate 10°C/min). The sintered ceramic microstructures were observed on gold coated fracture surfaces by scanning electron microscopy (SEM, JEOL 6300, Japan). Dry pressed samples were cold isostatically pressed (CIP) at 250 MPa from the as-received powders and were characterised and sintered in an identical manner to the slip cast samples.

The powders used were high purity γ-Al_2O_3 powders (CR 125, Baikowski , France) with specific surface areas of 105-110 m^2/g. α-Al_2O_3 could not be detected in X-ray powder diffraction of the as received powders or after 3 hrs attrition milling. The transmission electron micrograph (TEM) in Fig. 2 shows some typical aggregates and the primary particle size to be around 10-20nm. Particle size distributions (PSD) were measured using a photocentrifuge (Horiba CAPA700, Japan).

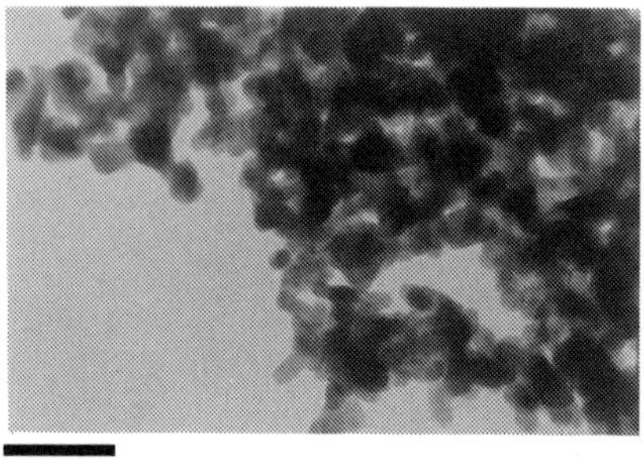

Fig. 2. TEM of γ-Al_2O_3 powders used in this study. Bar = 50 nm

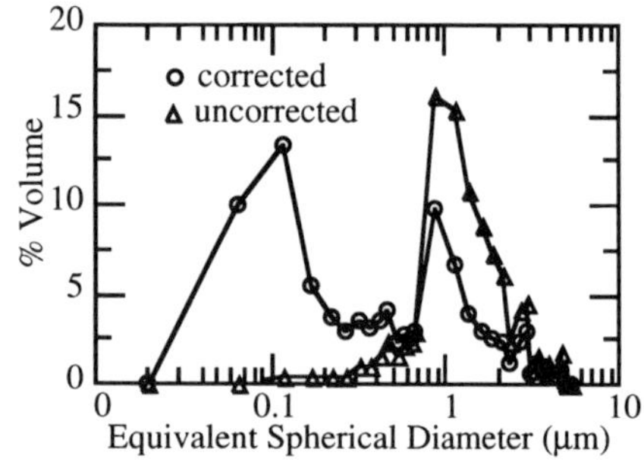

Fig. 3. PSD's for γ-Al_2O_3 (batch 2CR as received) with and without light scattering correction.

For sub micron particles it is essential to correct the photocentrifuge data for light scattering.[10] An example of how important an effect this correction has on photocentrifuge data is shown in Fig. 3 for an as received γ-Al_2O_3 powder. The change from a monomodal distribution centered around 1μm to a bimodal distribution with 47% of the particles less than 0.675 μm is striking.

Table I. Powder and slip characteristics for the γ-Al_2O_3 studied.

Sample	Powder Batch No.	Composition - Treatment	PSD modes	D_c / D_f (μm)	% coarse (>0.65 μm)	% volume solids- slip
B2 - sc	927N	as received	bimodal	1 / 0.12	60	14.8
C1 - sc	927N	milled 1 hr	monomodal	- / 0.07	8	14
D1 - sc	2CR	as received	bimodal	1 / 0.12	40	21.2
D3 - sc	2CR	70% as received 30% milled 3hrs	bimodal	1 / 0.07	25	20.3
ISO2*	927N	as received	bimodal	1 / 0.12	60	-
ISO3*	2CR	as received	bimodal	1 / 0.12	40	-

sc - slip cast, * isostatically dry pressed, D_c coarse and D_f fine fraction modes.

Unfortunately the computation of this correction, the first time it has been applied to nanosized powders[11], were in progress when the compositions for the bimodal distributions were made.

Consequently the ideal bimodal distribution of 70% coarse aggregates with a diameter (1 μm) about 10 times that of the fine aggregates (0.1 μm) has not been investigated. The characteristics of the PSD's, coarse to fine ratios for bimodal powders and sample nomenclature are recorded in Table I.

RESULTS AND DISCUSSION

The PSD's for the various powders and slurries used are shown in Fig. 4 after correction for light scattering. The two as received γ-Al_2O_3 powders already show bimodal distributions (B1, D1). The C1 slurry after 1 hr attrition milling shows a monomodal distribution and most of the coarse aggregates found in the as received powder have been broken. This powder and powders milled for longer times are good model powders for the fine aggregates in bimodal mixture. None of the powders were suitable for a coarse aggregate fraction, all had too many fines. Preliminary attempts to produce a better coarse fraction around 1 μm using centrifugal methods were not successful. A bimodal mixture D3 was prepared and although this mixture had 30% coarse and 70% fine (and not vice versa for the best expected packing), this ratio will however slightly increase the random close packing from 0.64 to about 0.70.[8] One could expect the same improvement from the fact that the powder is not monodispered and has a distribution in particle sizes.[12]

The green body characterization (Table II) showed average pore sizes in the 10-20 nm size range with very similar pore size distribution widths. The smallest pore sizes around 11 nm were seen with the CIP samples and the largest for the samples slip cast from the as received powders. The total pore volumes also followed the same rankings. From the total pore volumes measured from the N_2 isotherms a green density can be calculated and is always slightly lower than the measured green densities (Table II). The detection of pores >100nm by the N_2 isotherm method is poor and the discrepancies between the calculated and measured green densities may be attributed to small amounts of porosity >100nm. Mercury intrusion porosimetry (MIP, which is more sensitive to pores >100nm) measurements were made and showed similar values of porosity >100nm ie between 2 - 5 %. The low percentage of macropores makes the reproducibility of MIP difficult but the overall low value of macro porosity 2-5 % was confirmed. The green densities both measured and calculated also follow the same ranking as above from 45% for the isostatically pressed samples to 34% for the slip cast samples.

The sintered slip cast samples from the as-recieved powders have very low densities. The CIP samples (despite having the best green densities and smallest pore sizes) had lower sintered

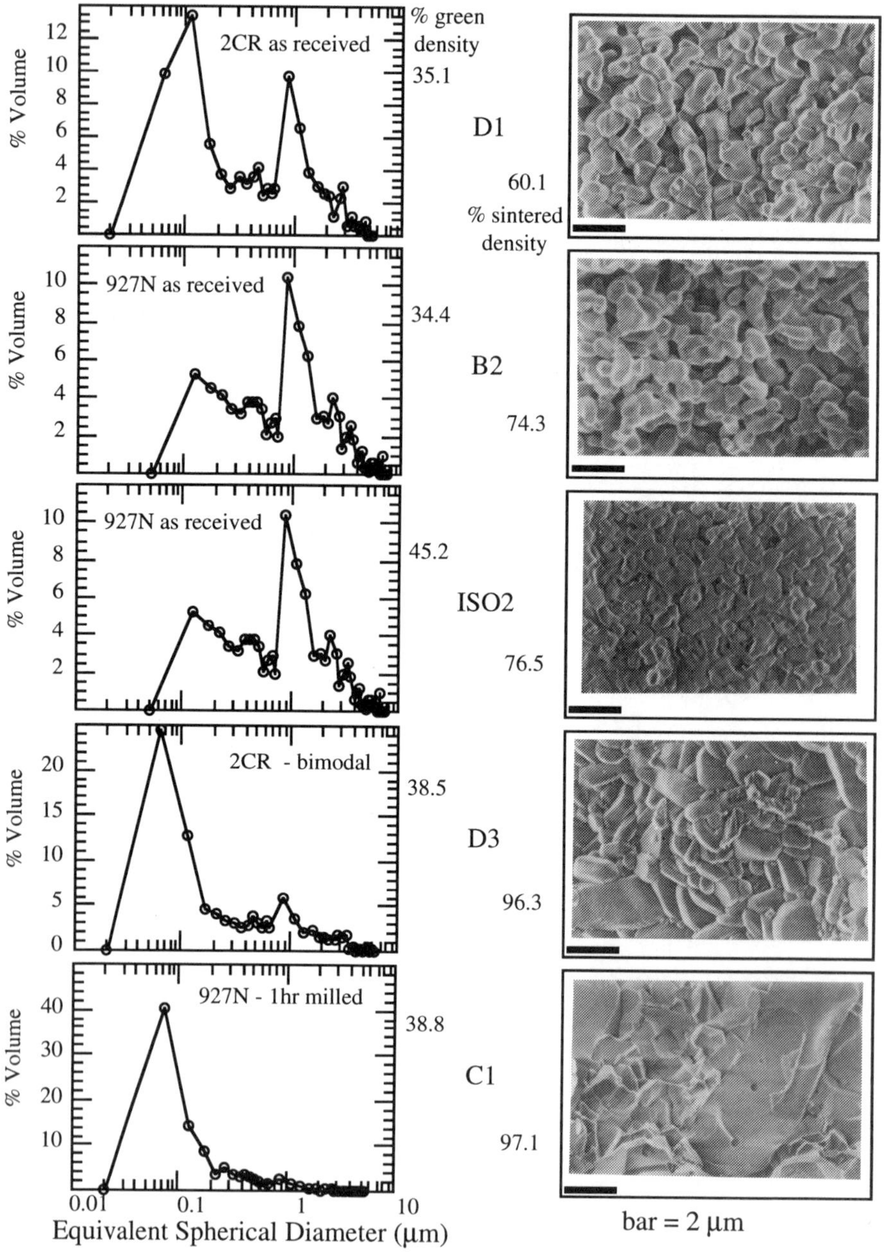

Figure 4. PSD's of cast powders and sintered microstructures (fracture surface).

Table II. Green body and sintered ceramic characteristics.

Sample	D1	B2	ISO2	D3	C1
Average Pore Size (nm) N_2	18.0	22.4	11.8	15.9	15.8
Relative pore size width£	0.65	0.81	0.56	0.78	0.71
Pore Volume (cm^3/g) N_2	0.57	0.70	0.39	0.52	0.51
Calculated Green Density (%)	34.0	29.6	43.1	36.1	36.6
Green Density (% ±0.6)*	35.1	34.4	45.2	38.5	38.8
Sintered Density (%)$	60.1	74.3	76.5	96.3	97.1

All results are ±5%, *density of γ-Al_2O_3 = 3.4 g/cm^3 and $ α-Al_2O_3 = 3.97 g/cm^3, £ average pore size / difference between 10% and 90% pore diameters in the cumulative distribution.

densities (ISO - 90%) than the best slip cast samples (D3, C1, 96-97%). The fracture surfaces of the sintered samples (Fig. 4) show grains of around 1μm. For the samples the slip cast samplesD3 and C1with densities above 90% there are some larger grains observed.

The highest sintered densities were achieved with slip cast samples with γ-Al_2O_3 that had undergone a milling step (C1 and D3). These have intermediate pore sizes but the smallest particle sizes. These facts suggest that is it the size of the aggregates that form the green body that help improve the densification. The grain sizes of the most dense samples seem larger and do suggest some anomalous grain growth. The milling step also has a small effect on the γ-α phase transition temperature (decreases by 10°C). Although no α phase was observed by XRD after 3hrs milling possible effects of seeding of the transformation and consequences on the sintering[13] can-not be ruled out. A more detailed analysis of the sintering of these aluminas has been and will be further presented elsewhere.[14,15]

CONCLUSIONS

Modification of the particle size distribution by attrition milling improves the green and sintered densities of slip cast samples when compared with samples cast from as received powders. The highest sintered density being acheived with the finest powder. The highest green density and smallest pore sizes were observed using isostatic pressing, but the sintered density was lower than the milled slip cast samples (80% cf 97%). Some of this improvement in density seems to have been at the expense of grain size but they remain around 1μm. Further work on reducing the aggregate size to give even fewer large aggregates than in the best dispersion in the current study and the addition of dopants is planned.

The attempt to use a bimodal distribution to improve packing and pore size homogeneity did not prove fruitful - mainly because the coarse powders contained too many fines and agglomerates. The production of the fine fraction by attrition milling was satisfactory but modification of the coarse distributions must be made to follow up the concept.

ACKNOWLEDGMENTS

The authors would like to thank, B. Senior - SEM, Ilford SA at Marly (Switzerland) for TEM, J. Castano ISO and SEM, J-C. Ebinger & L. Felberbaum - slip casting, L. Lemaire and S. Terrazzoni - sintering, and Baikowski for supplying the powders.

REFERENCES

1 R.D. Shull, "Viewpoint Update: Nanocrystalline and Nanophase Materials"NanoStructured Materials, 7 [1/2] 265-268 (1996).

2 J.R. Groza and R.J. Dowding, "Nanoparticulate Materials Densification," NanoStr. Mater. 7(7) 749-68 (1996).

3 J.M. McHale, A. Auroux, A.J. Perrotta and A. Navrotsky, "Surface Energies and Thermodynamic Phase Stability in Nanocrystalline Aluminas" Science, 277[5327] (1997).

4 R.M. German "Sintering Theory and Practice" Chp.5, J.Wiley, N.Y. (1996).

5 J.P. Smith and G.L. Messing "Sintering of Bimodally Distributed Alumina Powders" J.Amer.Ceram.Soc., 67[4] 238-42, (1984).

6 F.F. Lange, "Powder Processing Science and Technology for Increased Reliability", J.Am.Ceram.Soc., **72**(1) 3-15 (1989).

7 S. Taruta, N. Takusagawa, K. Okada and N. Otsuka,"Slip Casting of Alumina Powder Mixtures with Bimodal Size Distributions," J.Ceram.Soc.Jpn., 104(5) p447 (1996).

8 F. Zok and F.F. Lange, "Packing Density of Composite Powder Mixtures" J.Am.Ceram.Soc., 74[8] 1880-85 (1991).

9 S.J. Gregg and K.S.W. Sing " Adsorption, Surface Area and Porosity," 2nd Edition Academic Press, London, 1982.

10 P. Bowen, J. A. Dirksen, R. Humphry-Baker and L. Jelinek,"An approach to improve the accuracy of sub-micron particle size distribution measurement using the Horiba CAPA-700" Powder Technol., 74 67-71 (1993).

11 M. Staiger, P. Bowen, S.M. Sholz and R. Humphry-Baker "PSD Measurement and Assessment of Agglomeration of Sub-micron and Nanosized Particles", Proceedings of "2ieme Colloque, Science et Technologie des Poudres", 3-5 March, Lyon 1998.

12 G.T. Nolan and P.E. Kavanagh,"The size distribution of interstices in random packing of spheres," Powder Technology, 78 p231 (1994).

13 M. Kumagai And G.L. Messing, "Enhanced Densification of Boehmite Sol-Gels by α-Alumina Seeding," J.Am.Ceram.Soc., C-230-31 (1984).

14 P. Bowen, C. Carry, H. Hofmann, C. Legros,"Phase Transformation and Sintering of γ-Al_2O_3 - Effects of Powder Characteristics and Dopants (Mg or Y)" p.904-907, Euroceramics V, Versailles, Trans Tech Publications, Switzerland, 1997.

15 P. Bowen, C. Carry, H. Hofmann, C. Legros, to be submitted to J. Eur.Ceram.Soc.

FABRICATION OF PZT THICK FILM THROUGH CENTRIFUGAL SEDIMENTATION METHOD

Hirofumi Yamaguchi and Makoto Itoh,
NGK Insulators Ltd.
2-56 Suda-cho Mizuho-ku Nagoya 467-8530, Japan

Sin-ichi Hirano
Dept. of Applied Chem., Nagoya Univ.
Furo-cho Chikusa-ku Nagoya 464-8603, Japan

ABSTRACT

In recent years, a demand to fabricate thick films (approximately from 2 to 10 μm thick) has been increasing for an application of functional ceramics to such devices as an actuator, a chip-condenser and so on. Up to now, an appropriate method to form films with the above-mentioned thickness on a substrate has scarcely been developed.

In this work, a centrifugal sedimentation method was applied to form about 10 μm thick films of PZT at one step. Influencing factors in preparing suspension are investigated in order to form dense films, which are free from cracks. The addition of proper amount of both dispersing agent of sorbitan fatty acid and PVB binder was found to be a key for fabrication of dense films. The influence of binder content upon densification is discussed in terms of the viscosity behavior of suspension. The model for film formation process in centrifugal sedimentation is also proposed.

INTRODUCTION

For the electronic devices such as actuator and chip-condenser using functional ceramics, demands for fabricating 2 to 10 μm thick ceramics have been raised because of the miniaturization and improvement in performance. Self-sustained films, i. e. films formed without substrate, of about 10 μm in thickness can be produced by the modified doctor blade method. Question is still remained on whether the method is commercially applicable to form 2 to 10 μm thick film on substrate at one step. In Table 1, current techniques used for fabricating thick film on substrate are listed for comparison from viewpoints of possible thickness, productivity and cost. The screen printing method can not give films of less than 10 μm thick, although it has the highest productivity. The electrophoresis method[(1),(2)] can fabricate relatively thinner film, where a substrate has to be electrically conductive. A sol-gel dip-coating method provides thin films, by which many repetition of dipping-firing process is required to reach several μm thickness[(3)]. A dip-coating method using slurry affords a film of thicker than hundreds μm per one dipping. Therefore, a method for fabricating 2 to 10 μm thick film on a desired substrate has scarcely been developed.

Currently, a centrifugal phenomenon is widely applied as a method for solid / liquid separation in many fields. Though this technique is also utilized to characterize features of powders, slurry and so on, there are few works which have been concerned with fabricating thick films. As shown schematically in Fig. 1, a centrifugal method is considered to be adaptable to fabricate a thick film with desired thickness on a substrate, where a mixture of powder and solvent (i.e. suspension) centrifuges to form a thick film on a substrate placed at the bottom of container, followed by firing.

This work was carried out to examine the applicability of the centrifugal sedimentation method which may afford 2 to 10 μm thick films at one step. In this method, the control of the degree of dispersion in preparing the suspension is an essential key factor in order to prepare dense film. The purpose of present investigation is to clarify the influencing factors in preparing the suspension on the densification in a piezoelectric PZT-system as a subject material.

EXPERIMENTAL PROCEDURE

Powder Preparation

PZT powders used in this work were prepared by the conventional solid-state-reaction method. Weighed quantities of commercial-grade PbO, $SrCO_3$, ZrO_2, TiO_2, MgO, Nb_2O_5 with a composition of $(Pb_{0.95}Sr_{0.05})(Zr_{0.25}Ti_{0.375}(Mg_{1/3}Nb_{2/3})_{0.375})O_4$ were mixed by ball-milling, followed by being calcination and grinding. SEM photograph of powders is shown in Fig. 2. An average grain size and specific surface area were 0.4 μm and 7.4 m^2/g, respectively.

Preparation of Suspension and Film Forming

A procedure for the preparation of suspension to form film is shown in Fig. 3. The concentration of the PZT powders was determined to 0.6g in 100cm^3 solvent to adjust the thickness to 10 μm. A mixture of 75vol% xylene and 25vol% methylethylketone (MEK) was used as a solvent for suspension. Sorbitan monoleate (Span 80®) and polyvinylbutyral (PVB, type BLS, Sekisuikasei, JAPAN) were selected as a dispersing agent and binder, respectively. In this work, the effect of the dispersing agent on densification of formed films was investigated, considering the study on the effect of the addition of both organic substances. After mixing PZT powder (0.6g), solvent ($100cm^3$) and organic agent(s) (a few wt % to powder) using a ultrasonic homogenizer, the suspension was magnetically stirred for 5 hours at 30°C, followed by centrifugation. A thick film of PZT was formed on Pt foil placed on flat bottom of a Teflon container. The conditions of centrifugation were as follows; revolution: 2500rpm(1200G), duration: 20min. The film formed on Pt foil was fired in a closed MgO crucible at 1200°C for 3h after removing organic substances by calcining at 500°C in air.

The microstructures of the as-prepared and fired film were observed by scanning electron microscopy (SEM). Viscosity behavior was evaluated for the suspensions with various contents of binder. A small amount of suspension was frozen using liquid N_2 and then dried in vacuum in order to observe the structure of flocculated particles in suspension by SEM.

RESULTS and DISCUSSION

Effect of Addition of Dispersing Agent on Microstructure

SEM photographs of as-prepared films formed with various contents of dispersing agent; (a) no addition, (b) 0.5wt% to powder, (c) 1.0wt%, (d)5.0wt%, are shown in Fig. 4. A rough surface and non-uniform packing of particles were observed in the film without dispersing agent. The microstructure was remarkably improved with an increase of the content of dispersing agent. Films formed through the suspension with 1wt% and 5wt% contents had improved microstructures than that with 0.5wt%. The absorbed amounts of dispersing agent to powders measured by drying the supernatant after centrifugation were 8×10^6 mol/g and 11×10^6 mol/g for 0.1wt% and 1wt% content, respectively. The improvement of microstructure was considered to depend strongly on adsorption amount of dispersing agent. The main features of the films with 1wt% and 5wt% contents were; (1)flocculated particles, whose sizes were about 5 μm, (2)small void among flocculated particles. This structure reveals that the degree of dispersion might be insufficient.

Fig. 5 showed the microstructure of the film formed from the suspension with 5wt% content of dispersing agent. A dense film was not obtained due to the formation of many voids. This result is considered to be caused by the poor dispersion of particles in suspension which was prepared by the sole addition of dispersing agent as mentioned above.

Effect of Addition of PVB Binder on Microstructure

Both PVB binder and dispersing agent were added to improve a degree of dispersion because the added PVB binder bonding on the hydrophobic side of dispersing agent was considered to cause the steric hindrance[4]. Fig. 6 shows the microstructure of as-prepared film surfaces prepared through suspensions which contained 5-20wt% binder at 5wt% content of dispersing agent. The shape and the distribution of voids seem not to be influenced by the addition of binders less than 5wt%. Significantly dense packing structures of particles were realized by 10wt% and 15wt% contents of the binder. Conversely, porous structure was obtained through 20wt% content. Fired film surfaces through these four binder contents are shown in Fig. 7. Dense films were obtained through the suspensions from 10wt% and 15wt% binder contents in which as-prepared film had dense packing structure. In order to form a dense film by the centrifugal sedimentation method, it was found to be necessary that the degree of dispersion in suspension should be controlled by proper amount addition of both dispersing agent and binder.

Influence of PVB Binder on Suspensions

An apparent viscosity was found to change with a shear rate for suspensions including 5-20wt% content of binder as shown in Fig. 8. The apparent viscosity showed a maximum as a function of a shear rate for all suspensions, which resembles thixotropy. This result suggests that the particles in suspension were not well-dispersed but flocculated at some degree, and the flocculated particles may deform at a higher level of shear rate. The apparent viscosity increased with an increase of binder content at the shear rate range up to $0.3 \times 10^4\ s^{-1}$. On the other hand, the apparent viscosity increases in order of 10, 15, 5, 20wt% of the binder content above $0.7 \times 10^4\ s^{-1}$ shear rate, reflecting that the easiness of deformation of soft flocculates depends on the content of binder. Although the shear force generated during the centrifugal treatment is not clear, the relationship between the densification and the binder content is considered to depend upon the degree of deformation adjustable by the binder content.

Fig. 9 shows a SEM photograph of particles which were observed after freeze-drying of suspension with 15wt% binder content. This photograph demonstrates a structure of particles in suspension which is not well-dispersed but flocculated into about 1-2 μm in size as discussed above. The flocculated particles in suspension are considered to be deformed plastically to form dense film after deposited on substrates by centrifugal force.

Consequently, the process for a film formation in the centrifugal sedimentation would be proposed as following; (1)the formation of flocculated particles in suspension, (2)the plastic deformation of flocculated particles on substrate and (3)the bonding between the deformed particles to form dense film. In each step, the binder is considered to play important roles.

SUMMARY

The centrifugal sedimentation method was successfully applied as a simple method to form about 2 to 10 μm thick films. The main results obtained are shown below:

1. The centrifugal method is applicable to form dense and 10 μm thick film of PZT by one step operation.
2. The addition of proper amount of both dispersing agent and binder was found to be the key in order to form a dense film.
3. The particles have to be flocculated in suspension, which should deform plastically on substrate by centrifugal force to form films.
4. The binder was supposed to influence the formation of flocculated particles, its plasticity, and the bonding between deformed particles.

REFERRENCES

1 H. Nojima, M. Nagata, H. Shintaku, and M. Koba, "Fabrication of Y-Ba-Cu-O Superconducting Films on Cu Substrates by an Electrophoretic Deposition Technique," Jpn. J. Appl. Phys., 29[9] L1655-58 (1990)

2 T. Ishihara, K. Sato, and Y. Takita, "Electrophoretic Deposition of Y2O3-Stabilized ZrO2 Electrolyte Films in Solid Oxide Fuel Cells," J. Am. Ceram. Soc., 79[4] 913-19 (1996)

3 T. Yogo, Y. Takeichi, K. Kikuta, and S. Hirano, "Ultraviolet Patterning of Alkoxy-Derived Lithium Niobate Film," J. Am. Ceram. Soc., 78[6] 1649-52 (1995)

4 W. M. Sigmund, G. Wegner, and F. Aldinger, "Interaction of Organic Additives with Alumina Surfaces in a Ceramic Slurry," Mat. Res. Symp. Proc., vol. 407 313-18

Table 1 Comparison among current techniques for thick film forming.

	Possible Thickness (μm)	Productivity	Cost
Screen Printing	≧15	◎	◎
Dip Coating (sol-gel)	≦2	×	×
Dip Coating (Slurry)	≧100	○	○
Spin Coating	0.01～0.1	○	○
Electrophoresis	20～200	×	×

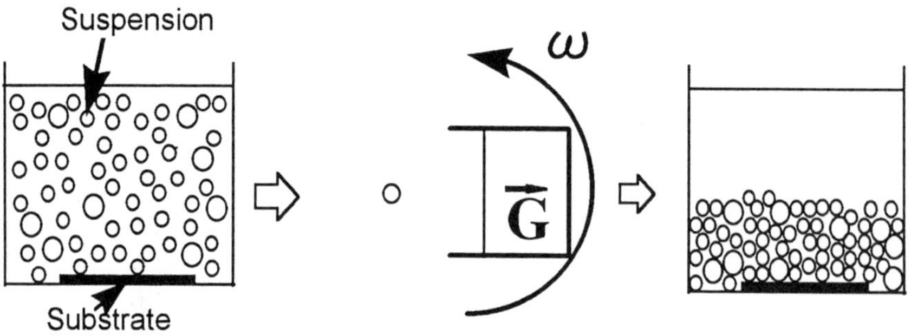

Fig. 1 A centrifugal sedimentation method for fabricating a thick film.

Fig. 2 SEM photograph of PZT-system powder.

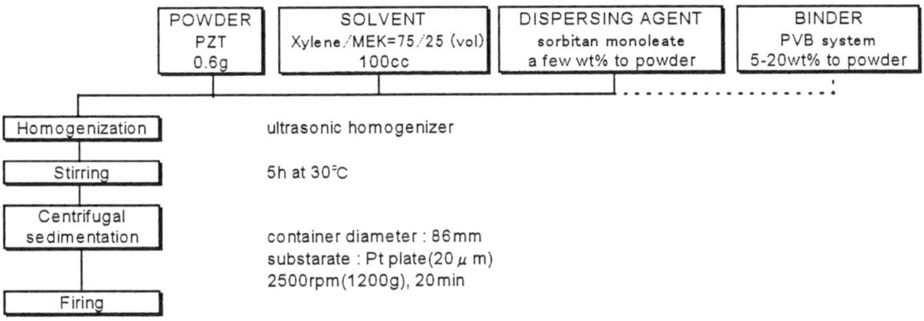

Fig. 3 Flow chart for preparation of suspension and film forming.

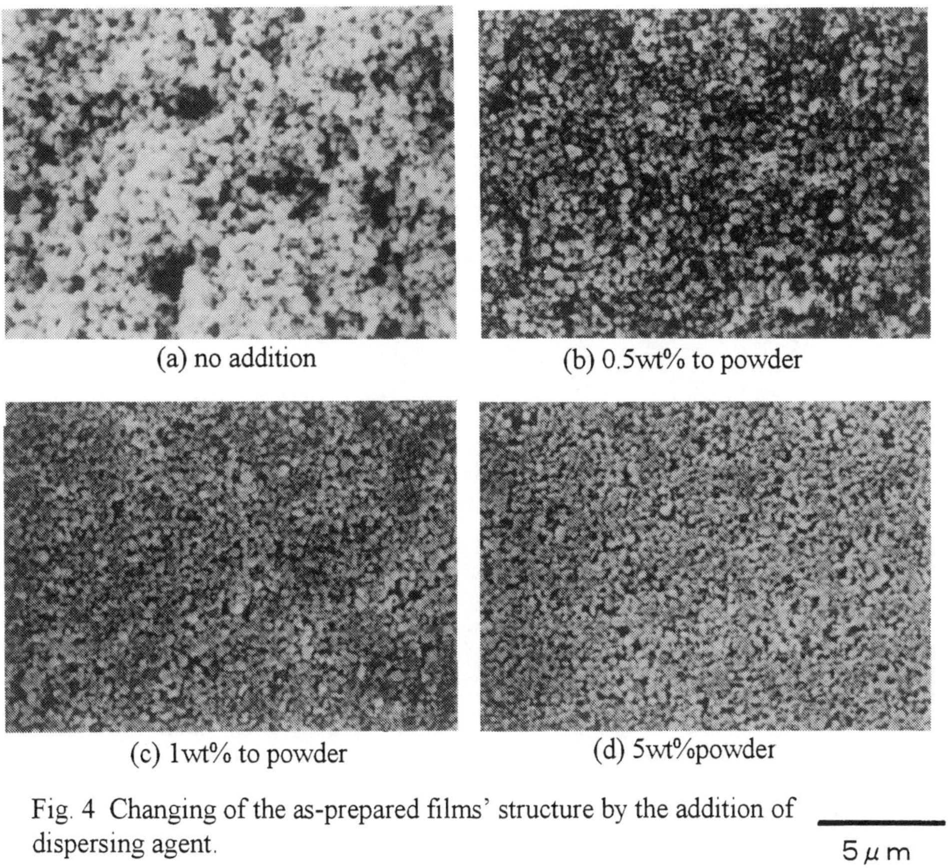

Fig. 4 Changing of the as-prepared films' structure by the addition of dispersing agent.

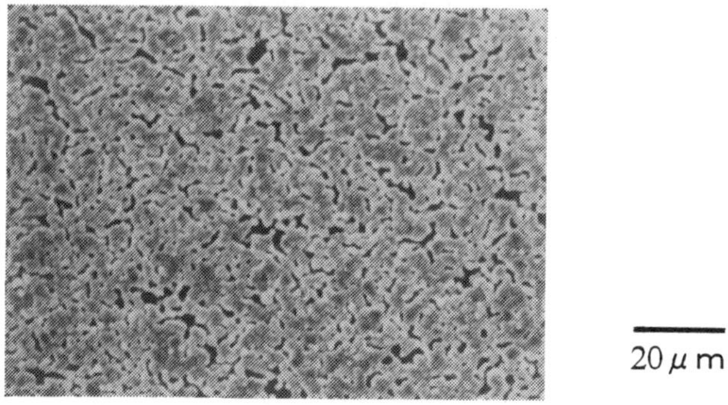

Fig. 5 Microstructure of fired film formed through addition of 5wt% dispersing agent.

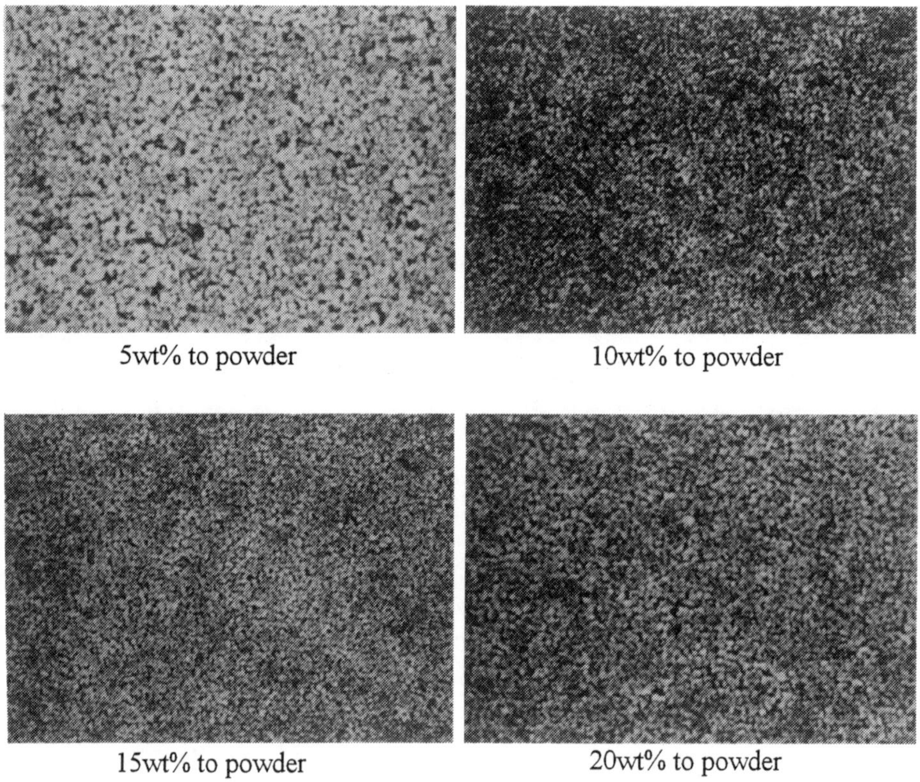

Fig. 6 Effect of the binder addition on the microstructure of as-prepared film.

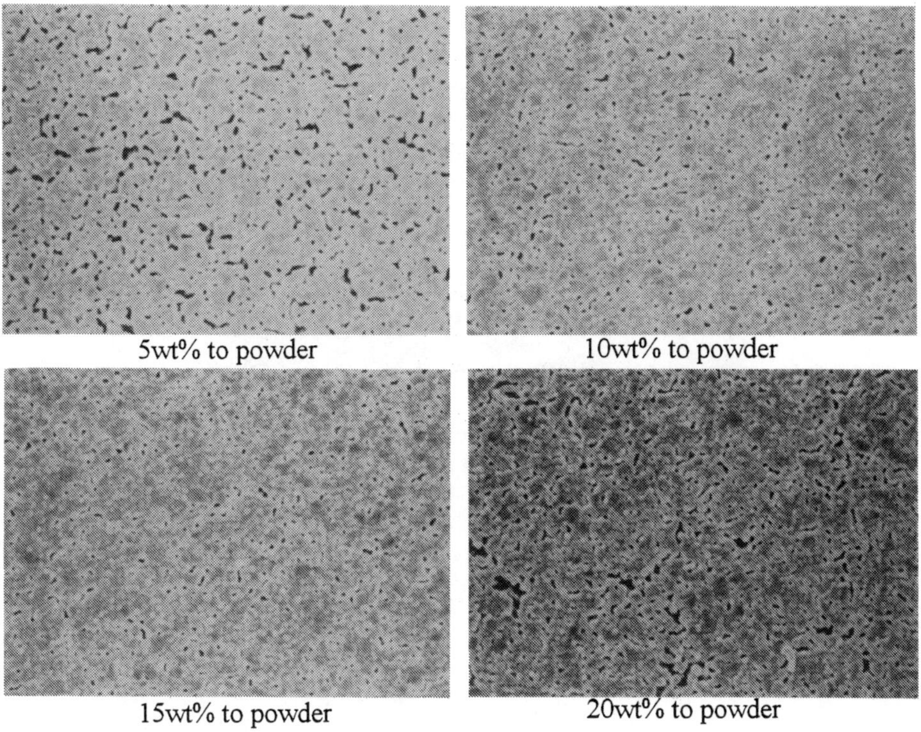

Fig. 7 Effect of the binder addition on the microstructure of fired film. 20 μm

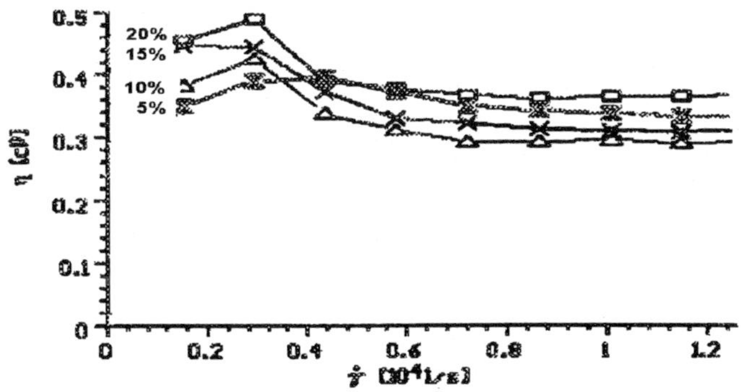

Fig. 8 Relationship between apparent viscosity and shear rate for suspensions with various contents of binder.

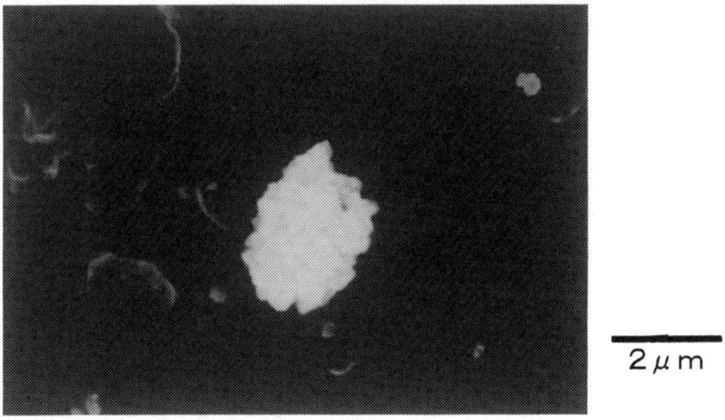

Fig. 9 SEM photograph of flocculated particles which were obtained by freeze-drying of suspension with 15wt% binder added.

CENTRIFUGAL FORMING OF PORE-GRADIENT MEMBRANES

Frank Müller, Chu-Wan Hong, and Peter Greil
University of Erlangen-Nurnberg
Department of Materials Science
Erlangen, Germany

ABSTRACT

Ceramic membranes gain in significance due to their excellent chemical and microbiological resistance, catalytic capability and mechanical stability up to high temperatures. Flat membranes for ultrafiltration (pore sizes ranging from 1 to 100 nm) were formed by a centrifugal casting process using multimodal suspensions of alumina powder. SEM and mercury-porosimetry were used to characterize the pore gradient structure. A gradient structure without interfaces was formed where the average pore size changes from 250 nm to 50 nm over a thickness of 5 mm. The effect of solid content and colloidal stabilization of the aqueous suspensions on the pore size distribution was examined. The experimental results are in good agreement with simulation of the pore formation process by discrete element method (DEM).

PRESSURE SLIP CASTING OF SiC AQUEOUS SLIPS

Ma. I. Nieto, A. Salomoni*, I. Stamenkovic,* and R. Moreno
Instituto de Cerámica y Vidrio
C.S.I.C. Crta. Valencia Dm. 24,300
28500 Arganda del Rey
Madrid, Spain

**Centro Ceramico Via Martelli, 26
40138 Bologna, Italy

ABSTRACT

A systematic study of the rheological behavior of aqueous -SiC slips is made in order to determine the best stabilizing conditions for slip casting operations. The effect of sintering aids (boron carbide and carbon black) on the rheology is studied. The dispersing state of the slip is improved by simultaneous addition of two stabilizing additives, one acting as deflocculant and the other allowing a complete wetting of the particles. Under these conditions, the viscosity of SiC slips containing concentrations of 2, 3 and 4 wt% C is nearly the same. Uniaxial pressure casting experiments are performed using rheologically optimized slips for applied pressures of 1.7, 3.4, 6.8 and 10.2 MPa. The pressure casting kinetics is determined taking into account the mass balance. The effect of the applied pressure is related to the casting parameters and to the characteristics of the green casts, such as density, wall thickness, drying shrinkage, etc. The obtained results demonstrate that when the slip rheology and the operating conditions are controlled, all these parameters are a linear function of the applied pressure.

PREPARATION OF FINE-GRAINED ZIRCONIA SYSTEMS BY COLLOIDAL PROCESSING

Y. Sakka, T. Uchikoshi, K. Ozawa, T. S. Suzuki and K. Hiraga
National Research Institute for Metals,
1-2-1, Sengen, Tsukuba, Ibaraki 305-0047, Japan

ABSTRACT

Slurries of monoclinic zirconia and CuO adsorbed tetragonal zirconia (3mol% Y_2O_3 doped tetragonal ZrO_2: 3Y-TZ) were prepared. Their dense compacts were obtained by pressure filtration follwed by CIP treatment. The compacts were densified by pressureless sintering in air to > 98%. Dense bodies of the undoped monoclinic zirconia and CuO doped tetragonal 3Y-TZ with average grain size under 100 nm are obtained.

INTRODUCTION

Polycrystalline zirconia polycrystal is one of the most promising materials for structural and functional applications. Pure zirconia transforms from monoclinic to tetragonal and cubic at high temperatures.[1] The transformation between monoclinic (*m*) and tetragonal (*t*) is accompanied by a large volume change, ca. 4.6%, and therefore cracks form during cooling. This problem makes it very difficult to sinter monoclinic zirconia powder to high densities, since the temperatures necessary for sintering of zirconia have always been well above the *m-t* transformation temperature, *i.e.*, above 1170 °C. To overcome this problem, dopants, such as Y_2O_3, CaO and MgO, have been commonly dissolved to stabilize the high temperature phases. Moreover, most of the previous studies on densification of undoped monoclinic zirconia were performed by hot pressing[3] or hydrothermal pressing[4].

To make dense monoclinic zirconia ceramics at lower temperatures than *m-t* transformation temperature without hot pressing, selection of raw material and processing method is very important. Required properties of the starting powder are high purity, fine particle size and no flocculation. Dense particle packing is also very important to reduce the densification temperature.

It is well known that yttria-doped tetragonal zirconia (Y-TZ) with fine-grained microstructure shows superplasticity.[4,5] Y-TZ with a grain size of ≈0.3 μm has been prepared by sintering chemically synthesized powder. However, to obtain a dense Y-TZ with grain size less than 0.1 μm, a special procedure is necessary. Kimura et al.[6] reported that liquid-phase sintering of Y-TZ with CuO is effective for densification at low-temperatures. Hwang and Chen[7] reported that the addition of CuO to Y-TZ can enhance superplasticity. Their powder processing method involves milling Y-TZ powder with oxide or acetate of Cu. Recently, Seidensticker *et al.*[8] demonstrated a new composite powder preparation involving adsorption of Cu ions on Y-TZ powder as a method of coating the surface. Homogeneous distribution of CuO in Y-TZ bodies can be obtained by this method, but during drying procedure of the coated Y-TZ powders the fine Y-TZ powders tend to agglomerate, resulting in large pore size distribution during compaction.[9,10]

Colloidal processing is a powerful method for controlling the density and microstructure of a consolidated compact.[9-12] The particle packing is affected by an interparticle force which can be changed from attractive to repulsive by adjusting the pH of aqueous suspensions. High repulsive potential among particles contributes to the dense particle packing. When the particle size is smaller than *ca.* 0.1 μm, however, re-dispersion of once flocculated particles becomes impossible even by either changing the pH or adding a dispersant. Therefore, it is necessary to prepare a well-dispersed suspension by wet processing and to consolidate the as-prepared suspension without powderization. This study is an approach to process the fine-grained (1) monoclinic zirconia polycrystal using monoclinic zirconia sol prepared by wet processing and (2) CuO-doped tetragonal zirconia using slurries of Cu adsorbed Y-TZ powers directly by a pressure filtration method.

EXPERIMENTAL AND RESULTS

(1) MONOCLINIC ZIRCONIA

Monoclinic zirconia sol prepared by hydrolyzing of zirconium oxychloride octahydrate at Sumitomo-Osaka Cement Co., Ltd. was used as the starting suspension. This preparation is accompanied by the production of hydrochloric acid, as follows:

$$ZrCl_2O \cdot 8H_2O \rightarrow ZrO_2 + 2HCl + 7H_2O$$

Since zirconia particles show a large positive zeta potential in strong acids, as-prepared fine zirconia particles are immediately dispersed into a strong hydrochloric acid solution and result in a well-dispersed aqueous zirconia suspension. The solid content of the sol was 2 vol% and the pH was 1.7. The morphology of as-prepared zirconia particles was spindle-shaped and average particle size was 80 nm by TEM observation. HREM lattice images of particles by showed good crystallinity. The single monoclinic phase was confirmed for the dried powder of the sol by ED and XRD. The *m-t* transformation temperature of the powder was confirmedby thermal analysis to be 1170°C.

It should be noted that the zirconia sol prepared by this method inevitably involves chloride ions, considered to inhibit the sintering of zirconia.[13] It is essential to eliminate chloride ions without changing the dispersion of the suspension. We applied an ion exchange method using anion exchange resin (Mitsubishi Chem. Corp., CA08P) for removal of the chloride ions from the suspension. By this method, most of the chloride ions were removed in an acid state (pH=3.6), which indicates that the particles retain a high positive zeta-potential.

Figure 1 shows the particle size distribution change before and after the removal of chloride ions from the suspension. Particle flocculation during the removal of chloride ions is small and most of the particles show good dispersion. Dechlorinated and concentrated (up to 10 vol %) zirconia sol was consolidated by pressure filtration at 10 MPa [14]. Before consolidation, the suspension was evacuated in a vacuum desiccator to eliminate air bubbles. CIP treatment at 400 MPa was conducted to improve the packing density to the relative density of 49%.

The compact was sintered pressurelessly at 1000 or 1100 °C in air. The relative density change versus sintering time is shown in Fig. 2. The compact is densified by pressureless sintering to the density of 5.72 g/cm^3 (98.2 % of theoretical

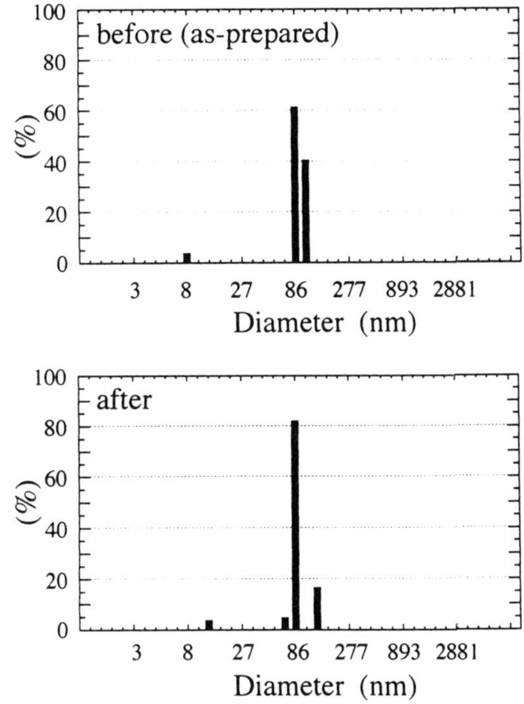

Fig. 1. Particle size distribution change before and after the removal of chloride ions.

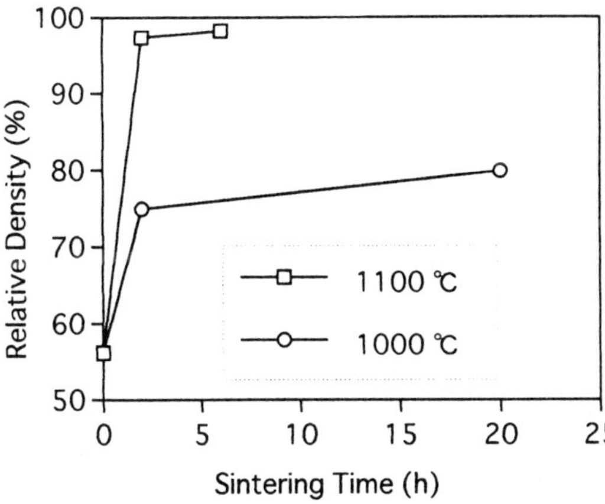

Fig. 2. Relative density as a function of sintering time at 1000 and 1100 °C.

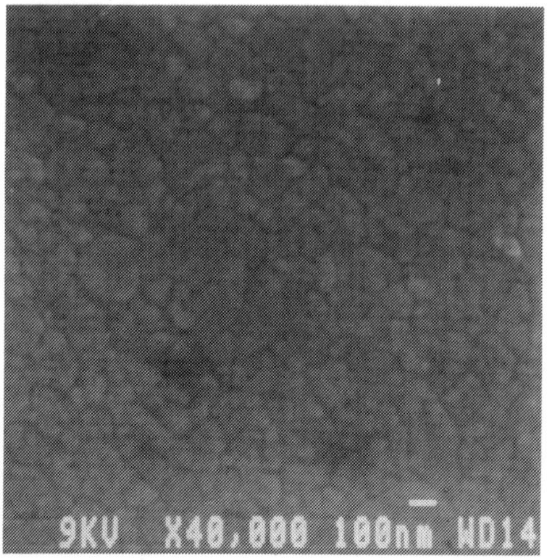

Fig. 3. Microstructure of the monoclinic zirconia ceramics sintered in air at 1100 °C for 6h.

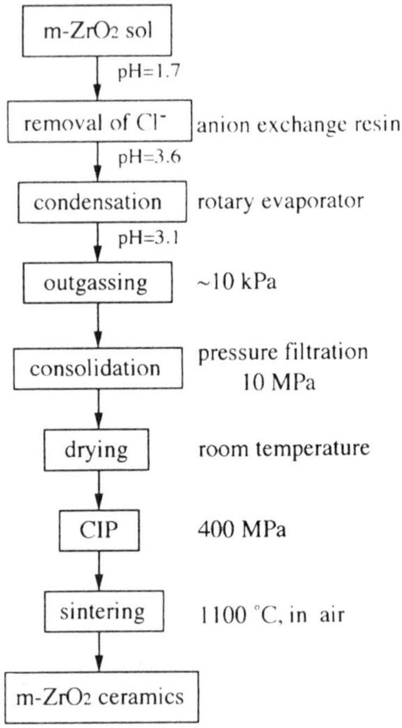

Fig. 4. Flow diagram for the synthesis of monoclinic zirconia ceramics.

density) in air at 1100 °C for 6 h. The microstructure of the sintered body observed by SEM using polished and thermally etched surfaces is shown in Fig. 3. The average grain size of the sintered monoclinic zirconia ceramics determined by the linear intercept method[15] was 92 nm. Figure 4 summarizes experimental procedures for synthesis of monoclinic zirconia ceramics. This simple and easy method is superior to the previous methods in the mass production of dense monoclinic zirconia ceramics.

(2) COPPER DOPED TETRAGONAL ZIRCONIA

The slurries of the Cu adsorbed Y-TZ powder were prepared as follows referring to the report of Seidensticker *et al.*[8] 3Y-TZ powder (Tosoh Corporation: grade 3Y-TZ) was dispersed with the aid of ultrasonic vibration and magnetic stirring in an aqueous solution of ammonium hydroxide and ammonium nitrate keeping the total $[NH_3+NH_4^+]$ equal to 0.1M at room temperature. Cu^{2+} was added to solutions by pipetting up to 10 mL of a 2.2 x 10^{-3}M solution of $Cu(NO_3)_2$. The initial pH was adjusted to 11. Since the zeta potential of the 3Y-TZ at pH=11 was -43.5 mV, which was determined by the Smoluchowski equation [16] from the electrophoretic mobility, dispersed slurry was obtained. The Cu^{2+} added slurry was stirred for 5 h. Slurry concentrated (up to 15 vol% solids) using a rotary evaporator was consolidated by pressure filtration (PF) [14] at 10 MPa.

The experimental CuO mol% adsorbed on 3Y-TZ powder was fixed as a function of initial concentration of Cu^{2+} solution because the Cu^{2+} adsorption on 3Y-TZ powder obeys the Langmuir type adsorption isotherm. Cu contents of compacts, determined by chemical analysis, were nearly equal to previous results.[8] Here, Cu is originally adsorbed onto 3Y-TZ powder as a coordination complex and CuO forms during drying.[17]

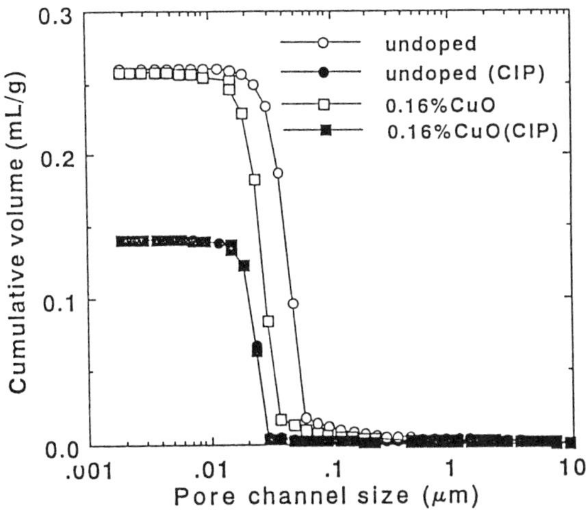

Fig.5 Pore channel size distributions of the pressure filtrated and CIPed compacts.

The green densities of undoped and 0.16 mol% CuO-doped 3Y-TZ after PF were 38.5 and 37.7 % of the theoretical, respectively. CIP treatment at 400 MPa was conducted to improve packing density. The densities of the undoped and 0.16 mol% doped 3Y-TZ increased up to 51.9 and 51.8 %, respectively. Figure 5 shows pore channel size distributions of PF and CIP treated compacts, which were determined by mercury porosimetry using standard values for mercury surface energy (0.48 N/m) and the contact angle (140°). A green compacts with a narrow pore channel size distribution with small pores could be obtained by PF and CIP treatment.

Compacts after PF and CIP treatment were sintered at fixed temperatures for 2h in air. Figure 6 shows sintering diagram of the two compacts. It is seen that CuO-doped compact densifies at lower temperatures. In CuO-doped 3Y-TZ, the following reaction has been reported[17] : CuO is transformed to CuO at 1031°C and consequently melts at 1134 °C in air. An endothermic reaction takes place around 1200 °C , which may be a reaction between CuO and 3Y-TZ. This phase is initially $Y_2Cu_4O_5$, but after extended holding at high temperatures, equilibrates to $Y_2Cu_2O_5$ composition that has a melting point of 1110 °C . Figure 7 shows SEM micrographs of polished and thermally etched

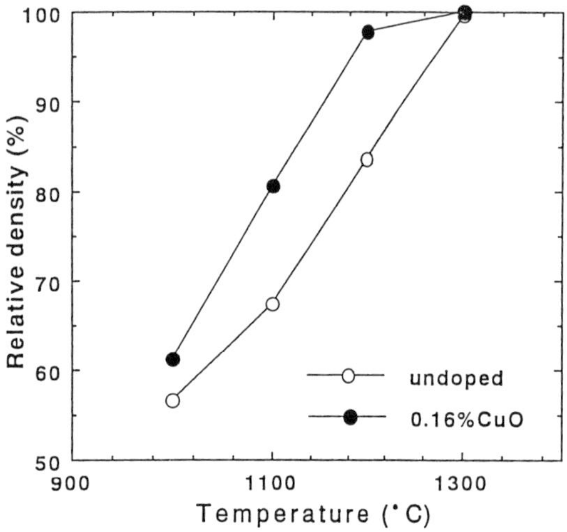

Fig. 6 Relative sintered density after heating at fixed temperatures for 2h in air.

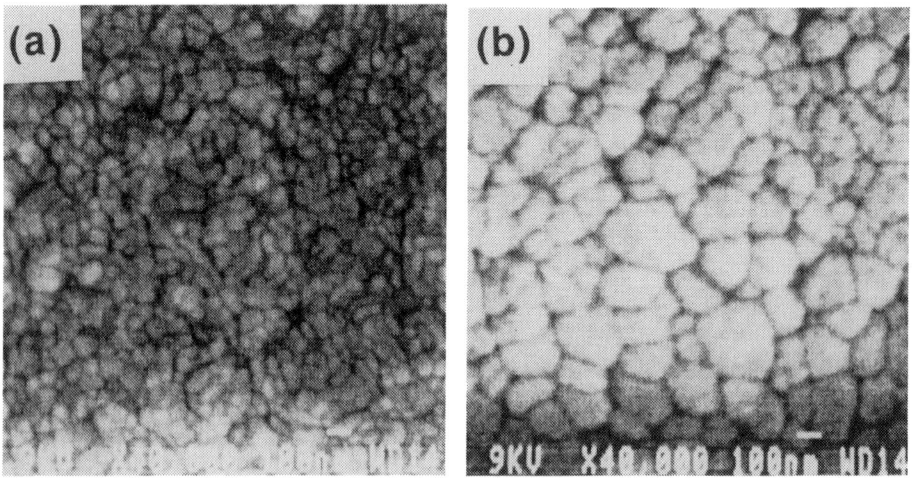

Fig. 7 SEM micrographs of the polished and thermal etched surfaces of 0.16 mol% CuO-doped 3Y-TZ polycrystals sintered at 1200 °C (a) and 1300°C (b).

surfaces of 0.16 mol% CuO-doped 3Y-TZ sintered at 1200 and 1300 °C. Mean grain sizes, determined by linear intercept [15], of CuO-doped samples sintered at 1200 and 1300 °C were 84 nm (fine) and 0.24 μm, respectively. The mean grain size of undoped 3Y-TZ sintered at 1300°C was 0.23 μm, which is nearly equal to CuO-doped sintered at the same temperature. It is noted that small amounts of CuO-doped Y-TZ with narrow pore size distributions densify at lower temperatures, resulting in fine-grained polycrystal. We have demonstrated that 3Y-TZ polycrystal prepared by colloidal processing exhibited enhancement in superplasticity comparable to that by conventional dry processing due to the elimination of large pores [18]. The fine-grained CuO doped 3Y-TZ sample is expected to show excellent superplastic properties and related experiments are now in progress.

CONCLUSIONS

We synthesized fine-grained dense monoclinic zirconia and copper doped tetragonal zirconia polycrystals by pressureless sintering in air. The key point is to consolidate the well-dispersed zirconia sol and slurries of copper adsorbed Y-TZ powder without powderization step by pressure filtaration.

ACKNOWLEDGMENTS

We thank Dr. S. Kuroda at NRIM for aid in the measurement of pore size distribution, and Dr. Y. Kaieda and Mr. N. Oguro for sample preparation by CIP. This work was supported by special coordination fund from the Science and Technology Agency of Japan.

REFERENCES

1 E. C. Subbarao, Advances in Ceramics, Vol. 3, Science and Technology of zirconia, pp. 1-24, Ed.A. H. Heuer and L.W.Hobbs, The American Ceramic Society, Columbus, Ohio (1981).

2 K. Sense, "A Method of Producing Very Dense ZrO_2," J. Am. Ceram. Soc. **44** [9], 465 (1961).

3 M. Yoshimura and S. Somiya," Fabrication of Dense, Nonstabilized ZrO2 Ceramics by Hydrothermal Reaction Sintering," Am. Ceram. Soc. Bull., **59** [10] 246 (1980).

4 F. Wakai, S. Sakaguchi and Y. Matsuno, "Superplasticity of Yttria-Stabilized Tetargonal ZrO_2 Polycrystals," Advaced Ceramic Materials, **1** [3], 259-63 (1986).

5 I-W. Chen and L. A. Xue, "Developement of Superplastic Structural Ceramics," J. Am. Ceram. Soc., **73** [9], 2585-2609 (1990).

6 N. Kimura, H. Okamura and J. Morishita,"Preparation of Low-Y_2O_3-TZP by Low-Temperature Sintering," Advances in Ceramics, Vol.24A, Science and Technology of Zirconia III, pp. 183-91, Ed. S. Somiya, N. Yamamoto and H. Yanagida, The American Ceramic Society, Columbus, Ohio (1988).

7 C-M. J. Hwang and I-W. Chen, "Effect of a Liquid Phase on Superplasticity of 2-mol%-Y_2O_3-Stabilized Tetragonal Zirconia Polycrystals," J. Am. Ceram. Soc., **73** [6],

1626-32 (1990).

[8] J. R. Seidensticker, M. J. Mayo and K. Osseo-Asare," Adsorption as a Method of Doping 3-mol%-Yttria-Stabilized Zirconia Powder with Copper Oxide," J. Am. Ceram. Soc., **76** [7], 1844-48 (1993).

[9] F. F. Lange, "Powder processing science and technology for increased reliability," J. Am. Ceram. Soc., **72** [1], 3-15 (1989).

[10] I. A. Aksay, "Molecular and Colloidal Engineering of Ceramics," Ceramic International, **17**, 267-74 (1991).

[11] C.H. Schilling and I.A. Aksay, Engineered Materials Handbook Vol.4, CERAMICS AND GRASSES, ASM international, pp.153-160 (1991).

[12] F. F. Lange and K. T. Miller,"Pressure filtration: Consolidation kinetics and mechanics," Am. Ceram. Soc. Bull. **66** [10], 1498-504 (1987).

[13] M. J. Readey and D. W. Readey, "Sintering of ZrO_2 in HCl atmospheres," J. Am. Ceram. Soc. **69** [7], 580-82 (1986).

[14] T. Uchikoshi, Y. Sakka, K. Ozawa and K. Hiraga,"Pressure Filtration and Sintering of Fine Zirconia Powder," J. Europ. Ceram. Soc., in press.

[15] M. I. Mendelson, " Average Grain Size in Polycrystalline Ceramics," J. Am. Ceram. Soc. **52** [8], 443-46 (1969).

[16] S. Ross and I. D. Morrion, Colloidal Systems and Interfaces; pp.205-63, Wiley, New York (1988).

[17]J. R. Seidensticker and M. J. Mayo, "Thermal Analysis of 3-mol%-Yttria-Stabilized Zirconia Powder Doped with Copper Oxide," J. Am. Ceram. Soc., **79**[2], 401-406 (1996).

[18] K. Hiraga and K. Nakano,"Cavitation Damage Mechanisms in a Superplastic Zirconia (3Y-TZP)," Mater. Sci. Forum, **243-245**, 387-92 (1997).

"ROBOCASTING": DIRECT FABRICATION OF CERAMICS AND COMPOSITES

Joseph Cesarano III and Paul Calvert*
Direct Fabrication Technologies Department
Sandia National Laboratories
Albuquerque, NM

*Arizona Materials Laboratory
University of Arizona
Tucson, AZ

ABSTRACT

A novel technique for the solid freeform fabrication of dense ceramics and composites has been developed. The "Robocast" process uses computer aided deposition of highly concentrated colloidal slurries to rapidly build complicated shapes without the use of molds or containers. Any conceivable two-dimensional pattern may be "written" layer by layer into a three-dimensional shape. Colloidal slurries containing upwards of 50 vol% solids are dispensed through the orifice of a piston driven syringe. Upon drying, the consolidated density is routinely greater than 60%. Crack-free alumina components have sintered densities greater than 95%. Components may be fabricated into bulk solids with large thicknesses unobtainable using slip casting; or, thin walled pieces made with aspect ratios greater than 20 and wall thicknesses less than 1mm. Several materials may be deposited at once for the fabrication of graded materials or hybrid components. Process feasibility and quality is dependent on the processing parameters of solids loading, slurry rheology, deposition rate, and drying rate. These interrelated parameters must be controlled so that sintering defects are prevented and shape tolerance is maintained. The Robocast process will be described and relations between processing parameters and deposition behavior will be presented.

FROM COLLOIDS TO COMPONENTS

Ludwig Gauckler
Department of Materials
ETH Zürich, Switzerland

ABSTRACT

Enzyme catalyzed reactions can be used to change the pH or the ionic strength in aqueous salt—solutions, sols, suspensions and emulsions. Thereby enzymes decompose a substrate by a time delayed catalytic reaction. The reactions can be controlled in the pH range from pH = 4 to 12 by the enzyme concentration and temperature. At low temperatures (~4°C) the enzymes are inactive, whereas around 40°C they reach their maximum activity. This gives the possibility to further delay or trigger the reaction by small temperature changes. Therefore enzyme catalyzed reactions are suitable to perform sol-gel transitions without disturbing the internal structure from the outside. These reactions can be used for sol-gel transitions of inorganic powder-suspensions, polymer suspensions and emulsions and combinations thereof. We describe examples of enzyme catalyzed reactions shifting pH from the acidic to the alkaline [1] and vice versa as well as those creating salt at a constant pH [2] and use them for coagulation. We then will give examples applying these reactions to a coagulation process of a high solids loading inorganic powder suspension and to some polymer sols. The process can be used in preparation of bulk ceramics, templating reactions and for surface patterning by stamping and capillary filling. Their application for a new ceramic green body fabrication method [3] is discussed for different powders.

[1] T. J. Graule, F. H. Baader and I. J. Gauckler, Enzyme Catalysis of Ceramic Forming, Journal of Materials Education, Vol. 16, No. 4, 1994, 243-267.
[2] T. J. Graule, F. H. Baader, L. J. Gauckler, Shaping of Ceramic Green Compacts Direct from Suspensions by Enzyme Catalyzed Reactions cfi/Berichte der Deutschen Keramischen Gesellschaft, 71, 6, 1994, 317-322.
[3] T. J. Graule, F. H. Baader, L. J. Gauckler, Casting Uniform Ceramics with Direct Coagulation, CHEMTECh 1995, 25(6), 31-37.

INTERNAL COAGULATION OF POLYMER/CERAMIC COMPOSITES

D. Kovacova, T. J. Graule, M. K. M. Hruschka, and L. J. Gauckler
Colloidal Chemistry-Ceramic Processing Group at the ETH
Zürich, Switzerland

ABSTRACT

Colloidal processing of stabilized ceramic suspension needs the transition from liquid to solid state. This may be achieved by drying, polymerization of monomers or thermal crosslinking reactions of binders. Two new processes which we investigated are the controlled destabilization of the suspension by minimizing the electrostatic repulsion and the internal gelation of polymer binders. In the present paper about the gelation of modified PAA and polysaccharides induced by internal change of pH or ionic strength. These gelling additives can be combined with ceramic suspensions to produce high strength, homogeneous ceramic green bodies.

THE RELEVANCE AND RELIABILITY OF DIRECT COAGULATION CASTING (DCC) FOR INDUSTRIAL FORMING PROCESSES

W. Burger, L. J. Gauckler,* G. Kiefer, and H. Ziegelbauer
CeramTec AG
P. O. Box 1149
D-73207 Plochingen, Germany

*ETH-Zürich
Department of Materials
Sonneggstr. 5
CH-8092 Zürich, Switzerland

ABSTRACT

The majority of structural ceramics is fabricated by pressing powder to green parts, followed by machining in the green or sintered state in order to achieve required dimensional tolerances and surface finish. Due to limitations of machining operations it is very difficult to produce complex parts with larger dimensions. In addition, the microstructure contains inhomogeneities up to 100 µm born in by the size of the spray granules used as feed stock. These inhomogeneities limit the attainable strength and reliability. Forming green parts by casting may be more economical in many cases compared to pressing and green machining. Direct Coagulation Casting (DCC) is an ideal technique for fabricating complex ceramic parts with high reliability. Ceramic parts have been produced under industrial conditions leading to excellent mechanical strength and to a high Weibull modulus.

MICROPOROUS MULLITE CERAMICS BY GEL-CASTING OF ALKOXIDE-DERIVED PRECURSOR POWDER

Hisao Suzuki, Masanori Shimizu, Hidehiro Kamiya*, Toshitaka Oto** and Minoru Takahashi**
Department of Materials Science
Shizuoka University
3-5-1 Johoku
Hamamatsu 432, Japan

*Graduate School of Bio-Applications and Systems Engineering
Tokyo University of Agriculture and Technology
Koganei, Tokyo 184, Japan

**Ceramic Research Laboratory
Nagoya Institute of Technology
10-6-29 Asahigaoka
Tajimi 507, Japan

ABSTRACT

Microporous fine powders of mullite with a stoichiometric composition were successfully prepared by controlling the agglomeration behavior of the sols prepared from molecular-designed precursor solution. Aging of the sols in the solvents with low dielectric constant, such as butyl alcohol, resulted in the coarse flocculation of the primary colloidal particles to form microporous fine precursor powders. These precursor powders crystallized into orthorhombic mullite with high surface area even after high temperature calcination. The microporous structure of the precursor powders was successfully maintained in the sintered body by using gel-casting as a forming technique. As a result, microporous mullite ceramic with a stoichiometric composition was prepared by the sintering at 1673 K for two hours. The true density of the microporous mullite ceramic reached more than 95% of theoretical and its surface area was as high as 25 m2/g.

GELCASTING AND AQUEOUS INJECTION MOLDING FOR SILICON NITRIDE STRUCTURAL CERAMICS

O. O. Omatete and J. P. Pollinger*
Oak Ridge National Laboratories
Oak Ridge, TN

*AlliedSignal Ceramic Components
Torrance, CA

ABSTRACT

The Gelcasting technology developed by Oak Ridge National Laboratory (ORNL) and the Agar Aqueous Injection Molding (AIM) technology developed by Allied Signal Ceramic Components as potential improved powder compact forming processes for use in silicon nitride structural ceramics. Both forming technologies are aqueous-based and thus have required extensive refinement and control of colloidal powder processing characteristics to be used effectively with submicron size silicon nitride and oxide sintering additive powders that are very reactive in water. The capabilities of Gelcasting and AIM as powder compact forming processes will be discussed and compared against current production/baseline processes such as slipcasting and isopressing/compaction pressing of silicon nitride. Capability improvements of Gelcasting and AIM discussed will include powder compact forming time, achievement of uniformity in powder packing density, and ability to form complex and large shapes. Limitations will be compared and discussed, including lower achievable powder packing densities, gel binder pyrolysis limitations, and the aging of the water-powder mixtures before use in forming due to hydroxylation. Finally, the potential of gelcasting and AIM will be discussed in light of the needs for current and future structural ceramic component production applications

ENZYME CATALYZED GELATION OF BOEHMITE CONTAINING SOLS

J. M. Helbig, Z. Xie, M. K. M. Hruschka, and L. J. Gauckler
Colloidal Chemistry-Ceramic Processing Group at the ETH
Zürich, Switzerland

ABSTRACT

Internal coagulation of boehmite containing sols was performed in order to derive fine grained alumina ceramics. We report about properties of wet green bodies and sintered ceramics in the system alpha-alumina powder (d50 >> 0.5 mm) and boehmite (d50 = 0.003 mm) after using a pH changing destabilization reaction. Additions of boehmite particles to α-alumina increases the wet strength of the green bodies. On the alpha-alumina rich side of the system alpha-alumina-boehmite the wet green strength increased from 0.01 MPa without boehmite to over 0.02 Mpa with 3vol% boehmite. The sintered parts had a density of >99.5% theoretical density and average grain size was about 3 mm. On the other side of the system small additions of alpha-alumina to the boehmite as seeding crystals did not change the wet green strength of less than 0.01 Mpa. After sintering the ceramic bodies were fully dense with a density of >99.4% theoretical density with a submicrometer grain size in the microstructure.

FORMING CERAMICS WITH BIOPOLYMERS

J. Brandt, E. Carlström, A. Kristoffersson, O. Lyckfeldt, and G. Lindsten*
Swedish Ceramic Institute
Box 5403
S-402 29 Göteborg, Sweden

*SKF Nova AB
S-412 88 Göteborg, Sweden

ABSTRACT

Biopolymers such as proteins and starches were used to develop new direct consolidation methods. Direct consolidation methods are powder forming methods where no removal of the forming medium prior to the consolidation of the body takes place. Direct consolidated bodies retain the high homogeneity of the original dispersion making better dimensional and microstructural control possible. Using biopolymers in these methods have the additional advantages that water and non toxic binders can be used. The polymer that are added are particle (globular) shaped, this makes the viscosity increase of the suspension much smaller compared to the addition of stretched out (fibrous) polymeric chains. In one of the methods a ceramic slurry was dispersed and a globular protein such as the egg white protein (albumin) was added to the suspension. The suspension was then poured into a mold and heated to about 60°C. The formed body then consolidated due to the coagulation of the protein. After drying and debinding of the protein the body was sintered. Prototypes of bearing balls and cutting tools of silicon nitride were manufactured by this technique. The foaming properties of the protein method could also be used to produce ceramic foams. The suspension was then vigorously stirred to create a foam. The foam was consolidated by heating, dried, and sintered. With this method porous sintered bodies with 50-80 vol% porosity were created. Another way to produce porous ceramic bodies was developed. In this method starch particles were used as the gelling agent. A ceramic suspension with additions of starch was poured into a mold and heated to about 60°C. The starch particles caused a consolidation of ceramic body by adsorbing water and swelling. The consolidated body could be removed from the mold, dried and sintered. The starch particles decomposed during sintering leaving pores of the same size as the original particles. Starch can be obtained from for example potato, corn, and rice in sizes ranging from 5 to 170 mym.

USE OF HYDROLYSIS ASSISTED SOLIDIFICATION (HAS) IN SLURRY FORMING Si_3N_4 BODIES

Kristoffer Krnel, Tomaž Kosmač
Jožef Stefan Institute
Jamova 39
1000 Ljubljana, Slovenia

ABSTRACT

The Hydrolysis Assisted Solidification (HAS) process was used in slurry forming Si_3N_4 ceramics. AlN powder was added to a highly loaded aqueous Si_3N_4 + Y_2O_3 suspension prior to shaping by aqueous injection molding. The $Al(OH)_3$ gel network formed by the AlN hydrolysis was first exploited for the provided increase in strength to the saturated green bodies. The thermal decomposition of the $Al(OH)_3$ later on served as a sintering additive. It was found, that the reactivity of AlN with water was suppressed or even prevented in the presence of dissolved silicic acid in the Si_3N_4 slurry. Therefore, the as received Si_3N_4 powder was leached in water before being used in the slurry preparation. With leached Si_3N_4 powder, green bodies were successfully produced and sintered to 95 % of theoretical density.

INTRODUCTION

The Hydrolysis Assisted Solidification (HAS) process is a novel Aqueous Injection Molding (AIM) variant which exploits the hydrolysis of AlN powder added to concentrated aqueous slurries[3]. During the hydrolysis reactions, water is consumed and aluminum hydroxides and ammonia are formed according to the reaction scheme[4]:

$$AlN + 2H_2O \rightarrow AlOOH_{amorph.} + NH_3 \quad (1)$$

$$NH_3 + H_2O \rightarrow NH_4^+ + OH \quad (2)$$

$$AlOOH_{amorph.} + H_2O \rightarrow Al(OH)_{3\ gel} \quad (3)$$

If a small amount of AlN powder is added to an aqueous suspension, the hydrolysis will not start immediately, since the surface of the AlN powder is passivated by a thin amorphous layer of hydrated alumina[5]. An incubation time, which depends on the thickness of the protection layer and its solubility in aqueous environments, is needed for water to dissolve this layer. Once it is expired, i.e. the surface layer is dissolved, the hydrolysis will start propagating resulting in a dramatic increase in the viscosity of the host slurry due to:

i) chemical bonding of water during Al(OOH) and $Al(OH)_3$ formation (with reference to Eqs. 1 and 3),

ii) gelling of the $Al(OH)_3$ reaction product to form a stiff network (Eq. 3),

iii)flocculation/coagulation induced by the increasing pH of the slurry initially set below the IEP of the host ceramic powder (Eq. 2).

In the present paper we report on the use of HAS in slurry forming Si_3N_4 bodies. The hydrolysis of admixed AlN powder was first exploited for solidification of the slurry in an impermeable plastic mold. Alumina, which is formed during thermal decomposition of aluminum hydroxide, later on served as a sintering additive to promote Si_3N_4 densification, in combination with Y_2O_3, though.

EXPERIMENTAL WORK

A nominal starting composition of 85 wt.% Si_3N_4, 8.6 wt.%Y_2O_3 and 6.4 wt.% Al_2O_3 was chosen for the preparation of sintered Si_3N_4 ceramics. Since it was intended to obtaine the Al_2O_3 from the thermally decomposed $Al(OH)_3$ reaction product formed during the HAS process, the starting powder mixture in the slurry consisted of 86 wt.% Si_3N_4 (LC-12, H.C. Starck), 8.7 wt.% of Y_2O_3 (99.99%, Ventron, Alfa products) and 5.3 wt.% AlN (Grade B, H.C. Starck).

To avoid premature hydrolysis of the admixed AlN powder the latter was added to the slurry after milling of the Si_3N_4 powder and its homogenization with Y_2O_3. These two powders were first ball milled for 2 hours in distilled water using alumina balls and then ultrasonicated for one hour with 20 sec intervals of ultrasonication and rest. Afterwards the slurry was filtered and the cake was washed, dried and redispersed in deionized water to remove silica from the Si_3N_4 powder surface. A commercial polyelectrolyte PC 33 (Zschimmer & Schwartz, FRG) was added as a dispersant, at a concentration of 0.5 wt.% of the total solids weight. The pH of the suspension was adjusted to a value of 9 by the addition of ammonia.AlN powder was then added to the slurry and ball milled for 15 minutes, followed by interrupted ultrasonication (20 sec intervals) for 15 minutes. The total solids loading in the slurry after AlN addition was 80 wt.% (55 vol.%) and its viscosity was 80 cPs at $D=100\ s^{-1}$.

Once homogenized, the slurry was poured into the container of a low pressure injection molding machine of our own construction. In the container, the slurry was evacuated under gentle stirring to remove trapped air. A plastic mold with a rectangular ($5\times5\times50\ mm^3$) cavity was heated to 80 °C by a water heating coil mounted on the mold. Molding pressure (1-2 bar) was provided by compressed air. Typical setting time was 3-5 minutes. After being solidified, the specimens were demolded, sealed in a plastic foil and kept in an oven at 50°C for one hour. Afterwards they were dismantled and dried in air at 80°C overnight. The samples were sintered in a graphite crucible at 1800°C for 2 hours in flowing nitrogen gas.

Solids loading, pH and viscosity measurements before and after AlN addition were used as standard tools for slurry characterization. The green density and shrinkage on drying of the HAS formed parts were measured geometrically. A three-point bending test (15 mm loading span) was used for the green strength determination. Distilled water was used as an immersion liquid for the sintered density measurements. Flexural strength was determined by four-point bending tests (loading spans 10 and 20 mm, respectively). Before the bending tests were performed, the tensile side of the test bars was ground and polished and the edges were removed. Sintered specimens were ground and polished for subsequent microstructural observations by SEM

RESULTS AND DISCUSSION

If the HAS process is to be applied in slurry forming Si_3N_4 ceramics, then the problem arises in obtaining high solids loading in the slurry which will roughly correspond to the fractional density of the dried green parts. In the case of aqueous Si_3N_4 slurries, containing additives to promote liquid phase sintering, relatively low solids loadings (45-55 vol.%) are commonly reported in the literature[6]. The high specific surface area of the powders used and their mismatching electrochemical behaviour in water require compromising solutions which result in lower solids loadings in aqueous Si_3N_4 slurries. Following reported recommendations in the literature[6], the as received Si_3N_4 powder was first heated for 8 hours at 550°C to oxidize the surface before it was mixed with yttria and dispersed in water. Once the corresponding amount of AlN (5.3 wt.% with respect to Si_3N_4 + Y_2O_3 premix) was added, the entire solids loading of the slurry was 54 vol.% which is well within the reported limits. The viscosity of the slurry was rather high, but still acceptable for low pressure injection molding. The AlN hydrolysis did not occur in this instance at room or elevated temperatures. The possible silicic acid formation due to the dissolution of the oxidized Si_3N_4 powder surface layer, may have adsorbed on to the AlN powder surface preventing the hydrolysis from starting. To assumption was investigated by collecting the supernat attained by centrifuging an aqueous Si_3N_4 slurry which contained 50 mg/l of silicon as determined by ICP-AES analysis. AlN powder was dispersed, in the above supernat and the change in pH of the suspension was monitored versus time at room temperature. The change in pH with temperature was also monitored (Figs. 1a, b). The increase in pH should indicate the proceeding hydrolysis reactions referenced by Eq. 2. No variation in pH was observed with the diluted AlN (2,5 wt.%) suspension confirming our previous assumption. However, in the more concentrated (25 wt.%) suspension the amount of dissolved silicic acid was obviously insufficient to prevent the hydrolysis. After a certain incubation period, the reaction started which was reflected in the pH rise of the slurry.

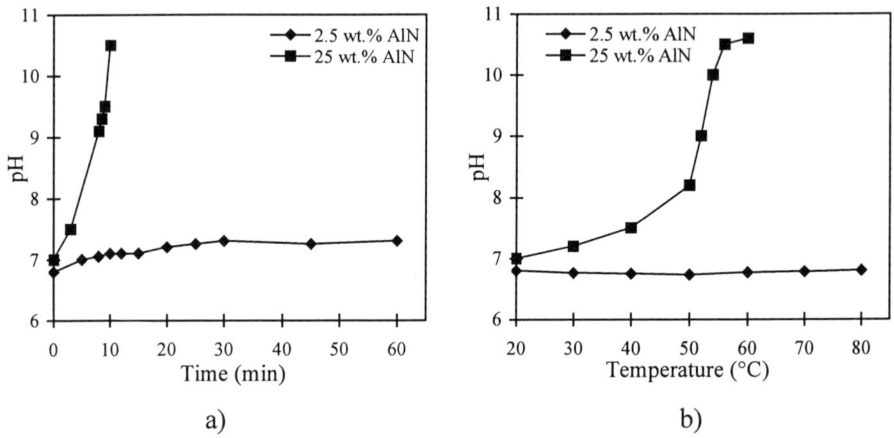

Figure 1. Variation in pH with time at RT (a) and temperature during constant heating (b) for Si_3N_4 supernat containing 2.5 wt.% AlN and 25 wt.% AlN, respectively.

In contrast to supernat, an equilibrium silicic acid concentration is retained in the aqueous Si_3N_4 slurry such that a larger amount of AlN powder (with respect to water) can be protected from hydrolysis.

If aqueous Si_3N_4 slurries are to be used in the HAS process the formation of silicic acid in the slurry must be prevented, possibly by chem-adsorptoin of the silane molecules on the powder surface[7]. Removal may be achieved by leaching the as received Si_3N_4 powder in water followed by filtration and washing. According to E. Laarz et al.[8] leaching has a beneficial effect on the dispersability of Si_3N_4 powder in water, therefore, this procedure was chosen for the slurry preparation for HAS. To check its effectiveness in preserving the reactivity of the AlN powder with water, an aqueous slurry consisting of leached Si_3N_4 and 5.3% AlN was heated at a constant rate while its viscosity was continuously monitored (Fig. 2). At approximately 50°C, the viscosity of the slurry started rising until the fluid character of the suspension was lost. This is a characteristic feature of the HAS process which was later on successfully used in slurry forming Si_3N_4 bodies. For comparison, the viscosity-temperature profile of an unleached Si_3N_4 slurry containing the same amount of AlN was also obtained in shown in Fig. 2. In this case, no change in viscosity was detected during heating up to 80°C.

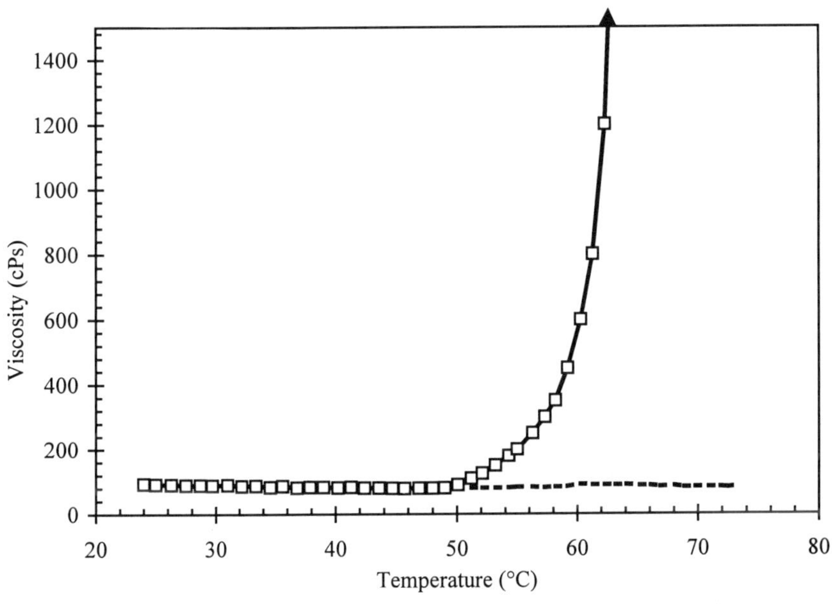

Figure 2. Variation in viscosity with temperature on constant heating for concentrated (55 vol%) aqueous Si_3N_4 suspension containing 5.3 wt.% of AlN.

The HAS formed parts were characterized before and after drying (Table I). As expected, the fractional density of the HAS formed green bodies corresponds well to the solids loading in the slurry. The bending strength of the green bodies is remarkable, significantly exceeding the values obtained with conventionally pressure cast green bodies containing alumina instead of AlN in the starting powder mixture (2 MPa). During solidification and subsequent drying, the AlN has almost completely transformed into $Al(OH)_3$ boehmite which gels forming a strong network between the particles. The presence of boehmite in the green body is one of the specific characteristics of the HAS process which increases the green strength but requires low initial heating rate during the sintering process to avoid cracking during its thermal decomposition.

Table I. Characteristic properties of the HAS formed Si_3N_4 bodies before and after drying

Drying shrinkage, %	0.6
Green density, % of theoretical	54
Green strength, MPa	12
Phases identified by XRD	α-Si_3N_4, Y_2O_3, $Al(OH)_3$, AlN (traces)

The density (94.6 % of theoretical) and average bending strength (350 MPa) of the sintered HAS formed Si_3N_4 samples are rather low. Among the 6 tested bars only one exhibited a highr bending strength value (640 MPa). However, the SEM micrograph of sintered Si_3N_4 (Fig. 3a) does not differ significantly from those reported in the literature for a dense strong Si_3N_4 ceramic. The dark elongated β-Si_3N_4 grains and white liquid glassy phase, which were revealed by SEM-BEI imaging of unetched polished surfaces are homogeneously distributed indicating that the presence of aluminum hydroxide in the green bodies had no significant influence on the densification and homogenous phase distribution. Based on the micrograph taken at high magnification (6000×) one would expect considerably higher fired density and correspondingly higher strengths than those reported above. The reason for this discrepancy becomes evident from the micrographs of the fracture surface taken at lower magnification (50×) (Fig. 3b).

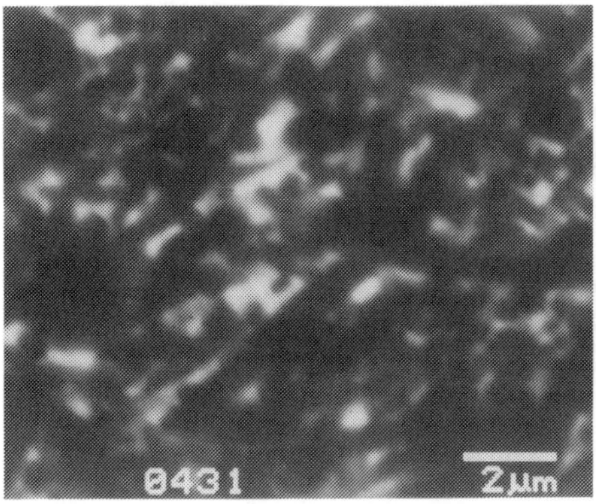

a)

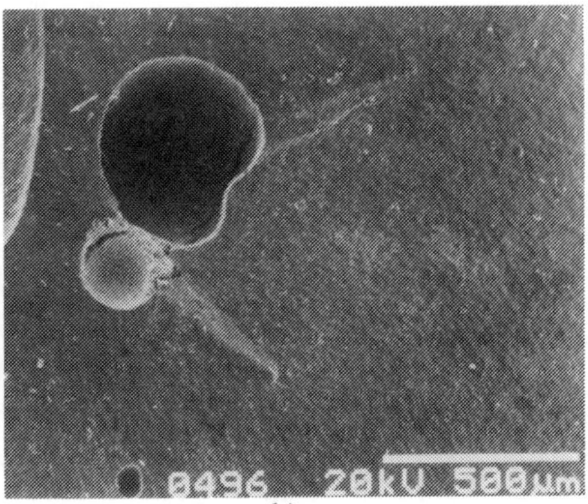

b)

Figure 3. SEM micrograph of sintered HAS formed Si_3N_4: a) back scattered electron image of polished surface (high magnification); b) image of fracture surface (low magnification)

Large, spherical pores (> 200 µm) which are present in the sintered ceramics not only contribute to a low sintered density but also act as stress concentrators. consequently reducing the strength. The porosity may result from released ammonia gas, but since such spherical pores were not observed with any other HAS formed ceramics, it is more likely that they result from trapped air which was not removed from the slurry due to its unfavorable rheological properties. Although the listed density and strength values might appear rather low, any conclusion regarding the mechanical properties of HAS formed Si_3N_4 would be premature at this point. Apart from large spherical pores, the microstructure of the sintered Si_3N_4 raises hopes that by an appropriate optimization of the rheological properties of an aqueous Si_3N_4 slurry the density and mechanical properties of HAS formed ceramics will be considerably improved.

CONCLUSIONS

The hydrolysis assisted solidification (HAS) process can be used in slurry forming complex-shaped Si_3N_4 ceramics, provided that the formation of silicic acid in the slurry is prevented or removed. Otherwise the silicic acid adsorbs on the AlN powder surface protecting it from hydrolysis. The fractional density of HAS formed green bodies roughly corresponds to the solids loading in the slurry which is an important process requirement. Gelling of the $Al(OH)_3$ reaction product results in remarkable strength of the dried green bodies. The presence of boehmite does not seem to exert any negative effect on the liquid phase sintering, and thus may be exploited as a sintering additive. However, the presence of trapped air in the highly concentrated Si_3N_4 slurry resulted in large spherical pores in the sintered ceramics. These pores contribute to a low sintered density, and also act as stress concentrators which reduce the strength of the sintered material.

REFERENCES

1 F. F. Lange, Powder Processing Science and Technology for Increased Reliability, *J. Am. Ceram. Soc.,* **72** (1989), 3-15.

2 T. Kosmač, Near-net Shaping of Engineering Ceramics: Potentials and Prospects of Aqueous Injection Molding, in proceedings NATO Advanced Research Workshop on Engineering Ceramics '96, Smolenice castle, Slovakia, 1996, editors: G.-N. Babini, M. Haviar, P. Šajgalik, Kluwer, Accademic publ., 1997.

3 T. Kosmaè, S. Novak, M. Sajko, Hydrolysis-Assisted Solidification (HAS): A New Setting Concept for Ceramic Net-Shaping, *J. Europ. Ceram. Soc.*, **17** (1997) 427-432.

4 P. Bowen, J.G. Highfield, A. Mocellin, T.A. Ring, Degradation of Aluminum Nitride Powder in an Aqueous Environment, *J. Am. Ceram. Soc.*, **73** 3 (1990) 724-728.

5 W.M. Mobley, Colloidal properties, processing and characterization of aluminum nitride suspensions, PhD thesis, Alfred University, Alfred, New York, 1996.

6 A.L. Nagel, Schlickergießen von sinterfähigen Si_3N_4 - Keramiken, PhD Thesis, Fakultät Chemie der Universität Stuttgart, Stuttgart, 1988.

7 M. Colic, G.V. Franks, M.L. Fisher, F.F. Lange, Improving the Aqueous Processing of Silicon Nitride with Chem-Adsorbed Organofunctional Silanes, to be published in the Journal of the American Ceramic Society.

8 E. Laarz, G. Lenninger and L. Bergström, Aqueous Silicon Nitride Suspensions: Effect of Surface Treatment on the Rheological and Electrokinetic Properties, Key Engineering Materials, Vols. 132-136 (1997) 285-288.

SHAPING OF CERAMIC BODIES FROM OXIDE/NONOXIDE POWDER MIXTURES BY ELECTROPHORESIS

K. Moritz, E. Müller
Freiberg University of Mining and Technology, Institute of Ceramic Materials
Gustav-Zeuner-Str. 3
D-09596 Freiberg, Germany

ABSTRACT

The topic of the investigations described in this paper was the simultaneous electrophoretic deposition of an oxide and a nonoxide powder from suspensions of a mixture of both powders. For these experiments, aluminium nitride and the sintering additive yttrium oxide were used as starting materials.

It was shown that the homogeneity of the Y_2O_3 distribution in the deposits is a function of the solid content of the suspension. From concentrated suspensions, it was possible to produce green bodies with a homogeneous distribution of the sintering additive. The formed bodies could be sintered to nearly theoretical density.

INTRODUCTION

Electrophoretic forming is known as a shaping technique which uses the force of an electric field acting on dispersed particles. Because the powder is directly consolidated from a stable suspension, this method offers the possibility of achieving a dense, homogeneous particle packing in the ceramic green body.

The electrophoretic deposition has been widely investigated as a forming technique for various powders like Al_2O_3 [1,2], ZrO_2, SiO_2 [3] and SiC [4]. Furthermore, laminated composites (e.g. ZrO_2/Al_2O_3 [5], SiC/C [6]) and fiber reinforced ceramic matrix composites[7] have been fabricated by electrophoresis.

The investigations described in the following text concern the application of electrophoretic forming to aluminium nitride. Due to the high percentage of covalent bonding in nitride materials, the addition of a sintering aid is necessary to obtain satisfactory densities by pressureless sintering. A common additive for sintering AlN is yttria. For this reason, the simultaneous deposition of AlN and Y_2O_3 by use of an oxide/nonoxide powder mixture was investigated.

EXPERIMENTAL PROCEDURE

The characteristic data of the AlN and the Y_2O_3 powder used for the experiments can be seen in table 1.

Table 1 Particle size and specific surface of the powders

powder	$d_{50,3}$ value of particle size distribution in microns	specific surface (BET) in m^2/g
AlN Grade C (H. C. Starck)	0.81	4.7
Y_2O_3 Grade C (H. C. Starck)	0.38	14.8

Aluminium nitride is highly sensitive to hydrolysis. Therefore, the powder was suspended not in water but in ethanol. The use of non-aqueous suspensions also avoided the formation of gas bubbles at the deposition electrode. By contrast, in aqueous media gas formation due to the electrochemical decomposition of water often leads to large pores in the green bodies.

The suspensions were prepared by ultrasonic agitation and by adding suitable dispersants. The influence of the organic additives (carboxylic acid, amine) on the surface charge of the particles in the suspension was characterized by measurements of the Electrokinetic Sonic Amplitude (ESA). By this method, suspensions of pure aluminium nitride or yttria powder, as well as slurries which contain a mixture of 97 wt.-% AlN and 3 wt.-% Y_2O_3 were examined. From suspensions of the powder mixture, planar green bodies (Ø 40 mm) were produced by electrophoretic deposition. The solid content of the suspensions was varied between 1.7; 3.5; 7; 14; and 28 vol.-%, respectively. The deposition was done in the direction of the gravitational force at an applied voltage of 120 V and an electrode distance of 30 mm.

After the drying step the organic additives (additive content < 1 wt.-%) were burned out from the deposits.

Small pieces of the upper and the lower side of the green bodies were cut off and the yttria content was analysed by ICP-OES. The green bodies were sintered at 1850 °C (2 h dwelling time) in nitrogen atmosphere.

RESULTS AND DISCUSSION

By adding a combination of a suitable carboxylic acid (for example polyacrylic or citric acid) and an amine (triethylamine or n-butylamine), a stabilization of ethanolic AlN suspensions can be achieved [8,9] For the investigations presented in this paper, the additive combination citric acid/triethylamine was used.

Fig. 1 shows the measured ESA signals of an AlN and an AlN/Y_2O_3 suspension, respectively, depending on the addition of citric acid (curves 1 and 2). Furthermore, the figure shows the ESA signal of suspensions which contain a constant amount of citric acid (0.65 mg per m² powder surface) as a function of the added amount of triethylamine (curves 3 and 4). The added amount drawn on the abscissa refers in the case of curves 3 and 4 only to the amount of amine, not to the sum of the citric acid and amine addition.

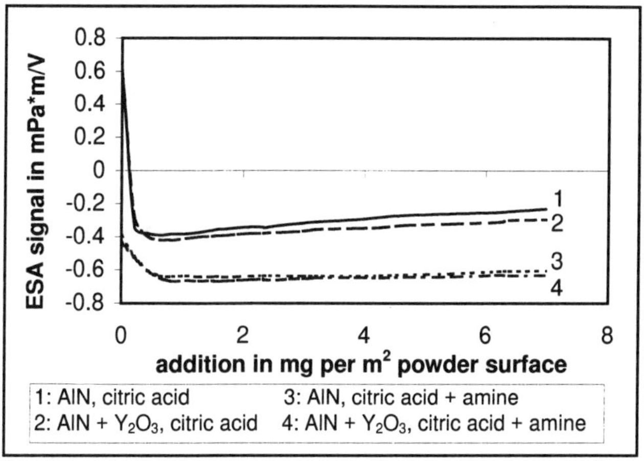

Fig. 1 ESA signal of AlN and AlN/Y_2O_3 suspensions, respectively, (solid content 3 vol.-%) vs. added amount of organic additives

The AlN particles are positively charged in the ethanolic suspension without organic additives. The positive particle charge can be explained by an acid-base-interaction between Bronsted-basic OH-groups on the partially oxidized or hydrolysed surface of the aluminium nitride on the one hand and small amounts of water in the suspension and the ethanol molecules, respectively, on the other hand. (An oxide-like surface of the AlN was proven by XPS.) By adding citric acid the sign of the surface charge reversed to negative values. In this case the surface charge is caused by the adsorbed carboxylic acid molecules which show a weak ionization in the dispersion medium. The ionization of the acid molecules can be increased by the simultaneous addition of the strong basic amine. In this way, an increase of the negative particle charge is obtained.

For the suspension of the pure AlN powder and that of the 97 wt.-% AlN / 3 wt.-% Y_2O_3 powder mixture nearly identical ESA curves were determined (Fig. 1).

A suspension of the pure Y_2O_3 powder was investigated by analogously performed ESA measurements. The results of these measurements are shown in fig. 2.

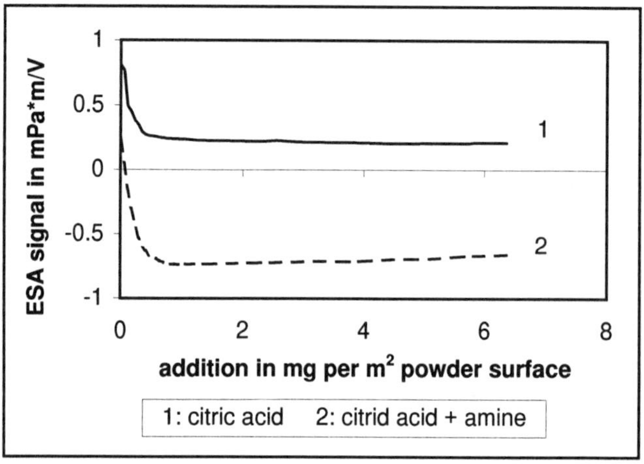

Fig. 2 ESA signal of a Y_2O_3 suspension (solid content 3 vol.-%) vs. added amount of organic additives

In the suspension without organic additives the surface of the yttria particles is positively charged, too. In contrast to the AlN suspensions, the exclusive addition of citric acid only reduces the positive ESA signal, but adding a combination of citric acid and triethylamine leads to a negative surface charge of the yttria particles.

By electrophoretic deposition ceramic green bodies were produced from suspensions of the AlN/Y_2O_3 powder mixture. The aim of the experiments was to investigate in what way it is possible to realize a homogeneous distribution of the oxidic sintering aid in the AlN green bodies by simultaneous deposition of the two powders. It can be assumed that differences in the magnitude of the electrophoretic velocity of the AlN and the Y_2O_3 particles under certain conditions cause the formation of a gradient in the yttria content. Besides different electrophoretic velocities of both powder components a contribution of sedimentation effects has to be taken into consideration as a possible reason for an inhomogeneous Y_2O_3 distribution in the deposits.

To investigate whether a compositional gradient had been formed or not, an unusually long deposition time of 60 min was chosen. Usually, depending on the desired thickness of the green bodies, deposition times in electrophoretic forming are in the range of only a few minutes.

In the above-mentioned experiments a complete deposition of the solid content was reached. The thickness of the green bodies and the yttria content at the upper and the lower side of the deposits depending on the solid content of the suspensions can be seen from table 2.

Table 2 Sample thickness and yttria content at the upper and the lower side of the green bodies

solid content of the suspension in vol.-%	thickness of the green bodies in mm	Y_2O_3 content in wt.-%	
		upper side	lower side
1.75	1.29	3.61	2.53
3.5	2.20	3.16	2.74
7	4.57	3.17	2.79
14	7.69	3.08	3.00
28	14.95	3.03	3.04

In the case of a low solid content of the suspension, analysis showed a higher yttria content on the upper than on the lower side of the deposit. By contrast, from concentrated slurries, green bodies with a homogeneous Y_2O_3 distribution over the sample thickness were obtained. The results suggest that with increasing solid content of the powder suspension a collective motion of the particles in the electric field dominates.

How can collective motions within a particle suspension be explained? For two different solid contents (1.75 and 28 vol.-%, respectively) the number of particles per unit of volume was estimated. It was assumed that the $d_{50,3}$ value of the AlN powder equals 0.8 microns and the $d_{50,3}$ value of the yttria powder equals 0.4 microns. In the case of the lower solid content, a concentration of $8.3 \cdot 10^{12}$ particles/dm^3 and, in the case of the higher solid content, a concentration of $1.33 \cdot 10^{14}$ particles/dm^3 was calculated. Hence, the median values of the distances between the centres of the particles could be achieved (5 microns at a solid content of 1.75 vol.-% and 2 microns at 28 vol.-%). Taking into account that the difference between the particle surfaces and their centres has to be calculated, the median value of the particle surface distances in the suspension with the lower solid content is 4.2 microns and in the case of the suspension with the higher solid content 1.2 microns. A collective motion of the particles exerts more and more influence, if the distances between the particles decrease to values comparable with particle diameters. The dimensions of the electrochemical double layer were not considered for these simplified estimations.

As a result of the stabilization of the suspensions by adding the dispersants, a dense, homogeneous particle packing was achieved in the deposits. The homogeneity of the particle packing is expressed by the monomodal pore size distribution shown in fig. 3.

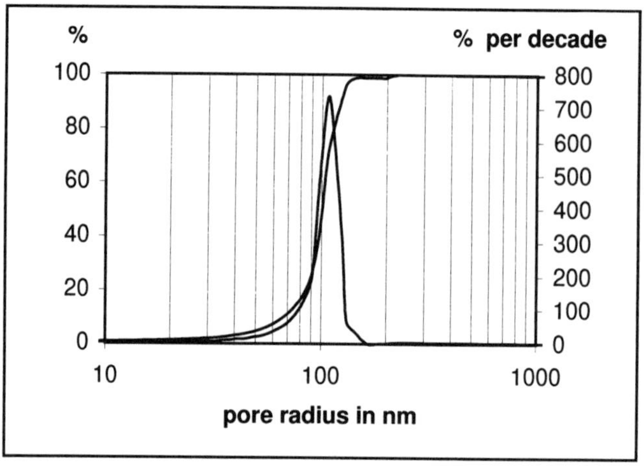

Fig. 3 Pore size distribution of an electrophoretically deposited green body of the AlN/Y_2O_3 powder mixture

The green densities of the formed bodies rose as the solid content of the suspensions increased. From slurries with 14 resp. 28 vol.-% solid content green bodies with an average density of 53.8 resp. 56.6 % of the theoretical density were obtained (theoretical density: 3.31 g/cm³).

Relative densities of the sintered bodies reached values of nearly theoretical density.

The SEM micrographs of sintered samples in fig. 4 illustrated the different yttria content at the upper and the lower side of deposits from suspensions with a low solid content (Fig. 4a) in comparison with the homogeneous distribution of the sintering additive in deposits from concentrated slurries (Fig. 4b). The light areas of the microstructure represent the secondary phase. As secondary phases both $Y_3Al_5O_{12}$ (YAG) and $Y_4Al_2O_9$ were detected by X-ray diffraction.

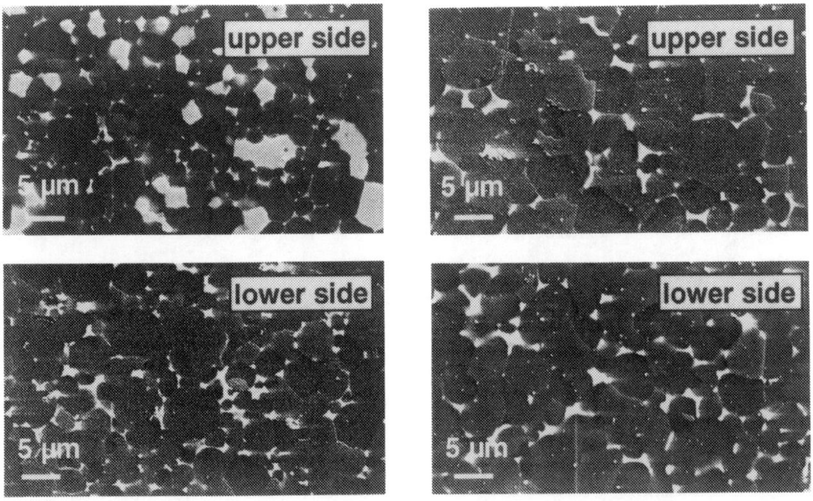

a) solid content of the suspension 1.75 vol.-%

b) solid content of the suspension 28 vol.-%

Fig. 4 SEM micrographs of polished sections of sintered samples

CONCLUSIONS

In suspensions of increased solid content, which are commonly used for electrophoretic shaping, a collective motion of the particles in the electric field dominates. Therefore, it is possible to realize a homogeneous distribution of oxidic sintering aids by electrophoretic deposition from oxide/nonoxide powder mixtures.

ACKNOWLEDGEMENTS

Financial support from the Deutsche Forschungsgemeinschaft (SFB 285) is gratefully acknowledged.

REFERENCES

[1] F. Harbach, R. Neeff, H. Nienburg and L. Weiler, "Zuverlässige keramische Bauteile aus kolloiden Suspensionen" ("Reliable ceramic components from colloidal suspensions"), cfi/Ber. DKG **67** [4]130 - 135 (1990)

[2] Y. Hirata, A. Nishimoto and Y. Ishihara, "Forming of alumina powder by electrophoretic deposition," Nippon Seramikkusu Kyokai Gakujutsu Ronbunshi (Japan) Int. Ed., **99** [2] 108 - 113 (1991)

[3] R. Clasen, S. Janes, Ch. Oswald and D. Ranker, "Electrophoretic deposition of nanosized ceramic powders"; pp. 481 - 486 in Ceramic Transactions, Vol. 51, *Ceramic Processing Science and Technology*. Edited by H. Hausner, G. L. Messing and S. Hirano, Am. Ceram. Soc., Westerville, OH, USA, 1995

[4] R. L. Meisel, "Elektrophoretische Abscheidung von α-SiC mit den partikulären Sinteradditiven C und B_4C" ("Electrophoretic deposition of α-SiC with the sintering additives C and B_4C") PhD thesis, Universität des Saarlandes, Saarbrücken, Germany, 1994

[5] P. S. Nicholson, P. Sarkar and X. Huang, "Electrophoretic deposition and its use to synthesize ZrO_2/Al_2O_3 microlaminate ceramic/ceramic composites," J. of Materials Science, **28**, 6274 - 6278 (1993)

[6] L. Vandeperre and O. Van Der Biest, "Electrophoretic forming of silicon carbide laminates with graphite interfaces"; pp. 359 - 366 in Fourth Euro Ceramics, Vol. 1: *Basic Science - Developments in processing of advanced ceramics - Part I.* Edited by C. Galassi, Gruppo Editoriale Faenza Editrice S.p.A., Italy, 1995

[7] P. A. Trusty, A. R. Boccaccini, E. G. Butler and C. B. Ponton, "Novel techniques for manufacturing woven fiber reinforced ceramic matrix composites. I. Preform fabrication," Materials and Manufacturing Processes **10** [6] 1215 - 1226 (1995)

[8] K. Moritz, T. Reetz, "Electrophoretic shaping of aluminium nitride," cfi/Ber. DKG **70** [7] 348 - 350 (1993)

[9] K. Moritz, T. Reetz, "Electrophoretic deposition of aluminium nitride"; pp. 425 - 430 in Third Euro Ceramics, Vol. 1, Edited by P. Durán and J. F. Fernández, Faenza Editrice Ibérica S.L., Spain, 1993

THE PRODUCTION OF ENGINEERING CERAMIC FOAMS FROM AQUEOUS SUSPENSIONS

JGP Binner, PAI Sepulveda[1] and RT Smith[2]
Department of Materials Engineering & Materials Design
The University of Nottingham
Nottingham, UK

ABSTRACT

A new manufacturing route for foam ceramics has been developed and is now being commercialised. The primary advantage of the process is its inherent flexibility. The foams can be near net shape manufactured in a variety of shapes and sizes and after production are simply dried and fired. Machining is easily performed in the green and/or fired state. In addition, the porosity and pore size distribution can be varied to suit the application and a wide range of ceramic materials can be foamed. The primary criterion is the ability to produce a stable ceramic suspension. Foams have been produced in alumina, partially and fully stabilized zirconia, mullite, cordierite and hydroxyapatite with densities ranging from 5% to 40% of theoretical. Foams are currently being developed based on silicon carbide and aluminium nitride. Applications are diverse and range from furnace linings, high temperature kiln furniture and crucibles, to catalyst supports, hot gas filtration and artificial bone.

INTRODUCTION

Over the past few years there has been a significant increase in interest in the production and use of highly porous ceramic materials. This is associated mainly with the properties such materials offer, viz. high surface area, high permeability, low mass, low specific heat and high thermal insulation. These characteristics are found to be essential for technological applications such as catalyst supports, filters for molten metals, hot gases and ion exchange, refractory linings for furnaces, thermal protection systems, heat exchangers and as porous implants in the area of biomaterials. Cell size, morphology and level of interconnectedness are also important factors that influence

[1] Now with Universidada Federal De São Carlos, São Paulo, Brazil.
[2] Now with Dytech Corporation Ltd, Sheffield, UK

the potential applications for these materials. Predominantly closed cell materials are needed for thermal insulation whilst open cell, interconnected materials are required for uses involving fluid transport such as filters and catalysts.

Historically, highly porous ceramics have been made by a variety of processes ranging from simple incorporation of a disposable additive, such as sawdust, wax or organic phases [1], through to more complicated techniques such as replication of polymeric foams via slurry coating [2] and chemical vapour infiltration of carbon skeletons [3]. The new process discussed in this paper utilises the *in-situ* polymerisation of an organic monomer to stabilise foams produced from aqueous ceramic powder suspensions and prevent them from draining and ultimately collapsing. The monomer has the added advantage of significantly improving the green strength of the foam to such an extent that it is readily machineable after drying.

Foams can be produced either by mechanical frothing, by injection of a stream of gas or by evolution from an exothermic reaction within a ceramic suspension. In most cases the addition of a surfactant is required to reduce the surface tension of the gas-liquid interfaces and hence stabilise the gas bubbles developed within the slurry. Such stabilisation only works for a limited period of time, however since several transformations in the bubble structure may subsequently occur due to a thinning of the lamella surrounding the bubbles. This necessitates a further mechanism to provide a longer term form of stabilisation since any changes in the foam structure prior to solidification influence the final cell size distribution, wall thickness and microstructure of the solid foams. These in turn have a major role in determining properties such as permeability and strength. For example, when the films surrounding bubbles remain intact until solidification, a closed-cell foam is formed. Open-celled foams are produced when the films partially rupture. In extreme cases, excessive film rupture can lead to foam collapse.

A number of approaches have been adopted for the setting of fluid suspensions, regardless of whether they have been previously foamed. They include the incorporation of gelling substances such as cellulose derivatives [4] and alginates [5] (or the use of compositions which can intrinsically form a gel [6]) and the *in situ* polymerisation of monomers [7]. The organic monomers used for the latter route must be soluble in water and retain a high reactivity. Commonly used materials include methyl methacrylate, butyl acrylate, acrylamide, and other acrylates. The polymerisation of these vinyl monomers can be brought about by a variety of initiating systems that have a marked influence on the rates at which the reaction proceeds. For example, chemical initiation using systems such as peroxide-amine, as initiator and catalyst respectively, can be very effective in producing fast polymerisation [8].

THE FOAMING PROCESS

The process for fabricating the ceramic foams, which has now been taken to pilot plant status, is given by the flowchart in figure 1. A stable, well-dispersed, high solids content, aqueous ceramic suspension is prepared which also incorporates an acrylate monomer together with an initiator and catalyst. The latter is used to provide *in-situ* polymerisation. After the further addition of a foaming agent, a high shear mixer is used to provide simple mechanical agitation which results in the formation of a wet ceramic foam. Process variables include the degree of mechanical agitation and the pressure in the system. The latter affects the degree to which the bubbles formed can expand and hence can be used to control pore size if required.

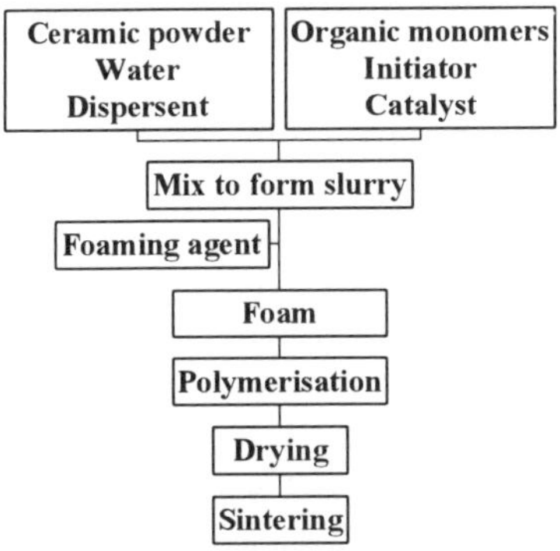

Figure 1: Process flow chart for production of ceramic foams.

One of the advantages of the *in-situ* polymerisation method is that it is common to observe a period of inactivity between the addition of reagents and the actual beginning of the polymerisation reaction. This is known as the induction period or idle time (t_i). The induction period is beneficial since it allows the casting of the fluid foam into a mould prior to polymerisation and stabilisation of the foam structure. However, on the negative side, the induction period also allows time for bubble enlargement and lamella thinning. These can result in the presence of flaws in the cell walls if excessive disruption of the films occur before polymerisation takes place. Good control of the induction time is primarily achieved by altering the concentration of initiator and catalyst, although other parameters are also very important in determining the induction time, such as the temperature and pH.

The manufacture of a ceramic foam begins with the selection of the appropriate ceramic powder. This depends on the end use of the foam and the physical and chemical properties required. However the heart of the process involves the creation of a stable ceramic slurry which necessitates the use of fine ceramic powders, typically D_{50}

< 3 μm. Materials have been processed with larger particle sizes but the quality of the dispersion is inferior and the ability of the powder to densify within the foam struts during sintering is reduced. Once selected, the powder is mixed with deionized water containing dispersants, initiator, catalyst and a premix solution containing ~6 wt% organic monomers. High shear mixing is used to destroy any agglomerates. The organic monomer solution provides a low viscosity liquid medium which can be readily polymerised to form a strong cross linked polymer-water gel. The polymerisation reaction traps both the ceramic powder and the water in the foamed structure. A foaming agent (surfactant) is then added to the slurry, the amount of which depends on the viscosity of the slurry and the final density of the foam desired, prior to transfer to a continuous foamer. This is then configured to produce a foam with the required density and cell size consistently and repeatably. Usually the slurry is foamed to between 2 and 7 times its original volume. The onset of polymerisation is controlled to allow enough time for the foam to exit the foaming unit and be cast into the appropriate mould. After polymerisation is complete the structure is strong enough to be demoulded and transferred to an oven for drying. The polymeric binder is burnt out as a controlled step in the sintering cycle to leave a mechanically strong, highly porous ceramic.

CHARACTERIZATION AND PROPERTIES

A wide range of ceramic materials can be produced as foams using the new process. Although the majority of research to date has focused on the engineering ceramic oxides, materials such as alumina, cordierite, mullite and zirconia, a large number of other ceramics have also been successfully foamed. Materials such as the bioceramic hydroxyapatite, the electroceramic lead zirconate titanate (PZT) and low thermal expansion sodium zirconium phosphate (NZP). Recently, work has begun on developing foams from non-oxide ceramics such as silicon carbide and aluminium nitride. In general, foams in the range 5% to 40% of theoretical density can be produced using the technique. For alumina and the other engineering oxides, three standard densities are now commercially available. These are 10%, 20% and 30%; for alumina foams (99.7% purity) their trade names are Por-Al 10, Por-Al 20 and Por-Al 30 respectively. Customised densities can be produced if required.

Although it is true to say that the lower the density of the foam the larger the cell or pore size, recent research has allowed a far greater degree of control to be achieved. Cell diameters as small as 20-50 μm or as large as 50-100 μm can be produced for the Por-Al 30 grade for example, whilst for comparison the Por-Al 10 grade typically has cell sizes of 300-400 μm. A typical micrograph of Por-Al 30 is shown in figure 2. Note how the pore walls and struts are solid and fully dense; this provides high strength and chemical resistance.

Figure 2: Typical micrograph of a 30% dense alumina foam.

Values for the modulus of rupture and crush strength of the Por-Al range are provided in table I. The former were measured by a four-point bend technique on samples measuring approximately 150 x 25 x 25 mm whilst the latter were measured on 25 mm diameter disks which were 25 mm thick. Not only are the values high for such porous materials, but they are also consistent and reproducible. Some of the thermal properties are illustrated in table II. Materials with a range of service temperatures from 1000 to 2000°C are available with thermal expansions ranging from zero to approximately 7 mm/mm/°C. The lower the thermal expansion the better the thermal shock resistance of the ceramic foam. After 140 hours at 1600°C, Por-Al 20 beams measuring 190x25x25 mm were found to deflect by just over 4 mm whilst the Por-Al 30 grade deflected only 2.5 mm illustrating long-term, high temperature stability. This conclusion is reinforced by the <1% linear shrinkage observed after holding any of the Por-Al range at 1750°C for 8 hours. The thermal conductivity is also relatively low, only about 1 $W\ m^{-1}\ K^{-1}$ for both Por-Al 20 and Por-Al 30 at about 1200°C. Although this is higher than some fibre-based products, ceramic foams offer a totally fibre- and dust-free working environment. Both zirconia and mullite also have lower thermal conductivities than the Por-Al range.

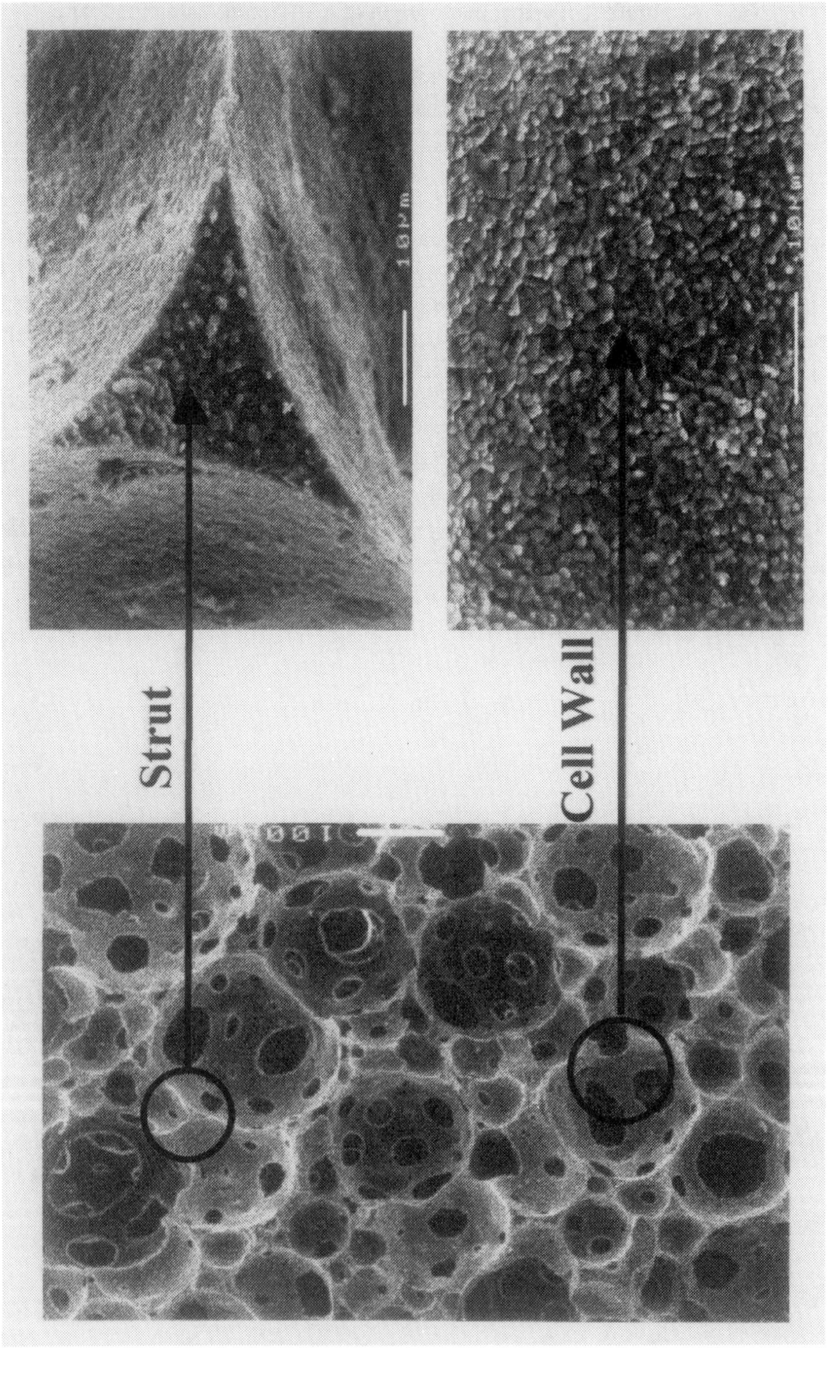
Strut
Cell Wall

Table 1: Modulus of rupture and crush strength for Por-Al ceramics

Grade	Modulus of Rupture / MPa	Crush Strength / MPa
Por-Al 10	2.5 ± 0.3	3 ± 0.5
Por-Al 20	10.4 ± 2.0	25 ± 6
Por-Al 30	25.4 ± 4.0	81 ± 15

Table II: Thermal properties of foam ceramics

Ceramic	Thermal expansion mm/mm/°C	Maximum service temp. /°C
Zirconia	-	2000
Alumina	7.08×10^{-6}	1900
Mullite	4.38×10^{-6}	1800
Cordierite	1.36×10^{-6}	1300
NZP	0	1500
Silica	-	1000

Figure 3: A small selection of shapes from Por-Al ceramics

A wide range of component shapes are available. The production route itself is intrinsically a casting process and hence tiles, tubes and a range of other custom shapes

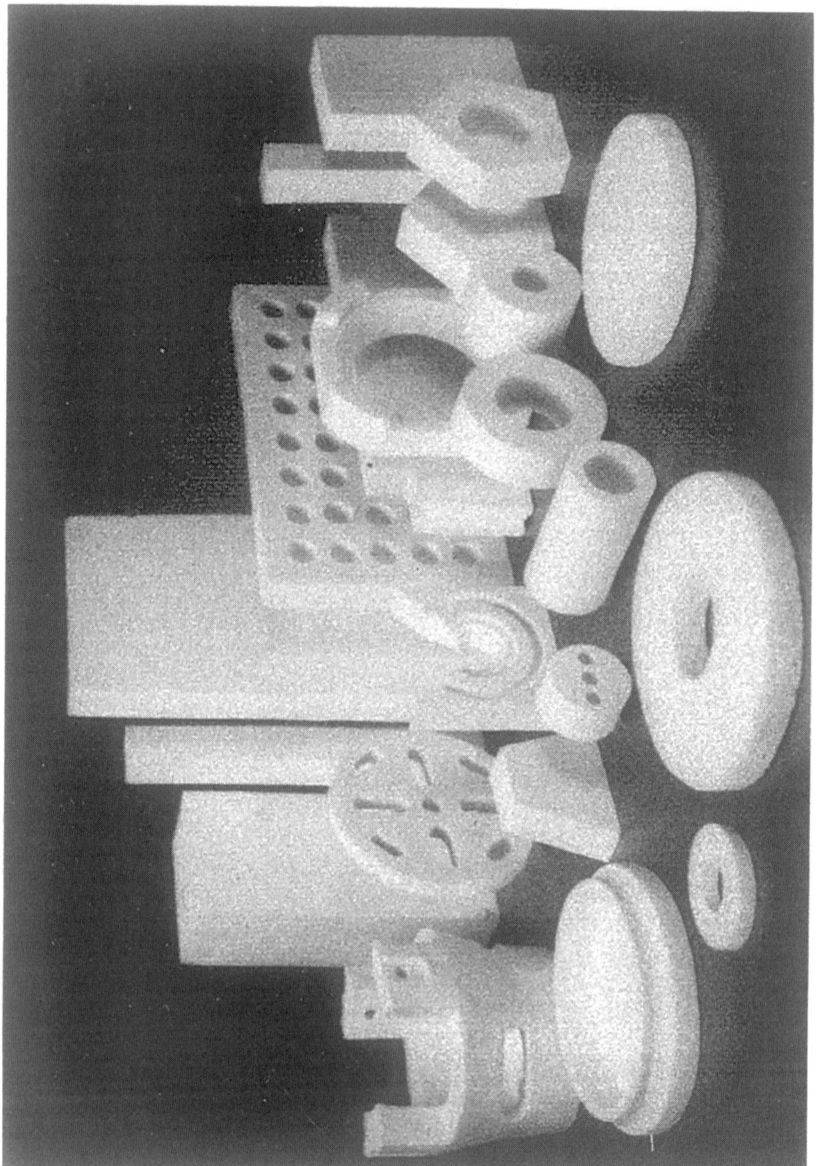

can all be produced very easily. In addition, both the green and fired foam may be readily machined, drilled, turned and slit opening up the possibility of producing some very complex shapes indeed. Figure 3 above illustrates just a small selection of the shapes which may be produced. In addition, it is a simple process to apply a dense surface coating either to eliminate permeability or to increase the mechanical properties of the surface layers. One application which could utilise this is the production of ultra-low mass ceramic crucibles. These can significantly reduce the thermal mass to be heated during processing of their contents, thus improving energy efficiency.

Applications for the engineering ceramic foams are increasing rapidly; to date they are being considered for thermal insulation, kiln furniture, crucibles, radiation shields, catalyst supports and where corrosion resistance is important. The hydroxyapatite foams have shown themselves to be successful both as potential bone substitute materials, bone cells grow into them readily, and also as a filter for removing heavy metal ions from aqueous solutions.

CONCLUSIONS

A new, novel and very flexible manufacturing route for producing advanced ceramic foams has been invented, developed, patented and commercialised as a result of a collaboration between The University of Nottingham and Dytech Corporation Ltd in the UK. Foams have been made from a wide range of ceramic systems and the results indicate that provided a stable aqueous suspension can be produced it is possible to manufacture a useful foamed product. Densities range from 5% to 40% of theoretical however for production purposes three grades are currently being developed; 10%, 20% and 30%. Cell size is dependent on the density of foam produced, however greater control of this aspect of the process is consistently being achieved. Dense coatings may be easily applied to the surfaces of the foam products to eliminate permeability and/or improve mechanical strength.

ACKNOWLEDGEMENTS

The authors would like to acknowledge Dytech Corporation Ltd, the Engineering and Physical Sciences Research Council and the Teaching Company Scheme in the UK and CAPES in Brazil.

REFERENCES

1 Kormarneni, S, Pach L, Pidugu, R, "Porous □-Alumina Ceramics using Boehmite and Rice Flour", *Mat Res Soc Symp Proc* **371** pp 285-290, 1995.

2 Minnear, W P, "Processing of Porous Ceramics", *Ceram Trans* **26** pp 146-156, 1992.

3 Lin, Y S, Burggraaf, A J, "CVD of Solid Oxides in Porous Substrates for Ceramic Membrane Modification", *AIChe Journal* **38** [3] pp 445-454, 1992.

4 Moreno, R, "The Role of Slip Additives in the Tape-Casting Technology", *Ceram Bull* **71** [11] pp 1647-1657, 1992.

5 Katsuki, H, Kawahara, A, Ichinose, H, "Preparation and Some Properties of Porous Alumina Ceramics Obtained by Gelatination of Ammonium Alginate", *J Mat Sci* **27** pp 6067-6070, 1992.

6 Bagwell, R B, Messing, G L, "Critical Factors in Production of Sol-Gel Derived Porous Alumina", *Key Engineering Materials*, **115,** pp 45-64, 1996.

7 Young, A C, Omatete, O O, Janey, M A, Menchhofer, P A, "Gelcasting of Alumina", *J Am Ceram Soc* **74** [3] pp 612-618, 1991.

8 Feng, XD, Guo, XQ, Qui, KY, "Study of the Initiation Mechanism of the Vinyl Polymerisation with the System Persulfate/N,N,N',N',-tetramethylethylenediamine", *Makromol Chem* 189 pp 77-83, 1988.

Preparation of Shirasu/alumina lightweight composites through a colloidal processing

Y. Sakka* and K. Sodeyama**
*National Research Institute for Metals,
1-2-1, Sengen, Tsukuba, Ibaraki 305-0047, Japan
**Kagoshima Prefectural Institute of Industrial Technology,
1445-1, Oda, Hayato-cho, Aira-gun, Kagoshima 899-5105, Japan

ABSTRACT

Fine hollow glass microspheres named Shirasuballoons with particle sizes under 20 μm and bulk densities above 0.5 g/cm^3 were prepared by heating rapidly milled Shirasu (vitric volcanic material) using a fluidized bed furnace. Dispersed slurry of fine Shirasuballoons and fine alumina powders were prepared by adding an appropriate amount of polyelectrolyte. The green compacts were sintered at 1200-1500°C for 2 h in air and preparation conditions of the lightweight Shirasu/alumina composites were discussed. Flexural strength and relative dielectric constants of the composites were measured. The dielectric constants obtained were interpreted using a modified Maxwell's equation.

INTRODUCTION

"Perlite" is a general term for a glassy, volcanic, rhyolitic rock, which is classified into three kinds, *i.e.* obsidian (water content: less than 2 mass%), perlite (2-5 mass%) and pitchstone (above 5 mass%) depending on their water contents[1]. "Perlite" is known to expand up to 20 times its original volume upon heating, which is called as expanded perlite. Because of its lightweight, expanded perlite is used in many fields, such as ceiling tile, pipe insulation, gypsum wallboard, cryogenic insulation and filter media, *etc.*[2]. However, the expanded hollow glass microsphere is vulnerable to shearing forces and breaks easily when mixed with resin and cement. One of the methods of solving this problem is reducing their sizes to increase their strength [3].

In Japan, hollow glass microspheres named as Shirasuballoons have been produced from vitric volcaniclastic materials named as Shirasu or Hakudo (a kind of the "perlite") since the first preparation in National Industrial Research Institute of Kyushu, Japan [4]. In attempts to produce fine balloons from finely milled particles under a conventional method, the water contained in the particles is evaporated in the process of heating; little remains at the glass softening point and fine particles are not formed. We have successfully produced fine Shirasuballoons of particle sizes under 20 μm and bulk densities under 0.5 g/cm^3 using a rapid heating in a sand-medium fluidized bed furnace [5-7].

The fine Shirasuballoons is expected to be a useful material for lightweight composites. In this study, conditions for preparing Shirasu/alumina lightweight composites were examined using the pulverized Shirasu or fine Shirasuballoons and mechanical and dielectric properties were examined.

EXPERIMENTAL

The powder named as Yoshida-Shirasu was supplied from Wakamatsu Shoten, and alumina powder is Sumitomo's powder with mean particle size of 0.5μm and specific surface area of 7.2 m^2/g (chemical composition: 99.5 wt% Al_2O_3, 0.05 wt% MgO, 48 ppm Na, 14 ppm Fe, 7 ppm Si). To pulverize the Shirasu into powders with the average particle size of 8.2 μm, a Jet Mill (IDS-2 type: Nippon Newmatic MFG. Co., Ltd.) was used. Foaming experiment of the pulverized Shirasu was conducted using a fluidized sand-bed furnace, whose apparatus is shown in Fig. 1. As for fluid bed division, the propane gas, supplied from the lower side, is ignited and burned out on the perforated dispersion plate by the highly preheated quartz sand of approximately a diameter of 1mm and the fluidization of the high-temperature heating medium is continuously maintained by the gas flames. The temperature of this fluid bed is controlled by the mixing ratio of propane gas to air and the feed quantity of the gases. The mixing ratio is controlled with an electromagnetic valve by monitoring the temperature of the fluidized reactor core using a thermocouple that is placed at the center of the furnace. In this experiment, heating time up to a maximum temperature of 1000 °C is less than 0.06 s, which is estimated by passage time of raw gas and temperature distribution in the furnace [7].

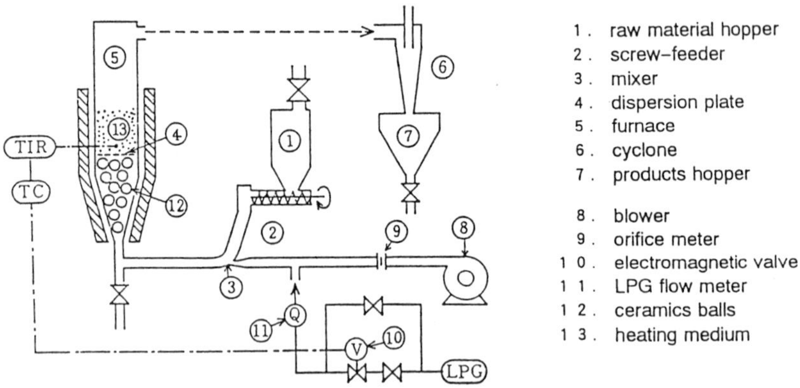

Fig. 1. Schematic of a fluidized bed furnace system.

The variations of zeta potential with pH for the pulverized Shirasu and fine Shirasuballoons were measured in order to adopt colloidal processing. The ratios of pulverized Shirasu or fine Shirasuballoons to alumina were 28.6 mass% for both, and the ratios of the powders to distilled water were set to 42.9 and 56.3 mass%, respectively. Ammonium polyacrylate (Kyowa Sangyo) was added as a polyelectrolyte, and acrylic emulsion (Mitsui-Toatsu Chemicals, Inc.) of 2.9 mass% was added as a binder. They were mixed for 1h with alumina balls in a plastic cylindrical container to prepare suspensions. After defoaming in vacuum, they were cast in a gypsum mold. The green bodies deried at 110°C were heated at 2.5°C /min in air to the temperature of 1300-1500°C. The particle density of starting materials and true density of sintered bodies were calculated from the weight and volume of specimens obtained by the He gas replacement method using a multipycnometer (MVP-1; Yuasa-Aionikusu). Here, the true density of the sintered bodies was measured after crushing in order to exclude the influence of closed pore.

The sintered bodies were cut into rectangular solids 3x4x35 mm in size. The strength of these samples was measured by the 3-point bending method with a span of 30 mm and crosshead speed of 0.05 mm/min. The relative dielectric constants were measured with sintered bodies planed into tiles about 50 mm square and about 3 mm thick by the electrode contact method using an LCR meter (4285A; Hewlett Packard) with a guard 38 mm in inside diameter, under the condition of 1 MHz in frequency.

RESULTS AND DISCUSSION

Fine Shirasuballoons was synthesized by the rapid heating the pulversized Shirasu at a firing temperature of 1000 °C . FigS. 2 and 3 show the particle distribution and SEM photographs of the pulverized and foamed fine Shirasuballoons, respectively. The pulverized Shirasu has a complicated angular shape. The fine Shirasuballoons show smooth spherical surfaces and consist of hollow glass spheres with outer shell thickness less than 0.5 μm. Table 1 shows chemical composition of them.

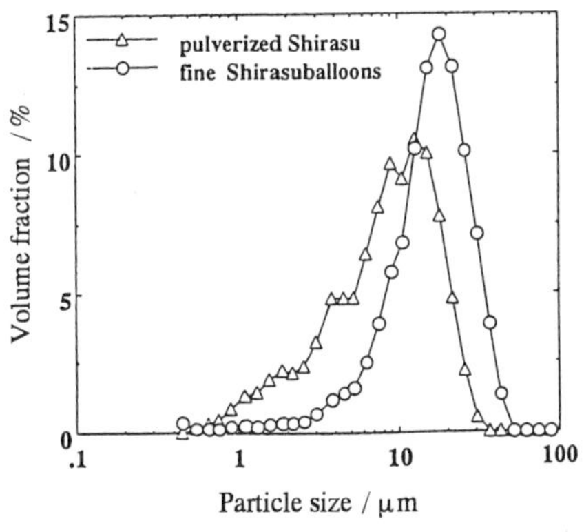

Fig. 2. Particle size distributions of pulverized Shirasu and fine Shirasu balloons.

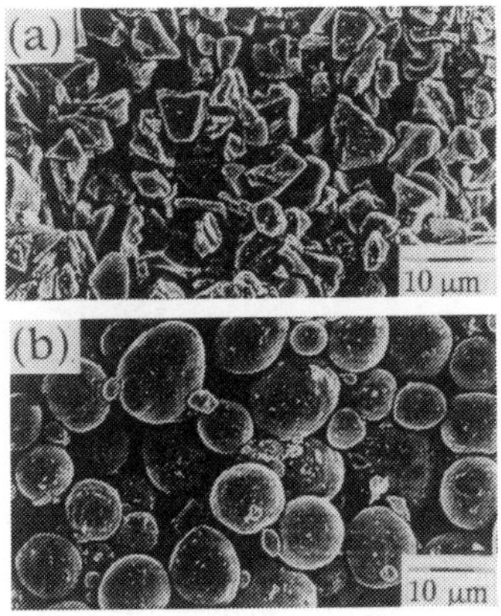

Fig. 3. SEM micrographs of (a) pulverized Shirasu and (b) fine Shirasu balloons.

Table 1 Chemical compositions of pulverized Shirasu and fine Shirasuballoons.

Sample Name	SiO_2	Al_2O_3	K_2O	Na_2O	CaO	Fe_2O_3	MgO	TiO_2	Ignition loss
pulverized Shirasu	71.4	12.7	2.8	3.4	1.7	1.7	0.5	0.3	5.5 %
fine Shirasuballoons	73.9	13.2	2.9	3.3	1.7	1.7	0.5	0.3	1.8 %

Control of electrostatic interaction has significant influence on the stability of the suspension. Fig. 4 shows the pH vs the zeta potential of the pulverized Shirasu and fine Shirasuballoons. Each has the same tendency as SiO_2 [8], takes large negative zeta potential at pH>9. The Shirasu or Shirasuballoons is stable at pH>9, where alumina is not stable (flocculated) compared to pH=3.5.[9] To solve this dilemma, we used a surfactant to stabilize the alumina by adding of 0.7 mass% ammonium polyacrylate. This quantity is sufficient to coat the surface of the alumina particles, where the pH value was in the range of 9.2-9.3. In this way, the suspension of Shirasu (or Shirasuballoons) and alumina is stabilized. The densities of the slip casted bodies were 3.22 and 2.65 g/cm^3 for pulversized Shirasu/alumina (hereafter abbreviated as SA green body) and fine Shirasuballoons/alumina (hereafter as BA green body), respectively.

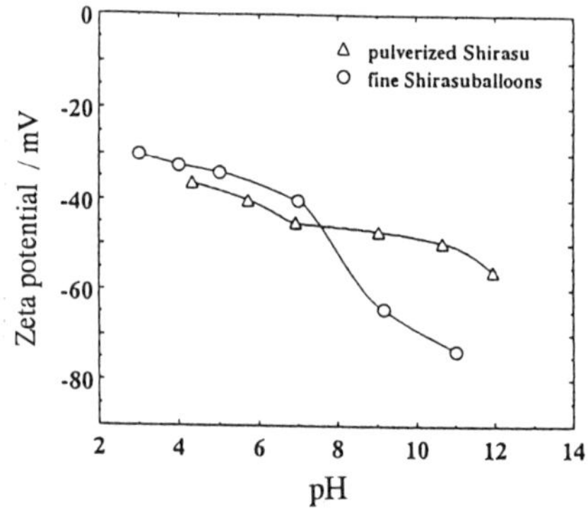

Fig. 4. Zeta potential vs pH for pulverized Shirasu and fine Shirasuballoons.

Shirasu/alumina light weight composites were prepared by sintering the SA or BA green bodies at 1300-1500°C for 1 h. Fig. 5 shows X-ray diffraction patterns after heating at fixed temperatures. In each case of SA and BA sintered bodies, intensity of alumina peaks decreased with heating at 1500 °C and the peak of mullite was generated; this is probably because the dissolution of alumina into the Shirasu phase occurred and then mullite was formed. Moreover, from the peak intensity of mullite, it is found that the use of pulverized Shirasu increases the amount of generated mullite. Below the heating temperature of 1400 °C, the peak of mullite was not observed.

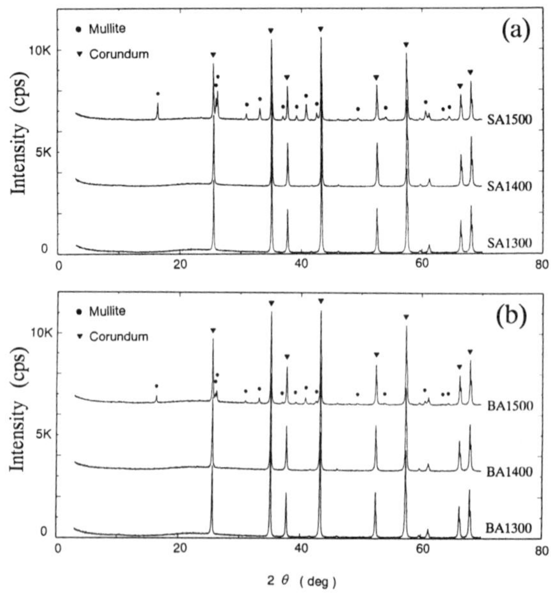

Fig. 5. X-ray diffraction patterns of (a) SA and (b) BA composites after sintering at 1300, 1400 and 1500°C for 1h.

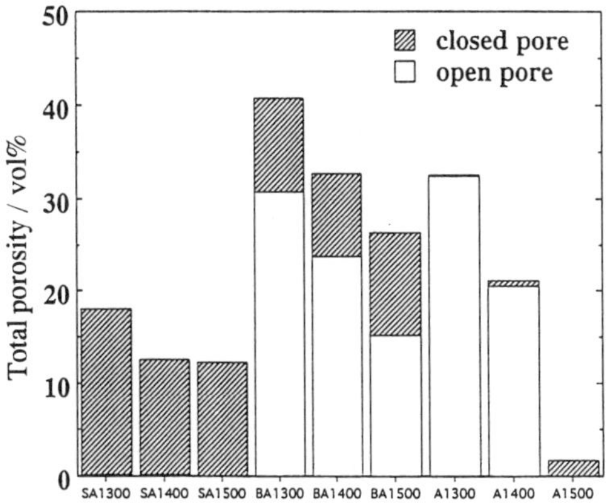

Fig. 6. Closed and open porosities of SA, BA and A samples after sintering at fixed temperatures.

Fig. 6 shows the relation between closed porosity and open porosity of SA, BA and A (alumina) sintered bodies heated at 1300-1500°C. The open porosity of SA sintered bodies is almost 0% at 1300-1500°C, and the closed porosity of SA1500 (SA heated at 1500°C) was 12% equal to that of SA1400. By contrast, in case of BA sintered bodies, open porosity decreased with increasing the sintering

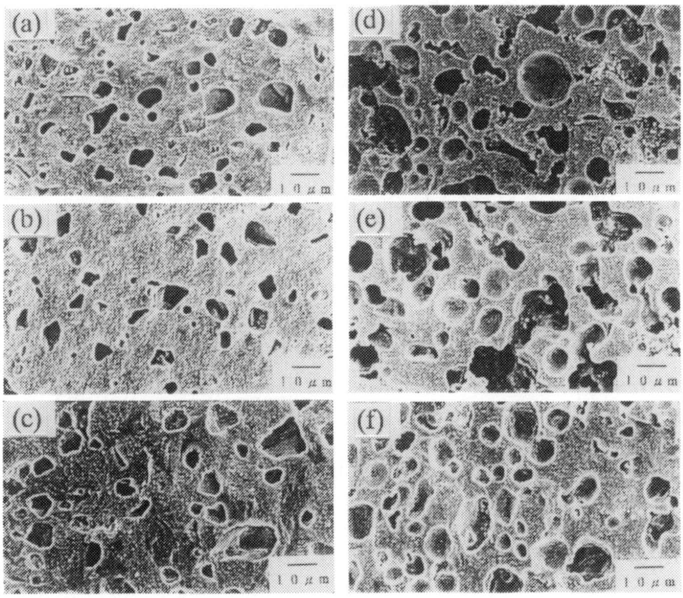

Fig.7. SEM micrographs of Shirasu/alumina composites, (a) SA1300, (b)SA1400, (c) SA1500, (d) BA1300, (e) BA1400 and (f) BA1500.

temperature, while even BA1500 had an open porosity of 15% and a closed porosity of 11%. Fig. 7 shows SEM micrographs of the farctured surfaces of SA and BA sintered bodies of 40 vol%, so that closed pores are dispersed sparsely, and their pores have complicated shape in many cases.

Fig. 8 shows the relation between the heating temperature and flexural strength. SA and SB sintered bodies showed maximum flexural strength that a heating tempearture of 1400°C; the SA sintered body had higher strength equal to that of A sintered body at 1400°C. Then at 1500°C, the flexural strength of both SA and BA sintered bodies became slightly lower than that at 1400°C. Generally, strength improves with a decrease in porosity, and yet the fact that strength decreased oppositely with 1500°C heating suggests that it is related with the mullite genertation.

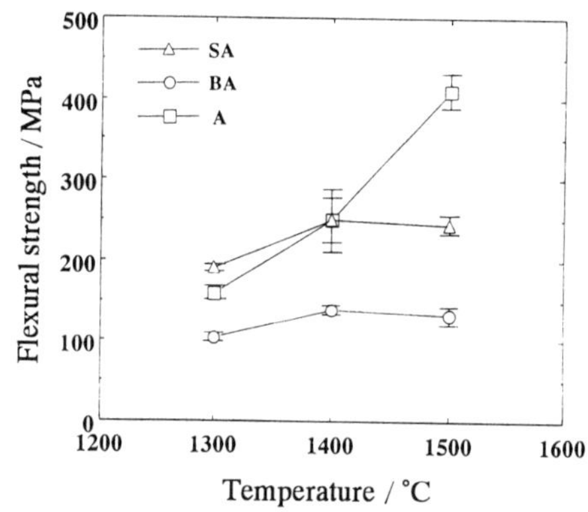

Fig. 8. Sintering temperature dependences of flexural strength of SA, BA and A samples.

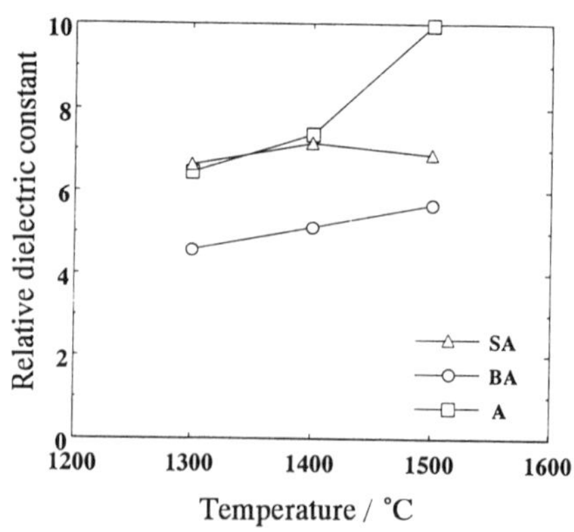

Fig. 9. Sintering temperature dependences of relative dielectric constants of SA, BA and A samples.

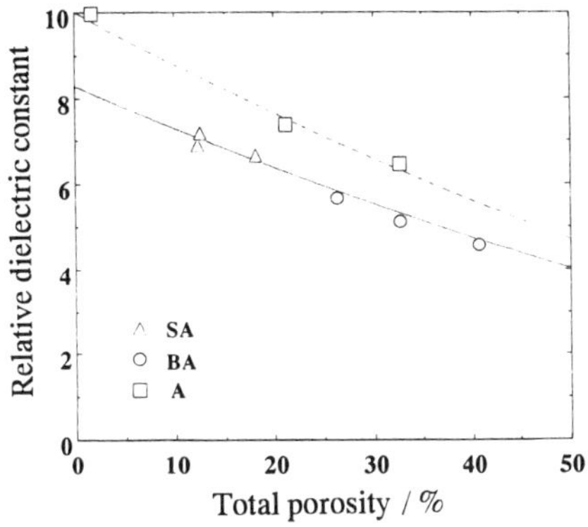

Fig. 10. Relationship between relative dielectric constants and total porosities.

Fig. 9 shows relative dielectric constant of each sintered body. The relative dielectric constant is greatly influenced by chemical composition and porosity, so that it agrees well with the tendency of bulk density. The relation between the relative dielectric constant and total porosity is given in Fig. 10. In each case, the relative dielectric constant lowers with an increase in porosity. The following modified Maxwell's equation is known with respect to the dielectric constant of a composite having dispersed particles [10].

$$K_{composite} = \{V_m K_m[2/3+K_d/(3K_m)]+V_d K_d\}/\{V_m[2/3+K_d/(3K_m)]+V_d\}$$

where V_d and V_m are volume fraction of spherical particles and matrix, respectively, and K_d and K_m are dielectric constant of a spherical particle and matrix, respectively. Caluculated results with spherical particles in alumina sintered bodies as pores are shown with a broken line. The calculated values are in good agreement with measured values. On the other hand, in case of Shirasu/alumina composites except for SA1500 and BA1500 which generated mullite, it was first assumed that they consist of two phases of pores and the matrix, the value of the matrix was obtained from the modified Maxwell's equation, and then the relative dielectric constant with each porosity was calculated from the former value. The result is shown with a solid line and the relative dielectric constant of the matrix 8.27 was obtained. The relative dielectric constant of Shirasu obtained from Maxwell's equation was 6.0 when the volume fraction of Shirasu was set to 40vol%, the same value as that of green body. This Maxwell's correction curve agreed well with the measured values of SA and Ba sintered bodies. SA 1500 with large amount of mullite showed the value slightly lower than modified Maxwell's curve for Shirasu/alumina composites without mullite generation. Shirasu/alumina composites provided lower dielectric constants than those of alumina sintered bodies with the same porosity.

In recent years, a technique of introducing pores (relative dielectric constant: 1.0) inside materials has been tried as a method of providing inorganic substrate materials for IC package with low dielectric constants.[11] The Shirasu/alumina lightweight composites are useful for their strength and low dielectric constants.

ACKNOWLEDGEMENTS

We wish to thank Y. Kamino, I. Tabata and K. Hamaishi (Kagoshima Prefectural Institute of Industrial Technology) for their helpful suggestions and discussions.

REFERENCES

[1] K. Hamano, "Natural glass and perlite," Yogyo-Kyokai-Shi, **73**, C8-C12 (1965).

[2] W. P. Bolen, "Perlite," Bull. Am. Ceram. Soc., **73** [6], 117-18 (1994).

[3] K. Sodeyama, Y. Sakka, Y. Kamino and K. Hamaishi, "Properties of Fine Shirasuballoons," J. Ceram. Soc. Japan, **105** [1], 79-84 (1997).

[4] K. Kimura, K. Jinnai and Y. Isayama, "On the Hollow Glass Micro-Sphere made from the "Shirasu" (Silas Balloons)," Yogyo-Kyokai-Shi, **80** [2], 84-91 (1972).

[5] K. Sodeyama, Y. Sakka, Y. Kamino and H. Seki, "Preparation of Fine Shirasuballoons from Vitric Volcaniclastic Materials," J. Japan Soc. Powder and Powder Metall., **42** [10], 1128-35 (1995).

[6] K. Sodeyama, Y. Sakka, Y. Kamino and I. Tabata, "Preparation and Characterization of Fine Shirasuballoons," J. Ceram. Soc. Japan, **104** [10], 963-68 (1996).

[7] K. Sodeyama, Y. Sakka, Y. Kamino, H. Seki, K. Nishimoto and M. Yamazaki, "The Manufacturing Process of the Fine Shirasuballoons Using a Fluidized Bed Furnace," J. Soc. Powder Technol., Japan, **34** [9], 697-703 (1997).

[8] T. Uchikoshi, Y. Sakka, K. Ozawa and K. Hiraga,"Preparation and Sintering of Silica-doped Zirconia by Colloidal Processing," Mater. Res. Soc. Symp. Proc., **457**, 33-38 (1997).

[9] J. Cesarano III, I. A. Aksay and A. Bleier, "Stability of Aqueous α-Al_2O_3 Suspensions with Poly(methacrylic acid) Polyelectrolyte," J. Am. Ceram. Soc., **71** [4], 250-55 (1988).

[10] Y. Chiang, D. P. Birnie and W. D. Kingery, "Physical Ceramics," John Wiley & Sons (1997) pp.466-73.

[11] S. Wu, L. C. DeJonghe and V. B. Dutta, "Development of Al_2O_3-Based Substrate with Low Dielectric Constant for IC Packaging," Ceramic Trans., **32**, 299-308(1993).

EFFECT OF WATER CONTENT ON CERAMIC FORMING BY COMPACTION

Junichi Nyuumura, Yao Zhang, Nozomu Uchida, and Keizo Uematsu
Department of Chemistry
Nagaoka University of Technology
Kamitomioka, 1603-1
Nagaoka, Niigata, Japan

ABSTRACT

Characteristics of ceramic powder granules and their compacts are affected critically by their water content, since the properties of organics added for controlling them depend on the moisture strongly. This paper examines the effect of water content on the granule strength and the diffusion coefficient of water in ceramic green body made by the compaction process. The granule strength was measured by a micro-compression testing machine. The diffusion coefficient was determined by following the weight change of the green body after step change of humidity. The granule strength was affected markedly by the PVA content and the humidity. The diffusion coefficient decreased as PVA concentration in compacts increased under a given humid condition.

DIRECT EXAMINATION OF THE STRENGTH VARIATION OF ALUMINA POWDER GRANULES WITH BINDER CONTENT AND RELATIVE HUMIDITY

Junichi Nyumura, Yao Zhang, Nozomu Uchida, Keizo Uematsu
Department of Chemistry, Nagaoka University of Technology
Kamitomioka 1603-1, Nagaoka, Niigata 940-2188, Japan

ABSTRACT

Stress-strain curves were examined on individual alumina powder granules with various PVA contents at controlled humidity at 25°C. The deformation behavior was affected markedly by the relative humidity (RH). Elastic deformation and brittle fracture occurred at RH=40%. Plastic deformation and fracture were noted at RH=60% and higher. The strength of granules increased with increasing PVA content under all conditions. The significant effect of RH on the deformation behavior was explained by the change of glass transition temperature of PVA with moisture content. The significance of strength variation on the processing of ceramics was also discussed.

INTRODUCTION

The powder compaction process is widely used in commercial production of ceramics. Granules made by spray drying are placed in a die and compacted uniaxially to form a green compact, which is then heated for densification. The characteristics of powder granules need to be fully controlled for high productivity as well as for high quality of product. Especially the mechanical properties of the powder granules need to be properly adjusted, since they govern the structure of compacts, and thus the structure and properties of the final products[1-4]. Die compaction is often applied to examine the stress-strain behavior of granules and also to discuss the characteristics of granules and their relevance to the structure of the compact[5-7]. These experiments, however, show only the overall behavior for a mass of granules. The local compaction behavior, rather than the overall compaction, governs the formation of processing defects in the compact. For minimizing the processing defects in final products, it is necessary to examine the characteristics of individual granules which directly affect the local structure of the compact. Kamiya et al[8] examined the characteristics of individual granules with micro-compaction equipment. They reported many interesting new results, but did not examined effects of some important parameters on the characteristics, such as binder and moisture content. Their influence must be examined, since a marked influence of humidity on compaction behavior was reported in a past study involving die compaction[9]. A variation of PVA content is present within a single granule[10] and affect the local deformation. In the present paper, the compaction behavior of single granule is reported under controlled humidity and temperature.

EXPERIMENT

The materials used in these experiments were industrial grade commercial alumina (Al160SG-1, Showadenko, Tokyo) and polyvinyl alcohol (PVA205, Kurare, Tokyo) as a binder. According to the supplier, the average particles size of alumina powder is 0.6μm. The degree of polymerization and the sophonification value of the binder are 500 and 88%, respectively. They were mixed with distilled water (solid/water=1) in a ball mill for 16h to prepare slurries of binder contents 0-3mass%. No dispersant was used in these slurries. A spray drier (SD-13, Mitsuikozan, Tokyo) was used to prepare granules at the inlet and outlet air temperatures 200°C and 100°C, respectively. The granules were stored in an air-tight container and exposed to the atmosphere at least for 1h before experiment by spreading them on a flat sheet to equilibrate the moisture content with the atmosphere. The relative humidity of the atmosphere was controlled within a few percent of specified values. Commercial micro-compaction equipment (MCTE-500, Shimadzu, Kyoto) was used to examine the compaction behavior of individual granules. Granules were placed on the sliding stage of the equipment, and one having a diameter around 50 μm was selected under a microscope. After its diameter was measured at two positions, the granule was moved to a loading device by sliding the stage slowly. The stress-strain curve was obtained at the loading rate 90 μN/s at 25°C. The tensile strength of the granule *St* was determined with the equation proposed by Hiramatsu et al[11].

$$St = 2.8Pf/\ \pi D^2 \qquad (1)$$

where *Pf* is the applied load at fracture and *D* the average diameter of the granule. The values obtained were only rough measures of strength, since the equation is only valid for small deformation. An optical microscope was used to observe the compaction process. SEM (JSM-5310LV, JEOL, Tokyo) was used to

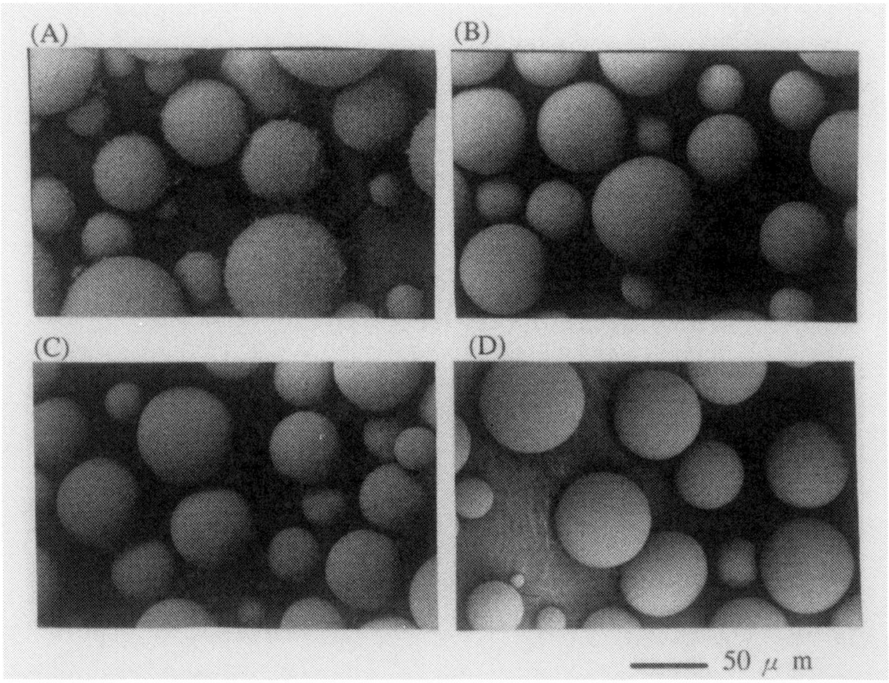

Fig. 1. SEM observastions for sprat-dried granule with the PVA content of (A)0mass%, (B)1mass%, (C)2mass%, (D)3mass%.

examine the morphology and structure of granules in detail. The liquid immersion technique was applied to examine the internal structure[12]. In this examination, the granules were immersed in methylene iodide to make them transparent. For observing the internal structure, an optical microscope in transmission mode (OPTIPHOT, Nikon, Tokyo) was used.

RESULTS

Fig. 1 shows the morphology of granules used in this study. Most of the granules shown in these micrographs have a near spherical shape. However, there were a small number of irregularly shaped granules present in a large mass of granules. The surface of granules containing PVA binder appears to be smooth. The surface of granules without organic additives appears to be rough.

Fig. 2 shows the liquid immersion photomicrograph of granules. Clearly the binder is distributed non-uniformly in granules. The dark layer at the circumferential region of granules corresponds to the segregated binder. The

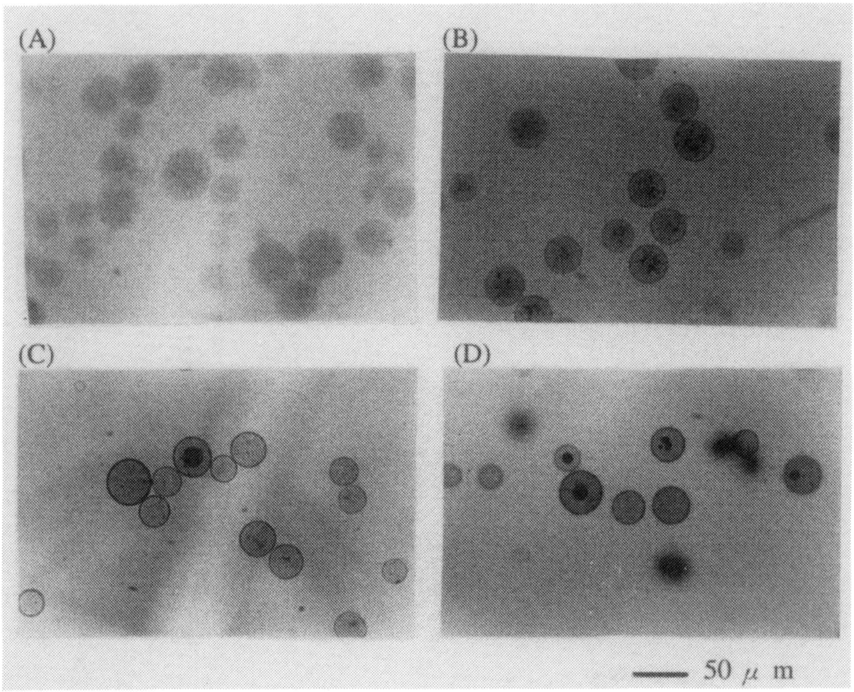

Fig. 2. The Liquid immersion photomicrograph of granules with PVA content of (A)0mass%, (B)1mass%, (C)2mass%, (D)3mass%.

darkness of the layer increased with increasing binder content. The reason the binder appears dark in the liquid immersion technique is that, due to the difference in refractive index, the light gets refracted away from the direction in which the light in all other areas of the sample is traveling. A separate experiment showed that the transmittance of light through granules increased significantly after the binder was removed by heating. A part of the binder added remains at the inner region of granules. The hollow granules and solid granules exist in the granules containing 3 wt% PVA. Granules are solid with some internal defects for PVA contents 0-2 wt%.

Fig.3 shows the effect of relative humidity on stress-strain curve for granules containing 3 wt% PVA at 25℃. Granules showed non-linear increase of strain with increasing stress at all conditions. Decrease of initial slope was noted in the curve with increasing relative humidity, showing that the granule was softened with moisture. At 40% relative humidity , an approximately linear relation was noted in the curve, showing nearly elastic behavior at this condition. The drastic change of strain at the stresses around 2N/mm^2 corresponds to the brittle fracture of the granule. At 60% relative humidity, the monotonic decrease of the slope was noted with increasing stress. At the stress around 1.4N/mm^2 a large change of slope was noted. This rapid deformation corresponds to the fracture of this granule. At the relative humidity 80%, the strain increased monotoneously with increasing stress, and no distinct fracture point was noted even with the strain over 20%. For this condition, the stress at the strain 20% was tentatively chosen as the fracture value. The fracture strength clearly increased with decreasing humidity. However, repeated experiments shows that a large variation of strength was present even at the same condition for these granules.

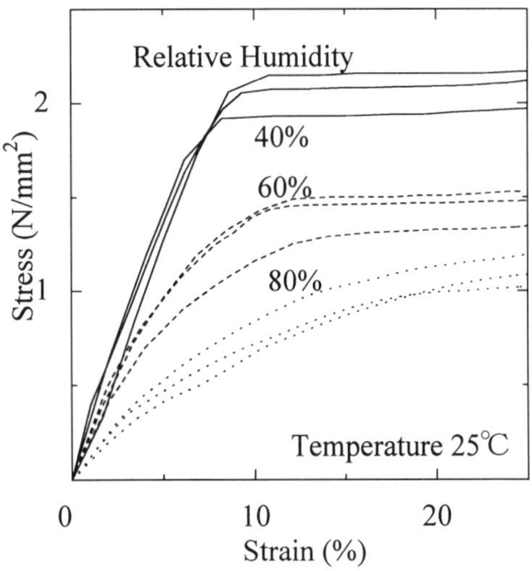

Fig.3 Effect of Relative Humidity on Stress-Strain Curve on Granule (3 mass% PVA)

Fig.4 shows the effects of PVA content and relative humidity on granule strength at 25℃. At the relative humidity 80%, the strength increased almost linearly with increasing PVA content. It increased more rapidly with increasing PVA content at the relative humidities 40 and 60%. The strength increased with decreasing relative humidity at all PVA contents. The strength was approximately the same for all relative humidities in granules containing no PVA. The effect of relative humidity on the granule strength increased with increasing PVA content of granule.

DISCUSSION

The stress-strain curves obtained for individual granules significantly deepen our understanding on the die compaction process. The marked change of granule strength was noted with binder content and relative humidity.

The latter can be easily explained in term of the glass transition temperature (Tg) of PVA[13]. Toyoshima reported the relationship for moisture content and sophonification value for PVA with degree of polymerization 1750, as well as that for Tg and moisture content[14]. Neglecting the slight difference in the characteristics of PVA, Tg for the present study is estimated. The values are 30°C and under 20°C for relative humidity 40% and over 60%, respectively. They suggest that elastic and plastic deformation should occur in the stress-strain curve for the relative humidity at 40% and over 60%, which agreed well to the experimental results.

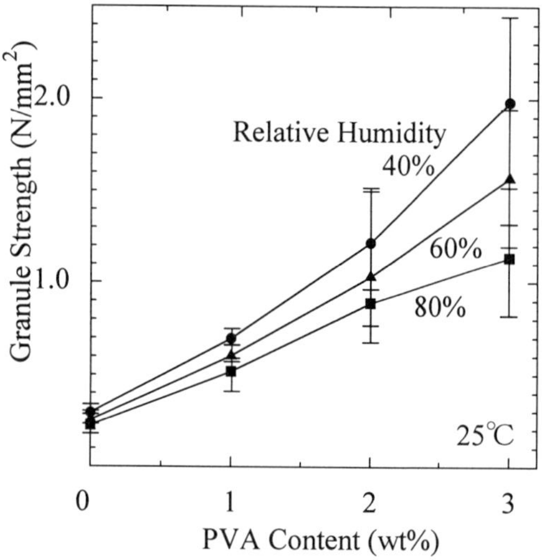

Fig.4 Effect of PVA Content and R.H. on Granule Strength

Significant variations were noted for granules with nominally the same composition at all conditions as shown in Figs 3 and 4. The similar change of strength is reported by Kamiya et al[8] for silicon nitride granules at a fixed condition. The source of the variation can be ascribed to the structure of the granules. Granules clearly have non-uniform structure as shown in Fig. 2, and this small difference can affect the mechanical behavior under stress. The variation is not reduced by the presence of binder nor its change in Tg.

The variation of strength should be very important for the fomation of defects in compact as well as in final product. Consider the strong granule among all granules in a die. At a certain high stress, deformation of granules is completed and rearrangement of primary particle starts to increase the relative density. The deformation, however, is not finished for the strong granule. At the end of the compaction process, the strong granule results in a region of low density. This region likely forms defect in the sintered ceramics.

In addition to the variation among granules, strength distribution should be present even within a granule. Surface region of a granule should be stronger than inner region due to high binder content. Formation of network defect structure was reported in our previous paper[12].

Kinetics for moisture content must be an important subject of discussion. In a separate paper, we estimated the moisture diffusion coefficient in a compact, an analogue of a granule, from the weight change experiment after step changes of relative humidity. The diffusion coefficient was about 3×10^{-5} cm^2/s in a spherical alumina compact containing 3 mass% PVA at 30℃ and 50% relative humidity. The saturation time was estimated for moisture diffusion into a granule with 50μm diameter and was below 1 second. At least a few minutes were needed to conplete a single run of granule compression test. Controlled temperature and humidity around the testing machine was again found to be indispensable for the present study.

CONCLUSION

The relationship between relative humidity and the characteristics of single granules were studied with the micro-compression testing machine. The test granules were prepared by spray-drying from slurries containing 0-3 mass% PVA and alumina powder. The stress-strain curve and granule strength was measured and the following result were obtained.

(1) The surface of granules without organic additives appears to be rough. The surface of granules containing PVA binder appears to be covered by primary particles uniformly. The binder is distributed non-uniformly in granules.

(2) Granules showed non-linear increase of strain with increasing stress at all conditions. Decrease of initial slope was noted in the curve with increasing relative humidity, showing that the granule was softened by moisture.

(3) The granule strength increased with increasing PVA content and decreasing relative humidity.

REFERENCE

1) H. Takahashi, N. Shinohara, K. Uematsu, J. Tsubaki, “Influence of Granule Character and Compaction on the Mechanical Properties of Sintered Silicon Nitride”, J. Am. Ceram. Soc., 79[4]843-48(1996)

2) S. J. Lukasiewicz, “Spray-Dried Ceramic Powders”, J. Am .Ceram. Soc., 72[4]617-24(1989)

3)Samir Baklouti, Thierry Chartier, Jean Francois Baumard, “Mechanical properties of dry-pressed ceramic green products: the effect of the binder", J. Am. Ceram. Soc., 80[8]1992-1996(1997)

4) B. J. Briscoe and N. Oezkan, "Compaction behavior of agglomerated alumina powders", Powder Technol., 90[3]195-203(1997)

5) C. W. Nies and G. L. Messing, "Effect of Glass-Transition Temperature of Polyethylene Glycol-Plasticized Polyvinyl Alcohol on Granule Compaction", J. Am. Ceram. Soc., 67[4]301-4(1984)

6) S. J. Lukasiewicz and J. S. Reed, "Character and Compaction Response of Spray-Dried Agglomerates", Am. Ceram. Soc. Bull., 57[9]798-805(1978)

7) R. G. Frey and J. W. Halloran, "Compaction Behavior of Spray-Dried Alumina", J. Am. Ceram. Soc., 67[3]199-203(1984)

8) H. Kamiya, K. Isomura, G. Jimbo, "Powder Processing for the Fabrication of Si_3N_4 Ceramics: I, Influence of Spray-Dried Granule Strength on Pore Size Distribusion in Green Compacts", J. Am. Ceram. Soc., 78[1]49-57(1995)

9) H. Tanaka, S. Fukai, N. Uchida and K. Uematsu, "Effect of Moisture on the Structure and Fracture Strength of Ceramic Green Body", J. Am. Ceram. Soc., 77[12]3077-80(1994)

10) Y. Zhang, N. Uchida and K. Uematsu, "Direct Observation of Non-Unifrorm Distribution of PVA Binder in Alumina Green Body", J. Mater. Sci., 30, 1357-60(1995)

11) Y. Hiramatsu, Y. Oka and H. Kiyama, "Determination of the Tensile Strength of Rock by a Compression Test of an Irregular Test Piece" J. Min. Eng. Jpn., 81, 1024-30(1965)

12). K. Uematsu, "Immersion Microscopy for Detailed Characterization of Defects in Ceramic Powders and Green Bodies", Powder Tech., 88, 291-98(1996)

13) T. Takigawa , H. Kashihara, K. Urayama, T. Masuda, "Critical Behabior of Modulus of Poly(vinylalcohol) Gels near the Gelation Point", J. Phys. Soc. Jpn., 59, 2598(1990)

14) K. Nagano, S. Yamane, K. Toyoshima, Poval, pp.148-170, Edited by Koubunshi-kankou-kai, Japan, 1970

HIGH PRESSURE PRECOMPACTION FOR SHAPE STABILIZATION OF SINTERED UO_2 PELLETS

K. Yanai
Nippon Nuclear Fueld Development Co., Ltd.

ABSTRACT

Cylindrical UO_2 pellets are commercially used for nuclear fuel in light water reactors. The pellets are fabricated by uniaxial dry pressing of UO_2 powder and pressureless sintering prior to centerless-grinding to meet the pellet diameter specification. Fluctuation of the diameter of as-sintered pellets should be decreased in order to minimize the grinding loss of UO_2 which raises the pellet fabrication cost. It is well-known that a cylindrical pellet fabricated by uniaxial dry pressing of ceramic powder deforms during sintering in a phenomenon which is called hourglassing. The hourglassing corresponds to the fluctuation of the shrinkage ratio of the pellet during sintering caused by density inhomogeneity in the green pellet and density homogenization during sintering. So the diametrical fluctuation of an as-sintered pellet can be decreased if the green density inhomogeneity is left in the as-sintered pellet. The green density inhomogeneity can be left in the as-sintered pellet by controlling the large pore distribution in the green pellet which influences the final sintered density. That is, the local volume of large pores should be increased with the decrease of the local green density. In this study, the UO_2 powder was precompacted under various pressures including pressures higher than the compaction pressure, and crushed prior to sieving of the UO_2 agglomerates to make hard aggregates, between which large pores are generated in the green pellets. After compaction by uniaxial dry pressing, the green pellets were sintered in the pressureless condition, and measurement of the diametrical distribution of as-sintered pellets was carried out using a laser micrometer. Also, the sintered density and the mechanical strength of as-sintered pellets were evaluated by the immersion method and biaxial flexure test, respectively. Using high pressure precompacted powder, large pores between hard aggregates were generated in the green pellets, especially in the lower density region. These large pores decrease the local sintered density in the lower green density region in comparison with that in the higher green density region. By changing the relationship between the local green density and the local sintered density, the diametrical fluctuation of as-sintered pellets was drastically decreased with the increase of the ratio of precompaction pressure to compaction pressure and it was scarcely dependent on BET specific surface area of UO_2 powder. The sintered density and the strength of sintered pellets were hardly decreased with the increase of precompaction pressure when the agglomerates made by precompaction pressure were less than 106 micrometer in size.

Films, Coatings, and Fibers

CHEMICAL SOLUTION DEPOSITION OF ELECTROCERAMIC THIN LAYERS AND THEIR PATTERNING BY MICROCONTACT PRINTING WITH SELF-ASSEMBLED MONOLAYERS

David Payne, Ralph G. Nuzzo, Paul G. Clem, Noo-Li Jeon, and Duk-Young Jung
Department of Materials Science and Engineering
University of Illinois at Urbana-Champaign
Urbana, IL

ABSTRACT

The solution deposition of electroceramic thin layers (10-2-10μ) from chemical precursors on a variety of substrate materials (Al_2O_3, MgO, ...) including semiconductors is described. The ability to integrate functional oxide materials with semiconductor substrates is considered critical for the development of future microelectronic technologies. Details are reported for the spin-casting of metallo-organic precursor solutions and their transformation into metal oxides (i.e., ceramic) thin layers, and the patterning of oxide devices. Characterization data are given for the thermal processing conditions, densification behavior, crystallization and interfacial stress development. Patterned devices are usually formed by uniform deposition followed by post-deposition ion-beam or chemical etching in a controlled environment (i.e., subtractive processes). A variety of patterning methods are reviewed. In the future, a new ambient atmosphere method is proposed which allows for the selective deposition of functional oxide thin layers without post-deposition etching(i.e., an additive process). In this method the substrate's surface is selectively functionalized by hydrophobic monolayers of octadecyltrichlorosilane (OTS) by microcontact printing (μ-CP) of the molecular mask. Solution deposition of chemical precursor over the functionalized patterns, followed by heat treatment and non-abrasive polishing, yields patterned oxide layers only on the unfunctionalized regions. A variety of micron-scale electrical, magnetic, mechanical and optical devices have been fabricated by this directed patterning method. Information is reported for μ-CP, solution chemistry, evolution of microstructure, and electrical properties, as well as the stress-related mechanism which enables selective de-adhesion and desired patterning. The ability to selectively deposit electroceramic materials on a variety of technologically important substrate materials suggests broad potential for integrated circuit and hybrid microelectronic applications. The feasibility of building-up multilevel structures is demonstrated.

PROCESSING OF CERAMIC FINE PATTERNS WITH EXCIMER LAMP

Koichi Kikuta and Shin-ichi Hirano
Graduate School of Engineering, Nagoya University
Furo-cho, Chikusa, Nagoya 464-8603
Japan

ABSTRACT

Ceramic processing with a novel light source, excimer lamp, was investigated for the preparation of ceramic fine patterns on substrates. Metal organic precursors chemically modified with alkanol amine were found to be as useful as photo-reactive resin, which is commonly applied in the production of electronic devices. Stereo selective UV irradiation of a homogeneous precursor film was achieved by an excimer lamp with a wavelength of 222 nm. Through this process, nitride ceramic fine patterns, such as titanium nitride, can also be prepared by changing the heat treatment atmosphere to the same as for oxide fine patterns.

INTRODUCTION

Processing of ceramic fine patterns is of interest for fabricating functional devices. Microcontact printing(μ-CP) with self-assembled monolayers(SAM)and some other processes have been studied for the preparation of ceramic devices such as $LiNbO_3$ waveguides[1-3] and the authors also reported stereoselective crystallization of $LiNbO_3$ by YAG laser beam scanning.[4] These processes, however, need laser equipment with a combination of designed UV sensitive precursors. The other patterning method is based on the system for microelectronics.[5] In the present work, a photo-reactive ceramic precursor was prepared by considering the chemical stability in an ambient condition to remove the unirradiated region after a photolithographic process. Through this process, ceramic precursor fine patterns can be synthesized the same as organic resist polymer. Photoreaction of metal organic compounds prepared by metal alkoxide and β-diketone was reported. This compound usually has the chemical bond with strong $\pi-\pi^*$ transition over 300 nm, which is easily decomposed with an ultra high pressure mercury lamp(HUV). Some of the β-diketonates don't result in homogeneous thin films due to the limited solubility in a solvent. Another approach for the photoprocessing is to select the wavelength of the UV light by considering absorption spectra of precursor films. We reported the pho-

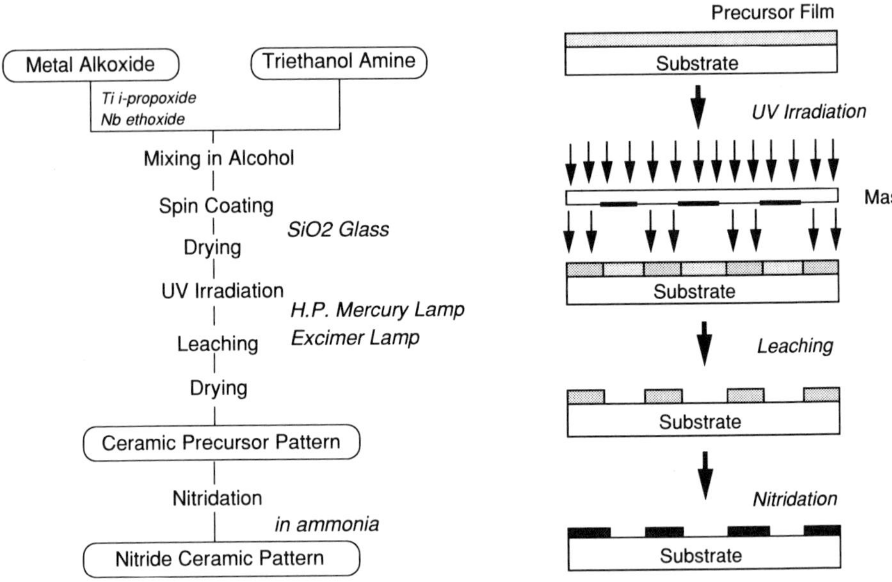

Fig. 1 Experimental Procedure

toreaction of chemical precursors synthesized by metal alkoxide and alkanol amine.[6] To enhance the reaction of a ceramic precursor, which does not have strong absorption over 300 nm, excimer lamps are a very promising light source, because of its wavelength tunability and out-put power controllability. Photoreactions of pure silicon alkoxide for protecting coat and oxidation of silicon surfaces were already reported.[7,8] There is no report on photochemical reaction of precursor materials for the patterning of functional nitride ceramics. In the present work, photoreaction of precursor films and nitridation of precursor patterns were investigated for titanium and niobium systems, which are very attractive materials for protecting coatings, electroding, and superconductors.

EXPERIMENTAL PROCEDURE

Ceramic precursor solutions were prepared by the chemical modification of metal alkoxide with chemical additives such as an alkanol amine. Among them, triethanol amine(TEA) was found to be valuable due to its high photoreactivity.[6] Preparation of precursor fine patterns were carried out as shown in fig. 1. Half of the alkoxy group of the metal alkoxide was exchanged by the TEA in titanium and niobium systems. A starting niobium precursor usually includes some precipitates in ethanol

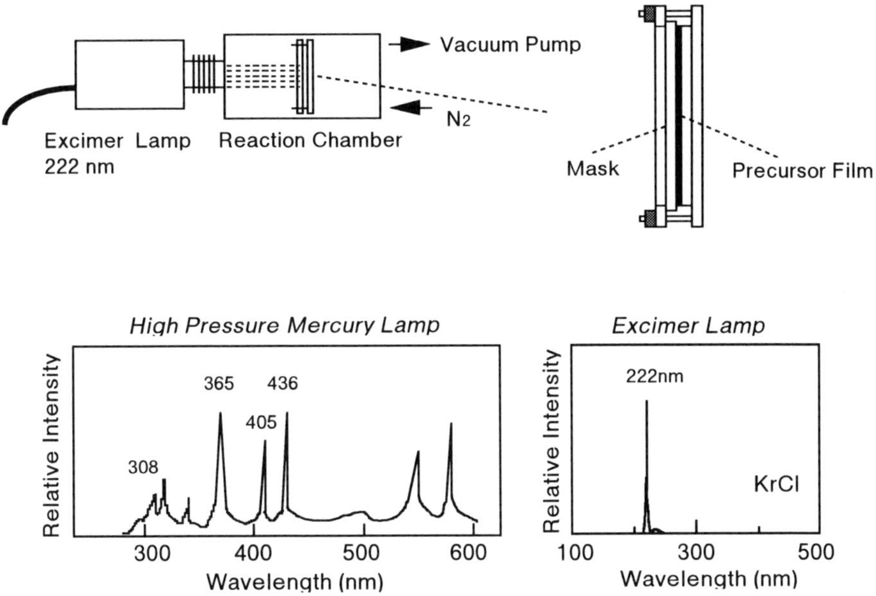

Fig.2 Excimer Lamp Equipment for UV Irradiation

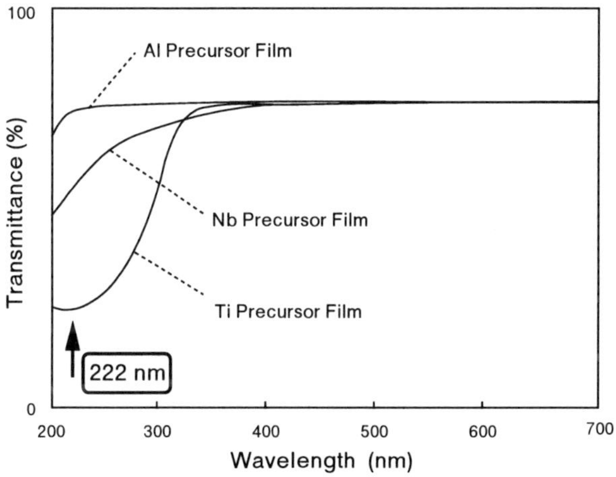

Fig. 3 VIS-UV Spectra of Precursor Thin Films

solution, so filtration of the precursor solution was done before coating. Homogeneous precursor films were prepared on a quartz glass substrate by spin coating of precursor solutions at 2000rpm. Ultraviolet light was irradiated through a photomask(Toppan Printing Co. Ltd.) as illustrated in fig. 2. Two light sources were used in this study, an ultra high pressure mercury lamp(Ushio, UI-502Q) and an excimer lamp (Ushio, UER20H) Then, the unirradiated region can be removed by leaching in ethanol solvent with a spin coater. The prepared ceramic precursor patterns were investigated by a photo spectrometer and a FT infrared spectrometer. Then, nitridation of the amorphous fine pattern was carried out in an ammonia atmosphere. The inside of tube furnace was first evacuated and then dry nitrogen was introduced. On heating, the nitrogen flow was exchanged with ammonia at 300°C. Nitridation of the precursor film was completed at 1000°C. Film thickness was measured with a surface profilemeter. Precursor films and crystallized nitride patterns were observed by an optical microscope. Crystalline phases were identified by X-ray diffraction (Rigaku, RINT 2100V).

RESULTS AND DISCUSSION

Processing of ceramic fine pattern

UV-visible spectra of the precursor films were investigated by a spectrophotometer as summarized in fig. 3. These films have absorption edges near the UV region, although the precursor films don't have strong absorption derived from π-π^* transition. Photoreaction of the ceramic precursor film was carried out with a high pressure mercury lamp(HUV) or an excimer lamp (222 nm). The UV-visible spectra indicated that these two lamps can be used for the patterning of titanium system, since the precursor films absorb the UV light efficiently in the spectra range of these light sources. For the niobium system, it is preferred to apply an excimer lamp with a wavelength of 222 nm. Even for the titanium system, photoreaction can be accomplished within a short time by the application of excimer lamp due to its high photon energy. The photochemical decomposition of organic groups could be observed in the IR spectra of the precursor films depending upon the UV irradiation time. The solubility in ethanol decreases remarkably in the irradiated region of the film. Therefore, the unirradiated region can be removed very easily by leaching with the solvent on a spin coater. After drying, the thickness of the patterned film was measured with a surface profile meter. The thickness of the prepared film from 0.2 mol precursor solution was around 70 nm per one coating cycle. Photoreaction of niobium precursor films were also achieved by an excimer lamp and the UV-visible spectra was changed by the irradiation. After 3 minutes irradiation, the irradiated region can be stabilized for leaching with ethanol solvent. These fine precursor patterns were subjected to the heat treatment described above.

This UV curing process can also be used for the preparation of thicker films by repeating the coating and photoreaction process.

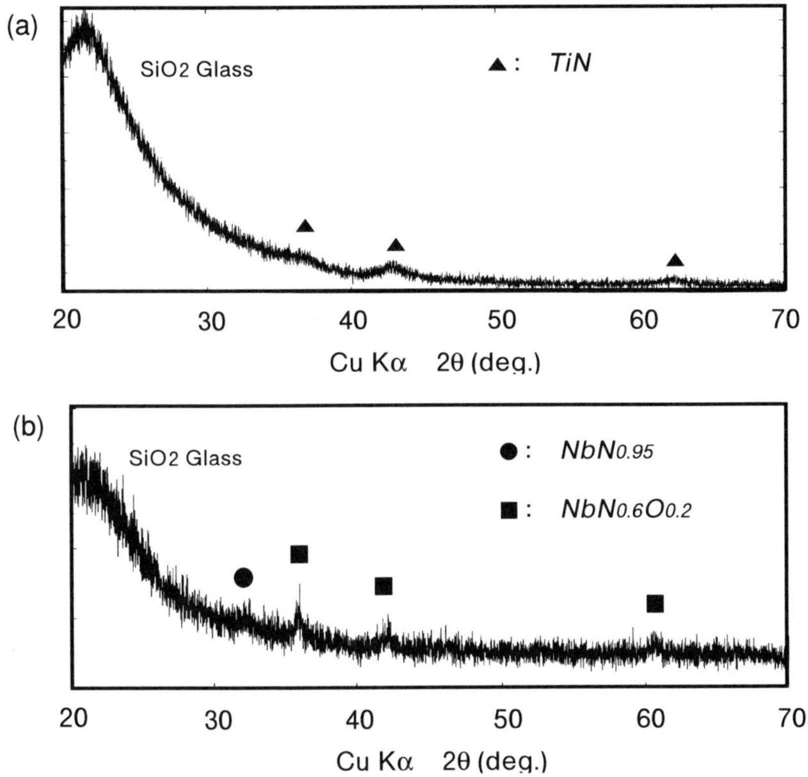

Fig. 4 XRD Profiles of Crystalline Films after Nitridation
(a) Titanium System (b) Niobium System

Heat Treatment of Precursor Film

Heat treatment of the precursor fine pattern was carried out in an ammonia atmospheres. The prepared film has a gold or black color according to the film thickness. The resultant film thickness is reduced to almost one third of that of the precursor film. In the titanium system, weak diffracted lines were observed and were assigned to stoichiometric titanium nitride. It is also confirmed that the prepared film is electroconductive by checking with a digital multimeter. The study on the relationship between the film thickness and conductivity is in progress. In the niobium system, X-ray diffraction revealed that two kinds of compounds, i.e. $NbN_{0.95}$ and $Nb_{0.6}O_{0.2}$, formed by this nitridation. Further improvement of the nitridation condition is required for the preparation of stoichiometric NbN films. One of the impor-

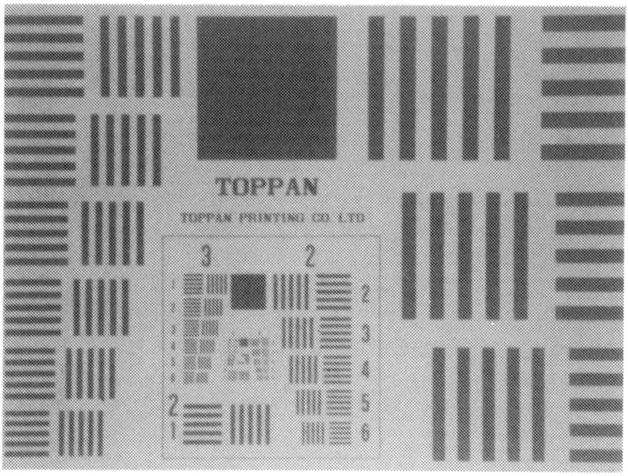

Fig. 5 Optical Micrograph of TiN Fine Pattern

200μm

tant factors influencing the nitridation is considered to be the amount of remaining organics in the precursor film, since this reaction is carbothermal reduction.

A prepared titanium fine pattern is shown in fig. 6. It is observed that this pattern is very smooth and the finest line width is around a few microns, and is in good match with the employed mask pattern.

CONCLUSION

(1) Preparation of photoreactive ceramic precursors was achieved by the chemical modification of metal alkoxide with triethanol amine for titanium and niobium systems.

(2) Photoreaction of the precursor film was successfully achieved by selecting a UV wavelength. As a novel light source, an excimer lamp with a wavelength of 222 nm was found to be applicable for niobium system.

(3) Crystallization of ceramic nitride fine patterns was attempted under nitridation condition using ammonia atmosphere at 1000°C. Electric conductive titanium nitride was prepared for the titanium system. Niobium nitride and niobium oxynitride phases formed in the niobium system by nitridation of niobium precursor films.

ACKNOWLEDGMENT

This work has been partly supported by Synergy Ceramic Project under the Industrial Science and Technology Frontier(ISTF) Program promoted by AIST, MITI, Japan. Under this program, part of the work was funded through NEDO. The authors are members of the Joint Research Consortium of Synergy Ceramics. The authors would like to thank to Mr. Yuji Iwamoto for his kind help.

REFERENCES

[1] N.L. Jeon, P.G. Clem, R.G. Nuzzo, and D.A. Payne, "Patterning of dielectric oxide thin layers by micro contact printing of self-assembled monolayers", J. Mater. Res., 10[12] 2996 - 2999 (1995)

[2] Katsuhide Shinmou, Noboru Tohge, and Tsutomu Minami,"Fine-patterning of ZrO_2 thin films by the photolysis of chemically modified gel films," Jpn. J. Appl. Phys. 33, L1181-1184 (1994)

[3] Toshinobu Yogo, Yoshikuni Takeichi, Koichi Kikuta, and Shin-ichi Hirano, "Ultraviolet patterning of alkoxy-derived lithium niobate film," J. Am. Ceram. Soc., 78 [6] 1649-52 (1995)

[4] Koichi Kikuta, Yasuhide Isobe, Toshinobu Yogo, Satomi Ono, and Shin-ichi Hirano, "Orientation control and laser-beam assisted crystallization of sol-gel derived, titanium-doped $LiNbO_3$ thin films," J. Am. Ceram. Soc., 79[9] 2289-92 (1996)

[5] Satomi Ono and Shin-ichi Hirano, "Patterning of lithium niobate thin films derived from aqueous solution," J. Am. Ceram. Soc., 80[10] 2533-40 (1997)

[6] Koichi Kikuta, Katsuhiko Takagi, and Shin-ichi Hirano, submitted to the J. Am. Ceram. Soc.(1998)

[7] Koichi Awazu, and Hideo Onuki, "Photo-induced synthesis of amorphous SiO_2 film from tertamethoxy-silane on polymethylmetacrylate at room temperature," J. of Non-Crystalline Solids, 215, 176-181 (1997)

[8] J-Y. Ahang and I. W. Boyd, "Low temperature photo-oxidation of silicon using a xenon excimer lamp," Appl. Phys. Lett., 71(20), 2964-2966 (1997)

CERAMIC THIN FILMS MADE BY PROPIONATE-BASED CHEMICAL SOLUTION DEPOSITION ROUTES

Ulrich Hasenko, Susanne Hoffmann, and Rainer Waser
Institut für Werkstoffe der Elektrotechnik II
Aachen University of Technology
52056 Aachen, Germany

ABSTRACT

Chemical solution deposition is an attractive preparation method for the deposition of oxide ceramic thin films, offering a variety of advantages such as flexibility in thin film stoichiometry, easy preparation technique and low costs. In our work, we demonstrate that metal propionates are versatile precursors for the preparation of ceramic thin films, fulfilling all the advantageous aspects named above. By the application of propionate-based precursor solutions, we deposited ceramic thin films of very different compositions and properties. The scope of our preparation technique comprises insulating and protective films of TiO_2 and ZrO_2, high permittivity dielectric thin films of $Ba_{1-x}Sr_xTiO_3$ ($0<x<1$), and conducting, magnetoresistive $La_{1-x}(Ca, Sr)_xMnO_3$ thin films. The routes can easily be extended to other materials systems. Thin films with very different microstructures can be prepared by modification of the described propionate-based CSD methods. Thin film properties like grain size and porosity can be influenced. Dense $BaTiO_3$ thin films with a columnar microstructure were deposited by a propionate-based chemical solution deposition method developed in our institute. The crystallization mechanism from wet film to crystalline ceramic thin film is investigated by different analytical techniques such as glancing incidence XRD, reflectance FT-IR and high resolution SEM.

A CHEMICAL SOLUTION ROUTE FOR PROCESSING THIN FILMS OF OXYGEN ION CONDUCTING BISMUTH VANADATE

S. Simner and B. Dunn
Department of Materials Science and Engineering
University of California
Los Angeles, CA 90095-1595

ABSTRACT

Copper-doped bismuth vanadate ($Bi_2V_{0.9}Cu_{0.1}O_{5.35}$), known as BICUVOX, possesses high oxygen ion conductivities at low temperatures. The ionic conductivity of this material at 300 °C (~ 1 x 10-3 S/cm) is 50 to 100 times greater than any other solid electrolyte in this temperature range. In order to utilize these materials in electrochemical applications, it is vital that they be prepared as high quality thin films. In this study we report the first preparation of BICUVOX thin films by using the decomposition of metallo-organic precursor solutions, a particularly attractive route for producing complex ceramic compounds. A chloroform solution containing bismuth and copper 2-ethylhexanoates and vanadium triisopropoxide was deposited onto polycrystalline MgO substrates by spin coating. Desired film thicknesses were achieved by repeated deposition and pyrolysis steps. Our work investigates the microstructural evolution of these films with respect to solvent removal, organic decomposition, crystallization and final densification using atomic force and scanning electron microscopy. Dense, crack-free films were produced after sintering at 800°C for 10 minutes. The phase purity of the films was established by XRD and EDS techniques and the ionic conductivity measured using a complex impedance method.

FORMATION OF EPITAXIAL $PbTiO_3$ ISLANDS ON $LaAlO_3$ AND $SrTiO_3$ SUBSTRATES BY CHEMICAL SOLUTION DEPOSITION

J. H. Kim, C. D. E. Lakeman, J. S. Speck, and F. F. Lange
Materials Department
College of Engineering
University of California
Santa Barbara, CA 93106

ABSTRACT

Formation of epitaxial $PbTiO_3$ islands on $LaAlO_3(001)$- and $SrTiO_3(001)$-oriented substrates by sol-gel method using a mixed Pb-Ti double-alkoxide solution precursor has been studied. Scanning electron microscopy (SEM), atomic force microscopy, and x-ray diffraction were performed to characterize microstructure instability and crystallinity of samples annealed at various conditions. Transmission electron microscopy (TEM) was also performed to determine the critical size of an island that has no defect in the interior and the facet planes of islands. $PbTiO_3$ thin films, formed by only one spin coat and annealed at 800 C for 1 h, show a continuous, a discontinuous, or an isolated microstructures depending on the initial concentration of solution precursors. SEM observations show that $PbTiO_3$ islands formed on $SrTiO_3$ substrates are well faceted but those on $LaAlO_3$ are not. Plan-view and cross-sectional TEM observations show that critical size of a defect-free island on $SrTiO_3$ is ~100 nm and it has a hemispherical shape. Critical sizes of defect-free islands on $LaAlO_3$ and its shape will be shown. Comparison of these experimental results to the theoretical calculated morphological stability diagrams will be discussed.

CHEMICAL SOLUTION DEPOSITION OF CMR EPITAXIAL THIN FILMS

A. D. Polli, F. F. Lange, M. Ahlskog, R. Menon, and A. K. Cheetham
Materials Department and the Institute for Polymers and Organic Solids
University of California at Santa Barbara
Santa Barbara, CA 93106

ABSTRACT

A simple process based on aqueous metal acetate solutions has been developed and employed to prepare $(La_{0.7}Ca_{0.3})MnO_3$ (LCMO) epitaxial thin films on (100) $SrTiO_3$ and pseudocubic (100) $LaAlO_3$. The films first crystallizes into a continuous, nanocrystalline network below 600°C. Upon further heat treatment, epitaxial grains grow and consume the nanocrystalline network. Heat treatment at 800°C for 1 hour is enough to complete this nanocrystalline-to-epitaxial transition, but surface roughness is eliminated only following treatment at elevated temperatures, ~1000°C. Four-circle x-ray diffraction shows that the epitaxial grains have a slight mosaic character (FWHM = 0.12°-0.35°). Temperature sweeps of electrical resistivity show that the conductivity and Curie temperature of the films increase with heat treatment temperature. Films fired at 600°C and 800°C for 1 hour show peak resistivities of ~5.0 W -cm and ~0.3 W-cm and Curie temperatures of ~190K and 270K, respectively. Films fired at 1000°C for 1 hour were found to have a Curie temperature near room temperature. For these films, the application of 1 and 8 tesla fields was found to drop the film's resistivity by ~10% and ~60%, respectively, at room temperature.

PROCESSING AND HARDNESS OF EPITAXIAL ALUMINA THIN FILMS CONTAINING NANOSIZED ZIRCONIA INCLUSIONS

Frank McNally and F. F. Lange
Materials Department and the Institute for Polymers and Organic Solids
University of California at Santa Barbara
Santa Barbara, CA 93106

ABSTRACT

Single crystal homoepitaxial alpha-alumina films containing nanoscale zirconia inclusions were grown on (006) sapphire substrates. These films were prepared from mixtures of aluminum nitrate and zirconium acetate aqueous precursors by spin coating and subsequent heat treatment at 1200°C. Powders were also prepared from these precursors by spray pyrolysis and subsequent heat treatment. XRD and TEM studies on both powders and films confirm the expected non equilibrium phase evolution from amorphous inorganic to gamma-alumina to gamma-alumina + tetragonal zirconia to alpha-alumina + tetragonal zirconia. Cross section TEM of films confirm homoepitaxy and random spatial distribution of spherical inclusions approximately 20 nm in diameter. Nanoindention studies show that these inclusions at 2, 5 and 15vol% do not significantly affect room temperature plasticity of alumina.

GROWTH AND CHARACTERIZATION OF TIN OXIDE THIN FILMS

L. Fu and X. Pan
Department of Materials Science and Engineering
University of Michigan
Ann Arbor, MI 48109

ABSTRACT

Tin dioxide thin films have been grown on different substrates using electron-beam evaporation deposition and metal-organic deposition (MOD). Microstructures and grown textures have been characterized by 4-circle X-ray diffraction and plane-viewing TEM. Interfacial atomic structure and growth defects have been studied by cross-sectional HRTEM. It has been found that the microstructure of the films strongly depends on the processing parameters such as the substrate structure, deposition temperature, post-deposition annealing temperature and time, and doping. Epitaxial films were grown on the R-surface of sapphire substrate using electron-beam evaporation deposition at high temperatures (500°–700°C) followed by annealing at the same temperature in air. Effects of metallic doping on the film growth and microstructure have also been studied. Ion-beam assisted e-beam deposition has been used to control the film growth, crystal defects, film stoichiometry and the texture. Data will also be presented on the electrical properties of the films. The relationship between the atomic structure and properties will be addressed.

PROCESSING EFFECTS ON STOICHIOMETRY OF HYDROTHERMALLY DERIVED $Ba_xSr_{1-x}TiO_3$ POWDERS AND FILMS

Ryan K. Roeder and Elliott B. Slamovich
School of Materials Engineering, Purdue University, West Lafayette, IN 47907-1289

EXTENDED ABSTRACT

Assessment of the effects of key processing variables is essential for stoichiometry control during hydrothermal synthesis of electroceramics, especially in the presence of multiple cationic species (e.g. Ba and Sr in $Ba_xSr_{1-x}TiO_3$).

$Ba_xSr_{1-x}TiO_3$ (BST) thin films were processed below 100°C by reacting hydrolyzed titanium metal-organic precursors in alkaline, aqueous solutions of $BaCl_2$, $SrCl_2$, and NaOH. Preliminary results from x-ray diffraction (XRD) qualitatively showed that Sr-rich BST films were formed from solutions containing equal initial concentrations of Ba and Sr. In order to provide bulk sampling for phase identification and quantitative composition analysis, BST powders were processed under similar experimental conditions as BST films.

BST powders were processed below 100°C by reacting nanosize TiO_2 powders in alkaline, aqueous solutions of $BaCl_2$, $SrCl_2$, and NaOH. The effects of processing variables (NaOH concentration, time, temperature, and the ratios of Ba, Sr, and Ti initially in the solution) on the resultant powder stoichiometry were examined. In particular, powders were processed in solutions containing either a large (5.0) or small (1.1) initial excess of Ba and Sr relative to Ti. Since metal-organic precursor films contain small amounts of Ti relative to the solution volume, powders processed in solutions with the larger initial excess of Ba and Sr relative to Ti were indicative of the expected processing effects on thin film stoichiometry.

All BST powders were characterized by XRD for phase identification and wavelength dispersive spectroscopy for quantitative compositional analysis. In all cases, Sr was more readily incorporated into BST powders than Ba, the extent greater for powders processed in solutions with a large (5.0) initial excess of Ba and Sr relative to Ti. BST powders processed in solutions with a large (5.0) initial excess of Ba and Sr relative to Ti consisted of a single phase solid solution. In contrast, BST powders processed in solutions with a small (1.1) initial excess of Ba and Sr relative to Ti exhibited a biphasic range of solid solubility corresponding to discrete Ba-rich and Sr-rich phases. The observed biphasic BST solid solutions suggest the presence of a metastable miscibility gap in the low temperature regions of the $BaTiO_3$-$SrTiO_3$ phase diagram.

Supported by the National Science Foundation grant DMR-9623744.

HYDROTHERMAL SYNTHESIS OF $Pb(Zr_x,Ti_{1-x})O_3$ POWDERS AND HETEROEPITAXIAL THIN FILMS

A. T. Chien, J. H. Kim, Y. S. Yoon, J. S. Speck, and F. F. Lange
Materials Department
University of California, Santa Barbara
Santa Barbara, CA 93106

ABSTRACT

Various perovskite heteroepitaxial thin films system have been successfully produced by hydrothermal routes at ambient pressure and low temperatures (90°-150°C). This processing method provides a simple low temperature route for producing a wide compositional range of $Pb(Zr_x,Ti_{1-x})O_3$ powders and epitaxial thin films on $SrTiO_3$ and $LaAlO_3$ single crystal substrates which has been applied to other ferroelectric materials. Growth mechanisms, structural development, and solution chemistry effects were studied by x-ray diffraction (XRD), scanning electron microscopy (SEM), and zeta potential measurements. Nucleation studies reveal that {100} faceted islands form along the surface steps on the $SrTiO_3$ substrate followed by growth and coalescence. EDS analysis of the $Pb(Zr_x,Ti_{1-x})O_3$ thin film shows that the Zr:Ti ratio is 44:56, nearly identical to the molar ratio of the precursors. The aqueous solution chemistry is found to control interfacial characteristics and plays an important role in controlling faceting. Faceting can be changed from {100} blocks to platelets by the introduction of additional cations (e.g. NMe^{4+}) during synthesis. Electrophoretic data shows that Pb is a potential determining counterion and adsorbs on $SrTiO_3$ surfaces. Electrical properties will also be discussed in relation to the aqueous solution chemistry.

Synthesis of Highly Oriented $(Pb_{0.85}La_{0.1})TiO_3$ Thin Films by Chemical Solution Deposition Method

W. Sakamoto, S. Yada, T. Kohigashi, K. Kikuta, T. Yogo and S. Hirano
Department of Applied Chemistry, Graduate School of Engineering,
Nagoya University, Nagoya, 464-8603 Japan

ABSTRACT

Crack-free $Pb_{0.85}La_{0.1}TiO_3$ thin films have been prepared by chemical solution deposition method. A homogeneous and stable PLT precursor solution was prepared by controlling the reaction of metal-organic compounds in ethanol with a key additive of acetylacetone. Highly oriented perovskite $Pb_{0.85}La_{0.1}TiO_3$ thin films were synthesized from the PLT precursor solution on $LaNiO_3/SrTiO_3/MgO(100)$ substrates without any formation of the pyrochlore phase. The dielectric constant of PLT films was about 600 at 25°C. The Curie temperature of PLT films was around 270°C at 100 kHz.

INTRODUCTION

Lanthanum modified lead titanate ($Pb_{1-1.5x}La_xTiO_3$, PLT) has been receiving great attention for pyroelectric sensors, nonvolatile memories and large variety of electro-optic devices because of its excellent pyroelectric, ferroelectric and electro-optic properties. The properties of PLT can be controlled by the amount of lanthanum doping. Recently, the demand for thin film processing has increased for integrated-device development.[1] (100) or (001) highly oriented PLT thin films are expected to give a large figure of merit and exhibit several excellent properties compared with other materials because of its polarization without poling treatment. The thin film processings of PLT have been investigated by several methods such as RF sputtering, laser ablation, chemical vapor deposition and chemical solution deposition (CSD) process. The CSD process developed for the fabrication of ferroelectric film provides several advantages, which include high purity, good homogeneity, ease of composition control, lower processing temperature and versatile shaping.[2,3] Hirano et al. reported processing and properties of $Pb(Zr,Ti)O_3$ thin films using metal-organic precursors.[2,4] Recently, materials for electrode are of quite interest for application in ferroelectric devices. Oxide electrodes such as RuO_2, IrO_2, $SrRuO_3$, etc. have been widely investigated for the improvement of the properties of ferroelectric films. Especially, $SrRuO_3$ and $LaNiO_3$ have a perovskite structure with good conductivity and good crystallographic matching with lead based ferroelectric perovskites.

This paper describes the synthesis of highly oriented PLT thin films by using $LaNiO_3$ as a bottom electrode. Crack-free layered $Pb_{0.85}La_{0.1}TiO_3/LaNiO_3/SrTiO_3$ thin films were successfully synthesized on MgO(100) substrates. The dielectric properties of the PLT thin film were also evaluated.

EXPERIMENTAL PROCEDURE

Anhydrous $Pb(OAc)_2$, $Ti(OEt)_4$ and $La(O^iPr)_3$ were selected as starting materials. The handling of chemicals and procedures were conducted in a dry N_2 atmosphere. Anhydrous $Pb(OAc)_2$ and $Ti(OEt)_4$ corresponding to $Pb_{0.85}La_{0.1}TiO_3$ composition were dissolved in absolute ethanol. This solution was refluxed for 18 h. Appropriate amount of $La(O^iPr)_3$ was added to the solution and then refluxed again for 18 h. Acetylacetone was mixed with the solution as a stabilizing agent. The molar ratio of acetylacetone to PLT precursor was 1.0. Finally, the PLT precursor solution was refluxed for 18 h and then concentrated to about 0.3 mol/L.

Films were fabricated using the precursor solution by dip-coating on MgO(100), Pt/MgO(100), $LaNiO_3/SrTiO_3$/MgO(100) substrates. A Pt(100) layer was deposited on MgO(100) by RF magnetron sputtering. $LaNiO_3$ thin films were prepared by chemical solution process with $Ni(OAc)_2 \cdot 4H_2O$, $La(O^iPr)_3$ and absolute 2-ethoxyethanol. $SrTiO_3$ buffer layer was also prepared from precursor solution with Sr metal, $Ti(O^iPr)_4$ and absolute 2-ethoxyethanol. The PbO top layer was fabricated using a 2-methoxyethanol solution of $Pb(OAc)_2$. Tani et al. reported the effect of the PbO coating on the surface of $Pb(Zr,Ti)O_3$ and $(Pb,La)(Zr,Ti)O_3$ thin films prepared by chemical solution process on microstructure and ferroelectric properties of the films.[5] The coating-crystallization process was repeated several times in order to increase the thickness of the film.

PLT powders were also prepared from the PLT precursor solution. The solution was hydrolyzed with deionized water which was beforehand diluted with absolute ethanol. The precipitate was dried at 100°C yielding a white solid, which was then heat-treated at various temperatures.

The prepared powders and films were characterized by X-ray diffraction (XRD) analysis using CuKα radiation with a monochromator. The films were observed by scanning electron microscopy (SEM) and atomic force microscopy (AFM). The electrical properties of films were measured using Au prepared by the vacuum deposition on the PLT films as a top electrode. The dielectric measurement was conducted using an LCR meter. The dielectric properties were measured for the films set in an electric furnace shielded using a Cu tube from 200°C to 400°C. The P-E hysteresis loops of the films were evaluated by a Radiant RT66A test system at 25°C.

RESULTS AND DISCUSSION

A homogeneous PLT precursor solution was prepared by controlling the reaction of $Pb(OAc)_2$, $La(O^iPr)_3$ and $Ti(OEt)_4$ in ethanol. The key point of the process was the prereaction of $Pb(OAc)_2$ and $Ti(OEt)_4$ in ethanol. The by-product of this reaction, ethyl acetate, was confirmed by GC and FT-IR analysis. The homogeneity and stability of the coating solution were found to be greatly improved by the addition of acetylacetone. The coordination of acetylacetone ligand to the PLT precursor was also confirmed by NMR spectroscopic analysis.

Figure 1 illustrates XRD profiles of PLT powders prepared from the PLT precursor solution heat-treated at 450°C and 500°C. The PLT powders are X-ray amorphous below 450°C, and directly crystallized to perovskite PLT at around 500°C without any formation of pyrochlore phase. According to the XRD analysis, the crystallized PLT powder is considered to be a cubic phase, although $Pb_{0.85}La_{0.1}TiO_3$ must have tetragonal structure at room temperature. The tetragonality of PLT powders decreases with

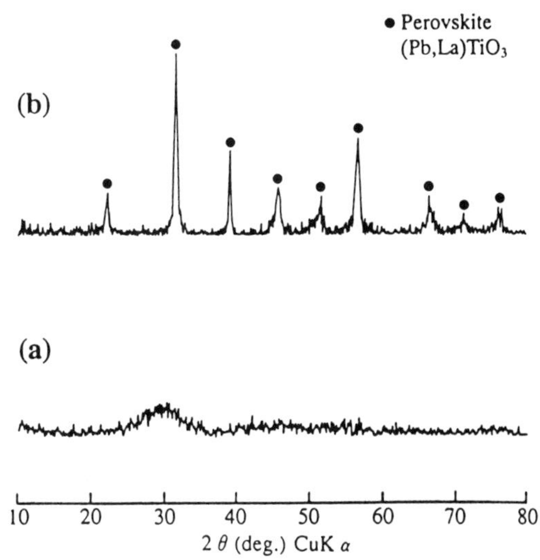

Fig.1 XRD profiles of $Pb_{0.85}La_{0.1}TiO_3$ powders heat-treated at (a) 450°C and (b) 500°C

increasing lanthanum in amount. Also, the current powder may have small grain size and poor crystallinity, therefore, the PLT powder crystallizes to the pseudo cubic perovskite phase.

In order to obtain highly oriented PLT thin films, MgO(100) and Pt(100)/MgO(100) substrates were selected. Highly (100) or (001) oriented PLT thin films were synthesized by selecting MgO(100) substrate as shown in Fig.2(a). However, PLT films on Pt(100)/MgO(100) include 111 reflection as shown in Fig.2(b). The (100) plane of PLT has a better lattice matching (lattice mismatch along a-axis; 0.2%) with the (100) plane of Pt than any other planes of PLT. However, the (111) plane is also easily formed because the (111) plane is the closest packed dense plane.

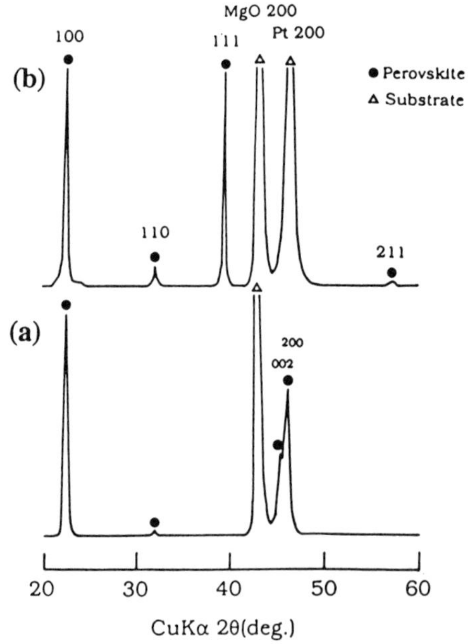

Fig.2 XRD profiles of $Pb_{0.85}La_{0.1}TiO_3$ thin films on (a) MgO(100) and (b) Pt(100)/MgO(100) substrates heat-treated at 650°C

To improve the ferroelectric property of PLT thin film and the degree of (100) or (001) orientation, $LaNiO_3$ was selected as a bottom electrode, because $LaNiO_3$ has a perovskite structure with good electric conductivity. Figure 3 shows the XRD profile of $LaNiO_3$ thin film on $SrTiO_3$/MgO(100) substrate. $LaNiO_3$ thin film with (101) orientation was found to form by

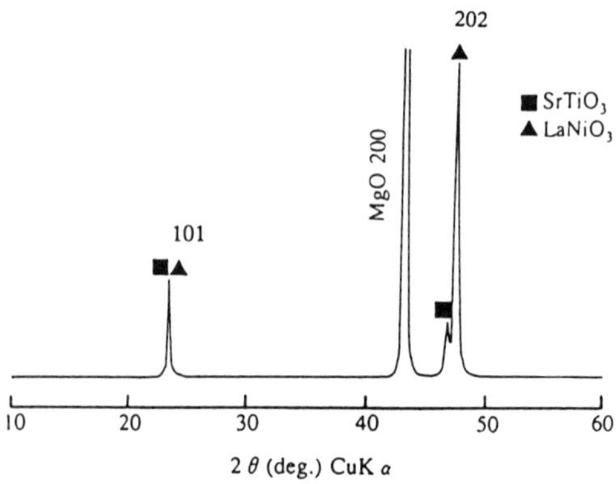

Fig.3 XRD profile of $LaNiO_3$ thin film on $SrTiO_3$/MgO(100) substrate heat-treated at 700°C

using $SrTiO_3$ buffer layer between $LaNiO_3$ film and MgO(100) substrate. The lattice mismatch between $LaNiO_3$ and MgO is large (about 8.9%). Therefore, $SrTiO_3$ was selected as a buffer layer, because $SrTiO_3$ has also the perovskite structure and the value of lattice parameter is suitable for both MgO and $LaNiO_3$. The electrical resistivity of the $LaNiO_3$ film was about 1.5×10^{-4} Ωcm measured by the four probes method.

Figure 4 shows the XRD profile of $Pb_{0.85}La_{0.1}TiO_3$ (PLT) thin film on $LaNiO_3/SrTiO_3$/MgO(100) substrate heat-treated at 600°C. From XRD analysis of the films, it turns out that PLT thin films on $LaNiO_3/SrTiO_3$/MgO(100) directly crystallized to the perovskite phase above 400°C with (100) or (001) preferred orientation. Thin films on substrate do crystallize to the perovskite PLT at lower temperatures compared with powders. This is due to the assistance of nucleation sites with atomic alignment of $LaNiO_3$(101). The direct crystallization to the perovskite phase was consistent with that of PLT powders. However, the film also seems to have a cubic-like structure, since the 100 or 001 and 200 or 002 X-ray diffractions are not separated as shown in Fig.4.

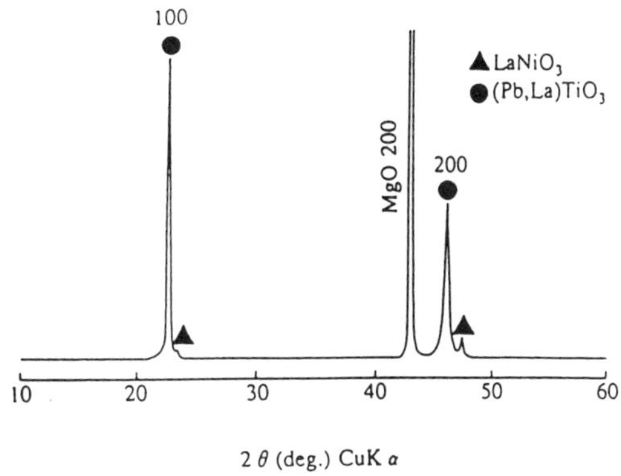

Fig.4 XRD profile of $Pb_{0.85}La_{0.1}TiO_3$ thin film on $LaNiO_3/SrTiO_3$/MgO(100) substrate heat-treated at 600°C

Figure 5 shows SEM photograph of PLT film on $LaNiO_3/SrTiO_3$/MgO(100) substrate heat-treated at 600°C. The films have smooth surface topology with no cracks and voids. The thickness of the crystallized film was about 0.6 μm after 20 cycles dipping. The roughness of the film was below 10 nm. The grain size of the PLT film was confirmed to range

Fig.5 SEM photograph of $Pb_{0.85}La_{0.1}TiO_3$ thin film on $LaNiO_3/SrTiO_3$/MgO(100) substrate heat-treated at 600°C

from 80 to 150 nm using AFM.

Figure 6 shows the temperature dependence of dielectric constant for the PLT film on $LaNiO_3/SrTiO_3/MgO(100)$ substrate heat-treated at 600°C. From this figure, the Curie point of the PLT film was found to be about 270°C. This value was a little lower than that of PLT bulk ceramics (about 330°C). The dielectric constant of the film at room temperature was about 600 at 100kHz. The internal stress and small grain size were considered to be responsible for the sift of the Curie temperature. The values of ε_r and tanδ were comparable with those of polycrystalline PLT films on $Pt/Ti/SiO_2/Si$ substrate.[6]

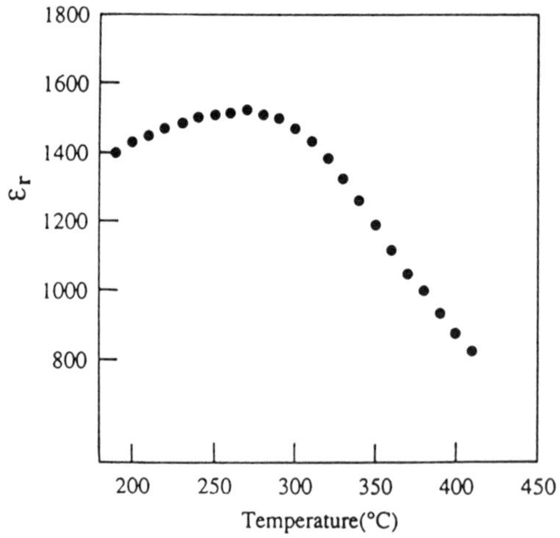

Fig.6 Temperature dependence of dielectric constant of $Pb_{0.85}La_{0.1}TiO_3$ thin film on $LaNiO_3/SrTiO_3/MgO(100)$ substrate heat-treated at 600°C measured at 100 kHz

Figure 7 shows a P-E hysteresis curve of the PLT film on $LaNiO_3/SrTiO_3/MgO(100)$ substrate heat-treated at 600°C. The remnant polarization and coercive field of the film were 6.0 μC/cm^2 and 98 kV/cm, respectively. The value of P_r was lower than that of PLT bulk ceramics. Since the synthesis of highly c-axis oriented PLT film is required to optimize its property, the improvement of synthesis conditions is now in progress.

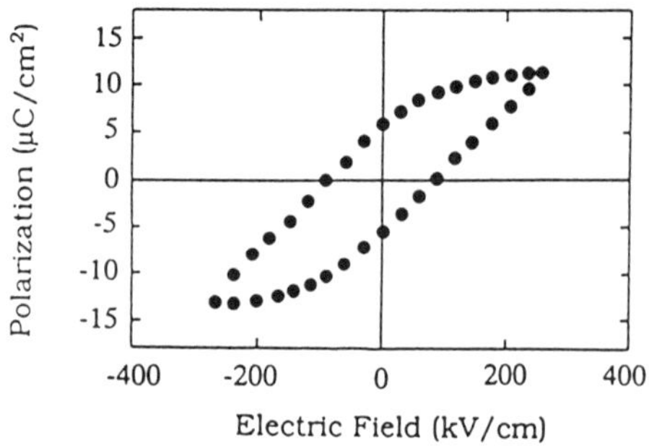

Fig.7 P-E hysteresis loop of $Pb_{0.85}La_{0.1}TiO_3$ thin film on $LaNiO_3/SrTiO_3/MgO(100)$ substrate heat-treated at 600°C

CONCLUSIONS

Highly oriented $Pb_{0.85}La_{0.1}TiO_3$ films were successfully synthesized on $LaNiO_3/SrTiO_3/MgO(100)$ substrates through metallo-organics. A homogeneous and stable PLT precursor solution was prepared from lead acetate, lanthanum isopropoxide and titanium ethoxide in ethanol. PLT precursor was stabilized by the coordination of acetylacetonate group to metals. Perovskite $Pb_{0.85}La_{0.1}TiO_3$ thin films with (100) or (001) preferred orientation were synthesized on $LaNiO_3/SrTiO_3/MgO(100)$ substrates above 400°C. The orientation of the PLT film was improved markedly by using $SrTiO_3$ as a buffer layer between $LaNiO_3$ film and MgO substrate. PLT thin films on $LaNiO_3/SrTiO_3/MgO(100)$ substrates showed a typical P-E hysteresis. The dielectric constant of the film was about 600 at 25°C. The Curie temperature of the film was found to be at around 270°C.

REFERENCES

[1]L. M. Sheppard, "Advances in Processing of Ferroelectric Thin Films", *Am. Ceram. Soc. Bull.*, **71** [1] 85-95 (1992).

[2]S. Hirano, T. Yogo, K. Kikuta, K. Kato, W. Sakamoto and S. Ogasawara, "Sol-gel Processing and Characterization of Ferroelectric Films", *Ceram. Trans.*, **25** 19-32 (1991).

[3]K. Kani, H. Murakami, K. Watari, A. Tsuzuki and Y. Torii, "Sol-gel Processing and Pyroelectric Properties of La-modified $PbTiO_3$ Thin Films", *J. Mat. Sci. Lett.*, **11** 1605-1607 (1992)

[4]S. Hirano, T. Yogo, K. Kikuta, Y. Araki, M. Saitoh and S. Ogasawara, "Synthesis of Highly Oriented Lead Zirconate-Lead Titanate Films Using Metallo-organics", *J. Am. Ceram. Soc.*, **75** [10] 2785-2789 (1992)

[5]T. Tani and D. A. Payne, "Lead Oxide Coatings on Sol-Gel-Derived Lead Lanthanum Zirconium Titanate Thin Layers for Enhanced Crystallization into the Perovskite Structure", *J. Am. Ceram. Soc.*, **77** [5] 1242-1248 (1994)

[6]W. Sakamoto, T. Kohigashi, K. Kikuta, T. Yogo and S. Hirano, "Processing and Properties of (Pb,La)TiO_3 Films by Sol-Gel Method", *The 7th US-Japan Seminar on Dielectric and Piezoelectric Ceramics, Extended Abstracts*, 81-84 (1995)

EVOLUTION OF POROUS MICROSTRUCTURES IN DOPED $PbTiO_3$ PYROELECTRIC THIN FILMS

A. Seifert, P. Muralt, and N. Setter
Laboratoire de Ceramique
Departement des Materiaux
Ecole Polytechnique Federale de Lausanne
1015 Lausanne, Switzerland

ABSTRACT

Calcium doped lead titanate (PCT) thin films were found to have excellent pyroelectric properties for applications in point-detectors and pyroelectric arrays. The properties can be further improved if porous, low-dielectric constant films are being used. Both 2-Methoxy ethanol and 1,3-Propanediol based solution precursors were synthesized and used for spin-coating platinized Si3N4/SiO2/Si wafers. The Ca content of PbTiO3 was varied from 0 to 30 percent and properties as well as microstructures of films processed from the different precursor systems were compared. The heating-rate during the final anneal was observed to be the controlling parameter for the evolution of the microstructures. Slow heating of the films at 1°C/sec leads to porous microstructures, whereas 98%-dense layers can be produced for fast heating at 60°C/sec. The evolution of the porous microstructures as well as the low temperature crystallization ($T<400°C$) of PCT-perovskite during fast heating was studied by RBS, XRD, SEM and TEM, and will be discussed. The influence of Ca on the polarization of the films as well as microstructure-property relationships will be presented.

ULTRAVIOLET RAY SHIELDING COATING ON GLASS BY SOL-GEL PROCESS

Hiroyuki Tomonaga, Akemi Mitani and Takeshi Morimoto
Research Center, Asahi Glass Company
1150 Hazawa-cho, Kanagawa-ku, Yokohama-shi
KANAGAWA 221-8755, JAPAN

ABSTRACT

Cerium oxide and Titanium oxide complex UV shielding coating deposited on glass via the sol-gel process was investigated. Addition of TiO_2 to CeO_2 changed the crystallinity and coordination status of CeO_2 and improved UV absorption. Formation of double-layered coating comprised of outer CeO_2-TiO_2 layer and inner TiO_2-SiO_2 layer diminished the interference coloring and improved the adhesion of the coatings to glass substrates. The double layer coated glass shows high UV-shielding properties without visible light absorption, and excellent durability against chemical and physical attacks, and has been widely used in automobile windows.

INTRODUCTION

Excessive exposure to ultraviolet ray (UV) in sunlight (6%) induces coloring and deterioration in paper, fabrics and resins. It also causes harmful damage to the human body such as premature aging of skin, skin cancer and cataract. Recent destruction of the ozone layer around the earth increases UV irradiation and its harmful effect on the human body[1)]. The necessity to shield from UV ray irradiation is of interest in many fields. For automobile applications, it is necessary to prevent the UV rays from damaging the interior materials and the skin of passenger. For this aim, we have developed an UV shielding coating on glass surface comprised of two metal oxides optical thin layers.

It is well-known that glass containing mixed oxides of cerium and titanium absorb UV effectively and exhibit yellow color[2)]. It is also reported that CeO_2-TiO_2 coatings by the sol-gel process show UV absorbance, high reflectance and interference coloring. Makishima et.al.[3)] studied the structures and phases of the coatings deposited on glass and aluminum foil. Sainz et. al.[4)] discussed the coloring mechanism of the SiO_2-TiO_2-CeO_2 coatings. However, their UV absorbance are not sufficient. Moreover, high reflectance and interference coloring are not preferable for automobile windows.

The following research was performed to address the above-mentioned problems. (1) maintaining CeO_2/TiO_2 ratio above 1.0 to increase UV absorption. (2) forming of double-layered coating on glass to diminish reflectance and coloring. This paper describes the materials, mechanisms and properties of the double layer coated glass.

EXPELIMENTAL

Coating solution for outer UV absorbing layer

Cerium nitrate [$Ce(NO_3)_3 \cdot 6H_2O$] and Titanium *n*-butoxide [$Ti(OC_4H_9)_4$, TBT] were used as metal precursors. TBT was dissolved in ethanol and stirred for one hour at 25℃. An aqueous solution of 6wt% HCl was added to the solution of TBT. The molar ratio of H_2O/TBT was 8.5. Concentration of this solution was 0.8M as Ti (Solution A). Cerium nitrate was dissolved in ethanol. Then, polyethyleneglycol (PEG, average MW=200) and acetylacetone (AA) were added as multidentate ligands. The molar ratio of Ce/PEG/AA was adjusted to 1/1/1. The obtained solution was heated up to 70℃ for 2 hours and cooled to room temperature (Solution B). Concentration of solution B was also 0.8M as Ce. Solution A and B were mixed in various ratios and aged at 35℃ for one day.

Coating solution for inner layer

Titanium isopropoxide [$Ti(OC_3H_7^i)_4$, TPT] and tetraethylorthosilicate [$Si(OC_2H_5)_4$, TEOS] were used as metal precursors. Various compositions of TPT and TEOS were dissolved in a solution of ethanol and acetylacetone (90/10 in weight ratio) and stirred for one hour at 25℃. Then 1wt% HCl aqueous solution was added to the solution under stirring to the molar ratio of H_2O/(TPT+TEOS)=2. Concentration of this solution was 0.4M as Ti+Si. Then the solution was aged at 35℃ for one day.

Coating procedure and analysis

Polished and cleaned soda-lime silicate glass (100x100x3.5mm) was used as the substrate. Solution was coated on substrate using the spin-coating process. After drying at 120℃ for five minutes, the coated glass was heated up to 700℃ and fired for 3 minutes.

The film thickness was measured using a profilometer (Sloan, Dektak 3030). The microstructure of the coatings was observed using a scanning electron microscope (Hitachi, S-900SEM). Crystal phases were determined using an X-ray diffractometer (Rigaku, RINT-2000). Transmittance and reflectance of the coatings were measured using a spectrophotometer (Hitachi, U-3500). State analysis was carried out by ESCA (Phi, type5500) and fluorescent EXAFS (Rigaku, R-EXAFS2000).

CeO_2-TiO_2 UV ABSORBING COATING

Figure 1 shows the UV transmittance of various CeO_2-TiO_2 binary compositions coated glass. Film thickness of all coatings was 160nm. The addition of TiO_2 to CeO_2 lowers UV transmittance and shows its minimum at around 40mol%. Although it was reported[4)] that CeO_2 precipitates in a solution of the molar ratio of $CeO_2/TiO_2>1$, no precipitation occurred in any ratios of CeO_2/TiO_2 solutions used in this experiment. Acetylacetone and polyethyleneglycol added to the solution seem to play important roles in the stabilization of the Ce ion in the solution.

Figure 2 shows XRD profiles of various compositions of CeO_2-TiO_2 coatings on glass. The coating containing only CeO_2 shows several peaks which have been indexed to the cerianite, CeO_2, crystalline structure. However, addition of TiO_2 to CeO_2 coating decreases the crystallinity of cerianite and forms a complete amorphous phase at the composition of TiO_2/CeO_2=40/60. Moreover, the peaks indexed anatase or rutile structure of TiO_2 do not appear by the addition of TiO_2 up to 40mol%. Therefore, addition of TiO_2 to CeO_2 increases disorder in the crystalline structure of cerianite.

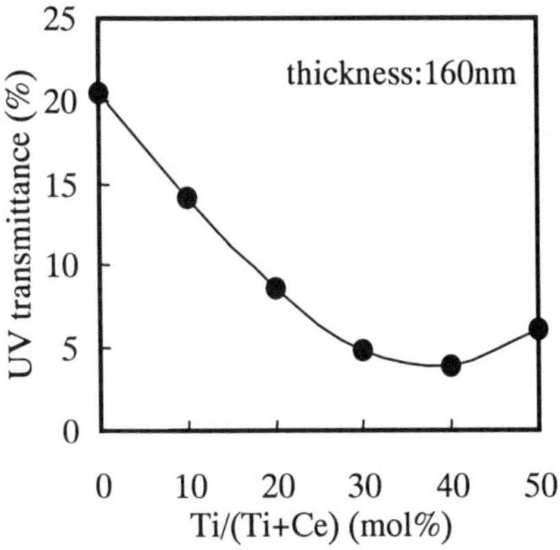

Fig.1 UV transmittance of CeO_2-TiO_2 coatings with various compositions.

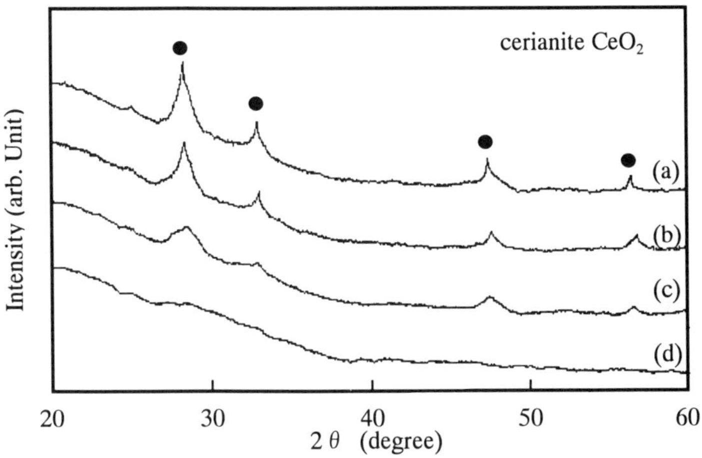

Fig.2 XRD profiles of CeO_2-TiO_2 coatings
CeO_2/TiO_2 molar ratio : (a)100/0 (b)80/20 (c) 70/30 (d) 60/40

Figure 3 shows Ce-3d region ESCA spectrum of CeO_2-TiO_2. The coating containing only CeO_2 is used as a reference. There are no significant differences in main peaks indexed as $Ce3d_{5/2}$ between TiO_2/CeO_2=0/100 and TiO_2/CeO_2=40/60 coatings. However, difference in the peaks around 898eV were observed. This peak is called shakeup satellite and indicates that a kinetic energy of photoelectron looses the energy corresponding to the excitation of a valence band electron because an electron release from $Ce3d_{5/2}$ orbital and an electron excitation from valence band occur at the same time. So the peak difference means that the excitation energy of valence band electron changes and indicates that TiO_2 addition to CeO_2 changes the excited state of Ce 3d electron into the lower level.

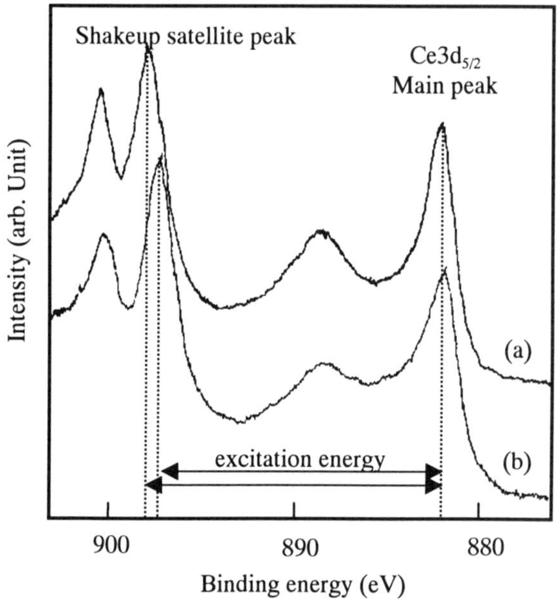

Fig.3 Ce3d region ESCA spectra of CeO_2 and CeO_2-TiO_2 coatings.
(a) CeO_2/TiO_2=100/0 (b) CeO_2/TiO_2=60/40

As mentioned above, addition of TiO_2 to CeO_2 changes crystallinity and excitation energy of the valence band electron. It is supposed that these changes are caused by some differences in the valence or coordination status of Ce ion. ESR measurements at 6K(liquid Helium temperature) showed that trivalent Ce^{3+} ion did not exist in both CeO_2 100% and CeO_2/TiO_2=60/40 coatings and it can be expected that only tetravalent Ce^{4+} ion is present in both coatings. Table I shows the results of EXAFS analysis on Ce-L orbital in the CeO_2 and CeO_2-TiO_2. Although the addition of TiO_2 to CeO_2 does not change the distance to neighboring O^{2-}, it reduces the coordination number of Ce ion to nearly 6. Considering that CeO_2 takes fluorite structure of which the coordination number of Ce is 8 and TiO_2 takes the anatase or rutile structure of which the coordination number of Ti is 6, it is expected that the structure of CeO_2 changes to anatase or rutile structure gradually through additions of TiO_2 to CeO_2.

Table I EXAFS results

sample	coordination number	distance to neighboring O^{2-}
CeO_2	8.0(standard)	0.234nm
CeO_2-TiO_2	6.2	0.233nm

INNER LAYER COATING

The refractive index of UV-absorbing layer is about 2.10 and higher compared to that of substrate glass (1.52). Therefore, the glass coated by UV-absorbing layer shows high reflectance and reflection coloring owing to optical interference. The inner layer was coated between the glass substrate and UV-absorbing layer in order to reduce the reflectance and reflection coloring. Refractive index and optical thickness required for the inner layer coating to reduce the reflectance and reflection coloring were calculated to be 1.7-1.8 and 110-150nm, respectively. Sol-gel derived SiO_2-TiO_2 binary coatings were used as the inner layer. Figure 4 shows the refractive indices of various compositions of TiO_2-SiO_2 coatings. The coating with molar ratio of TiO_2/SiO_2=33/67 had the refractive index of 1.76.

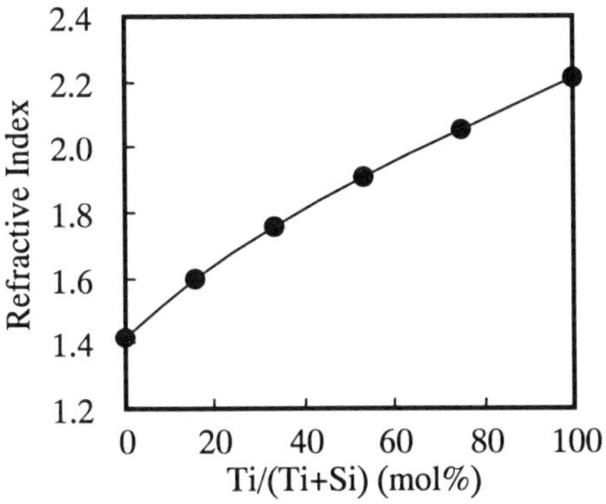

Fig.4 Refractive index of TiO_2-SiO_2 coatings with various compositions.

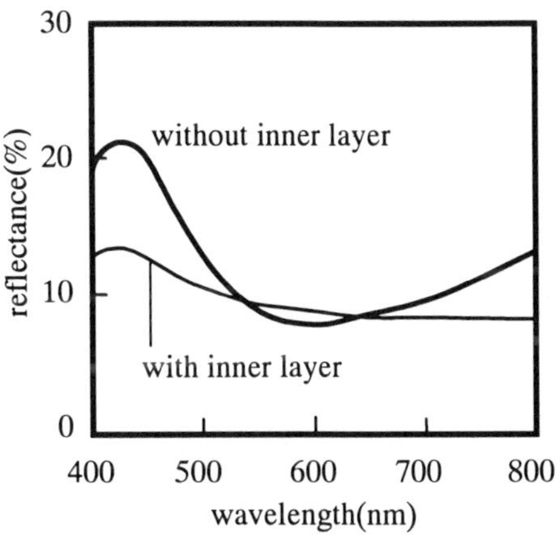

Fig. 5 Reflection spectra of UV absorbing layer coated glasses.

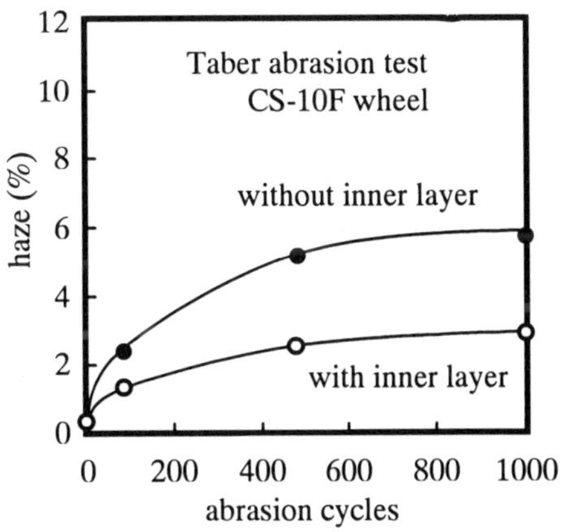

Fig.6 Abrasion resistance of UV absorbing layer coated glasses.

Figure 5 shows the reflection spectra of the glass coated UV-absorbing layer with and without an inner layer of which the refractive index and the optical thickness are 1.76 and 125nm, respectively. The inner layer reduces reflectance and prevents the glass from exhibiting a reflection color. Moreover, this inner layer plays a very important role for strengthening the adhesion force between the coating and the substrate as shown in Fig. 6. The insertion of the inner layer increases the number of surface active hydroxyl groups[5)] which can bond with the hydroxyl groups in the UV absorbing layer. Figure 7 shows the cross-sectional SEM image of the double layer coated UV-shielding glass.

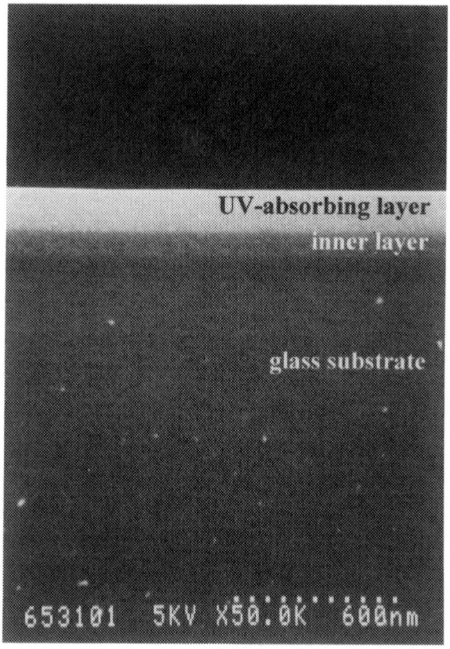

Fig. 7 Cross-sectional SEM image of double-layered UV-shielding glass

APPLICATION OF UV-SHIELDING GLASS TO AUTOMOBILE WINDOWS

Manufacturing procedure

Double-layered UV-shielding coatings have been applied to automobile windows. Several factors related to the coating solutions (e.g. solvents, ligands and additives) and the coating method have to be modified in order to apply the coating to large area glass substrate (max. ~2m^2). The coated glass was manufactured using following method. Polished and dried flat glass substrates are coated with inner layer and outer UV-absorbent layer. After drying, the coated glass is cut into the desired shape. Then the coatings are fired and hardened in the bending and tempering process of glass at around 700°C for a few minutes.

Performances of UV-shielding glass

Figure 8 compares transmission spectrum of the UV-shielding glass and substrate glass. The UV-shielding glass absorbs UV-rays effectively without significant absorption of visible light. The UV-shielding glass has the SPF (sun-protect factor) value of about 70, which is higher compared to commercial UV-protect cosmetic materials. As the coated glass has excellent stability against various chemical and physical attacks, they can be handled in the same way as normal glass. Results of the durability tests are shown in table II.

The UV-shielding coated glass has been practically used in side and rear windows of industrially produced automobiles for more than 4 years without any damage.

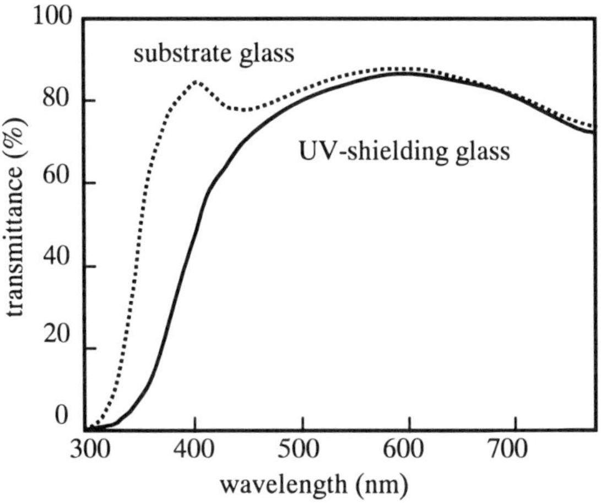

Fig.8 Transmission spectrum of UV-shielding glass compared with substrate glass

Table II Various stability test results of the UV-shielding glass

properties	test condition	results
abrasion	Taber abrasion 1000cycles*	haze=2.6%
acid	immerse in 0.1N H_2SO_4 240hrs	no damage
alkali	immerse in 0.1N-NaOH 240hrs	no damage
humidity	55°C 95%RH 240hrs	no damage
heat	350°C 240hrs	no damage
weathering	SWOM 10000hrs	no damage

*abrasion wheel : CS-10F

The efficiency of UV-shielding effects of the coated glass was examined by the following method. The glass was put on color charts, including cyan, magenta and yellow color printed on coated paper using a ink jet printer (Epson, typeMJ500C). The color charts were exposed to UV from Xenon lamp for 100 hours through the glass. Table III shows the change of color density by UV irradiation. Yellow and magenta colors are discolored completely and cyan color is changed considerably in the color chart exposed through the normal substrate glass. On the other hand, although yellow color is slightly discolored, no significant color change was observed in the color chart exposed through the UV-shielding glass.

Table III Changes of color densities by UV exposure test

color	initial	after exposure for 100hrs through	
		normal glass	UV-shielding glass
cyan	1.87	1.61	1.83
magenta	1.66	0.61	1.59
yellow	1.57	0.53	1.38

CONCLUSIONS

Ultraviolet ray shielding CeO_2-TiO_2 coatings on glass were prepared by the sol-gel process. UV absorption of CeO_2 was improved with the changes of crystallinity and coordination state by adding TiO2. The inner layer composed of TiO_2-SiO_2 between the substrate and CeO_2-TiO_2 coating prevented the glass from exhibiting interference coloring and improved the adhesion of coatings to the substrate. This double-layered UV-shielding coating glass is useful for protecting human and materials from the harmful effects of UV and shows excellent durability. The coated glass is widely used in the automobile industry.

REFERENCES

1. Executive Summary of the Environmental Assessment of Ozone Depletion-1991
2. C. A. Hampel, “Cerium in the Glass Industry”, The Glass Industry, **41** [2] 82-113 (1960).
3. A. Makishima, H. Kubo, K. Wada, Y. Kitami and T. Shimohira, “ Yellow Coatings Produced on Glasses and Aluminum by the Sol-Gel Process” Journal of American Ceramic Society, **69** [6] C127-C129 (1986).
4. M. A. Sainz, A. Durán and J. M. Fernández Navarro, “ UV Highly Absorbent Coatings with CeO_2 and TiO_2”, Journal of Non-Crystalline Solids, **121**, 315-318 (1990).
5. F. Gunji, T. Yoneda and T. Morimoto, “ Novel Functioning of Glass by Wet Coating Process: Part1. Water-repellent glass,” New Glass, **11** [4] 49-56 (1996).

THIN FILM SOLID OXIDE FUEL CELLS BY COLLOIDAL PROCESSING

C. Jacobson, S. J. Visco, and L. C. DeJonghe
Center for Advanced Materials, Materials Science Division
Building 62-113, Lawrence Berkeley National Laboratory
University of California at Berkeley
Berkeley, CA 94720

ABSTRACT

Thin film solid oxide fuel cells capable of delivering 2 Watt/cm2 at 800°C and below have been prepared and tested. The key process is one in which the solid oxide film is deposited by colloidal methods directly onto porous substrates. Control of the formation of the colloidal deposit and subsequent heat treatment leads to the formation of dense, crack and pinhole free electrolyte films with a thickness of up to 20 μm, in a single processing step.

DESIGN AND MICROSTRUCTURE CONTROL OF CERAMIC LAMINATES BY COLLOIDAL FILTRATION

A. J. Sánchez-Herencia and R. Moreno
Instituto de Cerámica y Vidrio, C.S.I.C.
Crtra. Valencia Km. 24,300
28500 Arganda del Rey
Madrid, Spain

ABSTRACT

Layered and functionally graded ceramics have received a great deal of attention for emerging applications in which higher mechanical performance or different functions are demanded. The development of laminated composites allow the processing of new materials with different properties to those of the present materials or to find new functions for existing materials. Colloidal filtration techniques provide a powerful tool for designing and fabricating these complex microstructures. The main requirement is the preparation of stable, well-dispersed slips, only possible from a good control of their rheological behaviour. A second aspect is the control of the casting kinetics, allowing the tailoring of desired microstructures with an accurate control of the thickness of each layer in the composite material. In this work, an evaluation of the colloidal filtration technique for the processing of layered ceramics in the system Al_2O_3/Y-TZP is made. The improvement of the rheology of either monophase or mixture slips and the study of the wall thickness formation rate make possible the preparation of multilayers, sandwich-like or functionally graded materials with very sharp interfaces between layers, a high dispersion in the case of both mono and multiphase layers and a very precise thickness control of each layer into the final tailored microarchitecture.

CHEMICAL METHODS FOR PROCESSING La-MONAZITE AND RELATED COMPOUNDS FOR CMC WEAK INTERFACES

P. E. D. Morgan, J. B. Davis, D. B. Marshall, and R. M. Housley
Rockwell Science Center
1049 Camino Dos Rios
Thousand Oaks, CA 91360

ABSTRACT

The economic coating of ceramic fibers and cloths etc. with stoichiometric La-monazite, which can provide a weak interface in ceramic matrix composites, requires new solution/and/or slurry chemistries. Some new solution methods (e.g. using dibasic acids such as phosphorous acid or methyl phosphonic acid with La nitrate) will be described. The progress of the conversion of precursors to monazite on heating and their topological relationships e.g. on sapphire will be reviewed. Some early prospects for this technology will be discussed.

DESIGN OF LAYERED ZrO_2/Y-TZP MATERIALS FOR CRACK BIFURCATION

A. J. Sanchez-Herencia and F. F. Lange
Materials Department
University of California at Santa Barbara
Santa Barbara, CA 93106

ABSTRACT

Sandwich like ceramic materials with thick outer layers of Y-TZP (= 5 vol.% of Al_2O_3) and thin inner layer of pure ZrO_2/Y-TZP composites with different composition ratios has been made by slip casting. After determining the kinetics of filtration, several materials have been obtained varying both the composition and thickness of the core layer and maintaining constant the outer layers. Due to the residual stresses state in these materials, crack bifurcation is observed in flexural strength determinations, when the amount of pure ZrO_2 (0% Y_2O_3) inside the thin core layer is higher than 50 vol.%. Also, due to the layered structure the amount of pure zirconia retaining the tetragonal phase is higher than that achieved in the monolithic compact.

EFFECTS OF STACKING STRUCTURES ON CRYSTALLIZATION AND PROPERTIES OF MULTI-LAYER $Pb(Zr_x,Ti_{1-x})O_3$ THIN FILMS

Hisao Suzuki, Yasuhiro Kondo, Shoji Kaneko, and Takashi Hayashi*
Department of Materials Science
Shizuoka University
3-5-1 Johoku
Hamamatsu 432, Japan

*Department of Materials Science and Technology
Shinan Institute of Technology
1-1-25 Tsujido-Nishikaigan Fujisawa
Kanagawa 251, Japan

ABSTRACT

Ferroelectric lead zirconate titanate, PZT, thin films with multi-layer structure were prepared at low temperatures by the dip-coating method of alkoxide precursor solutions. Stacking structures and heating schedules had large effect on the crystallization of the precursor thin films. Lead titanate thin film crystallized at lower temperatures around 673 K into perovskite phase. On the other hand, Lead zirconate thin film did not crystallize into perovskite phase at low temperatures. Therefore, a lead titanate layer acted as in-situ seeding layer and made the solid solution at higher temperature depending upon the thickness and the heating schedule. Well-crystallized lead titanate layer restricted the inter-diffusion between lead titanate and lead zirconate layers. As a result, a perovskite PZT thin films were prepared as low as 773 K from alkoxide precursor thin films with multi-layer stacking structure by choosing the suitable pyrolysis conditions before annealing.

PROCESSING OF SILICON NITRIDE BASED MULTILAYERED COMPOSITES

Y. Shigegaki
Ishikawajima-Harima Heavy Industries Co., Ltd. Tokyo 135 Japan
M. E. Brito, K. Hirao, M. Toriyama and S. Kanzaki
National Industrial Research Institute of Nagoya, Nagoya 462 Japan

ABSTRACT

Multilayered ceramic composites are promising candidate to meet multiple and improved performance, such as toughness, damage tolerance and strength. Two types of silicon nitride based composites were fabricated by stacking tape-casted sheets to tailor a layered structure consisting of layers with distinct mechanical properties. Discrete layered structures, consisting of β-Sialon/silicon nitride or dense/porous silicon nitride, were obtained by controlling starting powders and sintering additives. Neither crack nor delamination was observed within all layers or interfaces. Microstructure and interfaces were carefully characterized and processing parameters were clarified to obtain the aimed layered configuration. These composites presented high strength and high fracture toughness, which are closely related the layered structure..

INTRODUCITON

Multilayered ceramic composites have been widely studied in several systems, and most of the research and development have been done by keeping in major mind their potential implement on overcoming brittleness of ceramics materials. To date, an alternate layered composite composed of different materials has been reported to present fracture resistance and/or damage tolerance, as reported in ZrO_2/Al_2O_3[1,2], SiC/C[3], Si_3N_4/BN[4] and other systems[5-7]. The alternating layered structure gives the advantage for attaining the improved performance by laminating layers with distinct properties.

Beside affecting the composites properties, multilayered composites composed of the same material system should give a relatively easy and reliable processing way. If the aimed property is distinguished between the each layer, strengthening and toughening mechanism should arise even in the same material system. From this point of view, we have recently established the fabrication process of two types of multilayered silicon nitride, e.g., dense/porous silicon nitride system and β-Sialon/silicon nitride system. Silicon nitride has excellent mechanical properties (e.g., high strength and high fracture toughness), and porous silicon nitride and β-Sialon have lower elastic moduli, compared to dense silicon nitride. Therefore, we can make a distinction of elastic property in these layered structures. Furthermore, low elastic modulus is expected to improve thermal shock behavior[8].

This paper reports the fabrication process of two types of multilayered silicon nitride, dense/porous silicon nitride system and β-Silaon/silicon nitride system by laminating tape-casted sheets and sintering.

DENSE/POROUS SILICON NITRIDE

Fabrication Process

The microstructure of silicon nitride can be controlled by modifying sintering aid systems and/or raw powders. How to produce porous layers *in-situ* while densifying dense layers is a key issue to obtain a duplex dense/porous layered structure. A large addition of whisker is known to retard densification of composites by percolation of whisker. Therefore, distribution of sintering aids and incorporation of β-Si_3N_4 whisker in the porous layer would prevent the shrinkage at the porous layer. The effects of the larger addition of whisker on porosity in the porous layer and the bulk shrinkage during sintering have been evaluated to tailor the morphology and configuration of the pores.

Starting powders for the porous layer were α-Si_3N_4 powder and α-Si_3N_4 whisker with no sintering additives. Those for the dense layer were α-Si_3N_4, Y_2O_3, and Al_2O_3. The green sheets were formed by the tape casting. The thickness of the tape was adjusted to 50-100 μm. After alternate stacking, binder burnout and cold isostatic pressing, sintering was performed at 1850 °C under a nitrogen pressure of 1 MPa.

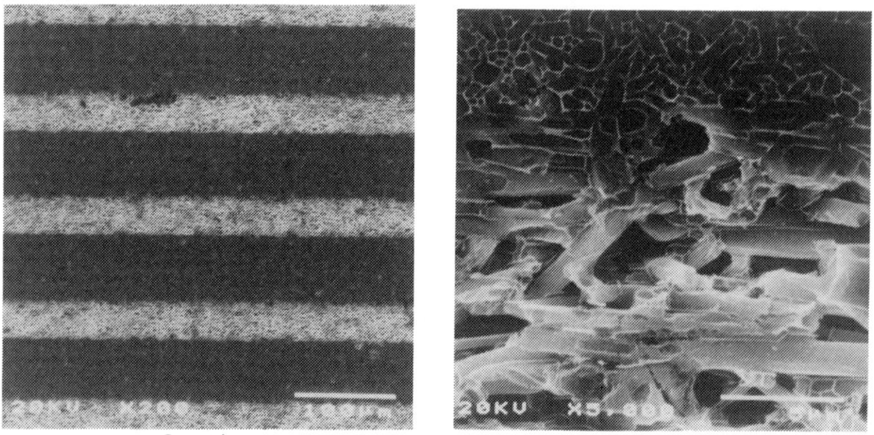

Overview Interface

Fig.1 SEM micrographs of cross section of dense/porous multilayered silicon nitride

After sintering, dense/porous layered structure could be obtained. Figure 1 shows scanning electron microscope (SEM) micrographs for the typical material with the porous layer, incorporating the 70 vol% whisker. The layers with sintering additives were sintered to nearly full density and no cracks or delaminations was observed, neither within the layers or at the interfaces. The porous layer presented the unique microstructure, where rod-like grains were aligned in the casting direction and anisotropic pores generated by tightly tangled elongated grains.

Figure 2 shows the bulk density and porosity in the porous layer. Porosity in the porous layer was calculated as the product of the bulk density and the volume fraction. The bulk density largely decreased by adding whisker of 30 vol% to the porous layer while porosity in the porous layer increased up to about 30 %. By a further whisker addition, bulk density and porosity in the porous layer slightly decreased.

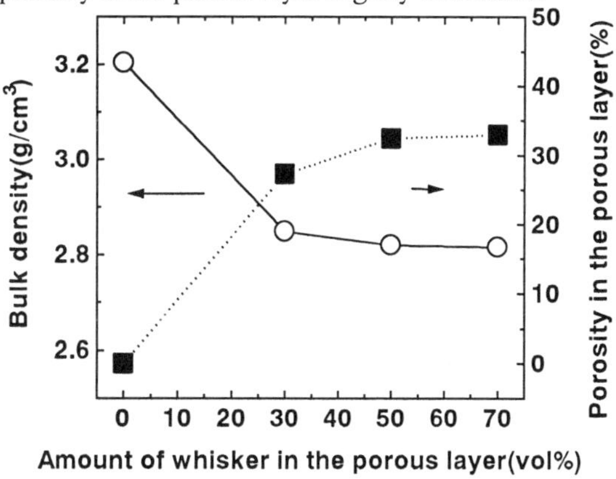

Fig.2 Variation of density and porosity in the porous layer with the amount of whisker

It can be easily thought that the difference in the shrinkage behavior of each layer during sintering significantly affects the obtained layered structure. Since layers are constrained each other, the shrinkage parallel to layers is expected to be small. On the other hand, the shrinkage normal to layers should be large to compensate its volumetric shrinkage. Figure 3 shows the linear shrinkages in three different directions. The arrow in Fig. 3 indicates the isotropic shrinkage value of the reference material. The shrinkage in the laminated direction (Z in Fig. 3) was relatively high in all layered materials, while the shrinkage in the casting direction (X in Fig. 3) was lower in materials incorporating whiskers in the porous layer.

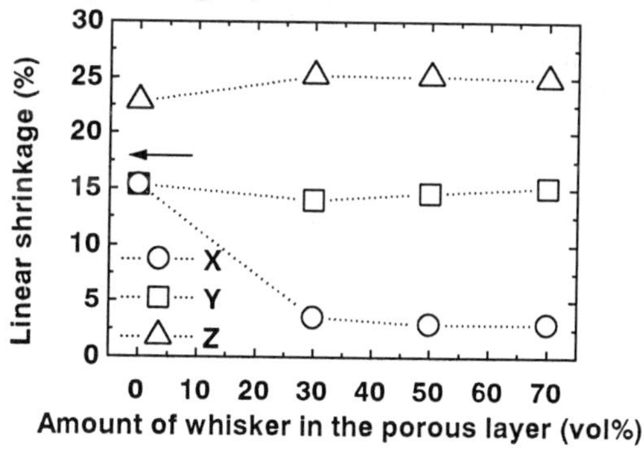

Fig.3 Variation of linear shrinkage with the amount of whisker

Considering nearly full density of the dense layers (Fig. 1), it is noteworthy that the dense layer compensates for the small shrinkage of porous layers in the untrapped laminated direction.

Mechanical Properties

Figure 4 shows the effect of span-to-depth ratio on the maximum tensile stress at failure in the case of the specimen with 70 vol% of whiskers in the porous layer. The fracture stress was especially high and the strain to failure in this material increased owing to its high fracture stress, while its elastic modulus decreased (230 GPa). Figure 5 shows SEM micrographs of fracture surfaces after bending test. Although the load-deflection curves always indicated a linear behavior, delamination cracks were often observed in central porous layers. In the high span-to-depth ratio region, the fracture stress was almost constant and a fracture origin was always confirmed in the dense layer on the tensile surface. As for three-point bending, the fracture was originated in the dense layer on the tensile surface by tensile stress followed by delamination. During four-point bending, a complicated fracture mode and local stepwise crack extension were observed. Delamination would play a dominant role in the case of four point bending.

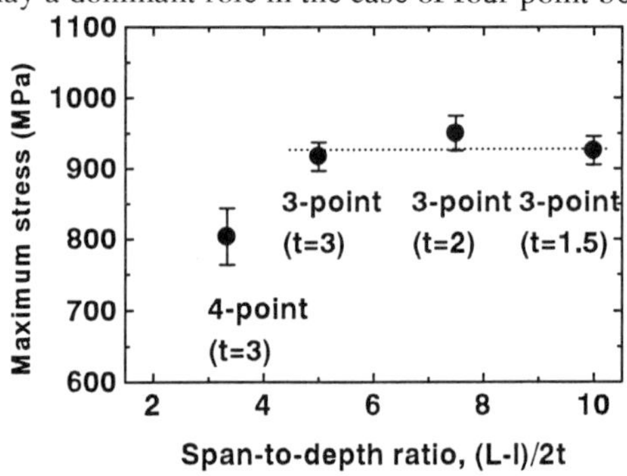

Fig.4 Effect of span-to-depth ratio on the maximum tensile stress at failure

3-point bending 4-point bending

Fig.5 SEM micrographs of fracture surfaces after bending test

Figure 6 shows R-curve behaviors[9]. The resistance markedly increased whenever a crack passed the porous layer, resulting in stepwise rising R-curve behavior. The higher fracture resistance would be attributed to crack deflection and pulling out caused by the aligned whiskers and the lower local stress intensity factor due to the differences in the elastic properties.

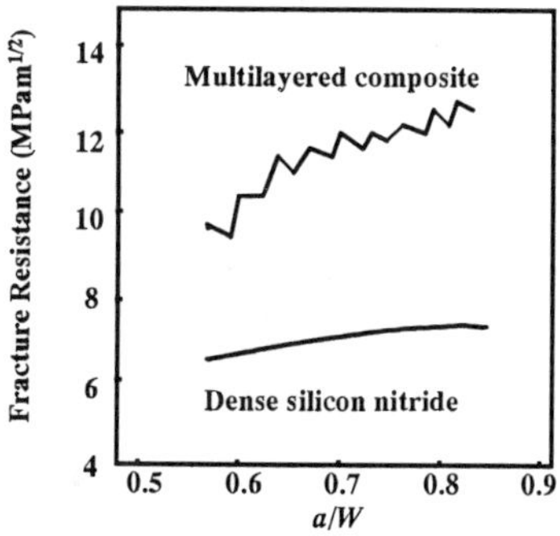

Fig.6 R-curve behaviors of multilayered silicon nitride and dense silicon nitride

β-SIALON/SILICON NITRIDE

Fabrication Process

Starting powders for the silicon nitride layer were α-Si_3N_4 with sintering additives of 5 wt% Y_2O_3. Those for the β-Sialon layer were α-Si_3N_4, AlN and Al_2O_3. The green compact was hot-pressed at 1800 °C for 1 hour under a pressure of 40 MPa in an ambient nitrogen atmosphere. Figure 7 shows SEM micrographs of polished and plasma-etched cross sections of hot-pressed specimens. The white layer is the silicon nitride layer. No cracks or delaminations were observed, either within the layers or at the interfaces. Xray diffraction pattern (XRD) phase analysis indicated no secondary phase, where clearly split peaks could be fully indexed as β-Sialon and silicon nitride (Fig. 8). At the interface, an intermediate region was detected with a thickness range of about 10 μm, where composition was functionally graded and the thickness of the graded region agreed well with the result of SEM observation.

β-Sialon and silicon nitride are not compatible in accordance with the Si-Al-O-N phase diagram[10]. Since appreciable diffusion of solutes (Al, N and O) from the β-Sialon layer towards the silicon nitride layer has occurred, the amount of solutes (z value) in the β-Sialon layers should have changed. The estimated z values from the lattice constants are summarized in Table 1, in conjunction with the β-Sialon monolith data. Considering errors in weighing raw powders and the amount of oxygen impurity in the powders, the β-Sialon layer in layered composites seemed to almost retain its initial composition (z=3). A relatively low z value in R_L of 1 is attributed to the large number of interfaces per unit cross-sectional area.

These results imply that a β-Sialon of graded composition was formed at the interface while retaining the two phase structure. Actually, the layered structure observed here could not be detected in a specimen sintered at 1850 °C for 3 hours under a nitrogen pressure of 0.9 MPa. The β-Sialon layer reaction in the early stage of sintering is considered to be helpful in tailoring the duplex structure, since the solid solution reaction occurred prior to densification during the sintering of β-Sialon[11].

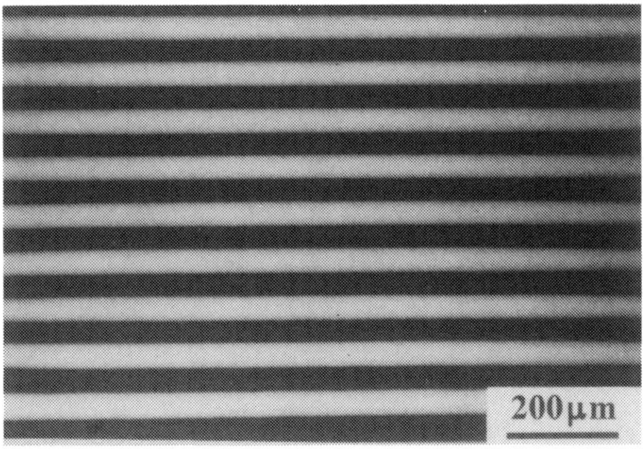

Fig.7 SEM micrographs of cross sections of β-Sialon /Si_3N_4

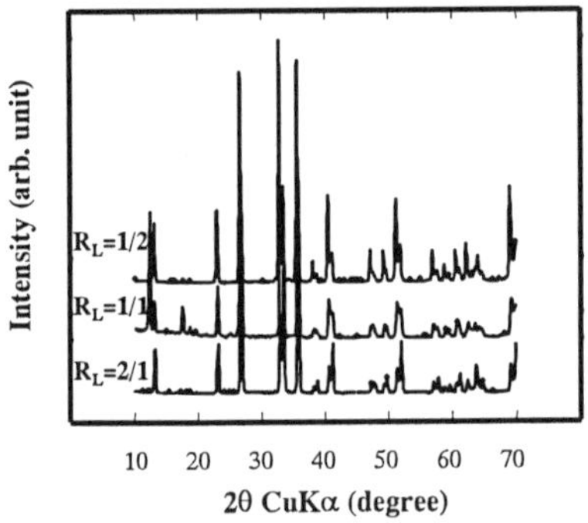

Fig.8 XRD patterns of specimens with lamination ratio ($R_L=N_{\beta\text{-Sialon}} / N_{\text{Silicon nitride}}$)

Table 1. Lattice constant and the estimated z value in β-Sialon layers.

R_L*	Lattice constant(nm)		Z value in $Si_{6-z}Al_zO_zN_{8-z}$	
	a	c	a-axis	c-axis
0.5	0.7688	0.2982	2.87	2.94
1	0.7682	0.2975	2.67	2.66
2	0.7690	0.2985	2.91	3.05
monolith	0.7691	0.2984	2.97	3.01

* $R_L = N_{\beta\text{-Sialon}} / N_{\text{Silicon nitride}}$

Mechanical Properties

Figure 9 shows the results of 4-point bending strength. The load-deflection curves during the bending tests always indicated a linear behavior. Fracture in the layered composites occurred in the β-Sialon layers. Judging from SEM observations, fracture always originated in the top or second β-Sialon layer. Strength of the β-Sialon layer, estimated by an equivalent strain model (the line in Fig. 7), increased appreciably compared to the monolithic β-Sialon. Two models based on residual compressive stress and local stress intensity factor could explain trends in strengthening behavior. The thickness of layers affects the strength of a discrete layer in the alternating layered configuration.

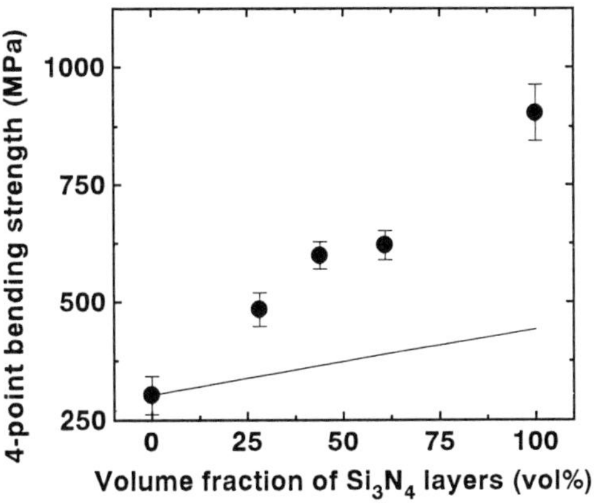

Fig.9 Variation of strength with volume fraction of β-Sialon layer

CONCLUSION

Two types of silicon nitride based composites, β-Sialon/silicon nitride and dense/porous silicon nitride, were fabricated by stacking tape-casted sheets and sintering process. An alternating layered structure was obtained. Neither crack nor delamination was observed within all layers or interfaces. The chemical compositions were functionally graded at the interface of β-Sialon/Si_3N_4. Density and Young's modulus of the composites were represented by the rule of mixtures. The composites presented low elasticity, high strength and high fracture resistance. The thickness of layers affects the strength of a discrete layer in the alternating layered configuration.

ACKNOWLEDGEMENT

This research carried out under the Synergy Ceramics Project of the ISTF program promoted by AIST, MITI, Japan. Under this program, part of the work has been sponsored by NEDO.

REFERENCES

1. D. B. Marshall, J. J. Ratto and F. F. Lange, "Enhanced Fracture Toughness in Layered Microcomposites of Ce-ZrO_2 and Al_2O_3," *J. Am. Ceram. Soc.*, **74** [12] 2979-87 (1991).
2. P. Sarkar, X. Haung and P. S. Nicholson, "Structural Ceramic Microlaminates by Electrophoretic Deposition," *J. Am. Ceram. Soc.*, **75** [10] 2907-909 (1992).
3. W. J. Clegg, K. Kendall, N. M. Alford, T. W. Button and J. D. Birchall, "A Simple Way to Make Tough Ceramics," *Nature (London)*, **347**, 455-7 (1990).
4. H. Liu and S. M. Hsu, "Fracture Behavior of Multilayer Silicon Nitride/Boron Nitride Ceramics," *J. Am. Ceram. Soc.*, **79** [9] 2452-57 (1996).
5. H. Katsuki and Y. Hirata, "Coat of Alumina Sheet with Needle-Like Mullite," *J. Ceram. Soc. Jpn.*, **98** [10] 1114-19 (1990).
6. C. J. Russo, M. P. Harmer, H. M. Chan and G. A. Miller, "Design of Laminated Ceramic Composite for Improved Strength and Toughness," *J. Am. Ceram. Soc.*, **75** [12] 3396-400 (1992).
7. P. Šajgalík, Z. Lenčéš and J. Dusza, "Layered Si_3N_4 Composites with Enhanced Room Temperature Properties," *J. Mater. Sci.*, **31**, 4837-4842 (1996).
8. Y. Shigegaki, M. E. Brito, K. Hirao, M. Toriyama and S. Kanzaki, "Strain Tolerant Porous Silicon Nitride", *J. Am. Ceram. Soc.*, **80** [2], 495-98 (1997).
9. T. Ohji, Y. Shigegaki, T. Miyajima and S. Kanzaki, "Fracture Resistance Behavior of Multilayered Silicon Nitride", *J. Am. Ceram. Soc.*, **80** [4], 991-94 (1997).
10. K. H. Jack, "Sialons and Related Nitrogen Ceramics," *J. Mater. Sci.*, **11**, 1135-58 (1976).
11. M. Mitomo, N. Kuramoto and Y. Inomata, "Fabrication of High Strength β-Sialon by Reaction Sintering," *J. Mater. Sci.*, **14**, 2309-16 (1979).

CERAMIC MEMBRANES FOR NANOFILTERS

H.Richter, A.Piorra, G.Tomandl
Freiberg University of Mining and Technology
Institute of Ceramic Materials
G.-Zeuner-Str. 3
09596 Freiberg, Germany

ABSTRACT

A polymeric sol-gel process was used to prepare ceramic nanofiltration membranes made of ZrO_2 and TiO_2. The covering of porous supports was possible by controlling of chemical processes responsible for gelation.
The prepared membranes have a thickness of < 50 nm which makes the further characterisation of the layers difficult. The average pore size of ZrO_2 investigated on sintered powders is 1.6 nm. The cut-off value of < 1 kD was determined by filtration tests with two colouring agents.

INTRODUCTION

Filtration is a very common separation technique because of its low energy requirement. Because of their higher chemical, thermal and mechanical stability than all other membrane materials ceramic membranes are of special interest.
The "Institute of Ceramic Materials" of the "Freiberg University of Mining and Technology" has been working in the field of ceramic membranes for several years. Ultrafiltration membranes (UF) with average pore sizes of 5 - 20 nm made of ZrO_2, TiO_2, γ-Al_2O_3 and TiN have been developed. The preparation of UF membrane takes place by covering a porous support with a colloidal sol and subsequently drying and firing.
For the production of ceramic layers with much smaller pores the average diameter of the precursor particles must be reduced. But a further reduction in the size of sol particles seems to be impossible. To prepare nanofiltration membranes (NF) the area of solid particles was abandoned and a polymeric sol-gel process developed.

By using the polymeric sol-gel route it was possible to deposit a further layer with much smaller pores on UF-membranes. Filtration tests with “Direct Red“ and “Orange G“ showed their applicability for NF.

PREPARATION

The metal alkoxides Zr-propoxide and Ti-ethoxide were partially hydrolysed by adding a small amount of water (eq. 1). The molar ratio H_2O : alcoxide was < 1. After hydrolysis a polycondensation reaction took place (eq. 2) /1,2/. During drawing-off water and alcohol a three-dimensional network grew. The resulting mixture is called “polymeric sol“.

Hydrolysis

$$Me(OR)_n + H_2O \rightarrow Me(OR)_{n-1}OH + ROH \qquad \text{Eq. 1}$$

Polycondensation

$$Me(OR)_{n-1}OH + Me(OR)_{n-1}OH \rightarrow (RO)_{n-1}Me\text{-}O\text{-}Me(RO)_{n-1} + H_2O \qquad \text{Eq. 2.1}$$

$$Me(OR)_{n-1}OH + Me(OR)_n \rightarrow (RO)_{n-1}Me\text{-}O\text{-}Me(RO)_{n-1} + ROH \qquad \text{Eq. 2.2}$$

The polycondensation starts immediately after the commencement of the hydrolysis. When using pure substances the reactions are very fast. The viscosity of the system increases so rapidly that no covering of a porous support is possible. For some alkoxides the affinity against water is so strong that every contact with water leads to a complete hydrolysis which prevents polycondensation.
The controlling of hydrolysis and polycondensation is possible by:

- Temperature
- Addition of organic additives
- Thinning with organic solvents

The influence of temperature was not tested. It is assumed that increasing temperature only leads to acceleration of the reactions.
The effect of organic additives like ß-dicetone, acetylaceton and acetic acid was tested. All additives reduce the velocity of the reactions. By adding Acetylaceton or Acetic Acid nearly constant conditions over a period of some hours were achieved. But only Acetic Acid finally allows the deposition of crack free ceramic layers.

By thinning of the alkoxides with ethanol or propanol the affinity against water decreases (Fig. 1). By extraction of solvent from a mixture of thinned alkoxide and water the sensitiveness against water increases and sol-gel transition can be initiated. Both the amount and the kind of solvent show an influence on the reactivity. By variation of water content, and variation of kind and content of the solvent the setting of viscosity for the covering of the porous support was possible.

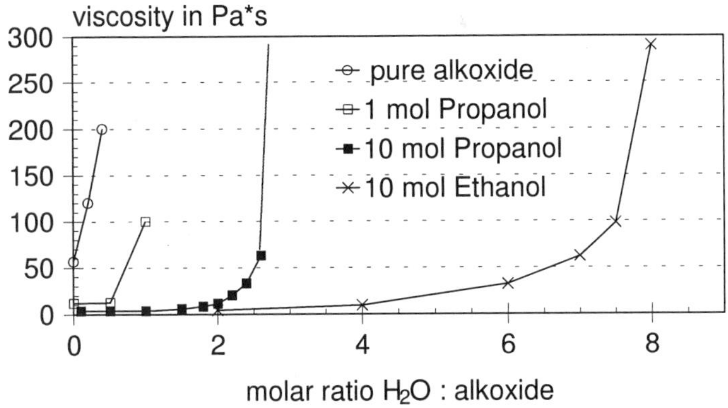

Fig.1: Viscosity versus water content depending on kind and content of alcohol

The polymeric sol was deposited on a porous support - a self-made ZrO_2 UF-membrane (Fig.2). Forced by the capillary forces a part of the solvent and liquid products of polycondensation reaction left the sol. This led to a strong acceleration of polycondensation reaction directly above the support surface. A polymeric layer was built and the excessive sol was poured off.
By drying over a period of three days and pyrolysis at 500 °C the transformation into ceramic took place.

RESULTS

Because of the extremly low thickness of the produced ceramic layers their investigation is very difficult. Only thick, but cracked membranes are visible by Scanning Electron Microscope (Fig. 3). Crack free ceramic NF-membranes are not detectable by using normal SEMs. The layer thickness is assumed to be < 50 nm.

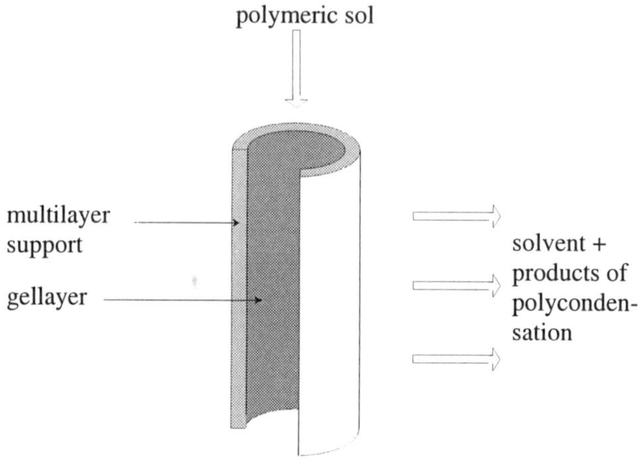

Fig.2: Covering the support with polymeric sol

Fig. 3: SEM picture of the cross-section of a cracked TiO_2-NF layer on TiO_2-UF-membrane

Investigation with Scanning Force Microscopy showed that the roughness of an UF-layer is 10 times higher compared to the same surface covered with a NF-layer. The NF-layer is possibly situated in pores and depressions of the support.

The determination of phase content and pore size distribution of supported membranes was not successful. Therefore polymeric sols were calcinated to powders. After firing at 500 °C pure tetragonal ZrO_2 and in case of TiO_2 anatas with a small amount of rutil was found.

In powder preparation probably a stronger shrinkage and combustion of the ceramics takes place compared to supported thin layer. The resulting pore volumes are very low. Therefore the determination of an average pore size of TiO_2 was not successful. By using N_2-adsorption/-desorption (Micromeritics ASAP 2000) a pore size of ZrO_2 of 1.6 nm (Dubinin-Asthakov) was found.

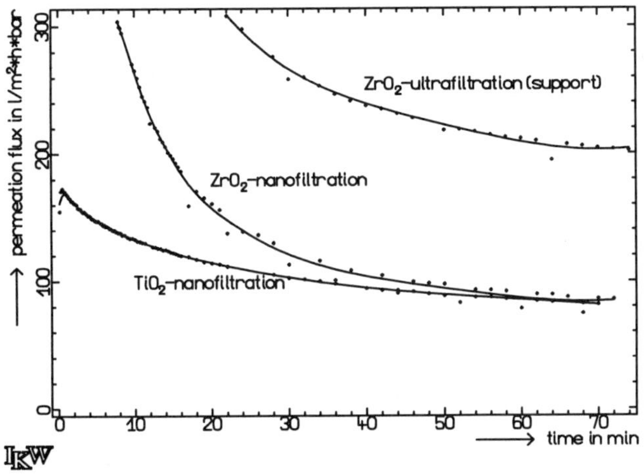

Fig.4: Permeation flux versus filtration time of ZrO_2 -UF (support) and ZrO_2 NF and TiO_2 NF membranes

The filtration properties of the membranes were determined by cross flow filtration with tubular membranes (filter area approx. 20 cm^2). For all NF-membranes a much lower water permeation flux compared to the supporting UF-membrane was found (Fig. 4). The reason for this is the smaller pore size. By using the equation of *Hagen-Poiseuille* the flux I grows with the pore size d^2 assuming a constant porosity ε, viscosity η, layer thickness x and transmembrane pressure Δp.

$$I = \frac{d^2 \cdot \varepsilon}{32 \cdot \eta \cdot x} \cdot \Delta p \qquad \text{Eq.3}$$

Table I: Permeation flux and rejection of ZrO_2 UF (support), ZrO_2 and TiO_2 nanofiltration after 60 min. crossflow filtration

	ZrO_2 UF (support)	ZrO_2 nanofiltration		TiO_2 nanofiltration	
	flux $l/(m^2 \cdot h \cdot bar)$	flux $l/(m^2 \cdot h \cdot bar)$	rejection %	flux $l/(m^2 \cdot h \cdot bar)$	rejection %
water	205	80	-	82	-
"Direct Red"	-	66	99,2	44	97,5
"Orange G"	-	45	30,0	41	47,3

In order to determine the rejection for "small" molecules after Larbot *et al.* /3/ two solutions in water were prepared (solution 1: 200 mg/l "Direct Red" (990,8 g/mol); solution 2: 200 mg/l "Orange G" (452,4 g/mol)). The flux decreased because pores were blocked by the test molecules. After 60 min. filtration ZrO_2 and TiO_2-membranes showed a very high rejection of > 97 % for "Direct Red" and of 30 % and 45 % for "Orange G". The cut-off value was less than 1 kD. By definition of Alam-Younsi *et.al.* /4/ a first condition for NF membranes was achieved. Therefore ceramic nanofiltration membranes made of ZrO_2 and TiO_2 were successfully prepared by using a polymeric sol-gel route.

CONCLUSIONS

A polymeric sol-gel process is practicable by partial hydrolysis and polycondensation of metal alkoxides. The controlling of the process is possible by the addition of Acetic Acid and alcohols. By covering a porous support and then drying and firing the preparation of ceramic NF membranes was carried out. Because of the extremely low thickness (< 50 nm) of the produced layers their investigation is very difficult. By crossflow filtration with test molecules the applicability for nanofiltration was proved.
The topic of future research is the determination of rejection of dissolved salts and other materials relevant for industry.

REFERENCES

[1] B.E.Yoldas, "Zirconium oxide formed by hydrolytic condensation of alkoxides and parameters that effect their morphology", J. Mat. Sci., 21, 1080-1086 (1986)

[2] K.Yamasa, T.Y.Chow, T.Horihata, M.Nagata,"Low temperature synthesis of zirconium oxide using chelating agents", J. Non-Cryst. Solids, 100, 316-320

[3] A.Larbot, S.Alami-Younssi, M.Persin, J.Sarrazin, L.Cot,"Preparation of a γ-alumina nanofiltration membrane", J. Membr. Sci. 97 (1994) 167-173

[4] S.Alami-Younssi, A.Larbot, M.Persin, J.Sarrazin, L.Cot, "Rejection of mineral salts on a gamma alumina nanofiltration membrane - Application to environmental process", J. Membr. Sci. 102 (1995) 123-129

ASSESSMENT OF THE CRITICAL THICKNESS AND FRACTURE TOUGHNESS OF THIN METAL-ORGANIC PRECURSOR FILMS

Ryan K. Roeder and Elliott B. Slamovich
School of Materials Engineering, Purdue University, West Lafayette, IN 47907-1289

ABSTRACT

A fracture mechanics approach for the study of cracking in thin films was demonstrated. The drying and cracking of thin metal-organic precursor films were examined by monitoring film thickness and evaporative mass loss. Experiments and calculations revealed two methods to prevent film cracking: (1) choosing a precursor to minimize evaporative stress during drying, and (2) adding elastomeric polymers to the precursor to increase the film fracture toughness.

INTRODUCTION

Sol-gel, metallo-organic decomposition, and hydrothermal routes to ceramic thin films depend on the mechanical integrity of the precursor film. The mechanical integrity of thin films has long been observed to be dependent on the film thickness. The critical film thickness for cracking is system and material dependent, reflecting the film fracture toughness. The critical film thickness has typically been observed in the range of 0.4 - 1.0 μm for a variety of materials and conditions.[1-6] Films above the critical thickness are observed to crack or decohere from the substrate. Cracking is caused by internal stresses that exceed the film's cohesive strength, and decoherence is due to stresses that exceed the adhesive strength to the substrate.[6-9]

Internal stresses in thin films arise from a material's response to constrained volume changes.[8,9] Constrained volume changes occur during many processes including drying, thermal expansion or contraction, pyrolysis, densification and curing. The internal strain induced by a constrained volume change is calculated as

$$\varepsilon = \frac{\Delta l}{l} = \left(1 + \frac{\Delta V}{V}\right)^{1/3} - 1. \qquad (1)$$

Assuming primarily elastic behavior (e.g. a small plastic zone), the internal biaxial film stress is given as

$$\sigma = \frac{E \cdot \varepsilon}{1-\nu} = \frac{E}{1-\nu} \cdot \left[\left(1 + \frac{\Delta V}{V}\right)^{1/3} - 1\right], \qquad (2)$$

where E is the film's elastic modulus and ν is Poisson's ratio.

Brittle films under a tensile stress are observed to crack or decohere from the substrate when above the critical thickness. The Griffith criterion for cracking in brittle solids shows that the stress required for fracture is inversely related to the film thickness, based on a macroscopic energy balance between the stress relieved and surfaces created during crack propagation. Thus, the Griffith criterion accounts for the critical thickness phenomenon, but is not alone sufficient. The stress concentration at the crack tip must also be considered because the cohesive strength of a material must be overcome on the atomic scale. The stress concentration required for crack propagation is described by the critical stress intensity factor (K_c), which is a geometry dependent material property. Evans and co-workers[10,11] used linear elastic fracture mechanics to predict the critical film thickness, h_c, in terms of K_c for a thin film as

$$h_c = \left(\frac{K_c}{\sigma \cdot \Omega}\right)^2, \text{ where } \Omega \propto \sqrt{\pi \cdot \left(\frac{E}{E_s}\right)}. \tag{3}$$

σ is the film stress and Ω is a semi-empirical non-dimensional parameter related to the ratio of elastic moduli for the film (E) and substrate (E_s). Eq. (3) predicts film cracking when the film thickness is greater than or equal to h_c. In this study, K_c for a thin film under mode I loading is expressed as "K_{Ic}" to denote *plane stress* and to differentiate from the commonly misused notation for *plain strain,* "K_{Ic}."

The above concepts of thin film fracture mechanics were used in the following experimental study of cracking in thin metal-organic precursor films. Rather than using Eq. (3) to predict the critical thickness for cracking, the critical thickness was measured and Eq. (3) used to examine the origins of film cracking. The drying and cracking of precursor films were examined by monitoring film thickness and evaporative mass loss. Internal film strains were calculated from the constrained volume shrinkages. Estimates of the normalized film stress and relative fracture toughness were used to identify processing effects on the cracking behavior. Cracking was prevented (increased critical thickness) by either minimizing the evaporative stress during drying, or toughening the precursor film via elastomeric polymer additions.

EXPERIMENTAL PROCEDURE

Two metal-organic precursors were investigated: titanium di(isopropoxide) bis(ethyl acetoacetate) (TIBE) and titanium dimethoxy dineodecanoate (TDD). Both have been used to process titanate thin films for microelectronic applications.[13-16] TIBE is known to crack readily,[13] while TDD does not. For spin coating, TIBE and TDD were diluted in 50 vol% toluene and 70 vol% p-xylene, respectively, such that the solution viscosities were nearly equal for shear rates of 1 to 1000 s^{-1}. TIBE solutions were further modified by dissolving 5.0 and 10.0 wt% (relative to TIBE) of an elastomeric styrene-butadiene-styrene block copolymer (Kraton® D1102C, Shell Oil Co.). The viscosity before and average film thickness after spin coating increased with polymer content. Also, two control solutions were prepared: as-received TIBE without solvent dilution and TIBE diluted in 50 vol% p-xylene. All films were spin coated in a forced convection fume hood at 5000 rpm for 15 s on Pt-coated glass substrates. The Pt coating provided reflectance for obtaining optical interference fringes and infrared spectra. Additional details of the TIBE solutions, spin coating procedure and viscosity measurements are available in Ref. 12.

Drying took place at room temperature and atmospheric pressure. The substrate mass was recorded beforehand so the film mass could be measured during drying. The film mass was monitored during drying using a microbalance (±0.005 mg), taking measurements 0.1, 0.5, 1.0, 1.5, 2.0, 2.5, 3.0, 4.0, 12.0, and 90 h after spin coating. The film morphology, including film thickness, crack initiation, crack growth, and crack arrest, was simultaneously observed using a standard optical microscope. Isochromatic color fringes resulting from optical interference of light reflected from the film and substrate were used to determine film thickness (Fig. 1). The critical thickness measurement technique is discussed in detail in Ref. 12.

Organic compounds present before and after spin coating were identified from infrared spectra obtained using a Fourier transform infrared (FTIR) spectrometer (Magna IR Spectrometer 55, Nicolet Analytical Instruments). The spectra of solvents, precursors, and precursor solutions alone were measured in transmittance by pressing a thin liquid film between NaCl plates. The spectra of films were measured in reflectance at time intervals beginning immediately after spin coating.

Films were subjected to a scratch test to determine the drying time when precursor films underwent a transition from behaving *primarily* as a viscous liquid to an elastic solid. After spin coating, films were lightly scratched with a pair of precision forceps at 5 min time intervals. The response of precursor films was observed in an optical microscope and micrographs were taken after 1 h of drying.

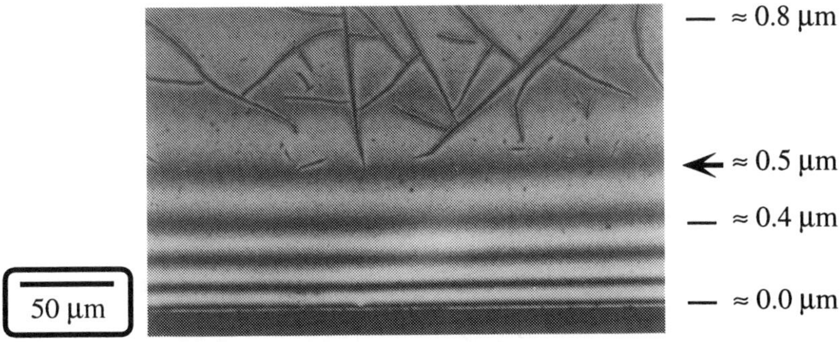

Fig. 1. Optical micrograph of a film made from the TIBE-50 vol% toluene solution with 5.0 wt% (relative to TIBE) polymer after drying 90 h. Cracks terminate at isochromatic color fringes corresponding to the critical film thickness. Note the substrate edge is located along the bottom side of the micrograph.

RESULTS AND DISCUSSION

Microstructural Observations

Observations of the cracking behavior and measurement of the critical thickness of TIBE films have been documented.[12] Crack growth in each film was observed to be complete after 90 h of drying. After 90 h drying, all TIBE films without polymer additions had largely decohered from the substrate; however, crack arrest was observed in the thinnest regions of the film (e.g. the outer edges of the substrate). TIBE films with polymer additions did not decohere. More importantly, cracking was much less widespread with increased polymer content, despite increased average film thickness as shown by color fringes. Unlike TIBE films, the TDD film did not crack, despite having the largest film thickness.

The critical thickness for films of each precursor solution was determined from micrographs taken where cracks initiating in thicker regions grew into thinner regions (e.g. Fig. 1). The critical thickness was measured as the thickness at which propagating cracks terminated. Ref. 12 discusses optical interference and explains the measurement of the critical film thickness from isochromatic color fringes. The critical thickness measured for each film is listed in Table I. The critical film thickness for cracking increased with increasing polymer content. Note the values reported in Table I contain up to 10% error inherent in the measurement technique. The critical thickness of TDD films could not be determined because these precursor films did not crack, even for films 50 μm thick.

Solvent/Precursor Evaporation

The measured mass of films during drying is shown in Fig. 2. All TIBE films exhibited similar behavior, while the TDD film lost relatively little mass. Decreases in the mass of the TIBE alone indicated that TIBE itself was volatile. An empirical exponential decay equation was used to fit the mass loss data, with film mass (M) as a function of drying time (t), $M(t) = A + B \cdot exp(-C \cdot t)$, as shown in Fig. 2.

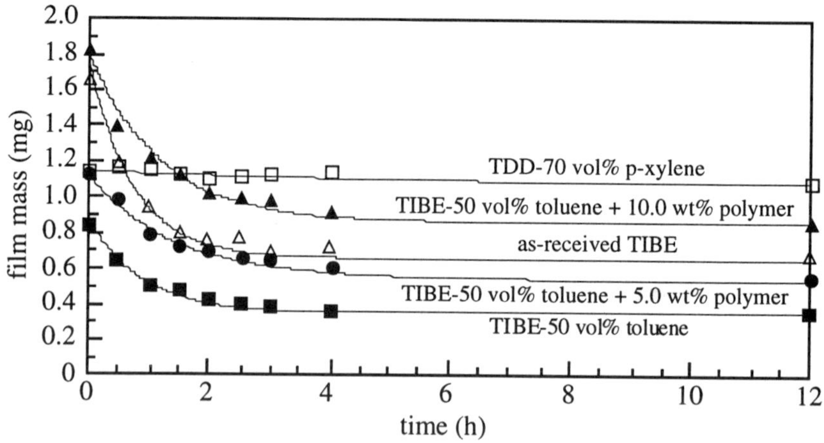

Fig. 2. Film mass for each precursor solution versus drying time, showing mass loss due to evaporation of volatile organics.

The evaporative mass losses were further investigated using FTIR. The FTIR spectra measured for films immediately after spin coating and after short drying times showed no presence of the identified solvent peaks. Thus, virtually all the solvent for each precursor solution evaporated during spin coating. Complete evaporation of solvents during spin coating was facilitated by the relatively high vapor pressures of toluene and p-xylene at room temperature, and the highly convective conditions while spinning at 5000 rpm in a forced convection fume hood. Thus, the origin of the mass loss in TIBE films remained uncertain, possibly due to the evaporation of TIBE itself, unreacted ethyl acetoacetone (5% in as-received TIBE[17]), or condensation products. In contrast, the negligible mass loss in the TDD film during drying indicated that TDD and its condensation products (if any) had a lower vapor pressure under the drying conditions.

Film Strain, Stress and Fracture Toughness

The scratch test was a simple and effective means to determine the glass transition* of a viscous liquid film to an elastic solid film. Shortly after spin coating scratches completely healed (Fig. 3, 13 min), as the precursor film retained a low viscosity. The response of the film became increasingly viscous for successive scratches, shown by film uplift alongside scratches (Fig. 3, 18 to 28 min). The transition into primarily elastic behavior was determined when no surface uplift was observed and scratch edges were serrated (Fig. 3, ≥ 28 min), indicative of elastic mixed-mode fracture. The approximate drying times to the glass transition (t_g) of precursor films are listed in Table I.

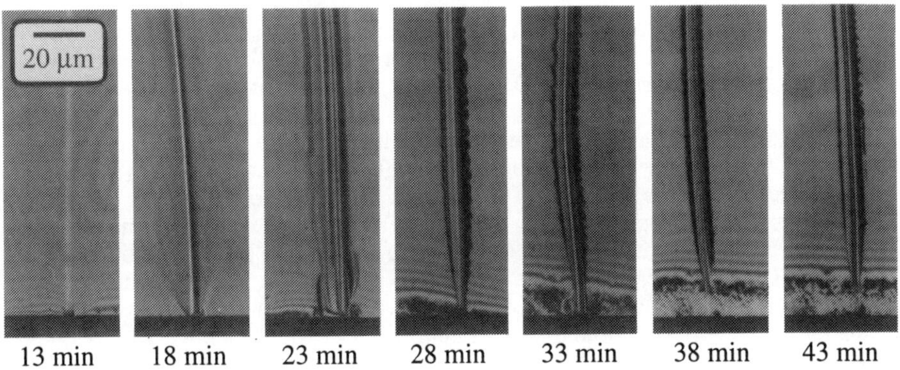

Fig. 3. Optical micrographs (taken after drying 1 h) of a film made from the TIBE-50 vol% toluene solution after performing the scratch test, showing the change in mechanical response from primarily viscous to primarily elastic with increased drying time. The substrate edge is located at the bottom of micrographs.

The measured mass loss and time to glass transition during drying were used to determine the internal strain and stress in each film. Film volume as a function of drying time, $V(t)$, was calculated from the mass loss data using the known densities and initial amounts of constituents in each precursor solution. Internal film strains were determined using Eq. (1) for the volume change that occurred after the glass transition at t_g. The volume change was determined from the film volume at the time of glass transition, $V(t_g)$, and after reaching a steady state, V_∞ (taken at 90 h). The internal linear strains are shown in Fig. 4a and Table I. The corresponding film stresses (Fig. 4a and Table I) were estimated using Eq. (2), normalizing to the unknown elastic modulus and assuming Poisson's ratio (ν) to be ≈ 0.5.

* In the above context, a glass transition refers to the transition of a primarily viscous material into a primarily elastic amorphous solid via molecular coalescence with drying.

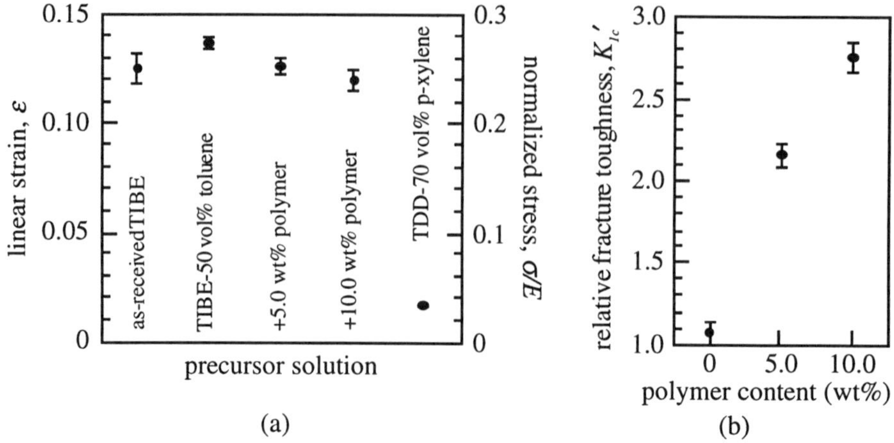

Fig. 4. (a) Linear strain and normalized internal stress calculated for each precursor film. (b) Relative (to that of TIBE alone) fracture toughness estimated for TIBE films with added polymer. Error bars correspond to the approximate range of error in determining the time to glass transition and measuring the critical thickness.

The internal strain and stress were nearly identical for all TIBE films, but were nearly ten times lower for the TDD film (Fig. 4a). Recall that the TIBE films cracked readily, and the TDD film did not crack at all. Also, recall that TIBE films lost significant mass during drying and the TDD film lost negligible mass. Since the TDD film showed relatively little evaporation, relatively little stress developed in the film. Thus, film cracking was prevented (increased critical film thickness) by choosing a precursor that minimized evaporative stresses during drying.

The film fracture toughness was calculated with Eq. (3) using: $\Omega = 1$ for the case $E << E_s$,[11] the measured critical film thickness, and the estimated internal film stress normalized to the unknown elastic modulus. To allow comparison between films, a relative fracture toughness (K_{1c}') was calculated by dividing the elasticity normalized fracture toughness by that of the as-received TIBE film, which exhibited the least resistance to cracking. The relative film fracture toughness calculated for precursor films is shown in Table I. Solvent choice and dilution showed little effect on the film fracture toughness. The relative fracture toughness for the TDD film could not be determined since there was no observed cracking.

The relative film fracture toughness increased with increasing polymer content (Table I). Fig. 4b shows a greater than two-fold increase in relative fracture toughness for TIBE films with small additions of polymer. Recall that the extent of cracking decreased and the critical thickness increased with increased polymer content in TIBE films. The polymeric molecules are speculated to have toughened films of smaller, monomeric metal-organic molecules by providing a more cohesive network via molecular entanglements. Moreover, film cracking was prevented (increased critical film thickness) by adding elastomeric polymers to the precursor to increase the film fracture toughness.

Table I. Experimental and Calculated Results for Precursor Films

Precursor Solution*	h_c (μm)	t_g (min)	ε	σ/E	K_{Ic}'
TIBE-50% toluene	≈0.11	25-30	≈0.14	≈0.27	≈1.1
TIBE-50% toluene+5% polymer#	≈0.51	25-35	≈0.13	≈0.25	≈2.2
TIBE-50% toluene+10% polymer#	≈0.92	25-35	≈0.12	≈0.24	≈2.8
TDD-70% xylene	>50	20-30	≈0.02	≈0.03	–
TIBE-50% xylene	≈0.17	25-30	≈0.12	≈0.25	≈1.2
as-received TIBE	≈0.10	30-40	≈0.13	≈0.25	1.0

* solvent in vol% relative to precursor and polymer in wt% relative to precursor

\# Kraton® D1102C, a styrene-butadiene-styrene block copolymer

Closing Remarks

The above experimental and analytical methods comprise a simple approach to study cracking in thin films. Admittedly, several shortcomings persist in the simplified analysis presented in this paper and are discussed below. However, the ability of the approach to compare behavior and elucidate the important effects on film cracking was demonstrated. The demonstrated means of preventing film cracking were consistent with those described in the literature.[1,6] Thus, the approach in this study is viewed by the authors as a foundation to build more refined and accurate analyses upon.

The analysis in this study included the following assumptions: biaxial stress, linear elastic fracture mechanics, isotropic and constant elastic properties, no substrate deformation, perfect adhesion, minimal viscoelastic stress relaxation, and constant stress through the film thickness. The random crack patterns observed in all films validated the assumption of biaxial stress. The use of linear elastic fracture mechanics is justified as long as the plastic (or non-elastic) zone is small relative to the crack velocity. However, other assumptions should be improved upon.

In the present analysis, the only governing factor for the magnitude of internal film stress was the volume change due to evaporation after the film began to behave primarily elastically. Viscoelastic stress relaxation (especially at a crack tip) was ignored in the calculations. However, the scratch test appeared to capture at least some of the effects of stress relaxation, shown by a decreasing internal film stress with increasing polymer content (Fig. 4b). Ideally, either the viscoelastic film properties or film stress should be measured. Several techniques have been used to measure film stress from substrate deflections.[18-21] With the addition of stress measurement, a better assessment of thin film fracture toughness could be made.

If the calculations in this study were carried out to exact numerical solutions assuming an elastic modulus of 1 GPa, the internal stress and fracture toughness in films of this study were on the order of tens to hundreds MPa, and tens to hundreds kPa·m$^{1/2}$, respectively. These values are believed to be an upper bound, because stress relaxation effects were not explicitly considered. Values reported in literature also suggest these stresses may be a slightly high estimate.[5,6,9] On the other hand, there are currently no known studies to compare thin film fracture toughness values. While exact calculations were not emphasized in this study, the use of simpler systems with well documented properties could facilitate such an objective.

ACKNOWLEDGMENTS

This work was supported by the National Science Foundation grant DMR-9623744. The authors thank Dr. Keith J. Bowman for helpful suggestions.

REFERENCES

1. C.J. Brinker, A.J. Hurd, P.R. Schunk, G.C. Frye and C.S. Ashley, "Review of Sol-Gel Thin Film Formation," *J. Non-Cryst. Solids*, **147-148**, 424-436 (1992).
2. G.W. Scherer, "Recent Progress in Drying of Gels," *J. Non-Cryst. Solids*, **147-148**, 363-374 (1992).
3. A. Atkinson and R.M. Guppy, "Mechanical Stability of Sol-Gel Films," *J. Mat. Sci.*, **26**, 3869-3873 (1991).
4. S.-Y. Chen and I-W. Chen, "Cracking during Pyrolysis of Oxide Thin Films–Phenomenology, Mechanisms, and Mechanics," *J. Am. Ceram. Soc.*, **78** [11] 2929-2939 (1995).
5. S. Bec, A. Tonck and J-L. Loubet, "Mechanical Properties of Pyrolytic "SiOCN" Thin Coatings," *Mat. Res. Soc. Symp. Proc.*, **308**, 577-582 (1993).
6. K. Sato, "The Internal Stress of Coating Films," *Progr. Org. Coatings*, **8**, 143-160 (1990).
7. M.D. Thouless, "Some Mechanics for the Adhesion of Thin Films," *Thin Solid Films*, **181**, 397-406 (1989).
8. G.P. Bierwagon, "Film Formation and Mudcracking in Latex Coatings," *J. Coatings Techn.*, **51** [658] 117-126 (1979).
9. S.G. Croll, "The Origin of Residual Internal Stress in Solvent-Cast Thermoplastic Coatings," *J. Appl. Polym. Sci.*, **23**, 847-858 (1979).
10. A.G. Evans, M.D. Drory and M.S. Hu, "The Cracking and Decohesion of Thin Films," *J. Mater. Res.*, **3** [5] 1043-1048 (1988).
11. M.S. Hu and A.G. Evans, "The Cracking and Decohesion of Thin Films on Ductile Substrates," *Acta Met.*, **37** [3] 917-925 (1989).
12. R.K. Roeder and E.B. Slamovich, "The Critical Thickness of Thin Metal-Organic Precursor Films: Effect of Polymer Additions," submitted to *J. Am. Ceram. Soc.*
13. E.B. Slamovich and I.A. Aksay, "Hydrothermal Processing of $BaTiO_3$/Polymer Films," *Mater. Res. Soc. Symp. Proc.*, **346**, 63-68 (1994).
14. E.B. Slamovich and I.A. Aksay, "Structure Evolution in Hydrothermally Processed (<100°C) $BaTiO_3$ Films," *J. Am. Ceram. Soc.*, **79** [1] 239-247 (1996).
15. A.S. Shaikh and G.M. Vest, "Kinetics of $BaTiO_3$ and $PbTiO_3$ Formation from Metallo-organic Precursors," *J. Am. Ceram. Soc.*, **69**, 682-688 (1986).
16. A.S. Shaikh, R.W. Vest and G.M Vest, "Dielectric Properties of Ultrafine Grained $BaTiO_3$," *IEEE Trans. UFFC*, **36** [4] 407-412 (1989).
17. Personal communication, Gelest Inc., Tullytown, PA.
18. J.H.L Voncken, C. Lijzenga, K.P. Kumar, K. Keizer, A.J. Burggraaf and B.C. Bonekamp, "New Method for the Measurement of Stress in Thin Drying Gel Layers Produced During the Formation of Ceramic Membranes," *J. Mater. Sci.*, **27**, 472-478 (1992).
19. K.E. Boggs, D.L. Wilcox, D.A. Payne and L.S. Allen, "Stress Development During High Dielectric Ceramic Thin Films Processing," *Proc. SPIE Int. Soc. Opt. Eng.*, **2256**, 350-355 (1994).
20. R.C. Chiu and M.J. Cima, "Drying Behavior of Granular Ceramic Films," *Ceram. Trans.*, **22**, 347-356 (1991).
21. T.J. Garino and M. Harrington, "Residual Stress in PZT Thin Films and Its Effect on Ferroelectric Properties," *Mater. Res. Soc. Symp. Proc.*, **243**, 341-347 (1992).

STRESS FORMATION DURING DRYING OF POROUS MEDIA

S. Lampenscherf, M. Bobeth, and W. Pompe
Max-Planck-Society
Research Group Mechanics of Heterogeneous Solids, Hallwachstr. 3
D-01069 Dresden, Germany

ABSTRACT

Stress formation and cracking during the dryin step are major problems in ceramic technologies. A model is presented, which describes densification and stress formation in a drying layer of a two-phase material containing a liquid and a solid phase.

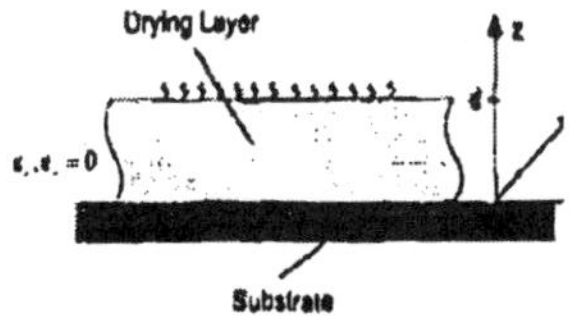

Stress formation occurs owing to capillary forces. The stress in the liquid phase casues it to flow to the drying surface of the body. Mass transport and stress distribution within the liquid phase are assumed to be coupled via Darcy's law. The mechanical behavior of the porous medium is decried according to Terzaghi' model (the intergranular stress model), where the compressibility, $K(c_o)$ of the porous medium depends in a nonlinear way on the volume fraction co(z,t) of the solid phase. It can be shown that the pressure $p_i(z,t)$ within the liquid is given by

$$\frac{\partial p_i}{\partial t} = - \frac{(1 - c_i)D(c_i)}{\eta_i}\left[\frac{\partial}{\partial t} p_i\right]^2 + \frac{k(c_i)}{\eta_i}\frac{\partial}{\partial_z}\left[(1 - c_i)D(c_i)\frac{\partial}{\partial_t} p_i\right]$$

where η_i depends on the viscosity of the liquid. The permeability $D(c_o)$ is determined by the pore size within the porous medium. The densification of the drying body is given by the evolution equation for the solid phase volume fraction, c_i

$$\frac{\partial c_i}{\partial_t} = - \frac{\partial}{\partial_z}\left[\frac{c_i(1 - c_i)D(c_i)}{\eta_i}\frac{\partial}{\partial_z} p_i(z,t)\right],$$

which result from the continuity equation for the solid phase and the Darcy law. Solutions of the two equations were calculated numerically. For limiting cases, analytical solutions are presented. The theoretical predictions for the overall stress in the layer are compared with stress measurements in thin drying layers obtained by recording the substrate bending with a Laser-Topography-Scanning System.

ELECTROPHORETIC DEPOSITION OF TZP ON POROUS SUBSTRATES

J. Will, L. Gubler, S. Al-Wakeel, M. Gödickemeier, M. K. M. Hruschka, and L. J. Gauckler
Colloidal Chemistry-Ceramic Processing Group at the ETH
Zürich, Switzerland

ABSTRACT

Dense thin films of zirconia, about 10-50 mm in thickness, were prepared via electrophoretic deposition using a suspension of tetragonal stabilized zirconia in ethanol. Porous Ni-cermet and $La_{0.85}Sr_{0.15}MnO_3$, respectively were used as substrates. These substrates act as anodes and cathodes in Solid Oxide Fuel Cells (SOFC) whereas the deposited and sintered zirconia film acts as electrolyte. A suspension of 25 m/o ZrO_2 (Tosoh, Tokyo, TZ-3YS) was prepared in ethanol and subsequently electrophoretically deposited on the porous substrates produced via tape casting. The composite was then sintered at 1350°C. Crack free co-sintering of the deposited film and the porous substrate is possible after adjusting the shrinkage of film and substrate by control of grain size and porosities. A kinetic equation for the correlation between deposition and current voltage is derived and verified by experiments.

ZIRCONIA COATINGS ONTO NICKEL ELECTRODES BY AQUEOUS ELECTROPHORETIC DEPOSITION

B. Ferrari and R. Moreno
Instituto de Cerámica y Vidrio
C.S.I.C., Crtra. Valencia Dm. 24,300
28500 Arganda del Rey
Madrid, Spain

ABSTRACT

Electrophoretic deposition is a powerful route to obtain thick coatings onto conductive substrates. The system ZrO_2/Ni is widely considered for technological applications, such as solid oxide fuel cells. Electrophoretic deposition (EPD) is usually performed using non-aqueous suspensions in which problems related to electrolysis are avoided. However, the possibility to use aqueous systems is receiving an increased interest due to economical and ecological benefits. In this work the processing conditions for the optimization of ZrO_2 electrodeposits from aqueous suspensions are studied. The dispersing characteristics of the slips are studied in terms of zeta potential and conductivity, both determining the optimization of a deposit. The kinetics of the process is analyzed for optimized dispersing conditions considering the deposited mass and thickness. The evolution of the slip conductivity and possible reduction in the solid concentration are determined and related to the voltage variation during the experiments. The characteristics of the green and the sintered deposits are also studied.

PATTERNING PROCESS USING MICROMETER PARTICLES

Hiroshi Fudouzi, Mikihiko Kobayashi, and Norio Shinya
National Research Institute for Metals
5th Research Group
1-2-1, Sengen, Tsukuba
Ibaraki 305, Japan

ABSTRACT

A novel method for fabrication of controlled microstructure has been proposed. In this method, an electrified pattern was formed on an insulating substrate by an electron or an ion beam, and then powder particles were attracted to the electrified pattern and adhered on the substrate. To demonstrate the potential of this processing, spherical silica particles with 5 µm diameter were used as a model, and a patterning structure of the particles was obtained on a calcium titanate ceramic substrate. Then, the substrate was heated at 1150°C for 30 minutes in air. A two-dimensional microstructure was fabricated by this method.

FABRICATION OF WOVEN NICALON™ (NL607) SiC FIBRE-YTTRIUM DISILICATE CMCs USING ELECTROPHORETIC DEPOSITION

P. A. Trusty* and C. B. Ponton

IRC in Materials for High Performance Applications, School of Metallurgy & Materials, University of Birmingham, Edgbaston, Birmingham B15 2TT, UK.

A. R. Boccaccini

FG Werkstofftechnik, TU Ilmenau, D-98684 Ilmenau, Germany

ABSTRACT

The processing problems involved in the fabrication of a Nicalon™ (NL607) woven fibre reinforced ceramic matrix composite using electrophoretic deposition (EPD) have been analysed and effectively solved. In this work, an yttrium disilicate precursor powder has been synthesised using hydrothermal processing and characterised subsequently in terms of particle surface properties in order to obtain a kinetically stable powder dispersion in water. EPD was performed under different constant d.c. voltage conditions using fibre preforms, both in the as-received condition and as pre-treated with a wetting agent. Infiltrated fibre mats were consolidated by hot-pressing using a schedule derived from hot stage optical microscopy studies. The resultant CMC samples were then characterised microstructurally using scanning electron microscopy.

INTRODUCTION

The synthesis of rare earth oxides, in particular oxides containing yttrium has been of great interest over the last 30 years [1-2]. Yttrium disilicate ($Y_2Si_2O_7$) occurs naturally as yttrialite and, with its high melting point of 1775°C and refractory character [3], it has potential as a material for high performance applications. Indeed, studies of silicon nitride using yttria as a sintering aid have highlighted the formation of a glassy yttrium disilicate phase at the grain boundaries. It has been shown that this amorphous material, upon crystallisation, exhibits improved high temperature properties [4].

To date, the published work on the synthesis of $Y_2Si_2O_7$ has involved a study of the phase evolution of the material system by means of X-ray diffraction, which gives an average macrocomposition. Much of this work has been on material produced by the solid-state reaction between Y_2O_3 and SiO_2 powders at temperatures in the region of 900°C - 1800°C [5]. In order, however, for the constituents to react completely and form a single phase, holding times at temperature in the region of 100 hours were necessary, a synthesis time which is obviously undesirable.

Recent work has shown, however, that a hydrothermally synthesised yttrium disilicate precursor powder produces single phase yttrium disilicate after calcination at 1050°C for 1 hour [6]. This paper aims to further this synthesis work by describing the stages required for the fabrication of woven Nicalon (SiC) fibre reinforced $Y_2Si_2O_7$ CMCs using electrophoretic deposition (EPD). The EPD technique is used to infiltrate the woven fibre preforms, prior to the subsequent hot-pressing stage. More details on the use of this technique to infiltrate 2D and 3D fibre preforms can be found elsewhere [7,8].

EXPERIMENTAL PROCEDURES

The nanosized yttrium disilicate precursor particles required to infiltrate the Nicalon fibre preforms were synthesised via hydrothermal processing using yttrium acetate ($Y(CH_3CO_2)_3.xH_2O$) and tetraethylorthosilicate (TEOS, $Si(OC_2H_5)_4$) as the chemical precursors. The synthesis procedure is described in detail elsewhere [6]. The resultant nanosized $Y_2Si_2O_7$ precursor powder was characterised subsequently in terms of its surface charge properties in water and its densification behaviour using the techniques described below.

Surface Charge Measurements

The surface charge properties of the precursor were measured using a DELSA 440 surface charge analyser. This involved suspending small powder samples in aqueous solutions of varying pH values and measuring their electrophoretic mobility as a function of pH. This information enabled the preparation of a kinetically stable sol of $Y_2Si_2O_7$ precursor particles for the EPD process.

EPD Trials

EPD trials were performed using an yttrium disilicate sol with Nicalon woven fibre mats in order to assess the infiltration parameters. EPD relies on the presence of nanoscale charged particles in suspension, which will migrate to an electrode of opposite charge under the influence of a d.c. electric field. The conducting carbon coating present on this particular grade of Nicalon SiC fibre (NL607) enabled the fibre mats to be used directly as an electrode in an EPD cell (see Figure 1). In such cases, where the fibres are conducting, on the application of an external field, the sol particles move towards, and actually penetrate/deposit onto the woven fibre preform which is functioning as the deposition electrode. When the fibre mats had been immersed in the sol, the fibre mat/sol system was ultrasonicated and vacuum degassed in order to promote particle/fibre contact and remove any air from inside the preforms. The fibre mats were used either in the as-received condition or after a surface pre-treatment, using polymethylacrylic acid (Versicol VK11) as a wetting agent to promote sol particle/fibre mat interaction. During EPD with this fibre system, it has been found that deposition is possible with applied voltages as low as 2 V d.c. [9]. However, higher voltages up to 10 V d.c. were also applied in an attempt to minimise the deposition time and optimise the material deposition density. Infiltrated preforms were examined using SEM to assess the extent of infiltration, and the infiltration parameters used for those preforms with the best level of infiltration then employed in the infiltration of several 50 x 50 mm fibre mats to be used for the hot-pressing studies.

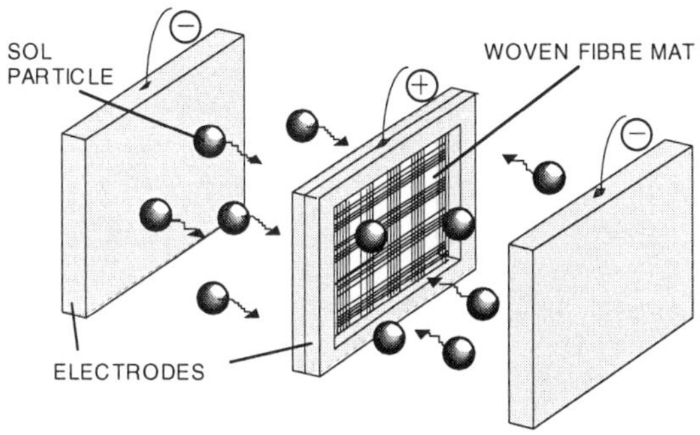

Figure 1 Schematic diagram of an EPD cell using a woven fibre mat as the deposition electrode

Hot-stage Optical Dilatometry

The flow behaviour of ceramic materials which are transiently viscous at high temperatures can be utilised, together with the application of pressure, in the densification of the matrix material surrounding the infiltrated fibre mats. The viscous flow of the matrix prevents the applied loads from being transmitted directly and non-uniformly to the fibres, which will in all likelihood cause fibre breakage. To this effect, if $Y_2Si_2O_7$ exhibits a viscous flow stage during sintering, this will be conducive to successful hot-pressing. Previous dilatometry studies on this system have shown that the viscous stage of $Y_2Si_2O_7$ can be enhanced by adding small amounts (ca. 3 wt %) of LiF dissolved in water to the precursor product [10]. The LiF acts like a flux by reducing the viscosity of the precursor during sintering. In order to gain insight further into the effect of LiF on the densification kinetics of $Y_2Si_2O_7$, the technique of hot-stage optical dilatometry was employed.

It has been shown that hot stage optical microscopy is a very useful experimental method for studying in-situ the pressureless sintering behaviour of powder compacts [11,12]. It enables quantitative monitoring of the densification process without the exertion of external loads (as in conventional dilatometry), which could influence the sintering kinetics.

Cylindrical compacts of the powders to be examined (5 mm in diameter by 5 mm) were prepared by uniaxial pressing at room temperature without the addition of any binder. Pressures of 100 MPa were applied in order to reach green densities of 1.36 g cm^{-3} (34% of theoretical). Constant-heating rate sintering experiments were performed in a standard heating microscope (Model 1750, LEICA, Wetzlar, Germany). The specimens were heated at 10°C/min up to the maximum temperature of 1500°C. Two specimens were used for each run and the results averaged. By taking photographs of the sample at pre-determined time intervals during the sintering process, the changes in length and diameter of the samples during sintering could be recorded "in-situ". These data enable the calculation of the axial and radial shrinkages as well as the density of the sample with enough precision at any temperature (and time).

Effect of LiF Addition on Densification Onset

From the results of the hot stage optical microscopy studies, the temperature ranges for the onset of densification in specimens with, and without LiF additions could be determined, as described later. Monolithic $Y_2Si_2O_7$ specimens were hot-pressed from the as-synthesised $Y_2Si_2O_7$ precursor powder, and then subjected to four point bend flexural strength testing. This was undertaken in addition to the composite work due to the fact that the available literature presents no mechanical property data on this material.

For the inter- (infiltrated) mat matrix material, however, the lower material viscosity produced by the addition of 3 wt% LiF to the $Y_2Si_2O_7$ precursor was exploited. A slurry of yttrium disilicate precursor suitable for applying to the infiltrated mats was made by dispersing 40 wt% of the LiF (3 wt%) doped yttrium disilicate precursor powder in water to which 10 wt% of polyethylene glycol (PEG) was added to act as a binder. These slurries were used in the lay-up of laminates comprising 6 infiltrated Nicalon fibre mats, which on drying in air, produced prepregs (with approximate dimensions of 50 x 50 x 10 mm^3) for hot-pressing. All the resultant hot-pressed specimens were sectioned and prepared for SEM examination.

RESULTS AND DISCUSSION

Electrophoresis

Figure 2 shows the particle electrophoretic mobility data for the undoped nanosized $Y_2Si_2O_7$ suspensions as a function of sol pH. From these data, it is evident that the point of zero charge for these particles occurs at pH 3. So, by using water (at pH 7) as the suspending medium, the $Y_2Si_2O_7$ particles will be negatively charged and hence deposit on the cathode during EPD.

Although the electrophoretic deposition of coatings on the fibre mats was possible at voltages as low as 4 V d.c., the best coating thickness was achieved when using 8 V for 1 minute. It is known that at voltages higher than 4 V d.c., gases are evolved at the electrodes during deposition, some of which become entrapped in the coatings and so produce pores. Whilst it would be ideal to eliminate such pores, their effect in this particular system is not too detrimental due to the fact that the final composite material will densify via a pressure-assisted viscous-led mechanism. It is envisaged that this stage will reduce the extent of this small scale porosity

The SEM micrographs in Figure 3 show the microstructures of the EPD-infiltrated fibre mats, which were used in both the as-received condition and wetting agent pre-treated condition. Figure 3a shows that, in the case of the as-received fibre mats, the intra tow regions appear not to have been fully infiltrated. This is due to the lack of sol/fibre wetting during the vacuum degassing and ultrasonication stages. Hence, any EPD deposited material remained only on the outer surfaces of the fibre mat and in

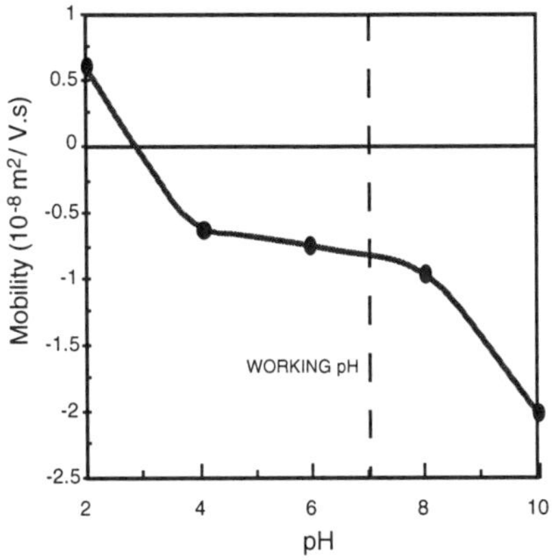

Figure 2 Electrophoretic Mobility data for $Y_2Si_2O_7$ powder in aqueous

the inter-fibre tow regions. With the pre-treated fibres, in Figure 3b, the wetting agent has, however, aided the ingress of the sol into the intra-fibre tow regions during the vacuum degassing and ultrasonication stages prior to EPD. These observations highlight the importance of fibre pre-treatment with this particular system in order to achieve a sufficient level of infiltration.

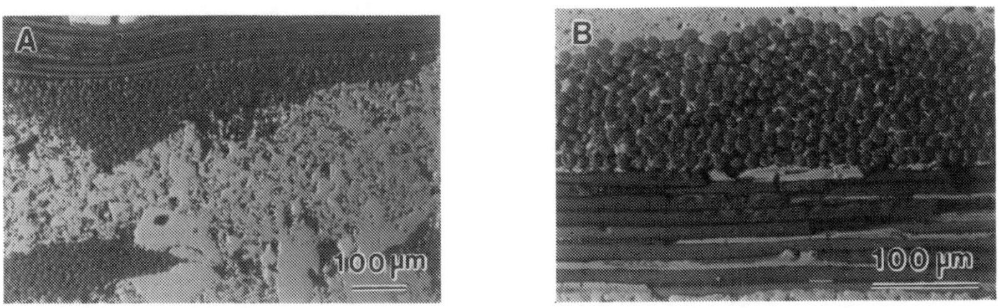

Figure 3: SEM micrographs of fibre mats infiltrated using EPD. (a) As-received fibre mat; (b) Fibre mat pre-treated with PMAA wetting agent

Hot Stage Optical Microscopy

Figure 4 shows the hot stage optical microscopy results for the $Y_2Si_2O_7$ precursor powder both with and without additions of LiF. The onset of densification for the LiF-free material occurs at 1050°C. In this temperature range, a sharp increase in the densification rate was evident. It is interesting to note that from previous DTA studies of this precursor powder, there is an exothermic reaction within this temperature range which has been attributed to the crystallisation of the amorphous precursor [10]. At 1500°C, the sample reached a density of 3.64 g cm^{-3} (90% theoretical). For the powder containing 10 wt% LiF, however, densification began at temperatures as low as 750°C and increased more slowly up to a value similar to that achieved with the LiF-free material. Thus, with an addition of LiF, it would be possible to apply a hot-pressing load in the lower temperature region of 750-850°C to aid densification, without transmitting the loads directly to the fibres.

Pressure-Assisted Sintering of Monolithic Yttrium Disilicate

It was decided to consolidate 25g of monolithic yttrium disilicate precursor powder using a hot-pressing schedule which incorporated dwell temperatures up to 1600°C due to the lack of densification via pressureless sintering during the hot stage optical microscopy tests. The hot-pressing schedule comprised an initial ramp to a temperature in the region of observed densification (1050°C) at a rate of 50°C/min, with a 30 dwell at temperature. During this dwell period, a load of 10 MPa was applied to the specimen. The load was removed subsequently and the specimen was then heated to 1600°C at a rate of 50°C/min, with a 2 hour dwell time. This was followed by furnace cooling.

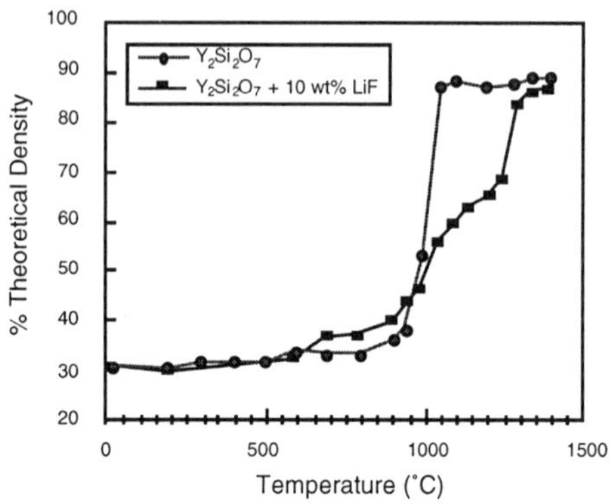

Figure 4 Hot stage optical microscopy results for the $Y_2Si_2O_7$ powder precursors

The monolithic specimen hot-pressed at 1600°C had a measured density of 3.72 gcm^{-3} (92% theoretical). This density is the highest achieved in this work so far, but it is envisaged that higher hot-pressing temperatures will be required to increase the density further. This view is validated by reference to work performed on the processing of pollucite (the most refractory known silicate with a composition $CsAlSi_2O_6$) which required a hot-pressing temperature of 1660°C in order for the material to exhibit a density of 91% theoretical [13]. Table 1 below shows the results for the 4-point bend testing of the monolithic yttrium disilicate specimens. As can be seen, the average room temperature flexural strength value is 219 MPa.

Table 1: Average 4-point Bend Test Results for Monolithic Yttrium Disilicate

Average Specimen Cross-sectional Area (mm^2)	Peak Force at fracture F_m (N)	Mid-point Displacement of Inner Span (mm)	Flexural Strength (MPa)	Coefficient of Variation (%)
5.062	40.32	0.35	219 (8 Specimens)	5.1

These values are comparable with the room temperature flexural strength data for mullite [14]. As far as the authors are aware, these are the first flexural strength data to be reported for hot-pressed monolithic yttrium disilicate. More property data is required for this material, however, in particular, high temperature strength tests and creep tests.

LiF-Doped Yttrium Disilicate Nicalon SiC Fibre Composite

The hot-pressing schedule for a LiF-doped slurry comprised an initial application of 3 MPa at room temperature, a ramp up to 820°C at 10°C/min with an increase in pressure to 10 MPa at this point and a dwell time of 20 minutes, followed by a ramp up to 1500°C at 10°C/min with a 15 minute dwell time at temperature, after which point the specimen was allowed to furnace cool. This lower final dwell temperature was used in order to minimise the well documented mechanical property degradation of these fibres with time at temperatures in excess of 1200°C [15].

It was found during hot-pressing that the LiF-doped matrix material became very fluid in the region of 800°C causing some material to be squeezed out from the prepreg. This result is important because it shows that a viscous flow stage during processing can aid the densification process without applying an excessive load on the fibres, thus reducing the propensity for fibre breakage during hot-pressing.

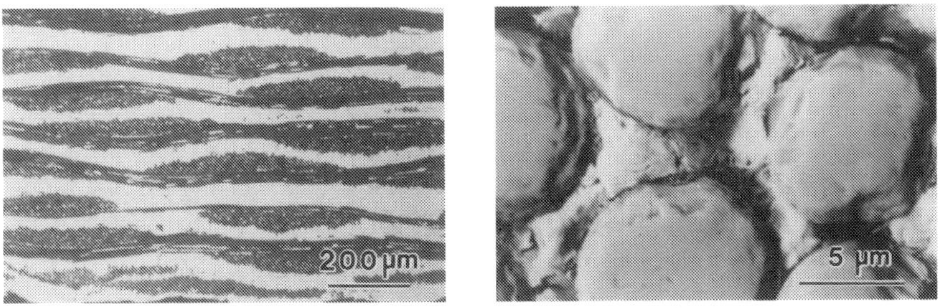

Figure 5 SEM Micrographs of hot-pressed composite material

Figure 5 shows a low magnification SEM micrograph of the composite material. As can be seen the intra-tow regions of the fibre mats are well infiltrated, and the fibre mats are closely spaced owing to the loss of matrix material during the viscous flow stage. This has the effect of increasing the fibre volume fraction in the final composite. It should be noted, however, that there are cracks in the matrix regions, which have been generated during the high temperature consolidation stage. The presence of these cracks is thought to be due to the fact that the applied load was maintained throughout the hot-pressing cycle. The presence of these cracks and their relationship to the hot-pressing cycle requires further investigation. As the addition of LiF enables a low temperature densification stage, composites could be hot-pressed at this temperature and then post-sintered at a higher temperature, allowing more control over the composite forming stage. Despite the need for further study, this iterative approach to CMC fabrication has nevertheless demonstrated the feasibility of fabricating woven fibre reinforced CMCs using EPD.

CONCLUSIONS

This work has shown that EPD is an effective, low cost technique for infiltrating conducting woven fibre preforms. In order to fabricate composites successfully from hydrothermally synthesised sols via EPD, optimisation of the processes employed at each stage is necessary. Results have shown that densification of pure yttrium disilicate precursor powders requires temperatures of 1600°C in order to reach a density in excess of 90% theoretical. The first ever reported flexural strength data have been generated from 4-pt bend specimens processed under these conditions, and it has been found that the material exhibits a mean room temperature fracture strength of 219 ± 11 MPa - comparable to that of mullite. However, more work is required on optimising the densification of the matrix regions in the composite specimens, involving taking advantage of the enhanced densification kinetics produced by the addition of LiF. It is envisaged also that a low temperature densification stage in conjunction with a high temperature pressureless sintering stage will lead to greater control over the final composite microstructure.

REFERENCES

[1]J.Ito and H. Johnson, "Synthesis and Study of Yttrialite", *Am. Mineralogist,* **53**, pp. 1940-1952, (1968).

[2]K. Liddell and D. P. Thompson, "X-ray Diffraction Data for Yttrium Silicates", *Br. Ceram. Trans. Journal*, **85**, pp. 17-22 (1986)

[3]E. M. Levin, C. R. Robbins and H. F. McMurdie, *Phase Diagrams for Ceramists 1969 Supplement*, The American Ceramic Society, Inc.; Fig. 2388 (1969).

[4]T. Hanada, T. Shinoda, S. Tanabe and N. Soga, "Preparation and elastic properties of rf-sputtered amorphous films", *J. Am. Ceram. Soc.*, **78**, pp.1383-1386 (1995) .

[5]J. Felsche, "Polymorphism and Crystal Data of the Rare earth Disilicates of Type R.E.$_2$Si$_2$O$_7$", *J. Less Common Metals*, **21**, pp. 1-14 (1970).

[6]P. A. Trusty, K. C. Chan and C. B. Ponton, "Synthesis of Sinteractive Single-Phase Microstructure Yttrium Disilicate Precursor Powder Using Hydrothermal Processing," *J. Mat. Res.*, (1998) In Press.

[7]P. A. Trusty, A. R. Boccaccini, E. G. Butler, and C. B. Ponton, "Novel Techniques for Manufacturing Woven Fibre Reinforced Ceramic Matrix Composites - I: Preform Fabrication", *Materials & Manufacturing Processes*. **10** (1995) 1215.

[8]P. A. Trusty, A. R. Boccaccini, E. G. Butler and C. B. Ponton, *Advanced Synthesis and Processing of Composites and Advanced Ceramics*. **2** (1996) 63.

[9]T. J. Illston, P. A. Doleman, E. G. Butler, P. M. Marquis, C. B. Ponton, M. J. Gilbert and R. Piramoon, UK Patent No. 9124822.9, November 1991.

[10]I. MacLaren, P . A. Trusty and C. B. Ponton, "A TEM Study of Hydrothermally Synthesised Yttrium Disilicate Powders : 1", Submitted to *Acta Mater.*, (1998).

[11]A. R. Boccaccini, "Anisotropic Densification During Sintering of Glass Powder Compacts," *J. Mat. Sci. Lett.*, **12**, 943-945 (1993).

[12]A. R. Boccaccini, P. A. Trusty and D. M. R. Taplin, "Anisotropic Shrinkage of Barium Magnesium Aluminosilicate Glass Powder Compacts during Sintering," *Mat. Lett.*, **24**, 199-205 (1995).

[13]R. S. Hay, T. A. Parthasarathy and J. R. Welch, "Creep and Stability of Pollucite", Wright Laboratory, Materials Directorate, WPAFB, Dayton OH 45433 (1994).

[14]T-I. Mah and K. S. Mazdiyasni, "Mechanical Properties of Mullite," *J. Am. Ceram. Soc.* **66**, 10, 699-703, (1983)

[15]G. Simon and A. R. Bunsell, "Mechanical and Structural Characterization of the Nicalon Silicon Carbide Fibre.", *J. Mat. Sci.*, **19** (11), 3649-3657 (1984)

ELECTROPHORETIC FILTRATION DEPOSITION (EFD) OF SAFFIL ALUMINA FIBRE-REINFORCED MULLITE MULTILAYER NANO-CERAMIC MATRIX COMPOSITES

+*C. Kaya, +P. A. Trusty, +*C. B. Ponton

+IRC in Materials for High Performance Applications, The University of Birmingham, Edgbaston, Birmingham, B15 2TT, UK

*School of Metallurgy & Materials, The University of Birmingham, Edgbaston, Birmingham, B15 2TT, UK.

ABSTRACT

Electrophoretic deposition filtration (EFD) and pressure filtration (PF) have been used to fabricate Saffil alumina fibre-reinforced mullite multilayer nanoceramic matrix composites from heterocoagulated nano-sized alumina sol particle and fumed silica powder particles dispersed in water. The surface charge properties of the particles in suspension were characterised by particle electrophoretic mobility measurements, while the effects of the short-range particle-particle interactions were observed using TEM. The heterocoagulated particle clusters were infiltrated into the multilayer Saffil fibre preforms using constant voltage EFD with varying deposition times, to establish the process parameters for the successful infiltration deposition of stoichiometric mullite. It was found that at a constant voltage of 10 V d.c., the increase in EFD deposit weight was almost a linear function of the EFD time, while the optimum electrode separation distance was 15 mm, producing full infiltration in 4 minutes. Furthermore, for the boehmite-silica system employed in the present work, the optimum alumina/silica particle size ratio for generating discrete heterocoagulated boehmite/silica particle clusters, and hence stoichiometric sintered mullite, was 3:1. Sintering at 1500°C for 3 hours produced sub-micron equiaxed mullite grains and no grain boundary glass phase, with fibre pull-out on fracture.

INTRODUCTION

Electrophoretic deposition (EPD) employs small charged particles in a liquid, i.e. a sol, which, on the application of an electric field, will move and deposit on an oppositely charged electrode. Conventionally, the electrode is an electrically conductive monolithic material but the application of EPD to the infiltration of porous reinforcing materials, e.g. 2D and 3D woven fibre preforms, has been patented recently [1]. When the woven fibre mat is electrically conductive, e.g. a Nicalon NL607 SiC fibre mat, it functions as the deposition electrode, i.e. the sol particles are infiltrated electrophoretically into the electrode and the term EPD is appropriate. However, when the woven fibre mat is electrically non-conductive, the infiltration and deposition occurs by the sol particles being electrophoretically drawn or filtered through the mat towards the adjacent electrode surface, and in the process becoming attached to the fibre surfaces. Thus, as in this paper, the process is one of Electrophoretic Filtration Deposition (EFD) rather than conventional EPD.

Subsequently, both EPD and EFD have been subject to on-going investigation by the authors et al. [1-3, 5-7] as high forming- and deposition-rate processes requiring relatively simple, low-cost equipment. They thus offer, in conjunction with pressure filtration (PF) where appropriate, new possibilities for the cost-effective design and manufacture of monolithic and woven fibre-reinforced ceramic components with more uniform and controlled microstructures. Of particular interest is the fabrication of material systems for high temperature structural applications in oxidising environments. For example, monolithic or fibre-reinforced mullite CMCs exhibit high thermal shock and creep resistance, low thermal expansion and thermal conductivity, good chemical and thermal stability, plus high refractoriness, making them an ideal material for high temperature applications [4]. These properties can only be attained, however, by controlling the very fine microstructure in conjunction with the amount and location of glassy phase in the sintered component. To this end, our studies using mixed silica and alumina sols of mullite composition have shown that preferential species migration during EPD/EFD leads to a loss of stoichiometry, and is influenced by the size ratio of the sol species involved [3, 5-7].

The fabrication of fibre reinforced mullite composites using EFD requires the matrix precursor ceramic particles to be nanosized and in stable suspension. The properties of the sintered matrix will thus be influenced significantly by the compositional homogeneity and stability of the suspension, as well as by the fabrication techniques. Suspension stability can only be achieved by controlling the short-range particle-particle interactions, which will in turn affect the rheological behaviour of the suspensions. To obtain a uniformly infiltrated green microstructure, EFD requires a kinetically stable, well-dispersed suspension having the highest possible solids-loading but a relatively low viscosity because this affects the particle electrophoretic mobility, and hence, deposition efficiency. Furthermore, the number of charged particles deposited per unit area of the electrode/fibre is influenced strongly by the process time, electrode separation and applied potential. In order to obtain higher heterocoagulated-particle EFD rates, a high dielectric constant and a low viscosity liquid (hence water is ideal) are necessary because the particle mobility within the suspension will be enhanced. To produce a homogeneous green infiltrated microstructure, it is, however, also crucial to optimise the solids-loading of the suspension without causing flocculation.

The present study examines the feasibility of fabricating Saffil alumina fibre-reinforced mullite matrix composites via a single-infiltration EFD process, followed by PF. The form of the heterocoagulated alumina/silica mullite precursor particles most appropriate for producing 'Saffil' alumina fibre-reinforced nano-mullite CMCs by means of EFD, PF and pressureless sintering has been determined.The relationship between the mullite precursors employed and the microstructure of the final sintered product has been correlated in terms of the location and amount of the glassy phase in the final sintered composite.

EXPERIMENTAL PROCEDURE

Mullite Precursor Preparation

Commercially available Saffil alumina fibre in the form of multilayer blanket (ICI Chemicals plc.) was used to make alumina fibre reinforced-mullite ($3Al_2O_3.2SiO_2$) multilayer nanoceramic matrix composites with alternate layers of variable porosity distribution-containing material and dense material, as the reinforcement material. The powders and sols used in this study are identified in Table I.

Table I. Physical and Chemical Property Data for the Powders and Sols (Manufacturers Data)

Material	Form	Average Particle Size (nm)	Particle Shape	pH stability Range
Remal A20*	Boehmite Sol	50	Hexagonal	4.5
Bindzil 30/220*	Silica Sol	15	Hexagonal	9.5-10
Aerosil 90	Silica Powder	20	Spherical	
Aerosil 130*	Silica Powder	16	Spherical	
Aerosil MOX80	Silica Powder	30	Spherical	

* Particle size confirmed using both Transmission Electron Microscopy and Particle Size Analysis.

To prepare the various alumina-silica mixed sols for the synthesis of stoichiometric 3:2 mullite (i.e. after sintering, 71.82 wt % alumina and 28.18 wt % silica), the base boehmite sol (i.e. the alumina source) was stirred vigorously whilst fumed silica powders were added to each alumina sol at the rate of 0.5 g/min; this low rate of addition was used in order to prevent the formation of large heteroflocculated clusters. Simultaneous ultrasonic agitation was used to enhance the powder dispersion, while the solids-loading was varied between 5 and 40 wt % of the dispersion liquid to determine its effect on the EFD behaviour. The pH of each suspension was adjusted using aqueous ammonia additions since according to the manufacturer, the dispersion stability of the alumina sol is maximised at this pH value. Our experimental observations confirmed this fact; the boehmite-silica mixed suspension gelled rapidly at pH values much higher or lower than 4. Kinetically stable, well-dispersed suspensions of boehmite and silica particles were thus obtained at a pH of 4. At this pH, the alumina particles are positively charged and surrounded by the negatively-charged silica particles. The resultant heterocoagulated particle clusters behave effectively as single negatively-charged sol particles. The surface charge properties of the single-species particles and the heterocoagulated particles in terms of their electrophoretic mobility were measured as a function of the sol pH and sol system employed, using a DELSA 440 surface charge analyser. Detailed information about the optimization of a heterocoagulated mullite suspension for the use in EFD can be found elsewhere [5,6].

Electrophoretic Filtration Deposition (EFD)

An EFD cell which allows the electrode separation to be varied was designed and constructed as shown in Figure 1(a).

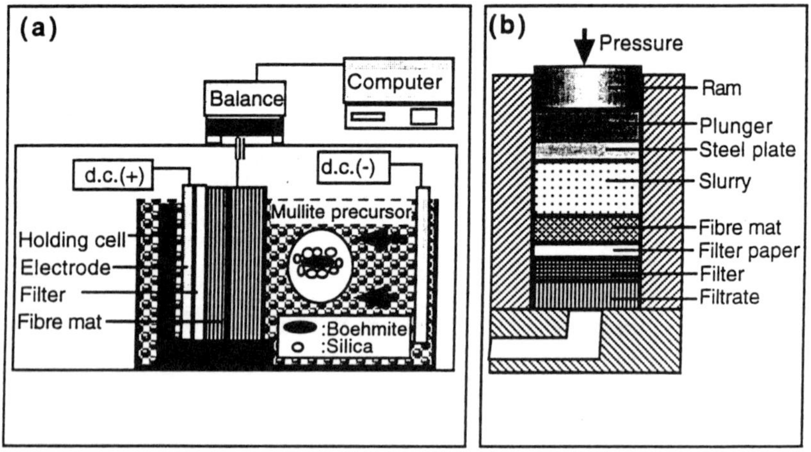

Figure 1. In-situ Electrophoretic Filtration Deposition Cell (a) and Cross-sectional view of the Pressure Filtration Apparatus (b) for Fabricating Fibre-Reinforced CMCs.

A filter membrane was placed adjacent to the positive electrode before applying the voltage in order to prevent gas bubbles being incorporated within the deposited ceramic matrix. A foldable holding-cell was used to hold the electrode, filter and fibre mat together. After the fibre mat was placed in the sol, the system was vacuum degassed to remove any entrapped air, and then the cell electrodes were connected to a 0-60 V d.c. power supply. The fibre mat was attached to the balance using a very thin wire. Weight measurements were recorded every second during EFD via a computer linked to the balance. EFD was performed subsequently under constant voltage conditions (10 V d.c.) using varying deposition times (2, 4, 6 and 8 minutes) and electrode separation distances (1, 1.5, 2 and 2.5 cm). Under the applied electric field, heterocoagulated composite particles, possessing a net negative surface charge migrated towards the positive electrode, infiltrating the fibre mat and being deposited out of suspension until a sufficient matrix thickness, which enveloped the fibre mat, was achieved. Following EFD infiltration, selected specimens were compacted further using pressure filtration (see Fig 1b). The resulting EFD/PF produced multilayer green body specimens were then freeze-dried before being sintered for 3 hours at 1500 °C and at 1600°C using a 3°C/minute heating and cooling rate.

RESULTS AND DISCUSSION

Figure 2 shows the particle electrophoretic mobility data for the nano-sized boehmite, fumed silica (Aerosil 130), and boehmite-silica aqueous sol suspensions as a function of sol pH. From these data, it is evident that the silica particles are negatively-charged over the whole pH range, whereas the boehmite particles are positively-charged below pH 8.5 and negatively-charged above pH 8.5. Hence, under the conditions used to obtain a stable mixed sol (pH 4), there is an electrostatic attraction between the positively-charged boehmite particles and the negatively-charged silica particles, enabling the resultant heterocoagulated sol particle clusters to move as single, negatively-charged composite particles.

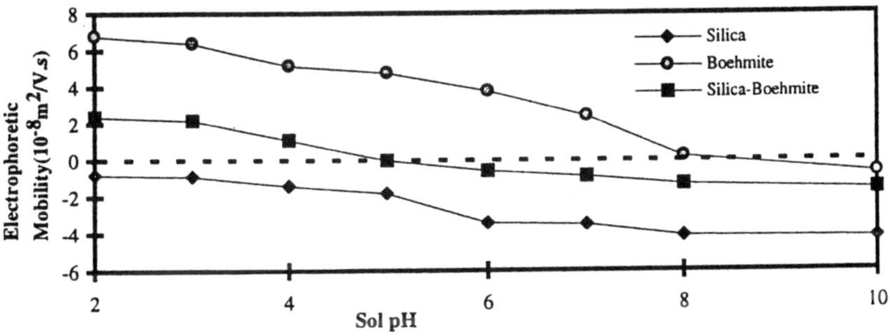

Figure 2. Particle electrophoretic mobility data for the individual Aerosil 130 amorphous silica (●) and Remal A20 boehmite (■) suspensions plus the resultant boehmite-silica (⊞) suspension as a function of sol pH.

Figure 3 shows Bright-Field TEM micrographs of (a) heterocoagulated clusters and (b) long-chained heteroflocculated clusters of silica/boehmite particles formed via mutual electrostatic attraction under the experimental conditions used, i.e. a pH of 4. The difference in the nature of the clusters formed in Figure 3(a) and Figure 3(b) is due to the relative size ratio of the precursor particles. With a boehmite to silica particle size ratio of 3:1 as in (a), more discrete heterocoagulated clusters are formed. However, when the size ratio decreases to 2.5:1, coverage of the boehmite particles by the silica particles is reduced, leading to less 'discrete' heterocoagulated clusters [7]. Thus, it has been observed experimentally that when the critical size ratio is much different from 3:1, the probability of forming long-chained heteroflocculated clusters is increased as shown in (b) where the size ratio is 1.25:1. Hence, for a given solids-loading, the particle size ratio also affects the viscosity of the suspension, and must therefore be taken into account when choosing the mullite precursor materials.

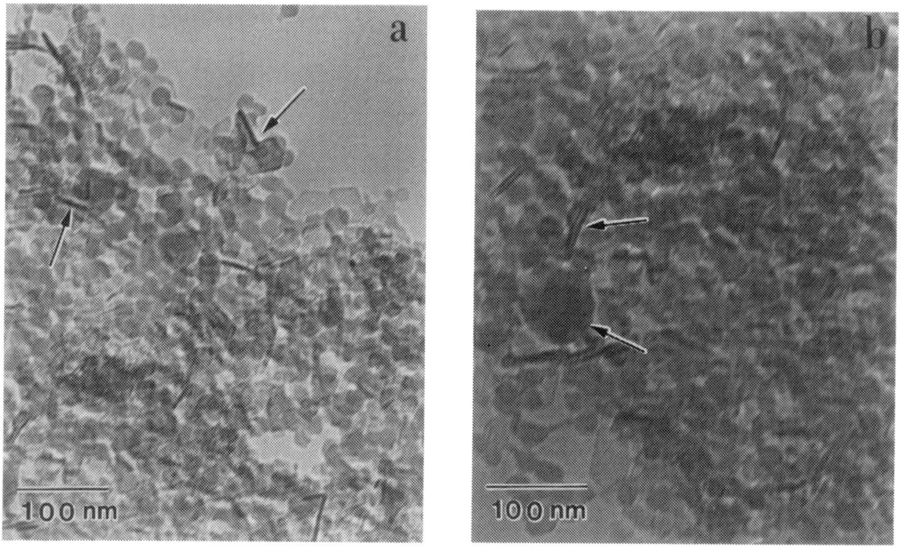

Figure 3. Bright-field TEM micrographs showing: (a) Heterocoagulated Remal A20 boehmite / Aerosil 130 fumed silica particle cluster as arrowed (Note the boehmite to silica particle size ratio is 3:1); (b) Formation of large heteroflocculated clusters (as arrowed) when the Remal A20 boehmite sol (50 nm) and Aerosil 0X 50 fumed silica particles (40 nm) are almost the same size.

The results of the EFD trials showed that the optimum electrode separation distance was 15 mm with an ideal constant applied voltage of 10 V d.c. Under these conditions, a deposition time of 4 minutes was found to be sufficient for full infiltration. This is shown by Fig.4(a) in which matrix particles are clearly attached to the fibre surface indicating infiltration right to the fibre surface. The microstructures of sintered specimens produced from a heterocoagulated sol (Remal A20 boehmite and Aerosil 130 fumed silica) under optimised EFD conditions are shown in Figure 4b. Sub-micron equiaxed mullite grains are observed in the samples sintered at 1500°C for 3 hours (Fig.4b). Further investigation of the grain boundary regions using TEM showed that whilst the samples sintered at 1500°C had clean grain boundaries and junctions (Fig.5a), those sintered at 1600°C possessed three-grain junctions entrapping a glassy phase (Fig.5b). This is attributed to the outward diffusion of SiO_2 from the mullite grains during sintering.

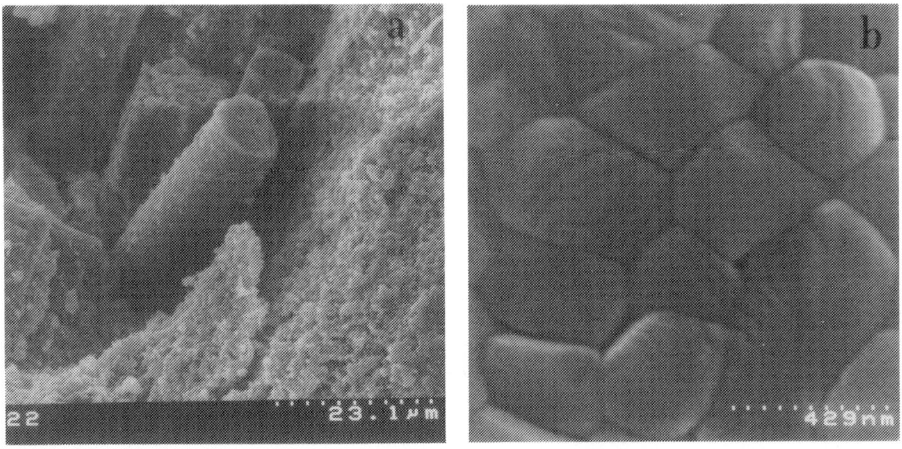

Figure 4. High Resolution FEG SEM micrographs of the fracture surfaces of the sample produced from a Remal A20 boehmite/Aerosil 130 fumed silica mixed sol (Note the boehmite to silica particle size ratio is 3:1) and sintered at 1500°C for 3 hours showing: (a) Infiltration right up to up fibre surface, plus fibre pull-out. (b) Mullite matrix microstructure showing equiaxed mullite grains.

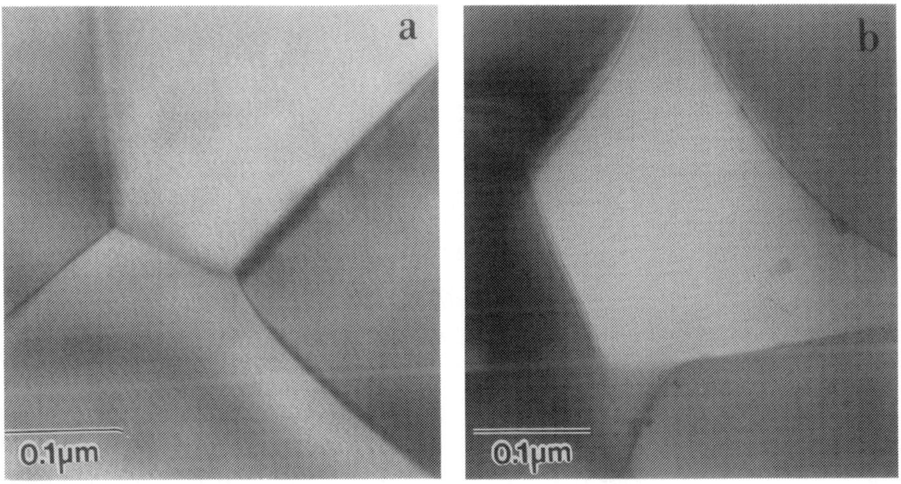

Figure 5. TEM micrograph of the crystalline mullite grains without glassy phase of a sample sintered at 1500 °C for 3 hours (a) and glassy phase at a triple-grain junction point in a mullite sample sintered at 1600°C for 3 hours (b).

CONCLUSIONS

It has been demonstrated that Saffil alumina fibre-reinforced mullite multilayer nano-CMCs can be fabricated using a single-stage EFD process and PF. Given the simplicity and very short processing time of this technique, as well as the low capital equipment costs, it has great potential as a cost-effective CMCs pregreg production route when compared with other processes such as thermal gradient CVI. Furthermore, the alumina/silica particle size ratio is shown to be a critical parameter in generating stoichiometric sintered mullite by EFD of mullite composition heterocoagulated nano-size alumina/silica sol particle clusters. For the boehmite-silica system employed in the present work, the critical ratio was found to be in the range between 2.5:1 and 3:1, with the latter being the optimum. It has also been found [5] that dense, homogeneous green mullite composition and sintered mullite microstructures can be prepared successfully only by controlling the short-range particle-particle interactions of the dissimilar particle species, and hence, the rheological behaviour of the suspension. Finally, the novel combination of EFD, PF and pressureless sintering is shown to be a successful route for fabricating fibre-reinforced multilayer nano-ceramic matrix composites.

Acknowledgements

Profs. M. H. Loretto and I. R. Harris are acknowledged for the provision of laboratory facilities in the IRC and the School of Metallurgy & Materials, respectively.

REFERENCES

[1]T. J. Illston, P. A. Doleman, E. Butler, P. Marquis, C. B. Ponton, M. J. Gilbert and R. Piramoon, "Method of Manufacturing a Composite Material", EP Patents No. 613455 (July 1997) and No. 613454 (September 1996). Original UK Patents Applied for in November 1991.

[2] P. A. Trusty, A. R. Boccaccini, E. G. Butler and C. B. Ponton, "Novel Techniques for Manufacturing Woven Fibre Reinforced Ceramic Matrix Composites - I : Preform Fabrication", Materials and Manufacturing Processes **10** [6] (1995) pp. 1215-1226.

[3]A. R. Boccaccini, P. A. Trusty, D. M. R. Taplin and C. B. Ponton, "Colloidal Processing of a Mullite Matrix Material Suitable for Infiltrating Woven Fibre Preforms Using Electrophoretic Deposition", J. Europ. Ceram. Soc. **16** [12] (1996) pp. 1319-1327.

[4]H.Schneider, K.Okada and J.Pask, " Mechanical Properties of Mullite Ceramics"; pp.199-228 in Mullite and Mullite Ceramics, edited by S.Somiya, H.Schneider, K.Okada and J.Pask. John Wiley&Sons, Chichester, UK, 1994.

[5]C. Kaya, P. A. Trusty and C. B. Ponton, "Preparation of Heterocoagulated Colloidal Suspensions for the Fabrication of Ceramic Matrix Composites by Electrophoretic Deposition," British Ceramic Transastions, 97, [2] (1998) pp. 48-54.

[6]C. Kaya, P. A. Trusty and C. B. Ponton, "Preparation and Characterisation of a Nano-size Colloidal Suspension for The Preparation of Multilayer CMCs Using Electrophoretic Deposition"; in Proceedings of Better Ceramics Through Processing, Ceramic Convention, Cirencester, UK (The Institute of Materials) (1998) p. 93.

[7]C. Kaya, P. A. Trusty and C. B. Ponton, " Electrophoretic Deposition (EFD) Forming of Saffil Alumina Fibre-Reinforced Mullite Multilayer Nano-ceramic Matrix Composites", J. Mat. Sci. Submitted (1998).

SYNTHESIS OF Al_2O_3 / YAG COMPOSITE FIBERS

Atsuya TOWATA, Hae Jin HWANG, Masaki YASUOKA, Mutsuo SANDO
National Industrial Research Institute of Nagoya,
1-1 Hirate-cho, Kita-ku, Nagoya 462-8510, Japan

Koichi NIIHARA
The Institute of Scientific and Industrial Research,
Osaka University, 8-1 Mihogaoka, Ibaraki, Osaka 567-0047, Japan

ABSTRACT

Polycrystalline α-Al_2O_3 fibers dispersed with YAG particles could be successfully fabricated modified sol-gel process including α-Al_2O_3 seed particles. YAG particles were synthesized in the Al_2O_3 fiber using reaction between aluminum isopropoxide and yttrium isopropoxide, and were dispersed homogeneously in Al_2O_3 matrix.

INTRODUCTION

Fiber-reinforced ceramic matrix composites (CMCs) have been developed as a promising class of structural materials with the significant improvements in strength and toughness over monolithic ceramics. Fibers used in these CMCs are required to possess strong heat resistant in high temperature oxidizing environments.

Oxide polycrystalline fibers have been studied for a number of years, especially Al_2O_3.[1-3] Al_2O_3 is a representative oxide material and undergoes a number of phase transformations. The structure of these so-called transition Al_2O_3 has been the subject of several investigations. The transition Al_2O_3 generally have a sintering behavior with a region of very fast sintering rate between 1000°C and 1200°C followed by a region of much lower sintering rates at higher temperatures.[4] The transition Al_2O_3 generally have a characteristic sintering behaviors relating to the phase transformation from γ to α-Al_2O_3.[5] The vermicular microstructure consisting of a network of large pores which develop during the transformation requires very high temperatures for achievement of high density. Because of densification difficulties caused by the transformation, Al_2O_3 bodies are normally fabricated by compaction and sintering of α-Al_2O_3 powders. However, it has been shown that seeding of boehmite gels with fine α-Al_2O_3 particles leads to an increase in transformation kinetics and to a lowering of transformation temperature.[6]

The creep resistance of polycrystalline ceramics is improved by the addition of hard second-phase particles.[7-10] The second-phase particles can affect the deformation rate of a polycrystal in two ways.[11] One is macroscopic modification of the continuum deformation mechanics, and the other is microscopic change of the interface -related deformation characteristics such as suppressing interface reactions (vacancy or interstitial creation /annihilation) or hindering grain boundary sliding. In addition, the second-phase particle may affect the microstructure.

YAG (yttrium-aluminum garnet, $Al_5Y_3O_{12}$) is a prime candidate for hardening Al_2O_3 due to its excellent creep-resistance. Corman[12] investigated the creep properties of single-crystal YAG.

This paper described new fabrication process of Al_2O_3 fibers dispersed with nano-sized YAG particles. The fibers were synthesized by a sol-gel method and addition of α-Al_2O_3 seed particles. Furthermore, effects of addition of the α-Al_2O_3 seed particles on the phase transformation of matrix Al_2O_3 and microstructure of fiber were investigated.

EXPERIMENTAL PROCEDURE

Four different types of Al_2O_3 fiber were prepared. These were (1) monolithic Al_2O_3 fiber, (2) monolithic Al_2O_3 fiber with α-Al_2O_3 seed particle, (3) YAG dispersed Al_2O_3 fiber (Call on "composite fiber") and (4) YAG dispersed composite fiber with α-Al_2O_3 seed particle.

In order to synthesis precursor fiber, aluminum isopropoxide (Nakarai Chemical Co., Japan) and yttrium isopropoxide (High Purity Chemical Co., Japan) were prepared. Aluminum isopropoxide and yttrium isopropoxide stabilized with 2-isobutoxyethanol were mixed in the isopropyl alcohol and ethyl 3-oxobutanate, according to the procedure described by Yogo et al.[13] In the case of fiber with seed particles, α-Al_2O_3 (Average particle size : 0.2 μm, TMDA, Taimei Chemical Co., Japan) were added to the solution. The solution was hydrolyzed by adding the hydrochloric acid, and was condensed, and then was spun to the precursor fiber. The precursor fiber was dried atmosphere, and then heat-treated at various temperatures in air.

The crystalline phase of the composite fiber was identified by X-ray powder diffraction (RU-200B, Rigaku Co. Ltd., Japan). Raman spectra were obtained with the excitation of the second-harmonic 532 nm line of a Q-switched continuous-wave Nd:YAG laser (CVI Laser, C-140). The scattered light was passed through a holographic notch filter (Kaiser Optical Systems, HSNF-532-1.0), analyzed with a single monochrometer (Jobin-Yvon, HR-320) and detected with an intensified photodiode array (Princeton Applied Research, 1421B-1024HQ). DTA (Differential Thermal Analyses) experiment were conducted on a DSC (Sinkuu-rikou,Japan). The fiber morphology and microstructure were observed by a scanning electron microscope (SEM, JSM-6320FK, JEOL, Japan). The ratio of Y/Al of the fiber was measured by inductively coupled plasma-emission spectrometry (ICP-ES, ICAP-1000S, NIPPON Jarrell-Ash, Japan).

RESULTS AND DISCUSSION

Figure 1 shows the Raman spectra of the aluminum isopropoxide and solution reacted with aluminum isopropoxide and ethyl 3-oxobutanate at 60°C for

12 h. Ethyl 3-oxobutanate converts to isopropoxy-bridge ($(OC_3H_7)^i$) in aluminum isopropoxide. In the case of the ratio of aluminum isopropoxide : ethyl 3-oxobutanate = 1 : 1, the peak of aluminum isopropoxide at 860 cm^{-1} was smaller than the unreacted and indicated the existence of the aluminum isopropoxide still. However, in the case of the ratio of aluminum isopropoxide : ethyl 3-oxobutanate = 1 : 2, the peak of the aluminum isopropoxide at 960 cm^{-1} almost disappeared. Then the ratio of aluminum isopropoxide and ethyl 3-oxobutanate was 1 : 1 in order to react with a part of isopropoxy-bridge.

As shown in Fig. 2, the Al_2O_3/YAG precursor fibers were successfully fabricated. The diameter of the fibers was from 15 to 150 μm and their length was below 30 cm. The surface of precursor fiber was smooth.

The X-ray diffraction patterns of the seeded and unseeded Al_2O_3 and composite (Y/Al = 0.62 mol%) fibers heat-treated at 900°C for 4 h were given in Fig. 3. As mentioned in a previous report,[14] the crystallization of α-Al_2O_3 was significantly accelerated by adding the seed particles. Although unseeded Al_2O_3 fiber (Fig. 3 (b)) started to crystallize to α-Al_2O_3, broadening of peaks which corresponded to amorphous Al_2O_3 were also observed at around 2θ=38° and 46° when heat-treated at 900°C for 4 h. On the other hand, in seeded Al_2O_3 fibers (Fig. 3 (a)) only crystalline peak of α-Al_2O_3 was observed. Although in the seeded composite fiber (Fig. 3 (c)) crystalline peak of α-Al_2O_3 appeared, the unseeded composite fiber (Fig. 3 (d)) was almost amorphous Al_2O_3. Much higher temperature was required to completely transform the unseeded fiber to α-phase.

Figure 4 shows the SEM photographs of the composite fiber heated at 1400°C for 4 h with various contents of α-Al_2O_3 seed particles. Figure 5 shows DSC curves of the composite fibers. The transformation temperature from θ to α phase Al_2O_3 was 1130 °C in the case of the seed particles of 0.2 wt %. The size of the Al_2O_3 matrix was 0.7 -0.8 μm and was much smaller than that of the unseeded fiber (Fig. 4 (o)). The pore almost disappeared. In the case of 3 wt % α-Al_2O_3 seed particles, the transformation temperature was 1110 °C and the size of the Al_2O_3 matrix was 0.5 μm. The grain size of the Al_2O_3 matrix decreased with increasing the content of the α-Al_2O_3 to 3 wt %. However the size of 10 wt % seed particles suddenly grew up to 5 μm. This is due to the sintering of the α-Al_2O_3 seed particles. In the case of Al_2O_3 / YAG composite fibers, similarly to Al_2O_3 fibers, the addition of α-Al_2O_3 seed particles enhanced transformation of Al_2O_3 matrix from θ to α phase. The temperature of the peak decreased from 1210 to 1090 °C when the amount of seeds reached 10 wt %. The critical quantity of seeds needed to shift the transformation amounts approximately to 1-2 wt %. The results were consistent with other reports.[15,16]

Another feature observed in Fig. 3 was the effect of the yttrium addition on the crystallization behavior of Al_2O_3 fibers. In comparison of Fig. 3 (a) with (c), the Al_2O_3 completely transformed to α-phase while amorphous Al_2O_3 still remained in the seeded composite fibers as seen from a broad peak 2θ=46°. This tendency was also found in the unseeded fibers heat-treated at the same temperature. It could, therefore, infer that yttrium suppressed α-Al_2O_3 crystallization. Although not given in this study, the temperature at which α-Al_2O_3 phase started to appear increased with increasing the volume fraction of Y_2O_3.

Figure 6 shows the SEM photograph for thermally-etched surface of the YAG dispersed composite (Y/Al = 1.07 mol%) fiber heat-treated at 1500°C for 4 h. In Fig. 6, white and spherical phases are YAG particles, and fine YAG particles with the size of several hundreds nm corresponds homogeneously dispersed within the Al_2O_3 matrix grains as well as at grain boundaries. Some earlier works[17-19] have demonstrated that yttrium segregated to Al_2O_3 grain boundaries and, beyond its solubility limit, YAG was precipitated to grain boundaries and pore surfaces. Whether all of the yttrium, in this study, was precipitated as YAG in the matrix grain boundaries was not known at this time, and it was probable that some yttrium was substituted within the Al_2O_3 grains and grain boundaries. The further investigation is on going.

All of the precursor fibers prior to heat-treatment were amorphous and consisted of weak-boned agglomerates, and the primary particle size was several nano-meters. As mentioned in Fig.3, the fibers transformed to stable α-phase at different temperatures, depending on the amount of seed particles and yttrium content. The transformation of an γ, θ to α-Al_2O_3 accompanies abrupt grain growth.[20] Microstructure development of each Al_2O_3 fiber, therefore, is significantly affected by their transformation behaviors. SEM photographs of the unseeded and seeded composite fibers (Y/Al = 1.07 mol%) heat-treated at 1200°C for 4 h were shown in Fig. 7. Both fibers are completely transformed to α-phase after heat-treating at 1200°C for 4 h. Similarly to microstructure of unseeded gel observed by Kumagai et al.,[16] the unseeded composite fiber showed vermicular grain structure. Agglomerates with particle size of several micro-meters and large voids that were on the same scale as the agglomerates were observed in the unseeded composite fiber. On the other hand, the seeded composite fiber showed homogeneous microstructure and small grain size compared to that of the unseeded composite fiber. Effects of the seeding on microstructure development as well as the phase transformation in Al_2O_3 fiber were obvious in this study.

CONCLUSIONS

Polycrystalline α-Al_2O_3 matrix composite fiber containing nano-sized YAG particles was successfully fabricated by sol-gel process containing α-Al_2O_3.The seed particle accelerated the phase transformation from θ to α-phase. The seeded composite fibers with YAG (Y/Al = 1.07 mol%) consisted of Al_2O_3 grains with size of 1- 2 μm and YAG particles with nano size. Dense and homogeneous microstructure was obtained by the simultaneous additions of the seed particles and yttrium.

REFERENCES

[1]M. H. Stacey, "Developments in Continues Alumina-Based Fibers, " Br. Ceram. Trans. J., **87** 168-72 (1988)

[2]Y. H. Chiou, M.T. Tsai, H.C. Shih, "The Preparation of Alumina Fibre by Sol-Gel Processing, "J. Mater. Sci., **29** 2378-88 (1994)

[3]V. Lavaste, M. H. Berger, A. R. Bunsell and J. Besson, "Microstructure and Mechanical Characteristics of Alpha-Alumina-Based Fibers", J. Mater. Sci., **30** 4215-25 (1995)

[4]D. I. Markin, W. Munro, and T. M. Valentine, "The Fabrication of α-Alumina by Reactive Hot Pressing," J. Mater. Sci., **6** 974-80 (1971)

[5]F. W. Dynys and J. W. Halloran, "Alpha Alumina Formation in Alum-Derived Gamma Alumina," J. Am. Ceram. Soc., **65** [9] 442-48 (1982)

[6]M. Kumagai and G. L. Messing, "Controlled Transformation and Sintering of a Boehmite Sol-Gel by α-Alumina Seeding," J. Am. Ceram. Soc., **68** [9] 500-505 (1985)

[7]K. Niihara, "New Design Concept of Structural Ceramics-Ceramic Nanocomposites," J. Ceram. Soc. Japan, **99** [10] 974-82 (1991)

[8]A. Nakahira, T. Sekino, Y. Suzuki and K. Niihara, "High Temperature Creep and Deformation Behavior of Al_2O_3/SiC Nanocomposites," Ann. Chim. Fr., **18** 403-408 (1993)

[9]T. Ohji, T. Hirano, A. Nakahira and K. Niihara, "Particle/Matrix Interface and Its Role in Creep Inhibition in Alumina/Silicon Carbide Nanocomposites," J. Am. Ceram. Soc., **79** [1] 33-45 (1996)

[10]L. C. Stearns and M. P. Harmer, "Particle-Inhibited Grain Growth in Al_2O_3-SiC : I, Experimental Results," J. Am. Ceram. Soc., **79** [12] 3013-19 (1996)

[11]I. W. Chen and L. A. Xue, "Development of Superplastic Structural Ceramics," J. Am. Ceram. Soc., **73** [9] 2585-609 (1990)

[12]G. S. Corman, "High Temperature Creep of Some Single Crystal Oxides," Ceram. Eng. Sci. Proc., **12** [9-10] 1945-66 (1991)

[13]T. Yogo and H. Iwahara, "Synthesis of Polycrystalline Alumina Fibre with Aluminum Chelate Precursor," J. Mater. Sci., **26** 5292-96 (1991)

[14]C. S. Oh, G. Tomandl, M. H. Lee and S. C. Choi, "The Effect of an Added Seed on the Phase Transformation and the Powder Properties in the Fabrication of Al_2O_3 Powder by the Sol-Gel Process," J. Mater. Sci., **31** 5321-25 (1996)

[15]E. Prouzet, D. Fargeot, and J. F. Baumard, "Sintering of Boehmite-Derived Transition Alumina Seeded with Corundum," J. Mat. Sci. Let., **9** 779-81 (1990)

[16]M. Kumagai and G. L. Messing, "Controlled Transformation and Sintering of a Boehmite Sol-Gel by α-Alumina Seeding," J. Am. Ceram. Soc., **68** [9] 500-505 (1985)

[17]A. M. Thompson, K. K. Soni, H. M. Chan, M. P. Harmer, D. B. Williams, J. M. Chabala and R. L. Setti, " Dopant Distributions in Rare-Earth-Doped Alumina," J. Am. Ceram. Soc., **80** [2] 373-76 (1997)

[18]R. C. McCune, W. T. Donlon and R. C. Ku, "Yttrium Segregation and YAG Precipitation at Surfaces of Yttrium-Doped Al_2O_3," J. Am. Ceram. Soc., **69** [8] C-196-C-199 (1985)

[19]P. Gruffel and C. Carry, "Effect of Grain Size on Yttrium Grain Boundary Segregation in Fine-Grained Alumina," J. Eur. Ceram. Soc., **11** 189-99 (1993)

[20]F. Oudet, P. Courtine, and A. Vejux, "Thermal Stabilization of Transition Alumina by Structural Coherence with $LnAlO_3$," J. Catal., **114** 112-20 (1988)

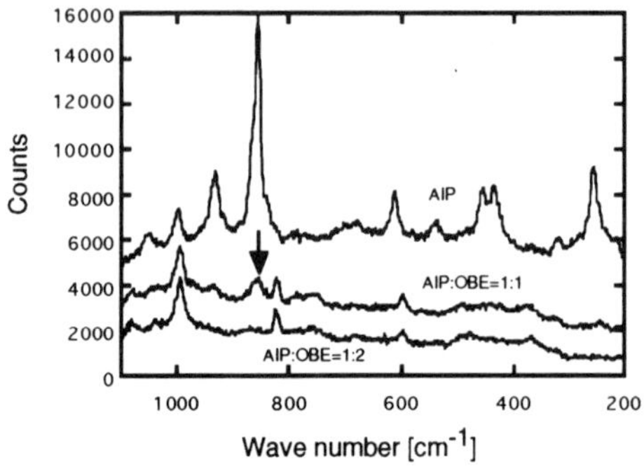

Fig. 1. Raman spectra of aluminum isopropoxide, solution mixed with aluminum isopropoxide and ethyl 3-oxobutanate in isopropyl alcohol at 60°C for 12h.

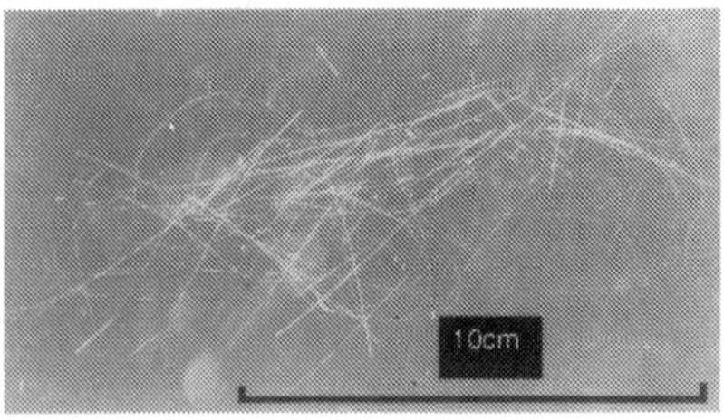

Fig. 2. Al_2O_3/ YAG composite precursor fiber.

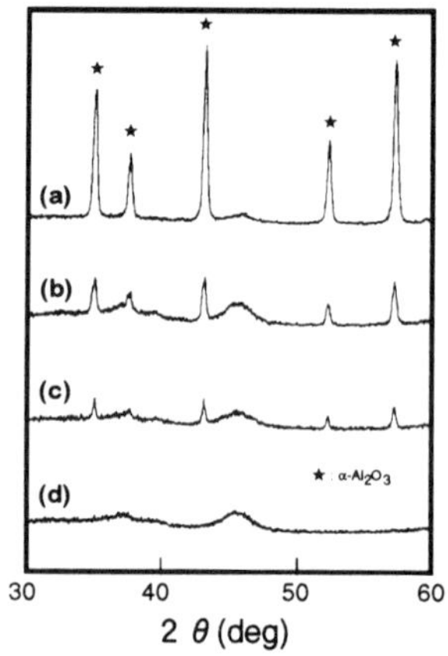

Fig. 3. X-ray diffraction patterns of the obtained fibers; (a) seeded Al_2O_3 fiber, (b) unseeded Al_2O_3 fiber, (c) seeded composite fiber and (d) unseeded composite fiber heat-treated at 900°C.

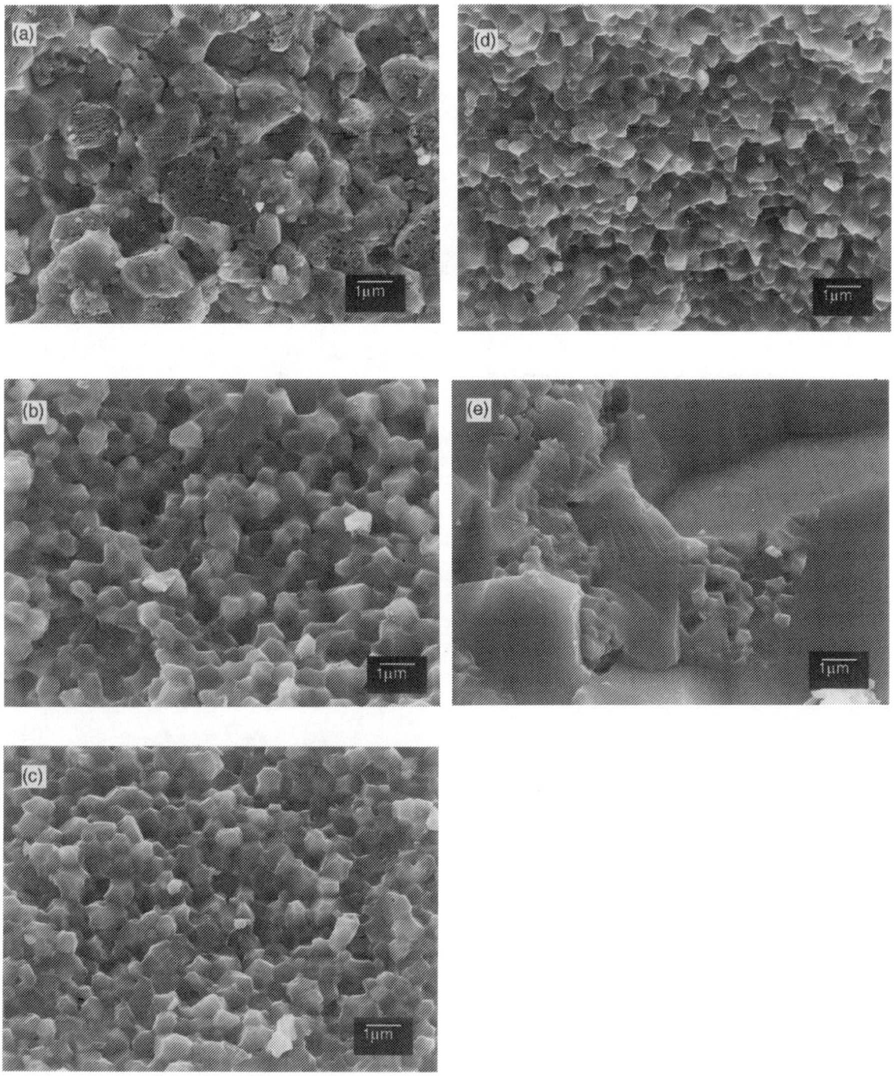

Fig. 4. SEM photographs of the obtained fibers with various seed contents.
(a) 0 wt% (b) 0.2 wt%, (c) 1 wt%, (d) 3 wt% and (e) 10 wt% heat-treated at 1400°C for 4 h.

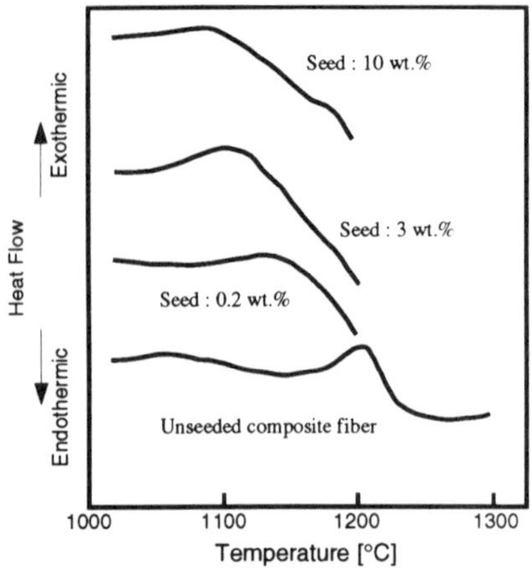

Fig. 5. DSC curves of the composite fibers. Samples with various seed contents were heated at 10°C/min in air.

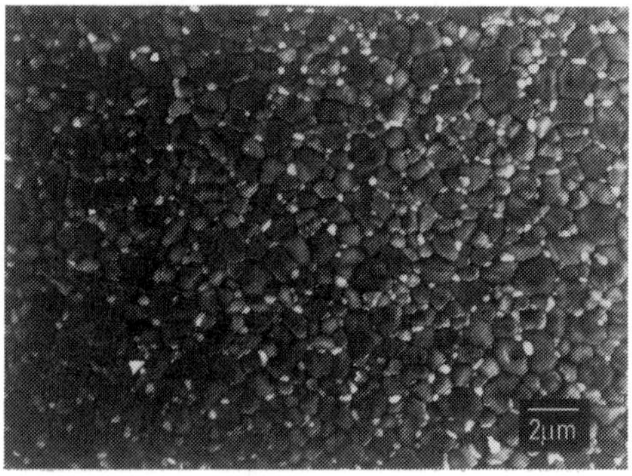

Fig. 6. SEM photograph of the Al_2O_3/YAG composite fibers (Y/Al = 1.07 mol%) heat-treated at 1500°C for 4 h.

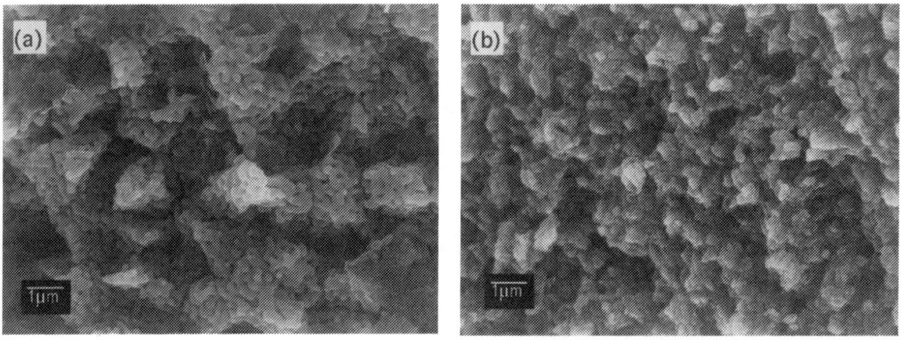

Fig. 7. SEM photographs of fracture surface of the Al_2O_3 / YAG composite fiber (Y/Al = 1.07 mol%) heat-treated at 1200°C for 4 h.
(a) Unseeded fiber and (b) seeded fiber.

SPINNING OF CERAMIC FIBERS WITH A SPECIAL STRUCTURE FOR MEMBRANE APPLICATIONS

Jan Luyten, Walter Adriansens, Inge Genné, Jos Cooymans, Roger Leysen
Process Technology, VITO
Boeretang 200
B-2400 Mol, Belgium

ABSTRACT

The gas permeation of microporous and dense ceramic membranes is rather low and limits industrial applications. In an attempt to solve this problem, a membrane module existing of ceramic fibers with a high surface/volume ratio is being developed for high temperature applications. In a first instance, porous Reaction Bonded Al_2O_3, with a high mechanical strength, and a dense O_2-conducting perovskite were envisaged as materials.

Based on a polysulfone/ceramic powder suspension, ceramic hollow fibers are synthesized by a spinning technique followed by a heat treatment. By using different compositions and by varying external parameters, like spinning rate, the air gap, the flow rates... ceramic hollow fibers with asymmetric as well as with homogeneous microstructures could be obtained.

INTRODUCTION

The gas permeation of microporous and dense ceramic membranes is rather low and limits industrial applications. In an attempt to solve this problem, a membrane module existing of ceramic fibers with a high surface/volume ratio is being developed for high temperature applications. With regard to the envisaged membrane applications [1,2] two specific hollow fiber structures receive special attention:

- The first one, a homogeneous porous structure for obtaining fibers of a porous Reaction Bonded Al_2O_3 (RBAO) [3,4]. Coating these fibers with different layers, mesoporous γ-Al_2O_3 (pore size 5 nm) and microporous top layers (pore size < 2 nm), gives us membranes with gas separative properties. By its special microstructure this porous RBAO is three times stronger than its conventional Al_2O_3 variant. Consequently, it solves the problem of brittleness of the fibers which was encountered by earlier researchers in this field [5].

- The second one, an asymmetric structure existing of a dense top layer with a sublayer with finger-like pores, is developed for dense membranes applications as N_2/O_2 separation using perovskite materials such as $LaSrFeCoO_{3-\delta}$. The objective here is to obtain sintered, self supporting fibers having a dense outer layer with the thickness smaller than the critical thickness for O_2-bulk conducting [6].

In this contribution, we describe a hollow fiber manufacturing route by spinning a polysulfone/ceramic powder suspension. Fibers with different microstructures could be synthesized by varying external parameters, such as the spinning rate, the air gap, the flow rate of the bore liquid, the composition and temperature of the suspension, coagulation bath, and bore liquid. The effect of the spinning parameters are already described elsewhere. Ph.D.-dissertation of S.Bergmans [7] gives a good overview of the state of art. However, the limited knowledge about the influence of the ceramic filler on the final membrane structure needs more research. The following parameters should be further investigated:

- the interaction of solvent-polymer and the ceramic filler
- the influence of the filling grade
- the influence of the particle size distribution
- the pre-treatment of the ceramic filler

Therefore different sets of spinning experiments were performed. Their microstructures were observed by using SEM and TEM and preliminary results of other characterizations e.g. pore size, porosity and mechanical strength can be shown.

EXPERIMENTAL PROCEDURE

First, polymer/ceramic suspensions were prepared for the spinning of hollow fibers. To obtain a homogeneous suspensions, a special mixing apparatus was used (High Speed Dissolver). The polymer, polysulfone (UDEL 1800 NT 11 supplied by Amoco), is frequently used for the synthesis of polymeric membranes. Two different kinds of powders were used. The first one consists of an Al/Al_2O_3 mixture to obtain finally a RBAO material. The Al powder was supplied by Baudier Poud. Met., and the Al_2O_3 powder by H.C.Starck [8]. The second powder $LaSrCoFeO_{3-\delta}$, a mixed conductor, was prepared from its component oxides (supplied by Merck). The compositions of the suspensions were optimized, taking into account the viscosity boundaries inherent to the spinning process.

In a lab spinning test rig with a nozzle with two concentric apertures, the suspensions are spun into hollow fiber shapes. By bringing the suspension into contact with a non-solvent the phase-inversion process of the polymer is initiated, resulting in a solidified network structure. Both at the in- and outside of the hollow fibers different non-solvent liquids were tested. The best results were obtained with a bore liquid (inside) consisting of a mixture of N-methylpyrrolidone (NMP), water and polyvinylpyrrolidone (PVP) and as coagulation liquid (outside) pure water.

At the end of the spinning process the "green" product is obtained. Figure 1 gives the flow sheet of the manufacturing route for the ceramic hollow fibers.

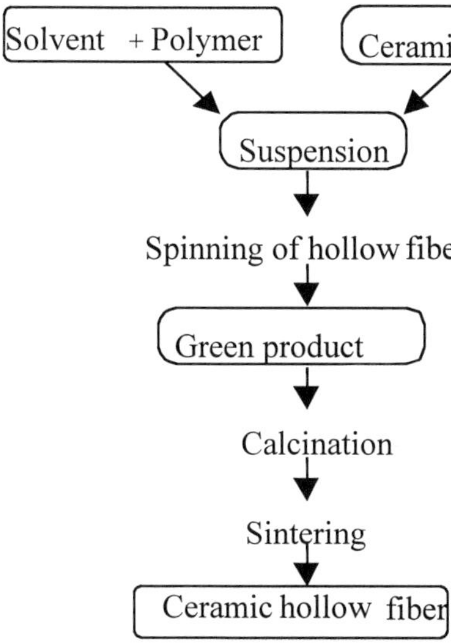

Fig. 1. Manufacturing route for ceramic hollow fibers, based on the phase inversion process

After pyrolysis, the hollow fibers are further thermally treated. In the case of Al/Al_2O_3 powder mixture, the fibers were first oxidized at 1100°C and then sintered at the temperature higher than 1300°C depending on the porosity or mechanical strength which were envisaged. In this way, hollow fibers (or capillaries) with an outside diameter of about 1 mm and an inside diameter of about 0.8 mm were produced.

RESULTS AND DISCUSSION

1. RBAO hollow fibers with a homogeneous microstructure

After optimization of the different spinning parameters like the spinning rate, the air gap, the flow rate of the bore liquid, the composition and temperature of the suspension, coagulation bath, and bore liquid, RBAO hollow fibers with excellent properties were obtained. The four point bend strength is a function of the porosity which could be varied by the final heat treatment. Figure 2 demonstrates this relationship and shows at the same time that our RBAO material is two to three times stronger then porous Al_2O_3 prepared by conventional powder techniques [9].

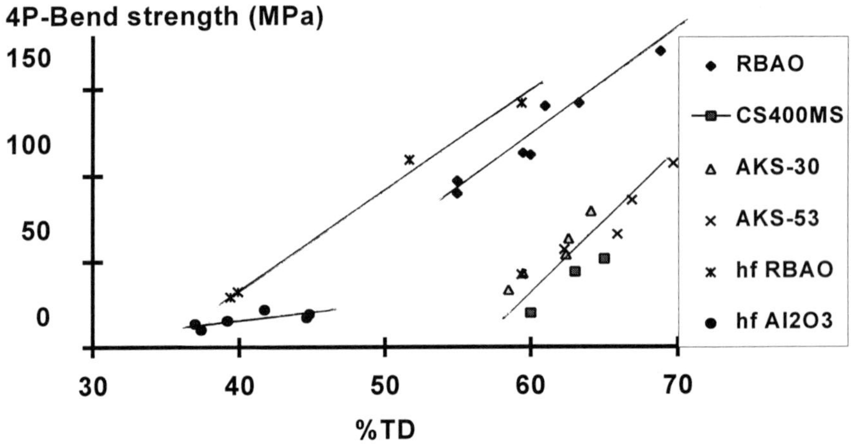

Fig. 2. Four-point bend strength of different porous supports in relation to their porosities
hf:hollow fibers, CS400MS, AKS-30, AKS-53 are spray dried powders from respectively Martinswerke and Sumitomo

It is also important to obtain the fibers with a sharp maximum in the pore size. A scanning electron microscopy (SEM) of the cross section, shows the nearly complete homogeneous microstructure, figure 3. Only very locally, a small form of a finger like structure, typical for phase inversion, could be observed. This kind of membranes can be easily coated with a second layer as e.g. γ-Al_2O_3 as is demonstrated in figure 4.

Table I summarizes the mechanical properties in function of the micro-structure, obtained from different spinning experiments , Table II.

Table I. Properties of RBAO hollow fibers (sintered at 1450°C)

Tests	Pore size (μm)	Pore Vol (%)	4p-bend strength (MPa)
1	0.4	31.4	163 ± 28
2	0.45	31.5	137 ± 25
3	0.8*	30.1	175 ± 16
4	2.0 *	30.9	143 ± 26
5	0.8*	31.6	149 ± 26

*No sharp maximum in the pore size.

Table II. Parameters of the different spinning experiments with RBAO

Tests	Bore liquid	Coagul. bath	Spinning rate n/min *	Air gap mm
1	50%NMP, 50% H_2O, 1 % PVP 360000	H_2O	25	5
2	50%NMP, 50% H_2O, 1 % PVP 360000	H_2O	50	5
3	H_2O	H_2O	25	5
4	H_2O	H_2O	50	5
5	H_2O	H_2O	25	-1 **

* n = rotational frequency,
** -1 = means under water

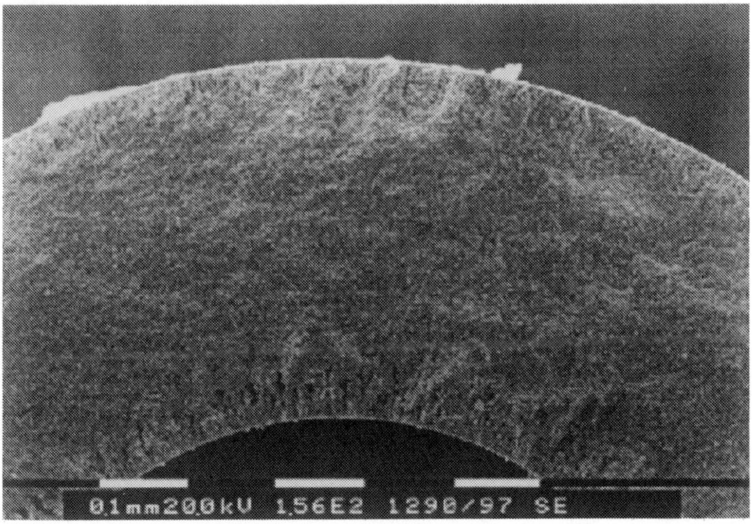

Fig. 3. Cross section of a nearly completely homogeneous microstructure

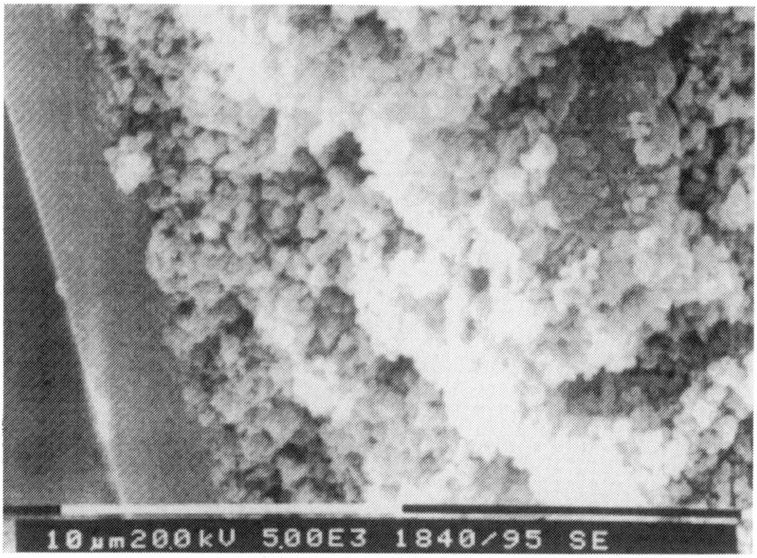

Fig. 4. A RBAO fiber coated with an γ-Al_2O_3 coating

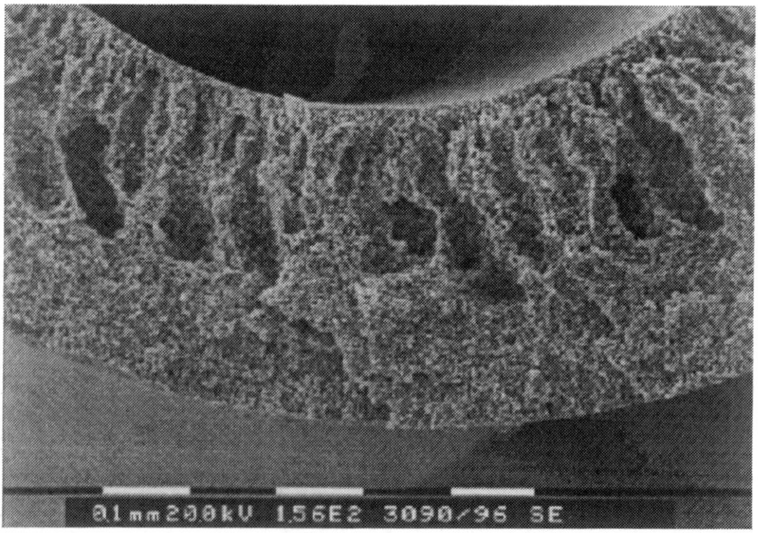

Fig. 5. Cross-section of an asymmetric type of hollow fiber
With a large quantity of finger-like pores

2. $LaSrFeCoO_{3-\delta}$ hollow fibers with an asymmetric microstructure

As mentioned in the introduction, hollow fibers with a finger like structure could be produced. The idea was to get a very porous membrane and thus a high gas flux. However, the obtained asymmetrical fibers were very weak. Figure 5 shows a cross-section of such a microstructure. Therefore, the parameters of the spinning process were changed to get a more homogeneous structure. The best value for the mechanical strength in four-point bend till now is 75 MPa. This relatively low value is due to the presence of some larger pores (fingers) near the surface of the fibers. Further optimization of the structure is ongoing. Figure 6 shows the surface of the $LaSrCoFeO_{3-\delta}$. The long sintering time has produced a thermal etching of the grains. Table III gives the mechanical strength obtained on fibers of four different spinning experiments, Table IV. The difference can be explained with a different quantity of finger-like fingers. The surface at the outside is dense.

Table III. Preliminary results of $LaSrCoFeO_{3-\delta}$ dense hollow fibers

Test	Pore Vol (%)	4-p bend strength (MPa)
1	8.2	75 ± 26
2	13.4	36 ± 21
3	11.0	41 ± 5
4	11.4	45 ± 5

Table IV. Parameters of the different spinning experiments with $LaSrCoFeO_{3-\delta}$

Test	Bore liquid	Coagul. bath	Spinning rate n/min	Air gap mm
1	50%NMP, 50% H_2O, 1 % PVP 360000	H_2O	20	10
2	50%NMP, 50% H_2O, 1 % PVP 360000	H_2O	9	5
3	50%NMP, 50% H_2O, 1 % PVP 360000	H_2O	9	15
4	50%NMP, 50% H_2O, 1 % PVP 360000	H_2O	9	0

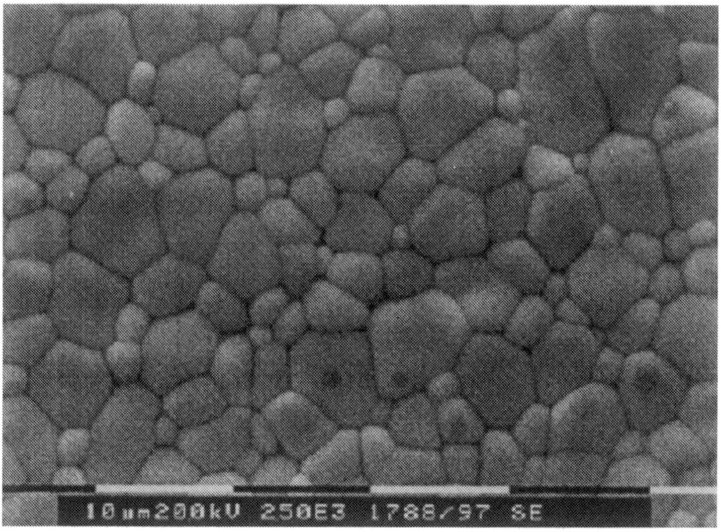

Fig. 6. The dense surface of a $LaSrCoFeO_{3-\delta}$

CONCLUSIONS

The spinning technique described here, allows to produce ceramic hollow fiber membranes.

Both asymmetric membranes with large finger like pores as well as symmetric ones with homogeneous pore distributrion could be obtained.

The mechanical strength of the asymmetric structure with large finger like pores is, certainly for the $LaSrCoFeO_{3-\delta}$, insufficient.

Therefore only fibers with a homogeneous structure will be produced in the future and further optimization will be necessary to go to hollow fibers that can be assembled into a membrane module. Of course, during that development step, special attention will be paid to the sealing problems.

ACKNOWLEDGEMENT

The authors want to express their thanks to L.Willems, P.Puts, V.Bukenbergs, R. Kemps and W. Hendrix for their technical assistance.

REFERENCES

[1]I. Genné, "Formation and characteristics of organo-mineral ultrafiltration membranes", Ph.D. Thesis, KU Leuven, 1996.

[2]H. Weyten, W. Adriansens, J. Cornelis and R.Leysen, "Preparation and characterization of high-Tc superconducting $Yba_2CU_3O_{7-x}$ filaments", Journal of Materials Science, 28, 1252-1256 (1993).

[3]J. Luyten, J. Cooymans, C. Smolders, S. Vercauteren, E.F. Vansant and R. Leysen, "Shaping of multilayer ceramic membranes by dipcoating", Journal of European Ceramic Society 17, 273-279 (1997).

[4]J. Luyten, S. Vercauteren, H. Weyten, E.F. Vansant and R. Leysen, "Synthesis of multilayer ceramic membranes", Proceedings 7th World Filtration Congress Budapest, pp. 380-384, 1996.

[5]K. Kusakabe, K. Ichiki and A.K. Fritzsche, "Polymeric gas separation membranes", John-Willey & Sons, inc., New York, 1993.

[6]A.J. Burggraaf, L. Cot, "Science and Technology Series", 4, Elsevier Science, Amsterdam pp. 435-515, 1996.

[7]S. Bergmans, "Spinning of hollow porous fibres", Ph.D. Thesis, KU Leuven, 1995.

[8]J. Luyten, J. Cooymans, P. Diels and J. Sleurs, "RBAO-used as part of a membrane configuration", Proceedings Third Euro-Ceramics V1, 657-662 (1993).

[9]J. Luyten, J. Cooymans and R. Leysen, "Shaping of a RBAO membrane support", Key Engineering Materials, 132-136, Euro Ceramics V, 3, 1691-1694, 1997.

BIOMORPHIC SILICON CARBIDE CERAMICS WITH CELLULAR MICRO-STRUCTURE

Peter Greil, Thomas Lifka and Annette Kaindl
University of Erlangen-Nuernberg, Department of Materials Science
Martensstr. 5
D-91058 Erlangen, Germany

ABSTRACT

Processing of cellular ceramics with anisotropic pore structures by liquid infiltration of Silicon into carbonized wood and subsequent reaction to SiC was investigated. Natural wood of different pore size distribution and composition was carbonized at 1800°C in inert atmosphere resulting in a one-to-one reproduction of the initial wood structure. The carbon template was converted to SiC by a rapid infiltration-reaction process at 1600°C. Spontaneous infiltration was achieved by using the continuous tracheidal cells in wood as a transportation path for liquid silicon when the cell diameter was less than 30 μm. Depending on the initial cellular microstructure of the various kinds of wood (ebony, beech, oak, maple, pine, balsa) ceramic materials of different anisotropic pore structure werde obtained.

INTRODUCTION

Material synthesis from biological structures has recently attained particular interest [1-4]. Due to the genetic evolution process, biological structures such as wood exhibit excellent strength at low density, high stiffness and elasticity, and damage tolerance on a micro as well as on a macro scale. Recent activities are directed to the question how to transform the hierarchical cellular structure of wood into inorganic materials with specific functional properties. Wood is a natural composite with cellulose, hemicellulose, and lignin as the major biopolymeric constituents (overall chemical composition of wood in wt.%: C 50, O 43.4, H 6.1, N 0.2, ash 0.3 [5]). Its unique hierarchical microstructural architecture exhibits a remarkable combination of strength, stiffness and toughness [6,7]. The microstructural features of wood range from mm (growth ring structures) via μm (tracheidal cell patterns, macro- and microfibril cell wall textures) down to nm scale (molecular fiber and membrane structures of cell walls), Fig. 1.

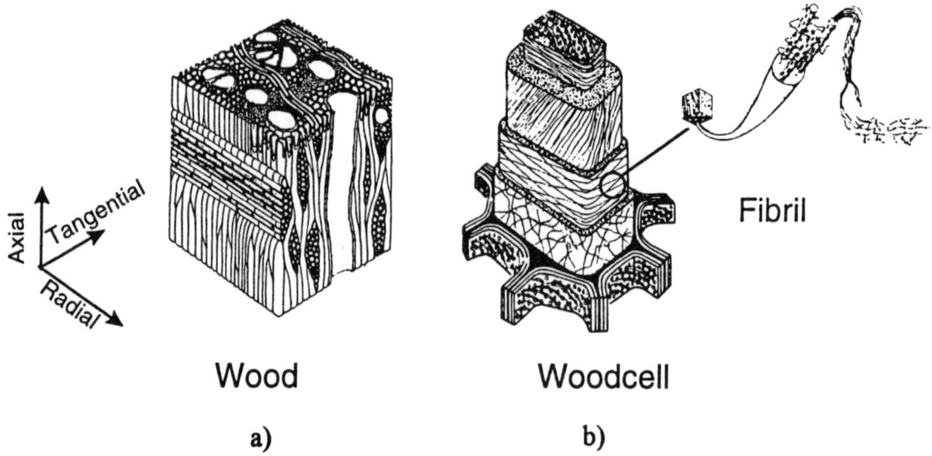

Fig. 1 Hierarchical microstructure of wood: a) macroscopic and b) microscopic cell structure.

Heating wood in non-oxidizing atmosphere at temperatures above 600°C results in decomposition of the polyaromatic constitutents to form a carbon residue which perfectly reproduces the originial cellular structure. SiC with highly anisotropic porosity was produced from wood by liquid infiltration of the carbonized preform with tetraethylorthosilicate (TEOS) and subsequent heating to 1400°C in Ar [8]. The carbon structure may also serve as a template for a variety of alternative reaction processes. In this work results on the use of various wood materials as microstructural templates for producing SiC-ceramic materials weith cellular pore structure by liquid Si infiltration and reaction process (LSIR) will be discussed.

EXPERIMENTAL PROCEDURE

Different kinds of natural wood were used to produce SiC-ceramics with different fractional densities and porosity structures: balsa, pine, maple, oak, beech, and ebony. The initial porosities ranged from 28% for ebony to 50-55% for beech, maple and oak up to > 90% for balsa, respectively. Mean diameters of the large tracheidal cells where in the range of 210 μm for oak down to 30 μm for pine.

Rectangular specimens (10 x 10 x 60 mm^3) of the carbon preform were prepared by pyrolyzing the dried (70°C, 15 h) wood in N_2-atmosphere in a carbon heated furnace. A slow heating rate of 1 °C/min was applied up to 500°C where the organic biopolymers have completely decomposed to carbon followed by an accelerated rate of 5°C/min up to the peak temperature. The specimens were hold at 1800°C for 4 h. The porous carbon preform was infiltrated with liquid silicon at 1600°C without pressure according to the conventional processing of SiSiC materials [9-11]. The specimens were kept for 4 h at this temperature to allow complete reaction of silicon with the carbon structure to form ß-SiC, Fig. 2. Unreacted silicon remained in part of the pore channels so that a cellular Si-containing SiC composite material finally was obtained. The conversion from the organic wood to the inorganic ceramic material was monitored by thermal analysis. The microstructure was characterized by scanning and transmission electron microscopy and X-ray diffraction. Density and porosity were measured by Helium pycnometry.

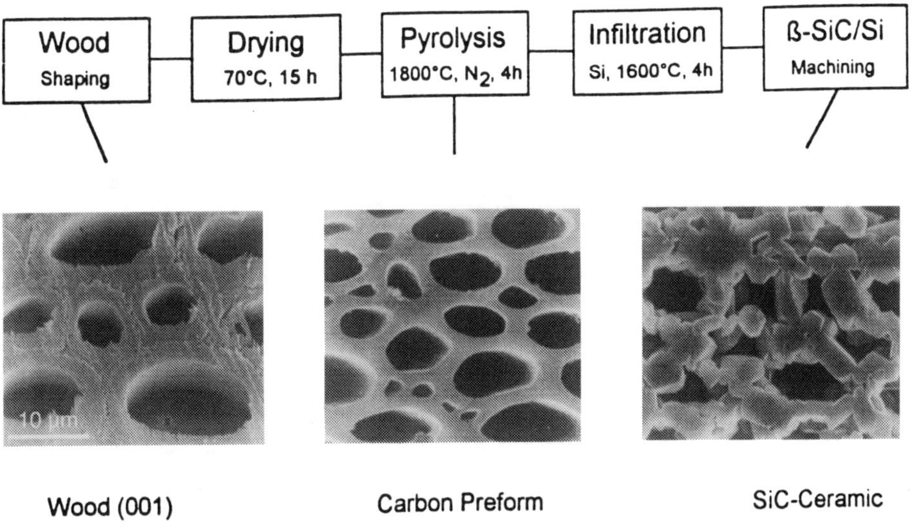

Fig. 2 Processing scheme of SiC from wood.

RESULTS AND DISCUSSION

Processing of porous carbon preform

Weight loss during heating of dried wood in inert atmosphere started below 200°C. Various maxima in decomposition rate at approximately 280°C, 340°C and 400°C, respectively, could be detected which are associated with the pyrolytic decomposition of hemicelluclose, cellulose and lignin [12]. At 500°C weight loss is almost completed. While the theoretical carbon content of wood is approximately

50 wt.% significantly higher weight losses up to 70...80 wt.% occurred due to evolution of volatile carbon hydrogen(hydroxide) species from fragmentation reactions of the polyaromatic constituents [13]. Different kinds of wood, however, may exhibit a wide range of weight loss due to the variation of molecular composition e.g. lignin to cellulose ratio and cellular pore structure. The aromatic structure of the lignin molecule which contains a high fraction of phenylpropane units combined through ether linkages or carbon-to-carbon linkages [14] gives rise to a weight loss of 55 wt.% whereas cellulose, which is characterized by oxygen bonded glucose units, shows a significantly higher weight loss of 80 wt.%. After break down of the -C-C- chains in the biopolymer macromolecules aromatic polynuclear carbon structure starts to form at temperatures above 500°C [15]. With increasing temperature as well as increasing lignin content of wood the degree of order of the graphitic like structural units increases significantly above 1400°C as indicated by XRD analysis of (0002) peak intensities [16] and TEM analysis (1800°C), Fig. 3.

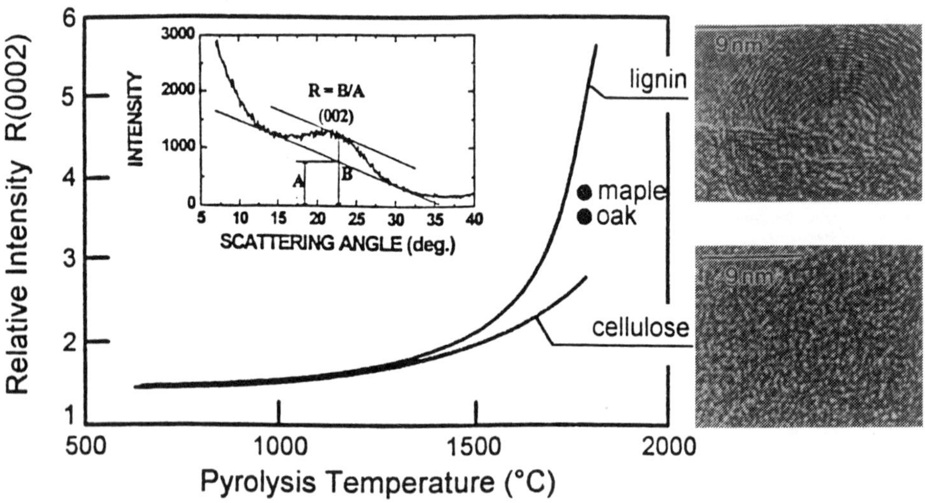

Fig. 3 Microstructural transformations during pyrolytic carbonization of wood.

Pyrolysis of wood into the carbon preform involves an anisotropic shrinkage of approximately 20 % in axial but 30 % in radial and up to 40. % in tangential directions. The magnitude of shrinkage anisotropy depends on the axial orientation of the cellulose fibers in the cell walls (macrofibrils) and the radially oriented ray parenchyma which provides a high stiffness of the cellular structure in axial and radial directions but allow a large strain in tangential direction during pyrolysis. Table I summarizes the processing characteristics of different kinds of wood.

Table I Microstructural characteristics of wood after pyrolysis at 1800°C and subsequent Si-infiltration at 1600°C.

Wood		Balsa	Pine	Oak	Maple	Beech	Ebony
Mean tracheidal diameter (μm)		150	30	210	60	50	-
Pyrolysis weight loss (wt. %)		73.5	73.8	70.4	74.9	74.2	64.6
Pyrolysis shrinkage (%)	axial	21	23	17	20	22	14
	radial	22	28	28	30	32	25
	tangential	22	31	33	40	38	30
Density (g/cm^3)	pyrolyzed	0.06	0.31	0.50	0.51	0.55	0.87
	Si-infiltrated	2.02	2.22	2.16	2.58	2.57	-
Porosity (%) (open/closed)	pyrolyzed	22/70	21/57	30/40	43/22	42/21	23/20
	Si-infiltrated	11/14	11/14	8/5	3/5	3/2	3/-

Si-infiltration and reaction to SiC

The tracheidal cell structure provides an easy access for the liquid Si to infiltrate spontaneously into the carbon preform. The wetting angle of liquid Si was found to be less than 10° on (0001) planes of oriented polycrystalline pyrolytic graphite but 50° on vitreous carbon at 1450°C [17]. Hence molecular structure of pyrolyzed carbon on the pore surface is of particular importance for the infiltration behavior and generally wetting and hence infiltration is facilitated with increasing crystallinity.

When the liquid Si comes into contact with the carbon preform reaction to ß-SiC occurs. Due to the specific volume expansion of 56 vol.% (equivalent to 16 % linear expansion) from C to SiC the open pore channel diameters of the carbon structure are reduced. Pore size analysis before and after Si-infiltration shows that all pores in the carbon preform with a diameter less than approximately 30 μm were completely filled up with Si whereas larger pores remained open. Despite the changes in porosity and anisotropic shrinkage the porous macro- and microstructure of the carbonized wood is retained after pyrolysis, Fig. 4 a-d.

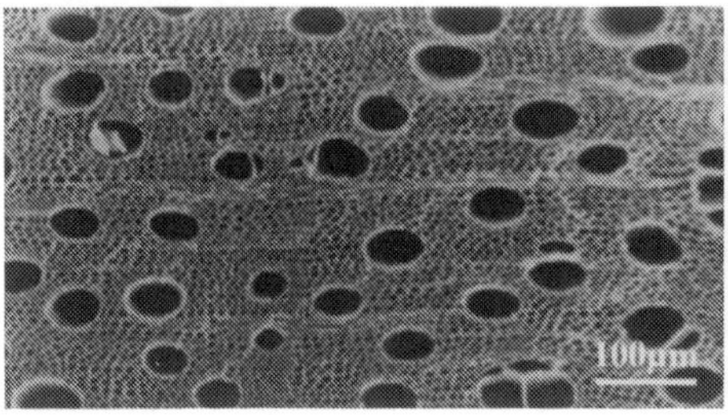

a) Pyrolyzed 1800°C

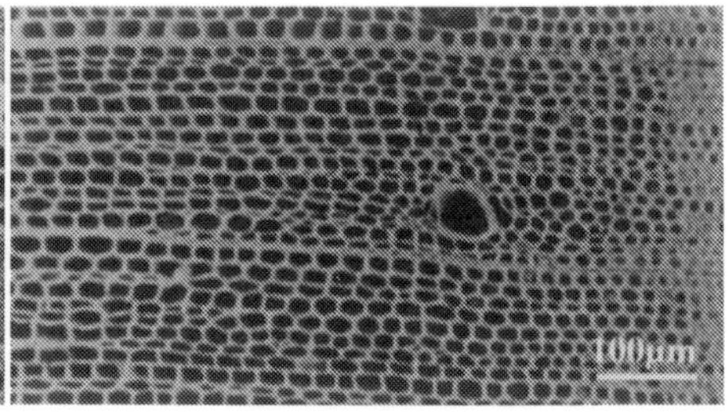

b) Pyrolyzed 1800°C

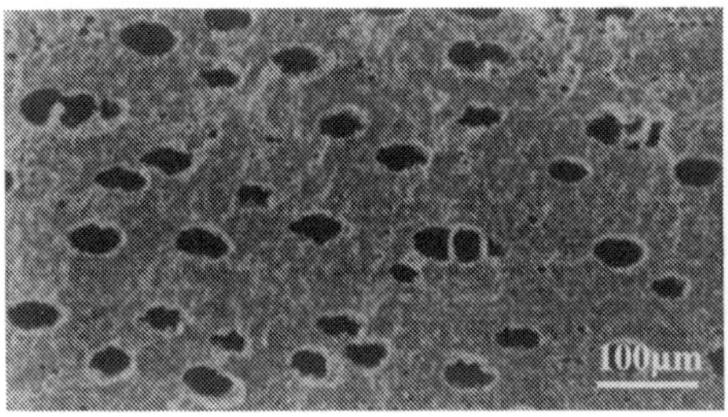

a) Si-infiltrated 1600°C

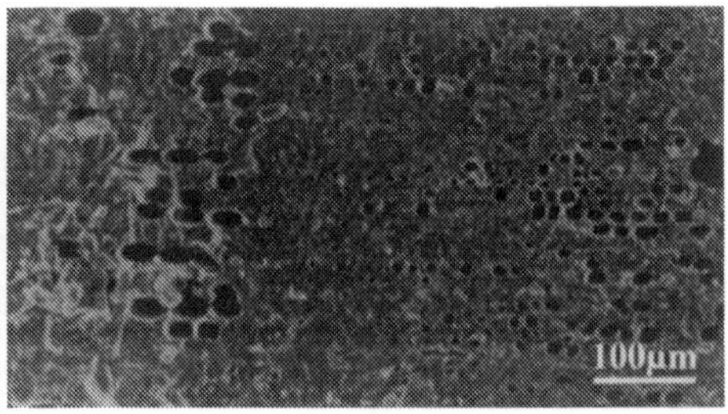

b) Si-infiltrated 1600°C

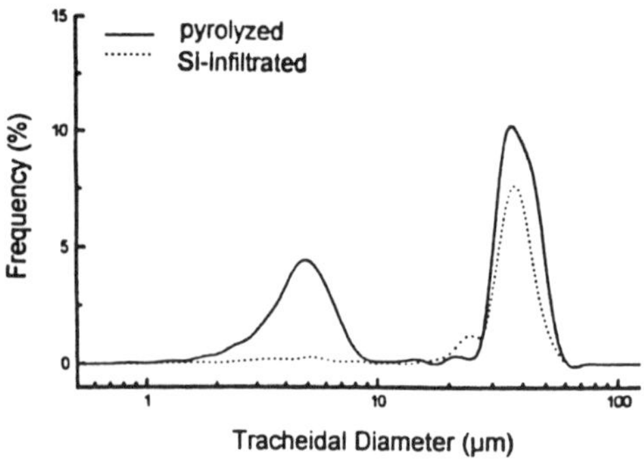

c)

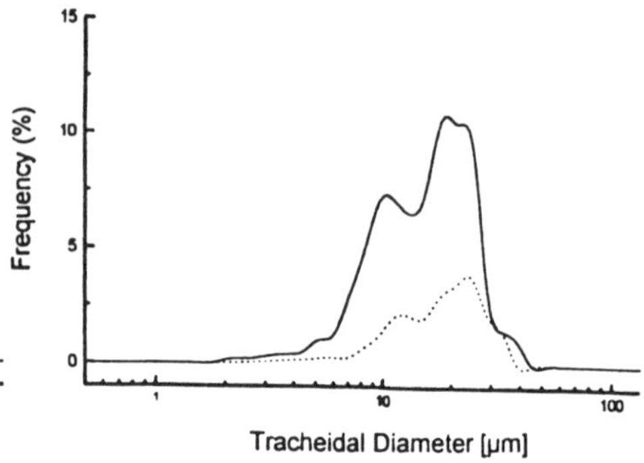

d)

Fig.4 Microstructures of maple a) and pine b) after pyrolysis and Si-infiltration; frequencies of tracheidal cell diameters of maple c) and pine d).

Other ceramics derived from the cellular carbon preform

The pyrolyzed porous carbon preform can be used for other synthesis reactions to form a variety of celullar ceramic composite materials, Fig. 5.

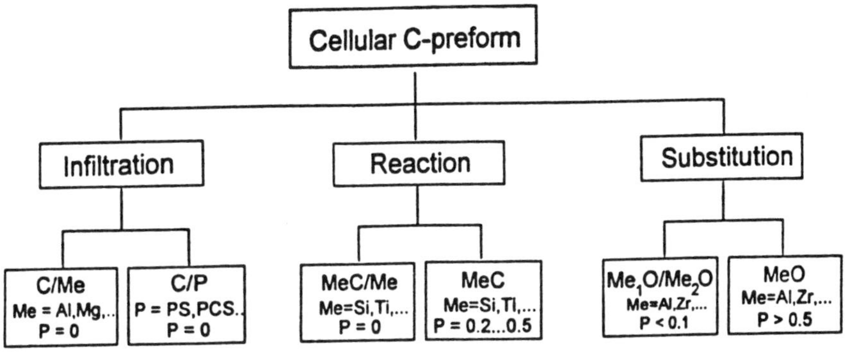

Fig. 5 Technology road map for manufacturing composite materials from wood as a biological template (P: porosity).

Infiltration of the porous carbon by non-reacting melts like Mg or Al (rapid infiltration under pressure avoids Al_4C_3 formation) or by low viscous polymers like conventional thermoplastics (PE, PVC, PA, etc.) or preceramic polymers (Polysilanes, -siloxanes, -carbosilanes, etc.) result in completely dense materials. Liquid infiltration with reacting melts like Si and Ti yield carbide/metal composites where the residual porosity is filled up with excessive metal. The reaction with gaseous infiltrants may yield single phase carbide ceramics with 20...50 % of open porosity. Infiltration with oxide forming sols/gels or suspensions and substitution of the carbon template by a second oxide forming infiltrant offers the possibility of producing celullar oxide ceramics of various composition.

CONCLUSIONS

Infiltration of the carbon preform derived from wood with liquid reactants such as Si or other metallic or non-metallic melts offers a versatile processing scheme to produce novel ceramics and ceramic composites with a unique oriented cellular microstructure. The anisotropy of their mechanical and other physical properties generally increases with porosity and great differences in strength, strain to failure, toughness, etc. may occur when loading is applied in axial, radial or tangential direction of the original wood structure. Anisotropic cellular ceramics of low density but high strength, however, could be of particular interest for applications in acoustic and heat insulation structures, as substrate, filter and catalyst carrier at high temperatures, as thermally and mechanically loaded ligth weight structures as well as for medical implant structures.

REFERENCES

[1] .A. Aksay, E. Baer, M. Sarikaya, D.A. Tirell (editors), Hierarchically Structured Materials, Mat.Res.Soc.Symp.Proc. **255**, Mat.Res.Soc. Pittsburg/PE (1992)

[2] P. Calvert, Biomimetic Ceramic and Composites, Mat.Res.Soc.Bull. **17** (1992) 36.

[3] M. Sarikaya, I.A. Aksay, An Introduction to Biomimetics: A Structural Viewpoint, in Microstructure of Materials, edited by K.M. Krishnan, San Francisco Press Inc., San Francisco/CA (1993)

[4] J.E. Mark, P.D. Calvert, Biomimetic, Hybrid and In Situ Composites, Mat.Sci.Eng. **C1** (1994) 159.

[5] R. Wagenführ, Anatomie des Holzes, VEB Fachbuchverlag Leipzig/GDR 2. Aufl. (1980)

[6] E.J. Gibson, Wood: A Natural Fibre Reinforced Composite, Metals and Materials **6** (1992) 333.

[7] P.W. Lucas, B.W. Darvell, P.K. Lee, T.D.B. Yuen, M.F. Choong, The Toughness of Plant Cell Walls, Phil.Trans.Roy.Soc.Lond. **B 348** (1995) 363.

[8] T. Ota, M. Takahashi, T. Hibi, M. Ozawa, H. Suzuki, Biomimetic Process for Producing SiC "Wood", J.Am.Ceram.Soc. **78** (1995) 3409.

[9] W.B. Hillig, Melt Infiltration Approach to Ceramic Matrix Composites, J.Am.Ceram.Soc. **71** (1988) C 96.

[10] Y.M. Chiang, R.P. Messner, C.D. Terwilliger, D.R. Behrendt, Reaction Formed Silicon Carbide, Mat.Sci.Eng. **A 144** (1991) 63.

[11] R. Pampuch, E. Walasek, J. Bialoskorski, Reaction Mechanism in Carbon-Liquid Silicon Systems at Elevated Temperatures, Ceram.Int. **12** (1986) 99.

[12] P. Ehrburger, L. Lahaye, E. Wozniak, Effect of Carbonization on the Porosity of Beech-wood, Carbon **20** (1982) 433.

[13] F.F. Byrne, H. Marsh, Indroductory Overview, in Porosity in Carbons, edited J.W. Patrick, E. Arnold Publ., London (1995)

[14] H.F.J. Wenzel, Chemical Technology of Wood, Academic Press, London (1970)

[15] F. Shafizadeh, Y. Sekiguchi, Development of Aromaticity in Cellulose Chars, Carbon **21** (1983) 511.

[16] J.R. Dahn, W. Xing, Y. Gao, The Falling Cards Model for the Structure of Microporous Carbons, Carbon **36** (1997) 825.

[17] T.J. Whalen, A.T. Anderson, Wetting of SiC, Si_3N_4, and Carbon by Si and Binary Si Alloys, J.Am.Ceram.Soc. **58** (1975) 396.

FASTER CERAMIC COMPOSITES VIA USE OF AN INVERSE TEMPERATURE PROFILE DURING CHEMICAL VAPOUR INFILTRATION

JGP Binner and Y Yin
Department of Materials Engineering & Materials Design
The University of Nottingham
Nottingham, UK

ABSTRACT

Conventional chemical vapour infiltration (CVI) of ceramic matrix composites suffers from the disadvantage that it is a very slow and expensive process. This is because when heated using standard radiant techniques the hottest part of a body is its surface. Hence during CVI deposition occurs preferentially at the surface blocking the interior from densification. Although many variations on the CVI theme have been investigated, none have succeeded in reducing production times to levels which will sustain the economic manufacture of low cost products. By using microwave radiation as the energy source it is possible to develop an inverse temperature profile, that is, the centre is the hottest part of the body. This allows deposition to occur from the inside out and hence production times to fall by at least an order of magnitude. This paper describes the results of initial experiments using both SiC/SiC and ZrO_2/ZrO_2 systems.

INTRODUCTION

Continuous fibre reinforced ceramic-matrix-composites (CMC's) are candidate materials for high temperature applications requiring high temperature strength, high strain to failure, chemical stability and thermal shock resistance. Examples include combustors, wear parts, radiant tubes and nozzles. The current world market is in excess of \$80m. Most applications require fully dense materials, however, the presence of the reinforcement inhibits full densification by simple pressureless sintering techniques. Alternative routes, such as hot pressing and hot isostatic pressing, utilise high pressure and temperatures, but these can reduce performance by degrading the fibre reinforcement. A current manufacturing solution relies on chemical vapour infiltration (CVI) to introduce matrix materials into highly porous preforms. In simple terms, this involves a solid material being deposited from gaseous reactants via thermal decomposition onto a heated substrate [1,2]. For example, methyltrichlorosilane (CH_3SiCl_3), the starting material to obtain silicon carbide, is decomposed between 1200-1400K:

$$CH_3SiCl_3(g) \rightarrow SiC(s) + 3HCl(g)$$

The process may be visualised as follows; the preform, usually a porous array of fibres, is placed in a high temperature furnace. Gaseous reagents flowing into the furnace penetrate the pores of the preform and deposit a solid coating onto the fibre surfaces. The deposition process continues until eventually the deposited material becomes the matrix. The advantages of this method include relatively low processing temperatures, the potential for near-net shaping and the ability to control matrix microstructures. Difficulties associated with CVI processing are the tendency to deposit preferentially near the surface leading to pore blockage or ‘crusting’, so inhibiting further densification, and hence, long processing times. Intermittent machining operations are required to remove the crusting which adds to the long times, which can exceed 6 months, for product manufacture. These problems stem from the conventional heating methods used to heat the preform, which rely on surface heating and conduction. Although many methods have been investigated to accelerate the process, such as forced flow, temperature and pressure modulation [3,4,5], the technical solution required is a cool surface with a hot interior.

Microwave heating is fundamentally different from conventional radiant element or induction techniques in that it couples energy directly into the component. This leads to heat being generated throughout the ceramic whilst the surrounding air remains relatively cool. Compared to conventional heating systems, an inverted thermal gradient develops as the bulk ceramic is heated whilst the surfaces lose heat by radiation and convection [6]. During microwave assisted chemical vapour infiltration (MACVI), cool reactant gases penetrate inward through the preform toward the hot interior region prior to the deposition reactions, in a sense, creating ceramic microstructures from the inside out which reduces crusting [7,8].

Many theoretical analyses and models have shown this inverse temperature distribution within the specimen [9,10]. It is influenced by the preform dielectric properties, porosity and the characteristics of the pores as well as external cooling conditions. However, systematic experimental studies of the temperature profile have not yet been reported. In this paper, the temperature distribution within a reticulated SiC-based preform heated by microwave radiation and the chemical vapour infiltration characteristics of two dielectrically different fibre-based preforms based on SiC and ZrO_2, respectively, are presented.

EXPERIMENTAL

Methyltrichlorosilane (MTS) was used as the starting material for SiC deposition. MTS carried by hydrogen decomposes into SiC and HCl according to the following formula:

$$CH_3Cl_3Si_{(g)} + nH_{2(g)} \rightarrow SiC_{(s)} + 3HCl_{(g)} + nH_{2(g)}$$

Reticulated SiC preforms supplied by Selee, USA, were used for preliminary and kinetic studies. The average pore size of the preforms was 20 pores per inch (ppi). Two preforms were stacked to make a total sample dimension of diameter 50 mm and 40 mm thick. Plain weave Nicalon fabric discs of a similar diameter were also used for SiC infiltration. About 30 discs were pressed and sewed together to form samples of ~7 mm thickness with an average porosity of about 60%.

ZrO_2 infiltration was accomplished by hydrolysis of $ZrCl_4$ with CO_2 and H_2 mixture, using an Ar carrier gas, at elevated temperatures. One of the advantages of using this system is that the overall reaction rate could be easily controlled by the water formation step. The hydrolysis process may be expressed as:

$$ZrCl_{4(g)} + 2CO_{2(g)} + 2H_{2(g)} \rightarrow ZrO_{2(s)} + 4HCl_{(g)} + CO_{(g)}$$

ZrO_2 fibre board (Zircar, USA), with a porosity of about 70%, was cut into cylinders approximately 50 mm in diameter and 50 mm thick and used as the preforms.

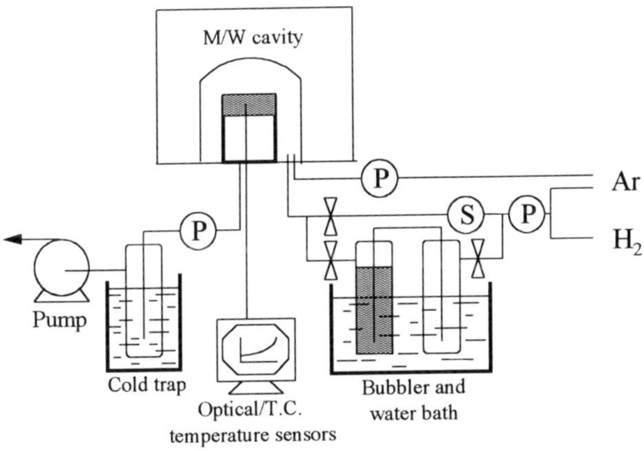

Figure 1: Schematic representation of the MACVI equipment for SiC/SiC composite infiltration.

Figure 1 shows the equipment built at the University of Nottingham and its experimental arrangement for SiC MACVI. MTS concentration, in the range of 2.0 to 4.5 mg per millilitre of carrier gas, was controlled by the water bath temperature and gas flow rate. The exhaust gas mixture was cleaned by a nitrogen cold trap. The total pressure of the reaction chamber was adjusted to around 95 to 97 kPa. In the case of Nicalon fabric preforms, the same total pressure was applied at the gas delivery side but gradually reduced at the exhaust side to keep a constant gas flow. The infiltration was terminated when the vacuum pressure reached the limit of the pump (about 7 kPa). For the ZrO_2 infiltration, the zirconium chloride was evaporated at a temperature between 250-310°C, i.e. slightly below its sublimation point, and was carried by argon. Hydrogen and carbon dioxide were pre-mixed in the ratio 100 (H_2) : 150 (CO_2) ml/min and further mixed with the $ZrCl_4$/Ar mixture within the reaction chamber. This procedure was intended to prevent a possible homogeneous reaction (sooting) in the gas stream.

Temperature measurement was performed by both optical thermometry (Accufiber/Lextron, USA) and sheathed thermocouples. In the latter case the thermocouple was calibrated under a microwave environment by the melting of pure aluminium and gold. Accuracy was found to be within ±10°C. The potential for possible electromagnetic interference caused by the insertion of sheathed thermocouples into the cavity was also checked and the results compared with the optical thermometry measurements. The measured temperature difference between the two systems was negligible.

The reaction chamber, which was made of quartz, was sealed onto the water cooled cavity platform via a PTFE fringe. Alumina based thermal insulation around the side and bottom of the sample was used to increase the vertical thermal gradient as well as reduce heat radiation into the cavity.

The phases of the infiltration products were identified by X-ray powder diffraction whilst the microstructure of the surface and interior were studied by scanning electron microscopy.

RESULTS AND DISCUSSION

Figure 2 shows the inverse temperature profile which was obtained with the reticulated SiC preform. The vertical temperature variation within the preform is plotted against a reference temperature, which was measured 5 mm from the top of the preform. A variation of approximately 100°C between the top and centre of the preform may be observed. A similar distribution occurred radially. The effect of the inverse temperature distribution is clearly seen in figure 3 where two specimens have been infiltrated at 1000°C for about 1 hour prior to being sectioned horizontally and vertically respectively. Concentric rings with different infiltration densities may be observed. The thickest deposits were to be found in the centre of the preform.

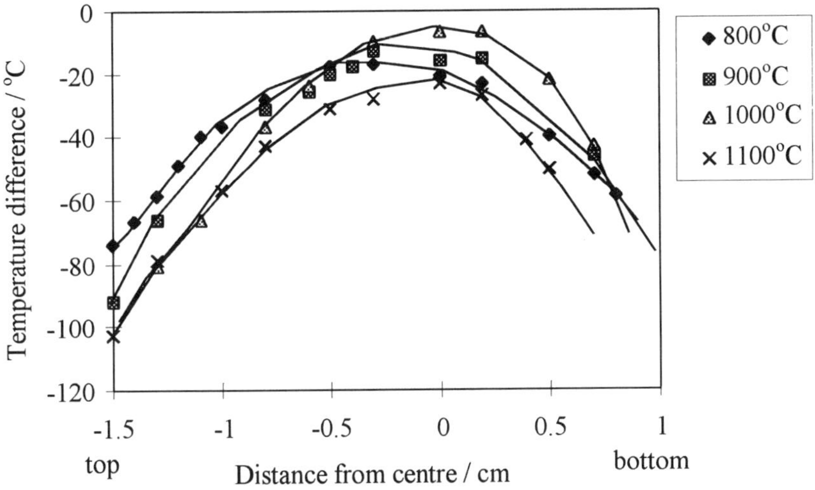

Figure 2: The measured vertical temperature distribution within a SiC preform.

Figure 3: Partly infiltrated SiC reticulated preform. Cross sectional views from side (left) and top (right).

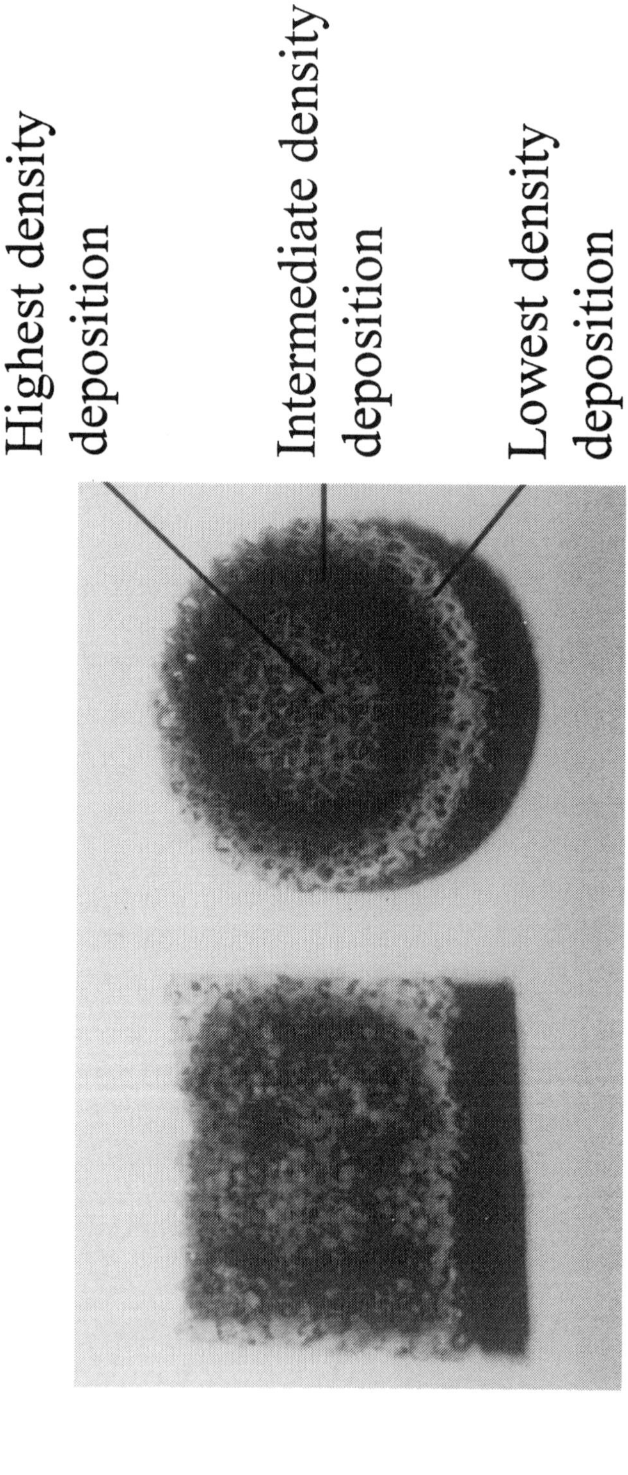
Highest density deposition
Intermediate density deposition
Lowest density deposition

The MACVI experiments using reticulated SiC foam were conducted in the temperature range between 800 and 1100°C. The hydrogen gas flow rate was 200 ml/min and the MTS concentration around 2.3 mg/ml. The effect of temperature on infiltration rate is shown in figure 4; the infiltration rate was observed to obey the Arrehnius Law, figure 5. The effective activation energy derived from the plot is 80 kJ mol^{-1}. This value is quite comparable with the widely quoted activation energy of 60-120 kJ mol^{-1} obtained from CVD experimental data [11,12].

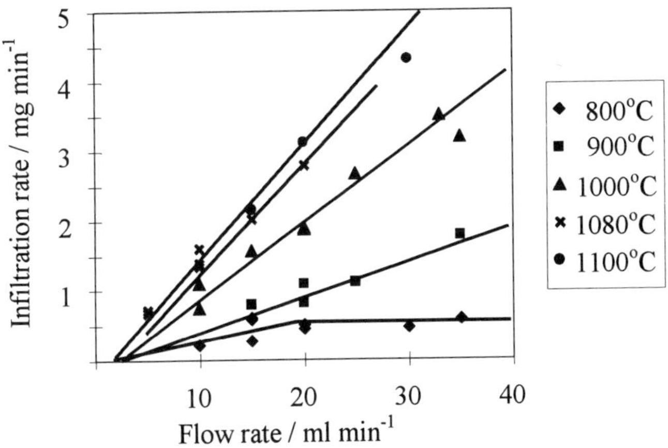

Figure 4: Infiltration rate as a function of flow rate at different infiltration temperatures.

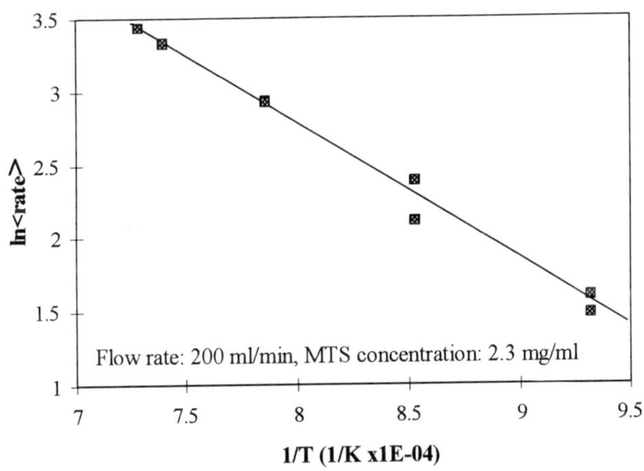

Figure 5: Influence of temperature on infiltration rate.

A micrograph of the centre infiltrated Nicalon preform is shown in figure 6. It shows a very high density of infiltration with only a few residual pores. The pressure drop across a 7 mm thick specimen reached 93 kPa after only 19 hours of infiltration at 1000°C and 250 ml/min gas flow rate resulting in an average of 70% theoretical density across the specimen as a whole. A similar level of infiltration would take several hundred hours using conventional CVI. A redesign of the experimental system to allow a larger pressure drop would result in more of the sample becoming fully dense. The maximum bulk density of the infiltrated preform was found to be dependent not only on the infiltration process but also the preparation method of the proform. The initial local porosity distribution affected both the microwave heating characteristics and the formation of non-accessible pores.

Figure 6: Infiltrated Nicalon preform. The Nicalon fibres shown on the top half of the micrograph are perpendicular to the ones at the bottom.

The influence of temperature and carrier gas flow rate on the infiltration rate of zirconia was also investigated and a very similar trend to the CVD work by Minet, Langlais and Naslain [13] was observed. However, the derived effective activation energy was 140 kJ mol^{-1} in the temperature range between 850 and 1000°C, compared to that of 418 kJ mol^{-1} for Minet et al. X-ray powder diffraction showed that the infiltrated phase was a mixture of both monoclinic and tetragonal zirconia. One interesting finding for the zirconia MACVI is that in the whole experimental pressure range, between 17 and 85 kPa, no homogeneous oxidation was found. In traditional CVI, the total pressure has to be controlled below 10 kPa to prevent sooting.

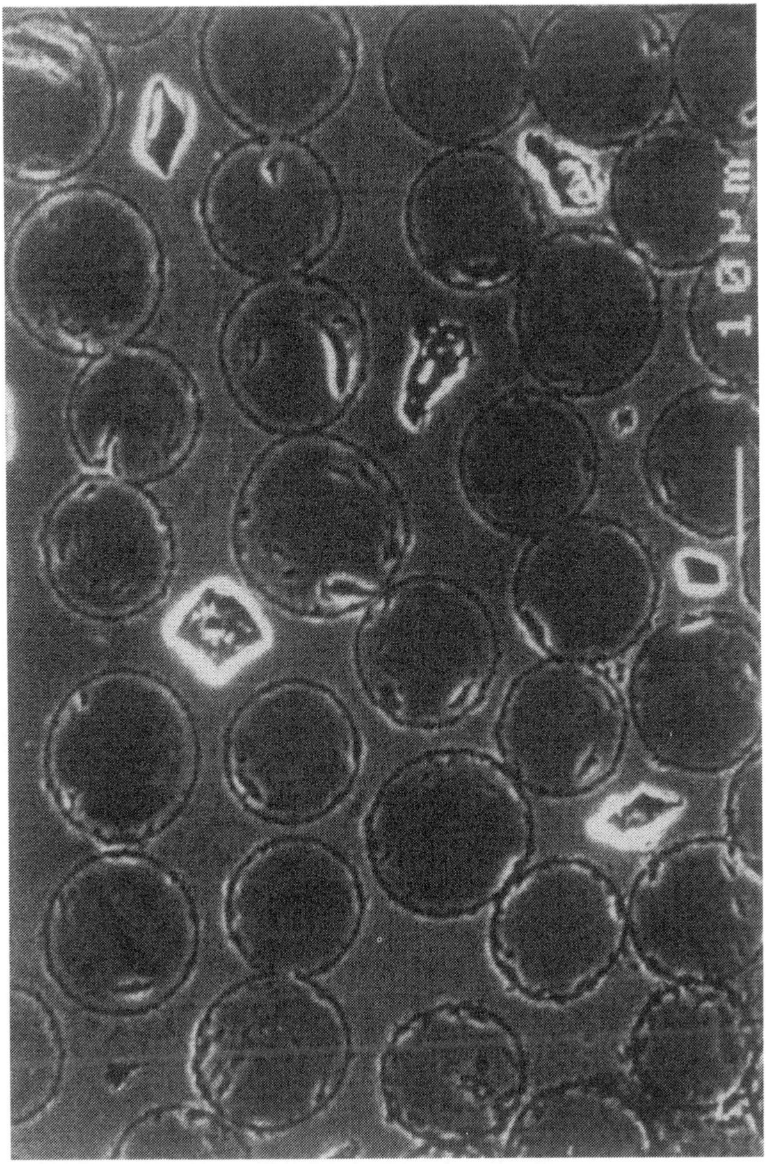
10μm

SUMMARY

In porous dielectric preforms an inverse thermal profile has been created and exploited using microwave heating. The temperature distribution has been found to reduce total infiltration time from hundreds of hours to less than 30 hours for an equivalent level of infiltration. Within the experimental temperature ranges, both SiC and ZrO_2 MACVI reaction kinetics could be described by Arrehnius plots.

ACKNOWLEDGEMENT

This project was supported by the ROPA programme of the Engineering and Physical Science Research Council, UK.

REFERENCES

1. Rossignol, J. Y., Langlais, F., and Naslain, R., *Proc. Int. CVD-9*, Cincinnati, ed. McD. Robinson, C. H. J. Van den Brekel, G. W. Cullen, J. M. Blocher and P. Rai Choudhury, The Electrochem. Society, Pennington, NJ, 1984, pp. 596-614.
2. Hannache, H., Langlais, F. And Naslain, R., *Proc. Euro CVD-5*, ed. J. O. Carlson and J. Lindstrom, University of Uppsala, Sweden, 1985, pp. 219-233.
3. Stinton, D. P., Caputo, A. J., and Lowden, R. A., *Am. Ceram. Bull.*, **65** [2] 347-50 (1986).
4. Besmann, T. M., McLaughlin, J. C., and Lin H.-T., *J. Nuclear Mater.*, **219** 31-35 (1995).
5. Gupte, S. M., Tsamopoulos, J. A., *J. Electrochem. Soc.*, 137[11] 3675-82 (1990).
6. Binner JGP and Cross TE, in Microwave Processing of Materials III, Materials Research Society Proceedings Vol 269, ed Beatty RL, Sutton WH and Iskander MF, 357-362 (1992).
7. Devlin, D. J., Currier, R. P., Barbero, R. S., and Espinoza, B. F., *Ceram. Eng. Sci. Proc.*, **14**[9-10] 761-67 (1993).
8. Spootz, M. S., Skamser, D. J., Day, P. S., Jennings, H. M., and Johnson D. L., *Ceram. Eng. Sci. Proc.*, **14**[9-10]753-60 (1993).
9. Morell, J. I., Economou, D. J., and Amundson N. R., *J. Mater. Res.*, 8[5] 1057-67 (1993).
10. Deepak, and Evans, J. W., *J. Am. Ceram. Soc.*, 76[8] 1915-23 (1993).
11. Besmann, T. M., and Johnson, M. L., Proceedings of the Third Int. Symp. on Ceram. Mater. and Components for Engines, Las Vegas, NV, Nov. 27-30, 1988, Ed. Tennery, V. J., American Ceramic society, OH, 1989.
12. Brenfleck, K. E., Fitzer, E., Schoch, B., and Dietrich, M., *Proc. - Electrochem. Soc.*, **84** [6] 649-62 (1984).
13. Minet, J., Langlais, F., and Naslain, R., *J. Less-Common Metals*, **132** 273-287 (1987).

Microstructures of Complex Ceramics

PHASE SELECTION DURING CRYSTALLIZATION

Carlos G. Levi
University of California
Santa Barbara, CA

ABSTRACT

The synthesis of ceramics by pyrolytic decomposition of solution-derived precursors is often accompanied by substantial departures from equilibrium, e.g. selection of metastable amorphous or crystalline structures, which can play a major role on the subsequent microstructure evolution. Pyrolysis typically occurs at low homologous temperatures, where long range diffusion is constrained, and hence many of the relevant transformations are partitionless. The phenomena share a common conceptual base with similar forms of metastability produced by other technologies, notably rapid solidification. A common thermodynamic foundation has been developed and used to assess the role of kinetics in phase selection and microstructure evolution in important ceramic systems. Phase hierarchy maps can be derived from phase diagrams and used to represent the menu of possible phases, and their relative stability, as a function of temperature and composition. Kinetic constraints which bias the phase selection away from the energetically most favored structure are usually the result of a requirement for ordering, complex atomic rearrangement, or partitioning during crystallization. Examples are given from ZrO_2 and Al_2O_3 based systems, including ternaries like PbO-ZrO_2-TiO_2 as well as some recent work on ZnO-based materials.

EVOLUTION AND HIERARCHY OF CRYSTALLISATION IN $CaO-P_2O_5-SiO_2-Al_2O_3-TiO_2$ GLASS-CERAMICS

S. Jordery, W. E. Lee, and P. F. James
Department of Engineering Materials
University of Sheffield
Sheffield S1 3JD, U.K.

ABSTRACT

The different levels of crystallization in a glass-ceramic can only be discerned using a range of microstructural analysis techniques including transmission electron and optical microscopes. Such a range of techniques is rarely applied to glass-ceramics. In this study, glasses in the tricalcium phosphate-anorthite ($Ca_3(PO_4)_2$-$CaAl_2Si_2O_8$ or TCP-CAS2) system with additional TiO_2 were crystallized. As the growth temperature increases, successive crystalline phases separate from the glass matrix. The phases present in the fully heat treated glass-ceramic at β-TCP, anorthite CAS2 and rutile TiO_2. Apart from TiO_2, these phases evolve from other polymorphs during the heat treatment. The metastable phases pseudo-orthorhombic CAS2 and α-TCP appear first around 880°C and transform into the stable phases triclinic anorthite CAS2 and β-TCP around 940° and 1000°C, respectively. The material crystallizes in stages. The first stage is the separation from the glass matrix of rutile TiO_2 and what is believed to be a calcium phosphate phase, whose sizes vary from 20 to 200 nm. This is followed by the appearance of larger crystals (1-2 μm), the metastable pseudo-orthorhombic CAS2, surrounding the already separated crystals. Finally, this phase disappears after the formation of anorthite (15-20 μm) spherulites.

LOW TEMPERATURE CONSOLIDATION OF METASTABLE PHASES IN THE Al_2O_3-ZrO_2 BINARY SYSTEM

Vikram Jayaram
Department of Metallurgy
Indian Institute of Science
Bangalore 560012, India

ABSTRACT

Spray pyrolysis of aqueous salts is used to produce amorphous oxide solutions of alumina and zirconia. These powders display greater instability than the equilibrium phases. Consequently, it is possible to consolidate them to high relative densities (95%) at 650°-700°C through the application of pressure. By this method, it has been possible to create in bulk form, new phases such as amorphous Al_2O_3-ZrO_2 and nano-crystalline t-ZrO_2 with gamma alumina. In addition, SiC particle reinforced alumina, which normally requires elevated temperature hot pressing in inert atmosphere, can be prepared in air at about 650°C.

PREPARATION OF ALUMINA-CHROMIA SOLID SOLUTION POWDER AND THIN FILM THROUGH PRECURSOR METHOD

Koji SATO and Yuji IWAMOTO
Synergy Ceramics Laboratory, Fine Ceramics Research Association,
2-4-1, Mutsuno, Atsuta, Nagoya 456-8587, Japan

Koichi KIKUTA
Department of Crystalline Materials Science, Nagoya University
1, Furo-cho, Chikusa, Nagoya 464-8503, Japan

Shin-ichi HIRANO
Department of Applied Chemistry, Graduate School of Engineering, Nagoya University
1, Furo-cho, Chikusa, Nagoya 464-8603, Japan

ABSTRACT

Processing of alumina-chromia solid solution powder and films was investigated through the chemical modification of metal-organic compounds. Solid solution film could be prepared with keeping starting composition at 1100°C through the addition of monoethanol amine. It is also found that the highly oriented solid solution film can be crystallized on sapphire C substrate. Plate-like crystals of around 1μm size were also observed on the surface of the film, which were identified as chromium-rich crystals due to the inhomogeneous decomposition. The number of the precipitates with chromium rich composition could significantly be reduced by the addition of monoethanol amine.

INTRODUCTION

Recently, ceramic film coating has been receiving great attention to improve the functionality and protecting the matrix. Alumina is one of the well known materials which can be used as a protecting film for corrosion and wearing. It is also known that such chemical and mechanical properties can be improved by the addition of chromia, since alumina can form a solid solution with chromia in the whole range of composition at elevated temperatures. Many studies on the system have been carried out intensively for the applications of catalysts and protecting film for non-oxide ceramics in oxidizing atmosphere[1]. Another interesting feature of this system is a spinodal decomposition of the solid solution below 950°C reported by Stubican[2]. In this study, preparations of proper precursor solutions and homogeneous solid solution film on the substrate were studied as a preliminary work for the microstructure control at lower processing temperatures.

EXPERIMENTAL PROCEDURE

Alumina-chromia solid solution has been commonly prepared by the heat treatment of mixed oxide above 1400°C, since the solid state reaction at lower temperature is very slow. These solid solutions with various compositions were synthesized as a standard for examining the composition of the prepared specimens as summarized in fig.1. It is clearly observed that the unit cell volume of those solid solutions linearly increased as increasing the chromium contents as shown in fig.2. In the present work, chromium contents of the specimens were determined by XRD based on this figure. In order to prepare solid solution film with an Al:Cr=50:50 composition, metal organic precursor solution could be prepared by the process of fig.3. It was necessary to select an appropriate nutrient compound for making this precursor without formation of precipitates due to the limited solubility and the lack of the stability. Precursor films were prepared by the spin coating of the solution on a sapphire substrate. Crystallization behavior and composition of the films were characterized by X-ray diffraction

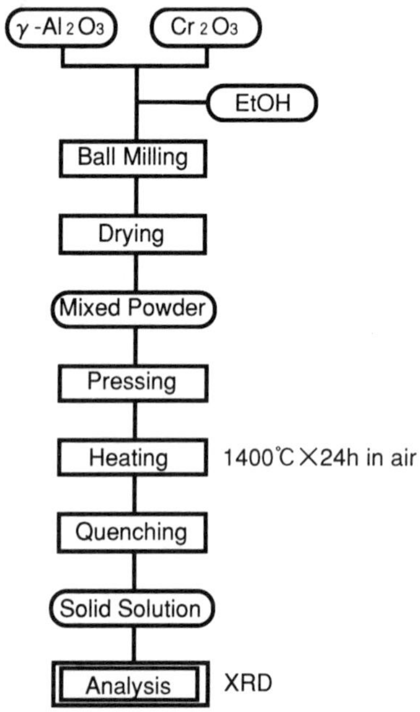

Fig1. Preparation of standard sample for determining lattice constant.

(RIGAKU, RINT 2500), and microstructure development of the solid solution film was observed by scanning electron microscopy (JEOL, JSM-6000F) with EDS system.

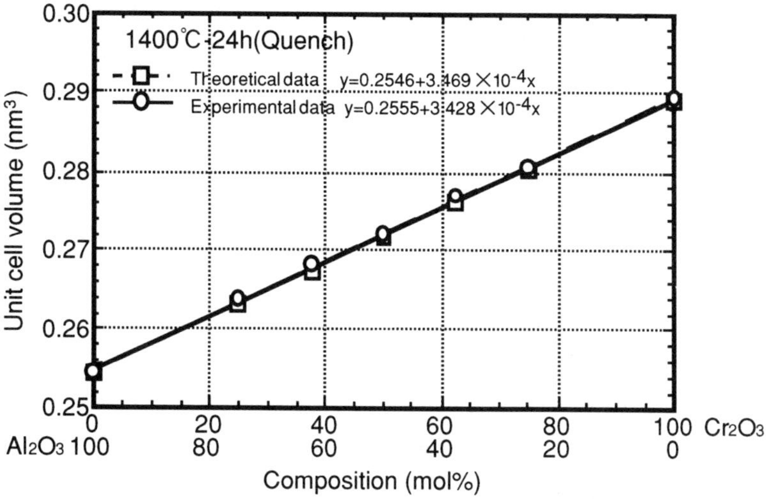

Fig.2 Unit cell volume in Al_2O_3-Cr_2O_3 system.

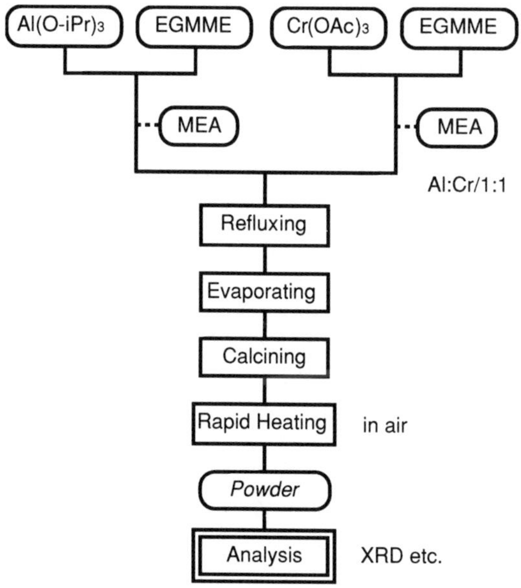

Fig.3 Preparation of Al_2O_3-Cr_2O_3 powder.

RESULTS AND DISCUSSION

Preparation of Precursor Solution

It is reported that alumina-chromia solid solution powder can be synthesized by the decomposition of inorganic salts and some other techniques[3,4]. It is, however, requested to synthesize homogeneous precursor solution to form a thin film on a substrate such as a sapphire single crystal. The number of available sources are limited for Al and Cr, so that we examined the solubility of each compound in a solvent and the homogeneity of the prepared film. Through this study, two available combinations were found as summarized in table1. Since some crystals with a rhombic shape were formed in films from methanol solution, ethylene glycol monoethyl ether (EGMME) solution was selected as a precursor to be crystallized to homogeneous film on a substrate. Further improvement of the solution was achieved by the addition of monoethanol amine (MEA) to suppress the in homogeneity in solution and preferential decomposition during calcination.

Table.1 Solubility of starting reagents in organic solvents.

	MeOH	EtOH	i-PrOH	EGMME
$Al(O\text{-}iPr)_3$	△	△	○	○
$Al(acac)_3$	○	△	×	△
$Al(OAc)_3$	×	×	×	×
$Cr(acac)_3$	○	△	×	△
$Cr(OAc)_3$	○	○	△	○

EGMME: ethylene glycol monomethl ether

○ :soluble
△ :slightly soluble
× :insoluble

Preparation of Alumina-Chromia Solid Solution

Preparation of alumina-chromia solid solution was firstly investigated for the powder specimen. Precursor solution with a starting composition of Al:Cr=50:50 was dried in advance of heat treatment, followed by the rapid heat treatment to 1100°C. A chromium rich phase was formed through the heat treatment as shown in fig.4. It is considered that the instability of starting precursor solution might be the reason for this behavior. Watanabe et al. reported that alumina rich and chromium rich phases were prepared in the composition range of Al_2O_3 30-70% by the heat treatment at 1300°C through hydrazine method[3]. In the present work, two processes were compared by the addition of monoethanol amine, to the system as summarized in fig.5.

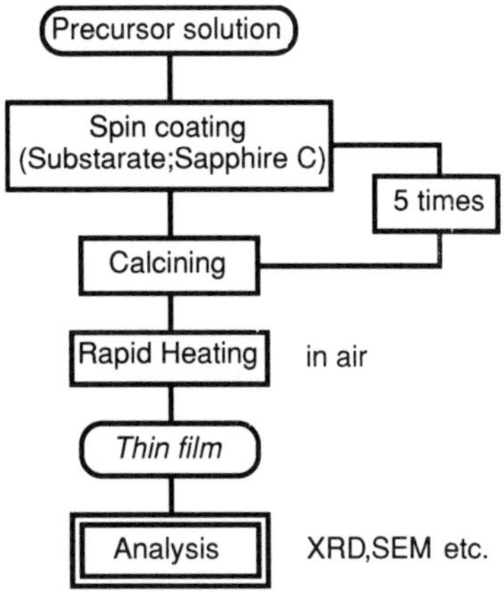

Fig.4 Preparation of Al_2O_3-Cr_2O_3 solid solution film.

After drying, these precursor powders were rapidly heated to 1100°C. XRD profiles in fig.6 clearly show the difference between these two processes. A homogeneous solid solution powder was successfully prepared by the modification of chromium acetate, which suggested that the instability of chromium acetate might be the reason for the partial phase separation of the chromium rich phase. This phase development behavior is similar to the result of Watanabe's work. It is still not clear whether an aluminum rich solid solution was formed or not in the sample. Processing of solid solution thin films was carried out using a sapphire substrate from these precursor solutions. Fig.7 shows the XRD profiles of the specimen after heat treatment at 1100°C. Very strong sapphire 006 and chromium rich phase were observed in this profile, indicating that an inhomogeneity was produced on the substrate during heat treatment. Even in this case, evidence for the aluminum rich phase could not be detected by XRD.

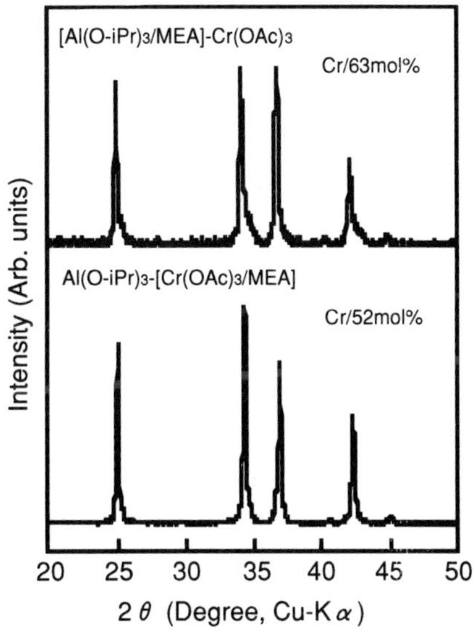

Fig.5 XRD profiles of Al_2O_3-Cr_2O_3 powder heat-treated in air.

Further detailed study using analytical electron microscopy is needed for revealing the crystallization process.

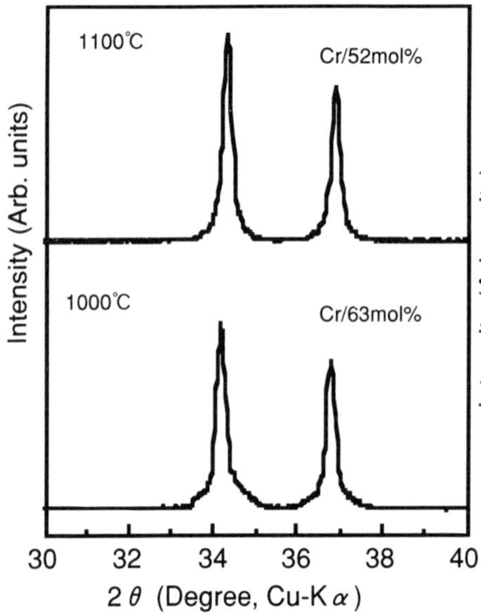

Fig.6 XRD profiles of Al_2O_3-Cr_2O_3 powder heat-treated in air.

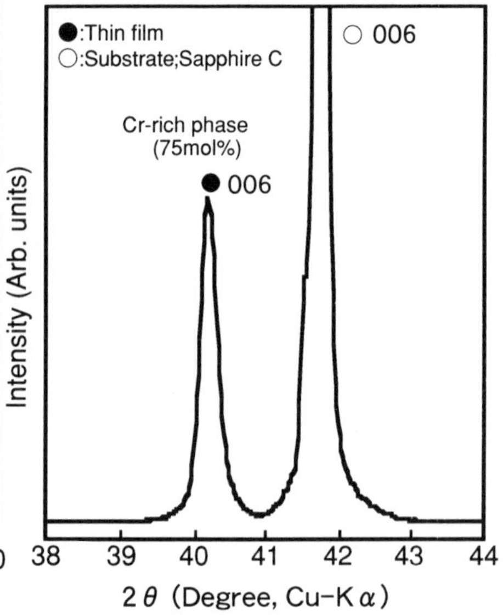

Fig.7 XRD patterns of the Al_2O_3-Cr_2O_3 thin film heat-treated at 1100°C in air.

Microstructure of thin films

SEM observation indicates the typical microstructure of the thin film as shown in fig.8. The film was composed of homogeneous crystallites around 50nm and some large precipitates with needle or hexagonal shapes as large as 1μm. Such hexagonal crystal was reported as the chromium rich solid solution prepared above 900°C[3)]. EDS analyses of these two parts exhibit that the precipitates have a relatively chromium rich composition. As mentioned in the previous section, chromium compound usually inclines to decompose at lower temperature compared to aluminum rich one. In this case, the crystallization of solid solution film could be influenced by the atomic alignment of substrate. Rapid heating was found to be required for the formation of homogeneous solid solution with a starting precursor, which was hardly achieved. This is due to the large heat capacity of the substrate or epitaxial growth of the solid solution film on sapphire. Improvement of the quality of the film was also examined by the modification of precursor solution with alkanol amine. These SEM photographs of fig.8 reveal that the number of the large precipitates was clearly reduced by this modification.

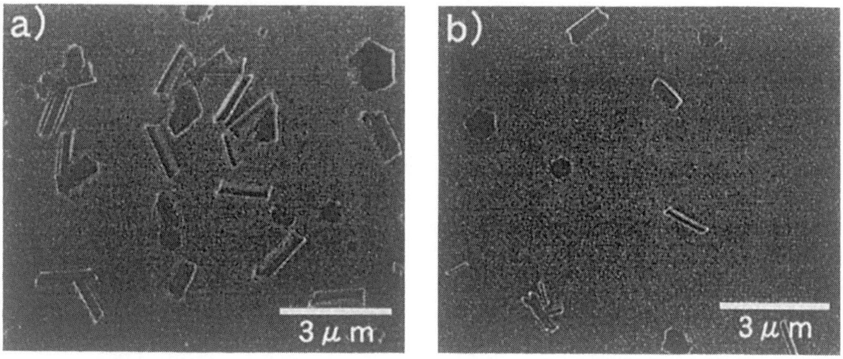

Fig.8 SEM photographs of Al_2O_3-Cr_2O_3 thin films.
a) Al(O-^{i}Pr)$_3$-Cr(OAc)$_3$, b) Al(O-^{i}Pr)$_3$-[Cr(OAc)$_3$/MEA]

CONCLUSIONS

(1) Ceramic precursor for alumina-chromia solid solution can be prepared by the combination of the selected organometallic compounds as a modifier.

(2) The precursor compound easily forms a chromium rich solid solution during heat treatment. Addition of monoethanol amine to chromium acetate was found to be very effective for suppressing the preferential decomposition of chromium rich phase.

(3) Alumina-chromia precursor can be crystallized on a sapphire substrate as a highly oriented film. This film had partly inhomogeneous microstructure, which is composed of very fine crystallites of 50nm and large hexagonal precipitates of micron size. The elemental analysis revealed that the large precipitate had relatively chromium rich composition. The microstructure was also improved by the addition of amine.

ACKNOWLEDGMENT

Work supported by NEDO as part of the Synergy Ceramics Project under the

Industrial Science and Technology Frontier (ISTF) Program promoted by AIST, MITI, Japan. The authors are members of the Joint Research Consortium of Synergy Ceramics.

REFERENCES

1) I. Zhitomirsky, R. Chaim, L. Gal-or, and H. Bestgen, "Electrochemical Al_2O_3-Cr_2O_3 alloy coatings on non-oxide ceramic substrates," J. Mater. Sci., 32, 5202-5213 (1997)

2) A. H. Schultz and V. S. Stubican, "Separation of phases by spinodal decomposition in the systems Al_2O_3-Cr_2O_3 and Al_2O_3-Cr_2O_3-Fe_2O_3", J. Am.Ceram.Soc., Vol.53, No.11, 613-616 (1970)

3) M. Watanabe, T. Hirayama, M. Yoshinaka, K. Hirota, and O. Yamaguchi, "Formation of Continuous series of solid solutions from powders prepared by hydrazine method: The system Cr_2O_3-Al_2O_3," Mater. Res. Bull. Vol.31, No.7, 861-868 (1996)

4) K. Shibata, M. Toshinaka, K. Hirota, and O. Yamaguchi, "Fabrication and mechanical properties of Cr_2O_3 solid solution ceramics in the system Cr_2O_3-Al_2O_3," Mater. Res. Bull. Vol.32, No.5, 627-632 (1997)

RECRYSTALLIZATION OF ALUMINA BY ADDITION OF FERRIC OXIDE

Yeong-Kyeun Paik, Ho-Yong Lee,* and Suk-Joong L. Kang**
Center of Interface Science and Engineering of Materials
Korea Advanced Institute of Science and Technology
Taejon 305-701, Korea

*An-Dong National University

**Sunmoon University

ABSTRACT

When Fe_2O_3 was dissolved into an (001) alumina single crystal through the vapor phase at 1500°C, chemically induced recrystallization (CIR) occurred. The shape of recrystallized grains were mostly plate-like. TEM analysis showed that low angle grain boundaries were formed by rearrangement of misfit dislocations and then transformed into high angle grain boundaries. The calculated dislocation density and the amount of strain required for the recrystallization were within the typical values for the recrystallization of cold-worked metals. The dislocation wall mechanism is, therefore, believed to be the nucleation mechanism of CIR in an alumina single crystal.

Preparation and Characteristics of Thermally Stabilized Transition Alumina

Osamu Yamanishi and Seiichi Hamano
Basic Chemicals Research Laboratory ,Sumitomo Chemical Company LTD.
5-1, Sobiraki - cho , Niihama - city, Ehime 792-8521,Japan

Abstract

Thermal stability of several transition aluminas, which were obtained by thermal decomposition of different aluminum salts, were compared. The transition alumina derived from solid sulfate showed the highest thermal stability. La doping conditions to the transition alumina were also investigated.

The highest thermal stability was obtained in the case when lanthanum salt solution and aluminium sulfate solution were mixed together homogeneously before the decomposition stage .

BET surface area of the La doped transition alumina was as high as 100 m2/g after holding at 1200 °C for 3 hours. Crystal structure of the alumina (La 3 to 15 wt% contained) showed gamma-type . SEM observation results showed that primary particles of the alumina were spheres with about 10 nm diameter and were well dispersed.

The alumina is considered to be suitable for catalyst carriers which are used over 1000°C .

Introduction

Recently , there have been variety of catalysts or catalyst supports used in the applications of chemical processes such as purification of automotive exhaust gases , high temperature steam reforming and catalytic combustion of hydrocarbons or hydrogen , as well as in the fields of catalytic reactions under high temperature gas turbines and boilers .

In these fields , generally transition alumina with high surface area ,comprising of primarily gamma-alumina , have been employed to utilize catalyst ingredient effectively . These catalyst supports are used at temperatures of 1000 °C or higher even at temperature exceeding 1200°C. Fore the catalyst supports , excellent heat resistance which exhibit a less specific surface area reduction even after used in such high temperature conditions , are needed . Many research works have been done on this field [1.2.3.4] .

We considered that retardation of sintering rate of transition alumina could be achieved by reducing of the average contact numbers among primary particles of the alumina and highly dispersing La atoms in each alumina particle .

We already reported that the thermally stabilized transition alumina [5] derived from thermal decomposition of aluminium sulfate showed higher thermal stability than that derived from aluminium hydroxide which was generally used as raw material to produce transition alumina .

In this paper we describe the preparation , characteristics and the mechanism of high thermal stability of the transition alumina derived from thermal decomposition of La doped aluminium sulfate .

Experiment

1 .Preparation

Several transition aluminas were obtained by thermal decomposition of different aluminium salts (Table 1). The following 3 type transition aluminas (LA1,LA2,LA3) were prepared as shown in Fig .1.

LA1 was the transition alumina derived from decomposition of the La doped aluminium sulfate . LA2 was the transition alumina whose La was doped after decomposition of aluminium sulfate . LA3 was the transition alumina whose La was impregnated on a received gamma-alumina .

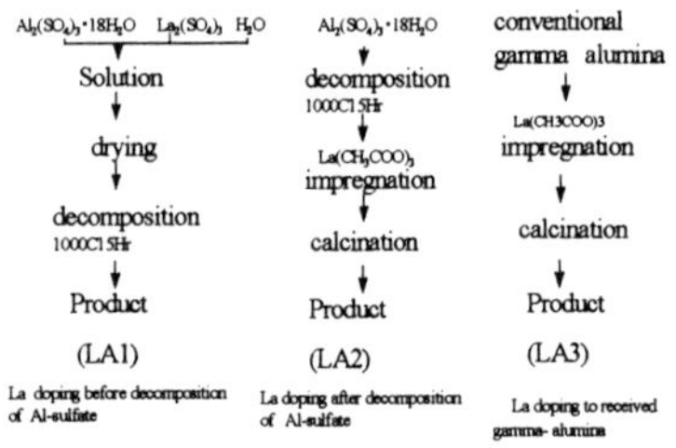

Fig.1 Preparation of La doped transition alumina

Table 1 Thermal decomposition of aluminium salts

Aluminum salts	BET surface area after 1100°C3hr m^2/g
$Al(OH)_3$	6
$Al(NO_3)_3 \cdot 9H_2O$	4
$AlCl_3 \cdot 6H_2O$	11
$Al_2(SO_4)_3 \cdot 18H_2O$	127

2. Measurement

For the prepared transition aluminas , specific surface area by BET method using nitrogen adsorption at 77K , transmission electron microscope (TEM) observation , crystal structure by powder x-ray diffraction using Cu K α radiation (XRD) and pore volume distribution were studied . Thermal stability was estimated by measuring BET specific surface area after heating . The pore volume distribution of LA1 was examined at molding conditions at 100, 500, 1000, 2000 kg/cm^2 and the thermal stability of every molded bodies of LA1 was estimated .

Pt was impregnated on LA1 and LA3 and sintering degree of Pt after 1100° C 3 hr soaking was observed .

Result and discussion

1. Thermal stability of the transition alumina derived from thermal decomposition of different salts

Table 1 shows the thermal stability of the transition aluminas which were obtained by thermal decomposition of different aluminium salts . The transition alumina derived from sulfate shows the highest thermal stability after heating at 1100° C for 3 hr .

The reason of such high thermal stability of the transition alumina derived from decomposition of sulfate is considered that removable weight ratio of sulfate is higher than other salts and released SO_3/H_2O gases make many pores in each secondary alumina particle .

2. Thermal stability of the La doped transition alumina

2.1 (Effect of the La doping process)

Table 2 and Fig. 2 show the thermal stability of the transition alumina which was prepared as shown in Fig.1 . LA1 and LA2 which were prepared by the decomposition of sulfate show better thermal stability than LA3 which was prepared by impregnation of La on the conventional alumina . The highest thermal stability is obtained in the case when La salt and Al sulfate solution are mixed before decomposition (LA1) . The BET surface area of LA1 after holding at 1200° C for 3 hr is over 100 m2/g .

Table 2 Effect of La doping on thermal stability

Alumina	LA1	LA2	LA3
La / Al_2O_3 ratio wt. %	3	3	5
BET surface area m²/g			
1100°C3hr	123	100	75
1200°C3hr	113	48	40

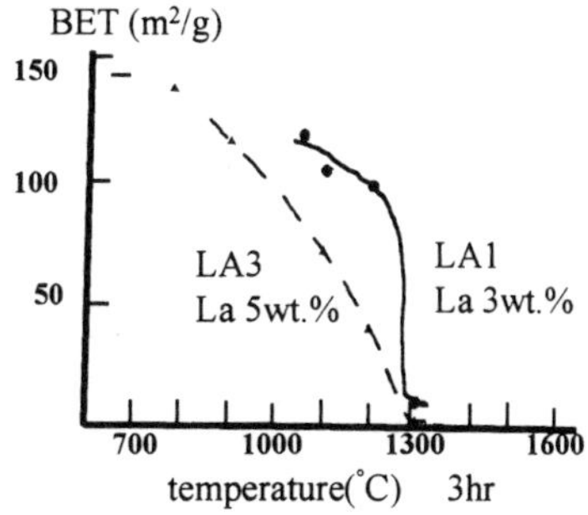

Fig. 2 BET versus temperature

2.2 (Long-run thermal aging estimation)

Transition alumina used as catalyst supports like automobile catalyst need strong durability at high temperature [(6)] .

Fig. 3 shows BET behavior during long time aging at 1100 °C of LA1 and LA3 . LA1 shows higher surface area (60m2/g) after 1100° C for 1000 hr than LA3 (10m2/g) .

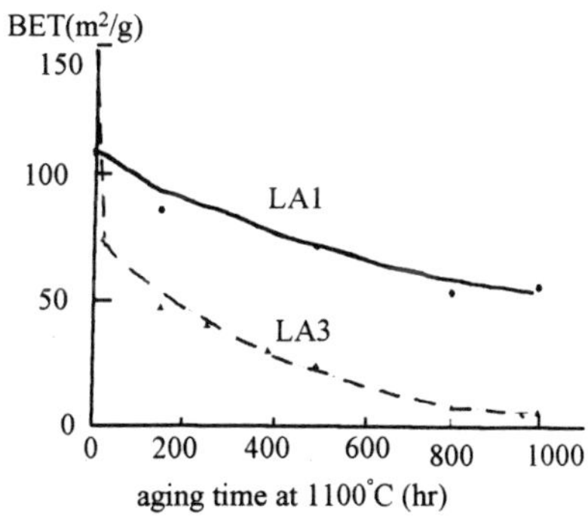

Fig. 3 BET behavior during long time aging

2.3 (Sintering behavior of Pt on the alumina)

The Pt catalysts supported on the alumina (Pt/Al_2O_3=3/100wt.) were prepared with LA1 and LA3 . Fig 4 shows the Pt sintering behaviors which were observed by field emission scanning electron microscope (FE-SEM) after heating at 1100° C for 3 hr . It is clear that Pt sintering rate in LA1 is slower than in LA3 .

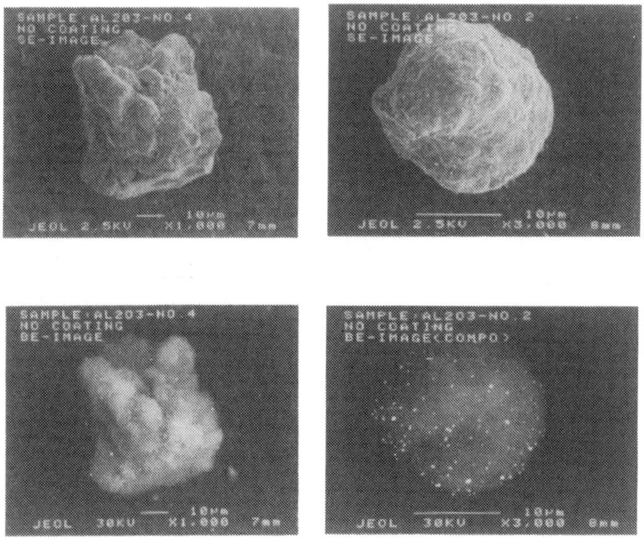

Pt/Al_2O_3=
3/100wt.
aging:
1100 °C3hr

LA1 LA3

Fig.4 Pt sintering behavior (FE-SEM)

3. Thermal stability versus La content

Fig 5 shows the effectiveness of La content in LA1 as for thermal stability . The highest BET surface area is obtained at La content between 1 to 3wt% after heating 1100° C 3 hr , 3 to 5 wt% after heating 1200° C 3 hr ,and over 15wt% after 1300° C 3 hr .

Fig. 6 shows the crystal structure of the transition alumina containing different La in LA1 . All samples are seemed to be detected as gamma- alumina type .This result shows that La atoms are dispersed in XRD detected range , but other micro analysis is necessary to discuss La atoms dispersion in detail .

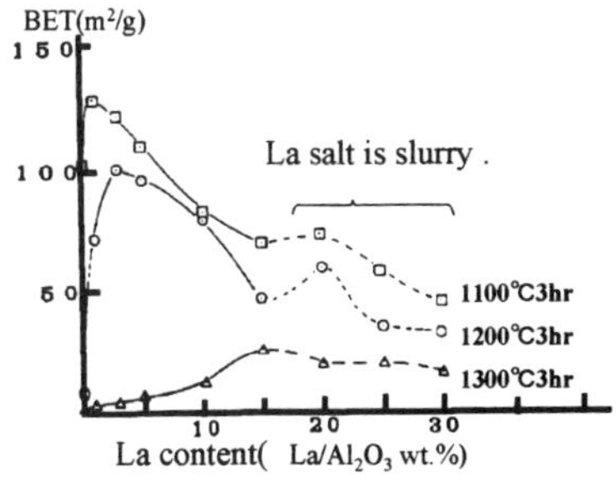

Fig. 5 Thermal stability versus La content in LA1

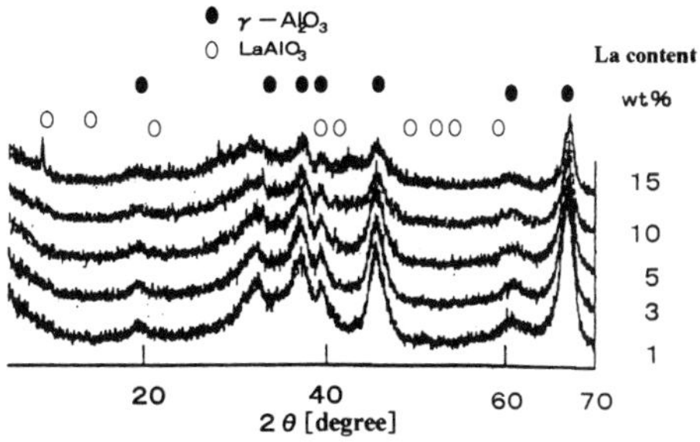

Fig. 6 **XRD of LA1**

4. Characteristics of the alumina

4.1 (TEM observation of each alumina)

Fig 7 shows the TEM observations of LA1 and LA3 when La content is 3 wt% .

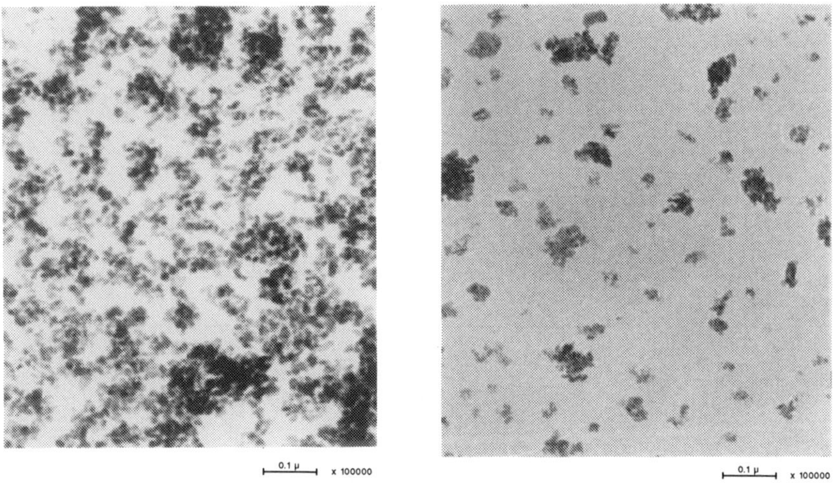

LA1 LA3

Fig.7 TEM observation

It is clear that primary particles of LA1 are spheres with about 10 nm diameter and are well dispersed although primary particles of LA3 are agglomerated .

4.2 (Pore volume distribution of the alumina)

Fig 8 shows pore volume distributions of LA1 and LA3 samples molded by pressing at 100kg/cm2 with La 3 wt% .

Total pore volume of LA1 is 1.29 cc/g and pore volume under 100 nm of LA1 is 0.72 cc/g . Total pore volume of LA3 is 0.61 cc/g and pore volume under 100 nm of LA3 is 0.30 cc/g .

Pore volume over 100 nm is considered as pores among secondary alumina particles and pore volume under 100 nm is considered as pores among primary alumina particles . Pore volume under 100 nm of LA1 is larger than that of LA3 .

The bulk densities of LA1 and LA3 are also measured , and the bulk density of LA1 is lower (0.30 g/cc) than that of LA3 (0.72 g/cc) .

From pore volume distribution and bulk density , it is considered that all primary particles of the alumina derived from decomposition of La doped aluminium sulfate (LA1) are loosely packed ,and average contact numbers among primary particles are less than another alumina which was prepared by conventional method (LA3).

Fig.9 shows pore volume distribution and thermal stability versus molding pressure of LA1 .

Thermal stability of LA1 decreases by reduction of pore volume.

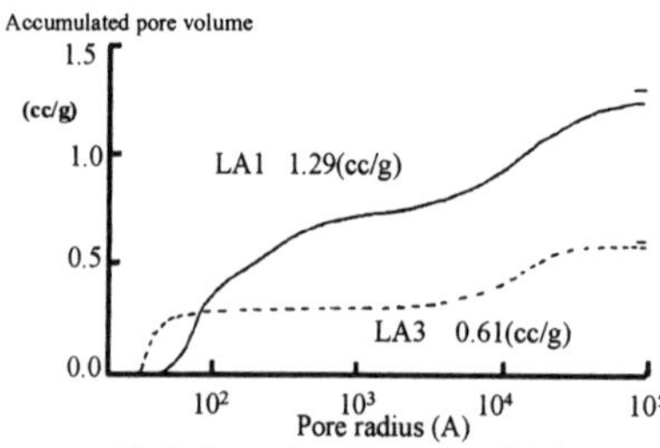

Fig.8 Pore volume distribution of aluimas

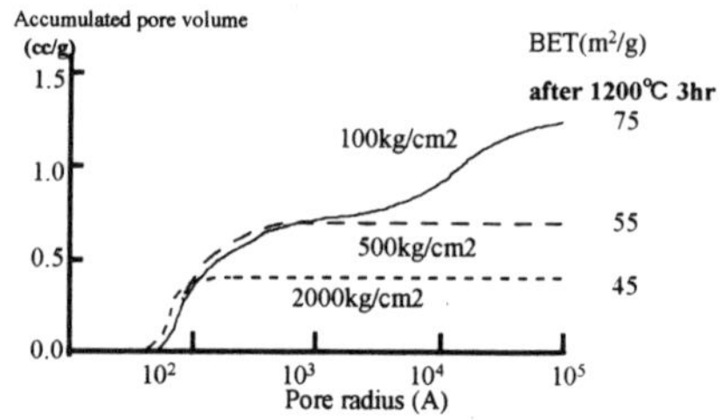

Fig. 9 Pore volume and thermal stability versus molding pressure of LA1

5. Speculation of the origins of high thermal stability of LA1

The origins of high thermal stability of this La doped transition alumina could be considered as follows :

Firstly ,removable weight ratio is very high and released SO_3 /H_2O gases make many pores in each secondary alumina particle . This speculation could be also expected from Kato's work (7) that explained the shape of transition alumina derived from decomposition of aluminium sulfate keeps the shape of hydrated aluminium sulfate .

Secondly , water which causes agglomeration in primary alumina particles through drying process does not exist after decomposition of sulfate. It was presented [(8)] that the morphology and thermal stability of the transition alumina from sol-gel methods which use the solvent include water are effected by conditions of drying .

Due to the first and second reasons, all primary particles of the alumina are loosely packed , and the average contact numbers among primary particles are less than other aluminas which are prepared by conventional methods. Therefore, the sintering rate of this transition alumina can be retarded.

Conclusion

We have developed thermally stabilized transition alumina using the decomposition process of La doped aluminium sulfate

1. The transition alumina derived from La doped aluminium sulfate shows higher thermal stability than impregnated alumina .(1200℃ 3 hr BET > 100m2/g)

   Long-run thermally aging tests and Pt sintering tests reveal that this alumina has better performance .
2. The origins of high thermal stability of this alumina could be considered as follows :

   a) The average contact numbers among primary particles are less than conventional alumina .

Reference

1) S. Matuda et al. : J. Am. Ceram. Soc. ,70[7]C-157-159(1987)

2) A. Ueno et al.: Sekiyu Gakkaishi,34,(6),477-485(1991)

3) F.Mizukami et al. : J. Materials Science Letters 9(1990)522-523

4) H. Arai et al. :Sekiyu Gakkaishi,36,(4),239-249(1993)

5) S. Hamano et al. : USP 5155085.

6) K.Ihara et al. : SAE Paper 902168

7) E.Kato et al. : Yogyo Kyokai Shi 85,(3),134-140(1977)

8) T.Fukui et al. : J. Materials Science Letters 13(1994)413-415

INFLUENCE OF Fe^{3+} ON THE SINTERING PROPERTIES OF CALCIUM HEXALUMINATE

C. Dominquez and R. Torrecillas
Instituto Nacional del Carbon
CSIC
Oviedo, Spain

ABSTRACT

Calcium hexaluminate ($CaO-6Al_2O_3$) forms extensive range of solid solution in the $CaO-Al_2O_3-Fe_2O_3$ system, but the influence of Fe_2O_3 in the sintering behavior and microstructure of calcium hexaluminate has not been properly studied. Specimens were prepared, in an air atmosphere, by reaction-sintering of homogeneous mixtures of CaO_3, $Al_2O_3$and Fe_2O_3 powders, XRD and dilatometry were used to study the phase development. The resultant microstructures were characterized by scanning electron microscopy and energy-dispersive X-ray microanalysis. It was found that the formation mechanism of the $CaO-6(Al,Fe)_2O_3$ solid solution, with the appearance of several intermediate phases (CA, CA_2 and CAF_2) during sintering, can be explained by means of the CaO- Al_2O_3- Fe_2O_3 equilibrium diagram. Two tendencies have been observed: up to 30-40% Al^{3+}-Fe^{3+} atomic replacement, addition of Fe enhances the formation and sinterability of CA6, whereas higher contents has the opposite effect. The amount of Fe3+ also affects the morphology and grain size of calcium hexaluminate on the final microstructure.

FORMING AND SINTERING OF ALUMINA WITH HYDRAULIC INORGANIC BINDER

T. Nagaoka, K. Kikuta,* and S. Hirano**
Fine Ceramics Research Association
Synergy Ceramics Laboratory
Nagoya, Japan

*NIRIN
Nagoya, Japan

**Nagoya University
Nagoya, Japan

ABSTRACT

Forming and sintering of alumina with hydraulic inorganic binder was studied in order to control the microstructure of alumina. In this study, barium aluminate, $BaO.Al_2O_3$ (BA) was selected as a hydraulic inorganic binder. BA is expected to form hydration bonds at room temperature and to form barium hexaluminate ($BaO.6Al_2O_3$) in an alumina matrix at high temperature by reacting with alumina. BA was prepared by heat-treatment at 400°C of barium aluminate hydroxide which was synthesized beforehand from a mixture of barium hydroxide and aluminum below 100°C in aqueous solution under atmospheric pressure. The hydration behavior of BA was studied to examine the capability as the inorganic binder. Microstructure and mechanical properties of the formed and sintered alumina will be discussed. Research promoted by AIST, MITI, Japan under the Synergy Ceramics Project of the ISTF Program. Under this program, part of the work has been supported by NEDO.

SYNTHESIS OF FINE GRAINED ALPHA ALUMINA VIA MECHANICAL SEEDING

M. L. Panchula and J. Y. Ying
Department of Chemical Engineering
Massachusetts Institute of Technology
Cambridge, MA 02139

ABSTRACT

Controlling the grain size of the ceramic materials has become increasingly important as these materials are used in more advanced structure applications. One of the chief goals over the last several years has been to reduce the grain size of the sintered ceramic to the nanometer range for improved strength and reliability. This has been particularly difficult for alumina, for although nanocrystalline transition aluminas are readily available, during sintering relatively few sites for α-alumina growth are created and the final grain size of the α-alumina is often in the micron to sub-micron range. Various researchers have investigated the effects of seeding the transition aluminas with external source of submicron α-alumina seeds or using other seeding agents (such as Cr_2O_3, Fe_2O_3, etc.) to affect the transformation with some success. However, it is also possible to affect the transformation via mechanical activation of the transition aluminas. The form of mechanical activation which we have investigated is high energy ball milling which has been shown to create a high density of α-alumina nucleation sites. This high density of nucleation sites results in a decrease of the transformation temperature (to as low as 950°C), a decrease in the final grain size of the material, and enhanced sinterability. Results of kinetic transformation studies as well as densification experiments will be discussed.

MECHANICAL AND FUNCTIONAL PROPERTIES OF TiO_2 MODIFIED POROUS α-Al_2O_3 LAYERS

Lars Grönroos, T. Mäntylä, A. B. A. Pettersson,* and J. B. Rosenholm*
Institute of Materials Science
Tampere University of Technology
P. O. Box 589
FIN-33101 Tampere, Finland

*Department of Physical Chemistry
Abo Akademi University
Porthansgantan 3-5
FIN-20500 Abo, Finland

ABSTRACT

The microstructural development of slip cast α-alumina layers doped with TiO_2 (rutile) have been studied. The effect of both sintering temperature and sintering additive TiO_2 concentration upon the determined material properties have been of major concern. The aqueous slips have been prepared according to a colloidal processing route. In order to optimize the stabilization and de-flocculate of particulate system an anionic polyelectrolyte was used. The total solids load was kept at ca. 76 wt%. Before casting the slips were monitored in respect to apparent viscosity. The castable slips were then poured in planar gypsum molds and dried over the night before firing. The TiO_2 concentration and sintering temperatures varied between 0–1 wt% and 1300°-1600°C, respectively. Microstructural changes have been determined in respect to apparent porosity and dynamic mean flow pore size. Mechanical properties were determined in respect to changes in Vickers hardness and functional properties were determined in respect to deionized water permeability. In the present work the following observations have been made. The apparent porosity (0–40%) is decreasing as a function of both sintering temperature and TiO_2 addition. The mean flow pore size (0.1–0.65 μm) is as well decreasing with increasing sintering temperature and TiO_2 addition. Vickers hardness (100–1000 kg/mm^2) is increasing as a function of both sintering temperature and TiO_2 addition. The water permeability (100–35000 $l/m^2/h$ at Δp=1 bar, layer thinness = 250 μm) of fired layers is decreasing as a function of both sintering temperature and TiO_2 additions.

FRACTURE OF Al_2O_3 WITH CONTROLLED PORES, EFFECT OF GRAIN SIZE

Andre Zimmerman, Mark Hoffman, J. Rödel, R. K. Bordia,* and B. D. Flinn*
Technical University of Darmstadt
Darmstadt, Germany

*University of Washington
Seattle, WA

ABSTRACT

The effect of grain size on strength of polycrystalline ceramics has been well documented. Over a wide range of grain size, the fracture strength is often represented in the form of the Orowan-Petch relationship. Although this behavior has been observed experimentally for many materials, the model based explanation of it has not been very satisfactory. In this study, a detailed investigation was conducted to obtain a sound explanation for this behavior. Samples with controlled grain size and strength limiting artificial pores were produced by controlled processing. Using these, strength was determined as a function of the grain size and fracture origin for each sample was identified by fractography. Finally, detailed micro-mechanics models were developed to correlate the measured strength to the observed fracture origin. Using this it is shown that, in the range of grain sizes investigated, the R-curve behavior of alumina is irrelevant for strength, the crack initiation, process depends on the grain size and that the critical defect is created by the interaction between processing defects (e.g. pores) and intrinsic defects (i.e. microcracks). Thus in order to improve the strength and reliability of ceramics, the processing defects should be controlled and microcrack initiation should be inhibited.

PROCESSING AND WEAR PROPERTIES OF Al_2O_3/ $AlPO_4$ – EPOXY COMPOSITE COATINGS

B. D. Craig and L. F. Francis
Department of Chemical Engineering and Materials Science
University of Minnesota

ABSTRACT

Ceramic-polymer composite coatings were prepared by fabricating porous ceramic coatings and then infiltrating with an epoxy resin and curing. The process leads to a coating microstructure consisting of interpenetrating ceramic and polymer phases. Porous ceramic coatings were prepared from aqueous dispersions of alumina particles (~0.3 μm average diameter) with phosphoric acid added. The dispersions were coated onto glass substrates using a doctor blade. Coatings were heated to temperatures between 200° and 500°C and reaction between alumina and phosphoric acid formed aluminum phosphate which served a binder. The pore content of the coating decreased from 50 to 25% as the relative amount of phosphoric acid increased from 0.0106:1 to 0.106:1 (H_3PO_4:Al_2O_3 weight ratio). The change in pore content is related to the formation of the lower density aluminum phosphate reaction product. X-ray diffraction, SEM and thermal analysis were used to characterize reactions and microstructure development. Composite coatings were prepared by infiltrating a cycloaliphatic epoxy resin into the porous ceramic coating. A combination of UV exposure and heating (~150°C) was used for curing. Coatings ranged from 10 to 100 μm in thickness and were completely filled with epoxy. A series of composite coatings with varying ceramic content (50 - 75 vol%) was prepared by changing the conditions used to make the ceramic coating. Another series was prepared with a constant ceramic content, but with varying amounts of polyol added to the epoxy to change its hardness. The microhardness and wear properties of the composite coatings were characterized by indentation and Tabor abrasion measurements, respectively. Wear resistance and hardness increased with the phosphoric acid content and temperature used in the fabrication of the ceramic phase. A sharp decrease in wear resistance with decreasing hardness of the epoxy phase indicated that the wear mechanism involved penetration of abrasive particles from the Tabor wheel into the coating followed by fracture. Comparison with coatings prepared with dispersed ceramic particles in an epoxy matrix showed that the interconnectivity of the ceramic phase was essential for achieving good wear resistance.

CONTROLLED MICROSTRUCTURES IN Ni/NiO COMPOSITES

Doris Clayton and D. S. Wilkinson
McMaster University
1280 Main Street West
Hamilton, Ontario L8S 4L7, Canada

ABSTRACT

The processing of Ni/NiO composites with controlled microstructures is currently under investigation. A porous ceramic preform is first prepared by tape casting, which is subsequently infiltrated with liquid metal. The porosity is controlled by the addition of fugitive agents to the ceramic slurry which are eliminated on heating. Nylon fibres of varying aspect ratios as well as polymer spheres can be used to tailor the size, shape, and quantity of porosity and hence the nickel phase of the composite. Of particular interest is the deformation behaviour of the composites which will be investigated with both characterization and fracture testing.

PROCESSING AND MECHANICAL PERFORMANCE OF MULLITE-Al_2O_3 FUNCTIONALLY GRADED CERAMICS

J. F. Bartolome, J. Requena, and J. S. Moya
Instituto de Ciencia de Materiales de Madrid (ICMM)
CSIC, Cantoblanco
28049 Madrid, Spain

ABSTRACT

Mullite/Al_2O_3 Functionally Graded Ceramics (FGC) have been obtained by sequential slip casting and subsequently firing at 1650°C in air. Functionally graded layered dense compacts with 99% density, with five different layers with a composition ranging from 100% mullite to 80% mullite/20% alumina and with thickness of 500 μm were obtained. Taking into account the thermal expansion coefficient and Young Modulus of each layer, the thermal stress distribution into FGC was determined by Finite Element Method (FEM). The bending fatigue test under cyclic loads has been carried out with both monolithic (80% mullite/20% alumina) and mullite/Al_2O_3 FGC. It has been found that the samples show a linear decrease in bending strength with the number of applied cycles of loading, being the slope of the monolithic 3 times the FGC one. It was found that in the FGC the residual stresses caused by thermal expansion mismatch between the layers are able to suppress the crack propagation due to residual compressive stress field around the crack tip. This fact can explain the better fatigue behavior of the FGC vs. the monolithic materials under low loads.

MECHANICAL AND THERMAL PROPERTIES OF LOW TEMPERATURE SINTERED AlN

Mitsuo Kasori, Hiroyasu Sumino, Akihiro Horiguchi, and Fumio Ueno
Research &Development Center, Toshiba Corp.
Komukai Toshiba-cho, Kawasaki, Kanagawa 210-8582, Japan

ABSTRACT

The effect of additives on low temperature sintering of AlN has been investigated. Simultaneous addition of Y_2O_3 and CaO lowered the densification temperature to a greater extent than separate Y_2O_3 or CaO addition. The most effective Y_2O_3/CaO ratio for densification temperature lowering was about 3/1 by weight. YF_3 was also effective in realizing low temperature sintering, but deteriorated the strength. The thermal conductivity was 110–150 W/(m·K) after 2 h–6 h sintering at 1600°C with Y_2O_3-CaO-LaB_6-W additives. The four-point bending strength was around 330–340 Mpa.

INTRODUCTION

Theoretical thermal conductivity of AlN is expected to be 319 W/(m·K) at room temperature[(1)]. However, the actual thermal conductivity of AlN ceramics is in the range of 70–270 W/(m·K) [(2)]. The highest value is almost the same as that of AlN single crystal, which is 285 W/(m·K) [(3)]. Development of high purity fine AlN powders and improvement of the sintering process contributed to this progress. AlN is an electrical insulator, having its resistance higher than 10 Ω^{13}·cm. Therefore, it is obvious that the heat carrier in this compound is phonons. Impurities or lattice imperfections deteriorate the thermal conductivity because of phonon scattering. Pure AlN easily reacts with water vapor in the atmosphere, and an oxide layer is formed at the surface[(4)]. Consequently, commercially available AlN powders contain around 1wt% oxygen. The oxygen atoms dissolve into the AlN lattice during sintering, and occupy the nitrogen sites, as shown by equation (1).

$$Al_2O_3 \xrightarrow{AlN} 3O_N + 2Al_{Al} + V_{Al} \qquad (1)$$

Substitution of three oxygen atoms for the nitrogen sites creates one Al vacancy for charge compensation, and deteriorates the thermal conductivity(1). Komeya *et al.* found alkaline-earth or rare-earth oxides were effective as additives for liquid phase sintering of AlN(5). The additives react with the aluminum oxide impurity in AlN powders during sintering, and the created aluminum-oxide containing phase generates a liquid phase. This liquid phase facilitates sintering. Shinozaki *et al.* found alkaline-earth and rare-earth additives were effective not only for sintering but also as an oxygen getter(6), that is, the additives prevent diffusion of oxygen into the AlN lattice and fix oxygen as a grain boundary phase.

The thermal expansion coefficient of AlN is 4.5×10^{-6} from room temperature to 300°C. This value is close to that of Si. Further, the mechanical properties of AlN are very close to alumina ceramics. These characteristics make AlN an excellent heat dissipating substrate. However, the higher cost compared to alumina has retarded rapid expansion of its application. One possibility to reduce the cost is sintering temperature lowering. The AlN ceramics are generally sintered at around 1800°C. If this sintering temperature could be lowered by 200°C, inexpensive furnace materials would be applied or a power-saving continuous sintering system could be introduced. In addition to the presence additives, particle size of the starting AlN powder also affects the densification temperature. Motoki *et al.* have achieved full densification at 1400°C for 96 h the YF_3 additive and ultra fine AlN powder which has an average diameter of about 60 nm(7). It is plausible that application of a finer powder is effective for densification temperature lowering, because of their larger surface energy. However, there are a lot of difficulties in dealing with ultra fine powders, particularly their poor packing density, shrinkage miss-match between the electrical conductor and the AlN matrix in the PKG use, and high oxygen concentration *etc.*

In this paper, novel additive combinations for low temperature sintering of widely-used AlN powder will be investigated.

2. EXPERIMENTAL PROCEDURE

The raw material of AlN was two types of carbo-thermal powders (Tokuyama F-type or Dow chemical XUS-35548, both having an average grain size of about 1.5 μm, and oxygen concentration of about 1 wt%). Additives, Y_2O_3, YF_3(8), $CaCO_3$, LaB_6(9) and WO_3(10), all of the reagent grade, were used singly or in combination.

The mixed powders were wet ball-milled. After 5 wt% acrylic binder (SERUNA-SG919, Tyukyo Oil) had been added, the powders were molded into 30 mm by 30 mm by around 8 mm compacts under a uniaxial pressure of 50 MPa. They were then de-waxed by heating to 700°C in nitrogen gas. The obtained mixed-powder compacts were sintered under atmospheric pressure in nitrogen gas. The temperature was raised at a rate of 1000°C /h to 1600–1700°C, and was then kept constant 2–100 h, in most cases for 2 or 6 h. Then the samples were allowed to cool in the furnace to room temperature with a rate around 700°C /h.

Powder X-ray diffraction was measured using Cu-Kα radiation (RU-200, Rigaku) after pulverizing the solid sample. The microstructure was observed using transmission electron microscopy (TEM) and energy-dispersive X-ray spectroscopy (EDS). Thermal diffusivity at 23±4°C was measured by the laser-flash method (TC-3000, Sinku Riko), in which a ruby laser (SLG-2006B, (wavelength, 590 nm, flash energy 10 J) Nihon-kousyuha) was the flash light source and an InSb cell (IV100C4, Fujitsu) was the infrared detector. Disk specimens for thermal diffusivity measurements having a diameter of 10 mm and a thickness of around 3.5 mm were prepared by cutting the sintered samples, and then coated with a thin gold film on the surface to prevent the flash penetrating into the ceramics. A carbon black film was applied over the gold film to enhance the absorption of the flash energy. Corrections on thermal conductivity regarding the flash duration time and flash inhomogeneity were carried out according to the Japanese Industrial Standard (JIS-R 1611). The thermal conductivity was calculated from the thermal diffusivity, the specific heat measured by differential scanning calorimetry, and the density. The four-point bending strength was tested with 3mm by 4mm by 40mm rectangular bars, using a 30-mm support span and a crosshead speed of 0.5mm/s.

3. RESULS AND DISCUSSION

3-1. LOW TEMPERATURE SINTERING WITH Y_2O_3 AND CaO ADDITION

We have found that simultaneous addition of Y_2O_3 and CaO lowers the densification temperature. As shown in Fig.1, the highest density for 1600°C sintering was obtained at CaO / (Y_2O_3+CaO)=0.25, in other words Y_2O_3/CaO was 3/1 in weight. The total content of (Y_2O_3+CaO) was fixed at 3 wt% and the sintering time was 2 h for all cases. The theoretical density of AlN is 3.26 g/cm^3. Actual density after full densification was dependent on the density of the grain boundary phases. In the case of sintering at 1700°C and 1800°C, almost complete densification was achieved except when Y_2O_3 was added alone.

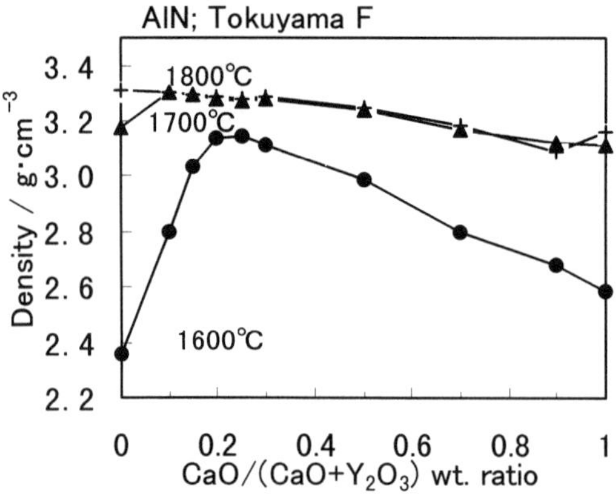

Fig.1. Relationship between density and CaO−Y_2O_3 weight ratio

It has been considered that AlN densification takes place by liquid phase sintering. In the case of Y_2O_3-CaO simultaneous addition, we have to consider the liquids which can be formed in the CaO-Y_2O_3-Al_2O_3 system. The lowest liquidus temperature in this system is found on the CaO-Al_2O_3 binary composition, being around 1360°C(11). Y_2O_3 addition to the CaO-Al_2O_3 eutectic raises the liquidus temperature. Densification temperature lowering by CaO-Y_2O_3 simutaneous addition indicates the sintering is controlled not only by the liquidus temperature but also by the solubility of AlN in the liquid.(12)

3-2. COMPARISON OF AlN SINTERABILITY AND BENDING STRENGTH WITH VARIOUS KINDS OF ADDITIVES

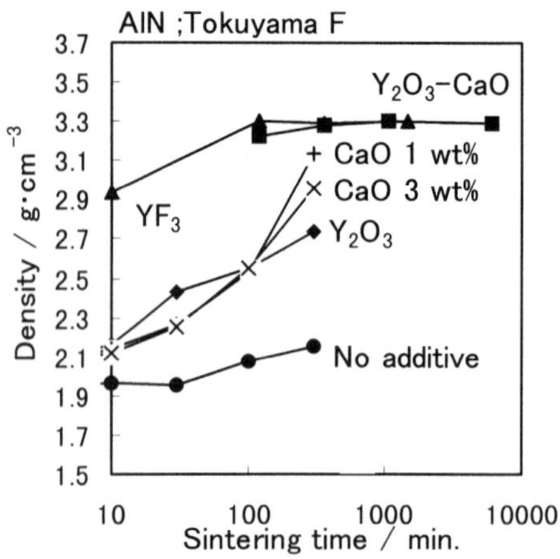

Fig.2. Density vs. Sintering time with various kinds of additives

Sintering at 1600°C for 10 to 6000 minutes have been performed to compare their AlN sinterability with various kinds of additives. The results are shown in Fig. 2. Addition of 3 wt% YF_3 and 3 wt% of Y_2O_3+CaO(3/1) was most effective for densification, followed by 1 wt% CaO, 3 wt% CaO and 3 wt% Y_2O_3. Judging from the XRD studies and weight changes during the sintering, it is considered that YF_3 reacts with Al_2O_3 impurity in AlN powder generating Y-Al-O phases, as the following equations show(8).

$$3YF_3+Al_2O_3 \rightarrow 3YOF+2AlF_3\uparrow \quad (2)$$
$$3YOF+3Al_2O_3 \rightarrow Y_3Al_5O_{12}+AlF_3\uparrow \quad (3)$$
$$3YOF+2Y_3Al_5O_{12} \rightarrow 9YAlO_3+AlF_3\uparrow \quad (4)$$
$$6YOF+3Y_4Al_2O_9 \rightarrow 5Y_2O_3+2AlF_3\uparrow \quad (5)$$

The four-point bending strength of the AlN ceramics with various additives are listed in Table I. To compare the strength in the fully densified state, the test pieces were sintered at 1800°C for 2 h for Y_2O_3 and CaO additives, and at 1600°C for 6 h for YF_3 and Y_2O_3-CaO(3/1) additives. As shown in the table, YF_3-added AlN ceramics has the lowest strength among the samples. Although the the amount of residual fluorine in YF_3-added ceramics was less than 500 ppm after sintering, the trace amount of fluorine might have deteriorated the grain-grain bond. The strength of the sample including 3 wt% CaO was weaker than that including 1 wt% CaO. The strength difference between the samples with 3 wt% added Y_2O_3, 1 wt% CaO and Y_2O_3-CaO(3/1) was not significant; the strength of the three samples was within the range of 280–300 MPa.

TableI. Four-point bending strength

	Y_2O_3	CaO	Y_2O_3/CaO=3/1	CaO	YF_3
	3wt%	1wt%	3wt%	3wt%	3wt%
	1800℃-2h	1800℃-2h	1600℃-6h	1800℃-2h	1600℃-6h
av. /MPa	302	292	284	219	186
So /MPa	326	315	304	236	195
m	5.7	5.4	6.4	5.6	9.4
n	25	25	19	25	20

av.; average, So; Scale parameter, m; Weibull modulus, n; Number of samples

3-3. LOW TEMPERATURE SINTERED AlN WITH Y_2O_3-CaO-LaB_6-W ADDITIVES

The effect of the added amount of (Y_2O_3-CaO) on sinterability and thermal conductivity has been investigated. The AlN powder used in this study was XUS35548. The (Y_2O_3/CaO) ratio was fixed at 3/1 in weight. Further, LaB_6 and W were added to facilitate sintering and blackening. Their content were fixed at 0.25 wt% and 0.3 wt% (0.38 Wt% WO_3), respectively. Sintering has been performed at 1600°C for 2 or 6 h. In the 2 h sintering cases shown in Fig. 3, around 4 wt% addition of (Y_2O_3-CaO) was optimal for lower temperature densification. Fig. 4 shows that an addition of around 3 wt% (Y_2O_3-CaO) was effective to optimize thermal conductivity. When the cooling rate was slowed down to 2°C/minute after keeping at 1600°C for 6 h, the thermal conductivity rose from 145 W/(m·K), which is obtained by the natural cooling, to 151 W/(m·K).

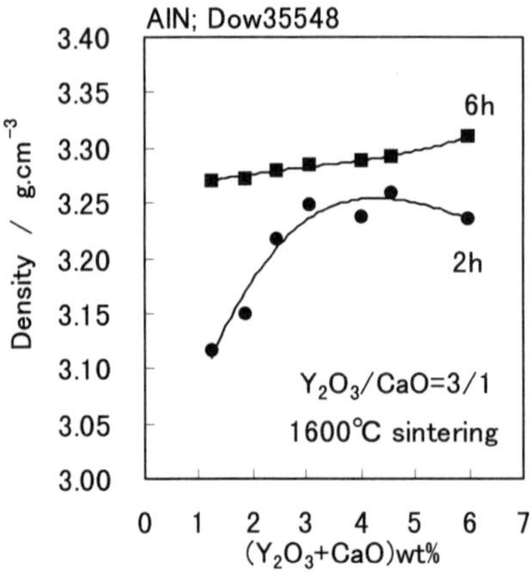

Fig.3. Relationship between density and amount of (Y_2O_3+CaO) addition

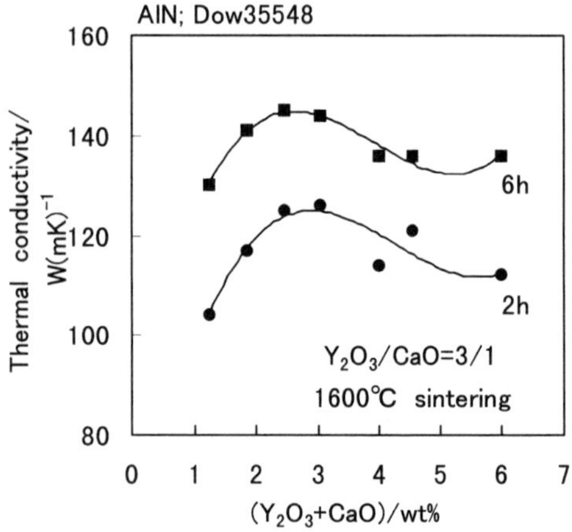

Fig.4. Relationship between thermal conductivity and amount of (Y_2O_3+CaO) addition

Table II. Secondary phases in Y_2O_3 –CaO–LaB_6–W additive system detected by XRD

Y_2O_3+CaO wt%	Secondary phases by XRD
1.24	$CaYAl_3O_7$? >$Y_3Al_5O_{12}$
1.85	$CaYAl_3O_7$? >$Y_3Al_5O_{12}$
2.45	$CaYAl_3O_7$?, $YAlO_3$, $CaYAlO_4$, >W_2B
3.05	$CaYAl_3O_7$?, $YAlO_3$, $CaYAlO_4$ >$Y_4Al_2O_7$, W_2B
4.00	$Y_4Al_2O_9$, $CaYAl_3O_7$?, $CaYAlO_4$>W_2B
4.54	$Y_4Al_2O_9$, $CaYAl_3O_7$?, $CaYAOl_4$>W_2B
5.99	$Y_4Al_2O_9$, $CaYAOl_4$>$CaYAl_3O_7$?, W_2B

Y_2O_3/CaO wt.ratio was 3/1, Sintered at 1600°C–6h

The secondary phases were examined by XRD. As listed in Table II, only when a small amount of Y_2O_3-CaO was added, a Ca-Y-Al-O compound ($CaYAl_3O_7$?) and $Y_3Al_5O_{12}$ were identified. As the amount of (Y_2O_3-CaO) addition was increased, phases containing less Al_2O_3, namely $CaYAlO_4$, $YAlO_3$ and $Y_4Al_2O_9$, have become detectable. The Al_2O_3 in the aluminate phases must come from the Al_2O_3 impurity of the AlN powder. It was considered that W_2B was formed as a result of a reaction between WO_3 and LaB_6. No La containing phase was identified by the XRD.

It should noted that, when 1.24 and 5.89 wt% (Y_2O_3-CaO) was added, the four-point bending strength of ceramics was 329 and 340 MPa in average, respectively. This means that there is no significant relationship between the strength and the (Y_2O_3-CaO) content.

Figure 5 shows a TEM photograph and EDX spectra of the ceramics to which 4.00 wt% (Y_2O_3-CaO) was added. At point B2, an AlN grain was formed. At point B5, O, Al, Y, Ca and La were detected. La seemed dissolved in the Ca-Y-Al-O or Y-Al-O phase. W peaks were found at point B6. From the

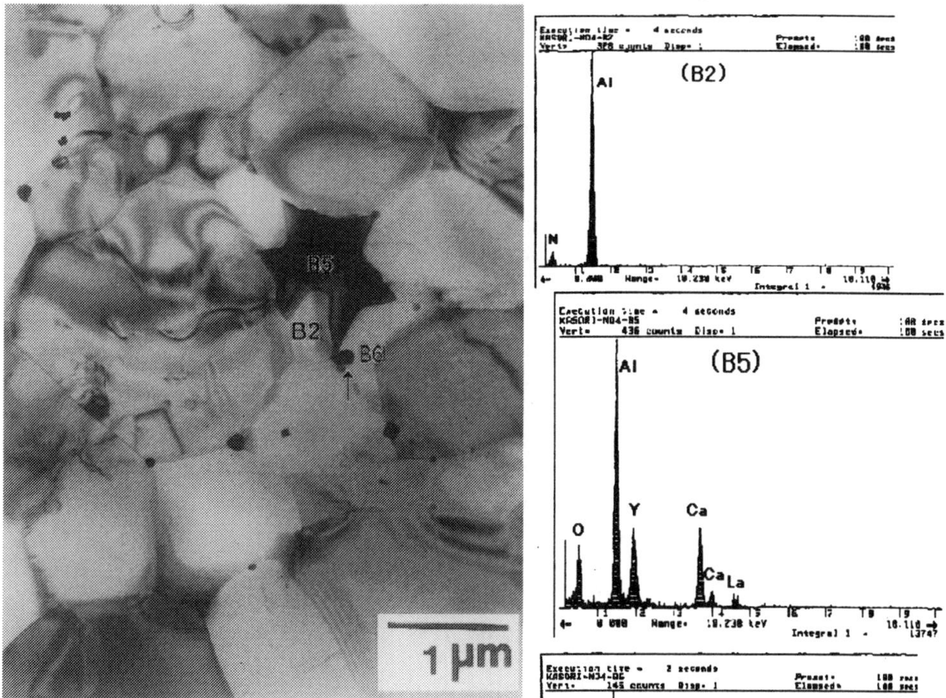

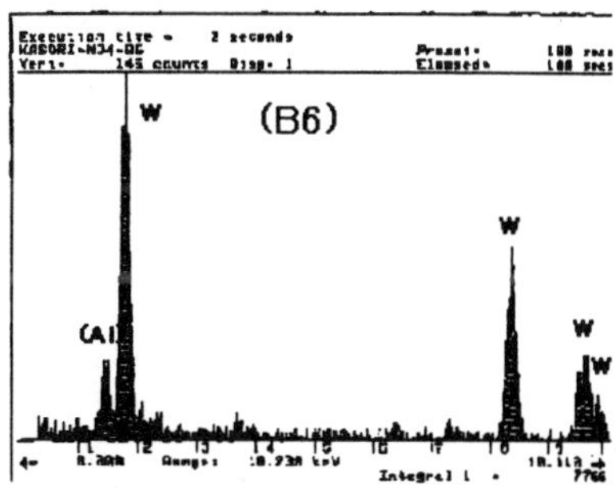

Fig.5. TEM photograph and EDX spectra of Y_2O_3-CaO-LaB_6-W added AlN ceramics

XRD results, W_2B seems to exist at B6. The Al peak at B6 seems to originate from surrounding AlN.

4. CONCLUSIONS

The influence of additives on densification temperature, thermal conductivity and mechanical properties of AlN ceramics has been investigated. The results are:

1.Simultaneous addition of Y_2O_3-CaO lowered the densification temperature. The most effective ratio of Y_2O_3/CaO was 3/1 by weight.
2.YF_3 was useful for low temperature sintering but deteriorated the strength.
3.Addition of 1 wt% CaO resulted in higher strength than addition of 3wt% CaO.
4.The thermal conductivity reached the range of 110–150 W/(m·K) after 1600°C, 2 h–6 h sintering with Y_2O_3-CaO-LaB_6-W additives. The four-point bending strength was around 330–340 MPa.
5.La seemed to be dissolved in the Ca-Y-Al-O or Y-Al-O phase.

This work has been entrusted by New Energy and Industrial Technology Development Organization (NEDO).

REFERENCES

[1]G. A. Slack "Nonmetallic Cristals with High Thermal Conductivity," J. Phys. Chem. Solids, 34 (1973) 321–335

[2]F. Ueno, A. Horiguchi "Grain Boundary Phase Elimination and Microstructure of Aluminum Nitride" Euro-Ceramics, (1989) 1.383–1.387

[3]G. A. Slack, T. F. Mcnelly "AlN Single Crystals," J. Crystal Growth, 42 (1977) 560-563

[4]E. Ponthieu, P. Grange, B. Delmon "Proposal of a Composition Model for Commercial AlN Powders," J. Europ. Ceram. Soc., 8(1991)233–241

[5]K. Komeya, H. Inoue, A. Tsuge "Effect of Various Additives on Sintering of Aluminum Nitride," Yogyo-Kyokai-shi(in Japanese), 89[6] 58–64 (1981)

[6]K. Shinozaki, K. Anzai, T. Takano, A. Tsuge, K. Komeya "Sintering Behavior and Thermal Conductivity of Y_2O_3-Added AlN Ceramic by Pressureless Sintering," p43 in Proceedings of 22nd Symposium on the Basic Science of Ceramics Society of Japan, Tokyo, Japan, 1984

[7] S. Motoki, S. Yokota, K. Kawasaki, K. Baba, N. Shohata "Higher Thermal Conductivity of AlN and Lower Temperature Sintering of AlN Ultrafine Powder Synthesized by Induction Thermal Plasma Process," p16 in Proceedings of Annual Meeting of Ceramic Society of Japan, Tokyo, Japan, 1991

[8] M. Kasori ;to be published

[9] T. A. Guiton ; Private Communication

[10]M. Kasori, F. Ueno, A. Tsuge "Effects of Transition-Metal Additions on the Properties of AlN," J. Am. Ceram. Soc., 77 [8] 1991–2000 (1994)

[11]Yu. P. Udalov, Z. S. Appen, V. V. Parshina; Zh. Neorg. Khim., 24[10]2786–92(1979); Russ. J. Inorg. Chem. (Engl. Transl), 24[10]1549-53(1979)

[12]Y. Yamasita, K. Shinozaki, N. Mizutani "The temperature Measurement of the Liquid Phase Formation in AlN-Y_2O_3(CO) system Using a Self-made High Temperature DTA," p601 in Proceedings of Annual Meeting of Ceramic Society of Japan, Tokyo, Japan 1993

CONTINUOUS AND LOW TEMPERATURE SINTERING ADDITIVE FOR ALUMINUM NITRIDE

A. Horiguchi, M. Kasori, H. Sumino, and F. Ueno
Materials and Devices Research Laboratories
Toshiba Corp.
Kawasaki, Kanagawa 210, Japan

Abstract

Due to the high thermal conductivity of aluminum nitride, its application to circuit substrates is increasing. Yttria and calcia were used as sintering additives of aluminum nitride. When either yttria or calcia was added alone, a densification temperature of about 1800°C was needed. However, the temperature decreased to about 1600°C when both additives were added at the same time. In addition, when a small amount of boron was added, the melting point of the grain boundary phase decreased by a further 50° to 100°C, reducing the densification temperature to about 1500°C. As a result, it was confirmed that a dense AlN sintered body having thermal conductivity of 150 W/mK is obtainable by sintering at 1600°C. Moreover, this low temperature sintering of a low cost manufacturing process is attractive to industry. AlN sintered sheets which had the above-mentioned characteristics were obtained by continuous low temperature sintering. This work has been entrusted by New Energy and Industrial Technology Development Organization (NEDO).

CONTROLLING GRAIN MORPHOLOGY FOR Si_3N_4 AND SiC MICROSTRUCTURES

Michael J. Hoffmann
University of Karlsruhe
Germany

ABSTRACT

Si_3N_4 and SiC ceramics can be densified by liquid phase sintering using oxides and nitrides as sintering additives. When the starting powders consist of a metastable or low-temperature modification, sintering is accompanied by a phase transformation. Since both materials exhibit a pronounced grain growth anisotropy, phase transformation could be used as a powerful tool to control the grain morphology by the number and size of the seeds of the stable modification in the starting powder. In case of Si_3N_4 a phase transformation from α-Si_3N_4 to β-Si_3N_4 occurs. It will be shown that the microstructural development is determined by the sintering additives as well as the initial β-content. The analysis is based on experimental results of pressureless as well as gas pressure sintered ceramics and nitrogen supersaturated oxynitride glasses, both containing different rare earth oxides. Grain growth during sintering is correlated to analytical TEM observations which indicate that the growth mechanism depends on the additive composition and can change from interface controlled to diffusion controlled. In order to understand the influence of the additives, the intergranular films have been investigated by HRTEM. Furthermore model experiments were presented to demonstrate the conditions of discontinuous grain growth. The growth anisotropy, which leads to the development of needle-like grains, is related to the crystal structure and the difference in the interfacial energy of prism and basal plane. Liquid-phase sintered SiC show a similar microstructural development as Si_3N_4, but phase transformation goes from β to α-Sic and the elongated grains exhibit a plate-let like morphology. The analysis of the densification mechanism and transformation kinetics show a much faster densification than transformation. SiC microstructures with high aspect ratio grains could only be achieved with β-rich starting powders, small amounts of α-seeds and after prolonged sintering times, ceramics prepared with 100% initial β-powder did not exhibit any transformation to α-SiC and the resulting grain morphologies were equiaxed similar to those with α-SiC starting powders.

TAILORING MICROSTRUCTURE DEVELOPMENT VIA TEMPLATED GRAIN GROWTH

G.L. Messing, M. M. Seabaugh, and S.-H. Hong, Department of Materials Science and Engineering, Materials Research Laboratory, The Pennsylvania State University, University Park, PA 16802

ABSTRACT

Seeding has been used extensively in materials manufacture and synthesis to regulate phase formation, crystallographic orientation, transformation kinetics and microstructure evolution. Only recently has seeding been used to directly influence the design of microstructures. Recent examples of seeded grain growth have opened new possibilities for *in situ* self-texturing and a high degree of architectural control through the number and placement of template particles. By orienting anisometric particles in a matrix, one can template subsequent grain growth processes and achieve highly textured microstructures. Processing schemes to achieve this goal involve the use of directional shear processes such as tape casting and extrusion. In addition, the initial physical environment for template particle growth strongly affects the path of microstructural evolution, final grain size and texture of the final ceramic. In this paper, we focus on three elements of this problem including (1) template particle alignment by shear forming processes and (2) anisotropic grain growth and texture development from the initial template particles in the oriented microstructure. Examples and data from our work on doped mullite and liquid phase sintered alumina are used to illustrate the underlying principles for the processing of textured ceramics by templated grain growth.

INTRODUCTION

The traditional goal of ceramic processing has been the development of dense, fine grained bodies of isotropic, randomly oriented grains. However, many ceramic crystal systems are inherently anisotropic, and processing must be tightly controlled to obtain the desired, fine grained isotropic microstructures. The objective of such processing is to produce dense bodies that exhibit isotropic physical properties, but in many cases, isotropic properties are not essential and represent an unnecessary compromise in performance. However, if a textured microstructure can be achieved which mimics that of the crystal structure, similarly anisotropic (and directionally improved) properties will also be obtained. Such improvements in directional properties could be desirable in conductive, superconductive, dielectric, piezoelectric and magnetic materials, for example

One method of obtaining such highly textured microstructures is templated grain growth (TGG). In templated grain growth, anisotropic template particles are aligned in a randomly oriented, fine-grained matrix. During sintering the larger anisotropic grains grow at the expense of the matrix grains, resulting in highly textured microstructures.

The TGG process is an outgrowth of prior investigations of anisotropic grain growth in sol-gel derived doped alumina and mullite systems. Seeded sol gel derived alpha alumina has been shown to form dense microstructures with extremely fine grain size at low sintering temperatures. When seeded sol-gel derived alumina is doped with 0.15 to 0.40 wt% TiO_2 a fine grained microstructure initially develops, but then platelet-shaped anisotropic grains nucleate and rapidly consume the finer matrix grains.[1] Doping above 0.6 wt% results in exaggerated grain growth. Anisotropic grain growth of platelike grains is also observed in oxide derived alpha alumina systems sintered in the presence of an aluminosilicate liquid phase.

In mullite derived from diphasic gels, doping with 3 wt% B_2O_3 resulted in a microstructure of highly anisotropic interlocking needle-shaped grains.[2] A similar microstructure can be obtained from mullite doped with 3 wt% TiO_2.[3] In both the mullite and alumina systems, sintering initially produces a dense, fine-grained microstructure, then anisotropic grains nucleate and consume the matrix. The growth of the anisotropic grains is dramatic in all three systems, due to the fine grain size of the matrices.[4]

The preceding studies showed that appropriate doping would develop microstructures with grains of controlled morphology, but it was impossible to texture by doping alone, since the number of nuclei and their orientation remained uncontrolled. In TGG, morphology is controlled by doping, and texture development is controlled by introducing a population of well-oriented anisotropic seed particles which grow to dominate the microstructure.

Three systems were investigated, sol-gel derived mullite doped with TiO_2, sol-gel derived alumina with $Ca(NO_3)_2$ and SiO_2 liquid phase formers, and mixed submicron powders of alpha alumina and $CaCO_3$ and SiO_2 liquid phase formers. Initial work in the sol-gel derived systems sought to maximize the size difference between matrix and template particles. Later work in alpha-alumina derived systems was initiated to increase sample size and to simplify processing.

Experimental Procedure

Sol-Gel Derived Mullite[5]

Mullite was prepared from diphasic precursors, which consisted of boehmite and colloidal silica. The mullite was doped with a titania with a mean diameter of 0.2 μm consisting of anatase and rutile (10%). Mullite whisker seeds were made from a 3 wt% titania-doped mullite by the chemical leaching method. To obtain mullite whiskers, sintered pellets of TiO_2-doped mullite were (1) dissolved in hydrofluoric acid for 12h, (2) washed with distilled water to remove the HF, (3) treated with an ammonium hydroxide solution to remove adsorbed HF, and (4) rinsed with distilled water to remove ammonia.

After heterocoagulating the boehmite, silica, and titania at pH 3, they were combined with a 25 wt% solution of 1:1 PVA to glycerol in water. The slurry viscosity was adjusted by evaporating the water. Slurries were cast at a speed of 5 cm/sec with a double doctor blade of between 500 and 1500 μm height. Tapes were dried in air, cut into 10 by 15 mm, then calcined at 500°C for 2h. Samples were sintered between 1450°C and 1650°C for times between 1 min and 300 min. Microstructural observation were performed using SEM in the fired surfaces of the mullite tapes and polished sections either parallel or perpendicular to the casting direction.

Sol-Gel Derived Liquid Phase Sintered Alumina[6]

The samples in this study were produced from slurries derived from seeded boehmite sols. Colloidal boehmite sols (15 wt%) were prepared from commercially available boehmite powder. Seed crystals of alpha alumina were prepared by dispersing high purity powders in water at pH 3 and centrifuged to remove particles larger than 0.1 μm. 1.5 wt% seeds (dwb of the boehmite) was added to the boehmite to provide nucleation sites for the γ-α-Al_2O_3 transformation.

Liquid phase formers were added to the sol to provide a $CaO:SiO_2$ ratio of 1:1. CaO was added to the sol in the form of $Ca(NO_3)_2$. SiO_2 was added in the form of a commercially available colloidal dispersion. Liquid phase formers were added in 0.5, 1.0, and 5.0% (dwb boehmite) amounts. 4 wt% (dwb) of hexagonal α-Al_2O_3 platelets (5-10 μm) were added to the mixed sols to provide substrates for anisotropic grain growth. Finally, glycerol was added as a plasticizer to aid tape removal after casting.

After the components were mixed, the viscosity was adjusted by evaporation of the solvent. The tapes were cast on a glass surface and allowed to dry, then cut and laminated at 1000°C for 15 minutes. The laminated samples were fired at temperatures between 1200°C and 1600°C, for times ranging between 0.1 minute to 8 hours. Fired samples were cut and polished, etched with dilute HF to remove the glassy intergranular phase, then examined by SEM.

Mixed Oxide Liquid Phase Sintered Alumina [7,8]

Submicron α-Al_2O_3 and $CaCO_3$ and colloidal SiO_2 were dispersed in water using a polyelectrolyte dispersant. Hexagonal platelets of α-Al_2O_3 were added to the dispersion, which was then sonicated to aid dispersion and allowed to stir overnight. After the mixture was well dispersed, polyvinyl alcohol and glycerol were added as a binder and plasticizer, respectively, Butanol was added as a surfactant and defoamer. The slurry was tape cast at 200 s^{-1} and 500 μm thickness and dried at 80% relative humidity for two days. The tape was then cut into sections 18 mm wide and 24 mm long, then stacked in groups of six and laminated at 30 MPa for three minutes. Samples were calcined at 1000°C for 3 hours, then sintered at temperatures between 1250 to 1650°C for 30 minutes.

RESULTS AND DISCUSSION

All three systems investigated for TGG developed strong textures, as shown in Figure 1, indicating both the promise of this technique for other systems. In addition to demonstrating the effectiveness of TGG as a processing method, these results also provide insight into the important parameters of TGG. The orientation of the template particles in all the samples shows the utility of shear

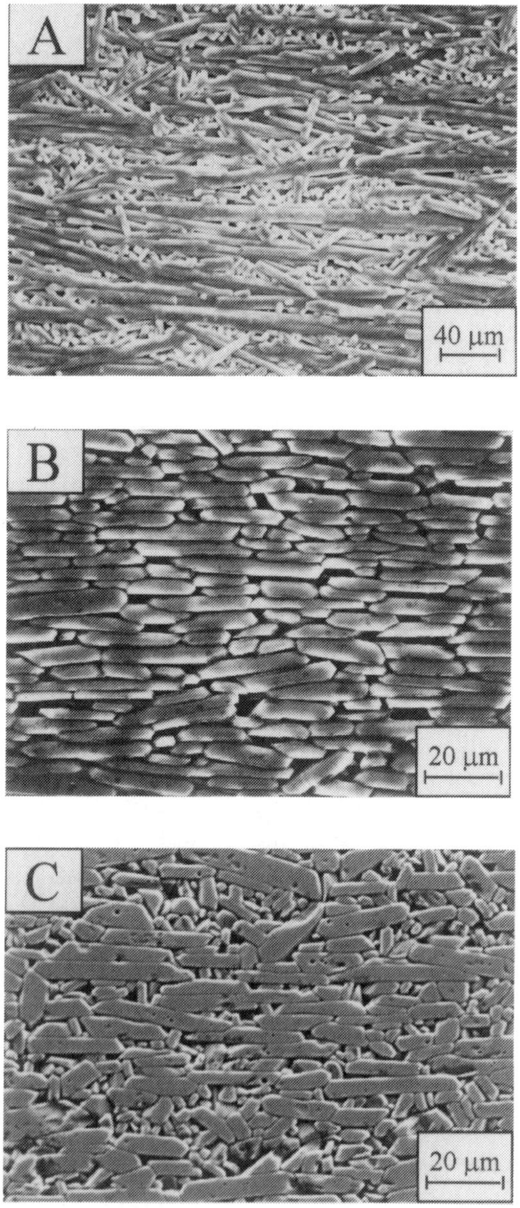

Textured Microstructures of A) Sol-Gel Derived Mullite[5], B) Sol-Gel Derived Liquid Phase Sintered Alumina[6] , and Mixed Oxide Liquid Phase Sintered Alumina[7]

forming methods in aligning both platelet- and needle-shaped templates. The strong textures in all three systems also indicates the growth of template particles dominate the microstructure development, which can be explained in terms of Ostwald ripening theory by their initial size advantage and the high driving force for their growth supplied by the matrix.

Matrix characteristics also play an important role in texture development. The extremely fine matrices of sol-gel derived systems result in a large template to matrix size ratio, which in turn results in a high driving force for template growth. This high driving force is reflected in the extremely textured microstructures that result. However, the oxide-derived alumina system shows that sol-gel processing is not essential to the process, and sufficiently high driving forces for anisotropic template growth and texture development can be obtained from powder processing. Two important differences are apparent between the two alumina samples, the size of the matrix grains and the degree of faceting. Samples grown from sol-gel derived powders typically show less coarsening of the matrix and a higher degree of faceting of the template grains.

From examination of these and other microstructures, the growth of the template particles appears to be a two step process. In the first step, growth in the fast-growing direction proceeds rapidly, while prominent facets develop in the slower-growing directions. Eventually, template particles impinge upon one another, and growth in the fast direction slows considerably. It is also possible in systems with few templates or a relatively coarse matrix that matrix grains could cause impingement. In the second step, after impingement, growth occurs predominantly in the slow-growing direction, until matrix coarsening limits the driving force for any further growth.

SUMMARY

Templated grain growth is a powerful approach to regulate microstructural development and to obtain unique microstructures. We have successfully adapted the process for producing textured $Sr_2Nb_2O_7$ and $Bi_4Ti_3O_{12}$, as well[9,10]. Although TGG is a seemingly simple process, the need for chemically compatible seed crystals, the availability of such materials in anisotropic form, and the need to learn how to induce anisotropic growth are sometimes formidable problems. Some materials, especially cubic systems are intrinsically resistant to this process; especially is the goal is texture development. Clearly, we are just scratching the surface. With a concerted effort TGG promises to add significantly to the tools available to the processing community for tailoring the microstructure/property relations in ceramic materials.

REFERENCES:

1. D. Horn and G. L. Messing, "Anisotropic Grain Growth in TiO_2-doped Alumina", J. Mat'l Sci and Eng. **A195** 169-78 (1995).
2. S-H. Hong, W. Cermignani and G. L. Messing, "Anisotropic Grain Growth in Seeded and B_2O_3-Diphasic Mullite Gels", J. European Ceramic Society, **16** 133-141 (1996).

3. S-H Hong and G. L. Messing, "Anisotropic Grain Growth in Diphasic Gel-Derived Titania-doped Mullite ", J. Am . Ceram. Soc. **81** [5] 1269-77 (1998).
4. M. Seabaugh, S. H. Hong and G. L. Messing, "Processing of Textured Ceramics by Templated Grain Growth" in Ceramic Microstructure: Control at the Atomic Level, eds. A. P. Tomsia and A. Glaeser, Plenum Press, New York, 303-310.
5. S-H Hong and G. L. Messing, "Development of Textured Mullite by Templated Grain Growth", in press J. Am. Ceram. Soc.
6. M. M. Seabaugh, I. H. Kerscht and G. L. Messing, "Texture Development by Templated Grain Growth in Liquid Phase Sintered α-Alumina" J.Am Cer. Soc. **80** [5] 1181-88 (1997).
7. M. M. Seabaugh and G. L. Messing, "Modeling Texture Development by Templated Grain Growth in Liquid Phase Sintered α-Alumina"
8. E. Suvaci, M. M. Seabaugh and G. L. Messing, Processing of Textured Reaction Based Alumina Ceramics by Tempalted Grain Growth" submitted to J. Am. Ceram. Soc for publication
9. B. Brahmaroutu, G. L. Messing, S. Trolier-McKinstry and U. Selvaraj, "Templated Grain Growth of Textured $Sr_2Nb_2O_7$ ", edited by B. Kulwicki , IEEE (1996)
10. J. Horn, S. C. Zhang, U. Selvaraj, G. L. Messing, and S. Trolier-McKinstry, "Templated Grain Growth of Textured $Bi_4Ti_3O_{12}$" accepted J. Am. Ceram Soc.

CHARACTERIZATION OF MICROSTRUCTURAL EVOLUTION AND TEXTURE DEVELOPMENT IN TEMPLATED LIQUID PHASE SINTERED ALUMINA

Matthew M. Seabaugh, Gary L. Messing, Mark D. Vaudin,* and James P. Cline*
Materials Research Laboratory
The Pennsylvania State University
University Park, PA 16802

*NIST
Gaithersburg, MD 20899

ABSTRACT

Liquid phase sintered α-Al_2O_3 has been demonstrated to develop highly textured microstructures via anisotropic grain growth of oriented template particles. Using Rietveld x-ray texture characterization techniques, the effects of processing parameters and sintering conditions on texture development have been evaluated, and the kinetics of texture development estimated. The microstructural evolution of these materials has also been characterized. The grain size has been measured parallel and perpendicular to the thickness direction, and compared for various processing and sintering conditions. Growth rates in different crystallographic directions were also measured. Both sets of data were compared with the data obtained from the Rietveld characterization and values from grain growth and crystal growth theories. The importance of interfacial reactions to growth rate in different crystallographic directions is also examined.

SINGLE CRYSTAL-LIKE HIGH TEMPERATURE FERROELECTRICS BY A TEMPLATED GRAIN GROWTH APPROACH

Bhaskar Brahmaroutu, Gary L. Messing, and Susan Trolier-McKinstry
Materials Research Laboratory
The Pennsylvania State University
University Park, PA 16802

ABSTRACT

Grain-oriented polycrystalline mixed metal oxides which exhibit single crystal-like properties can be fabricated by a templated grain growth (TGG) approach. The polycrystalline body can be made to exhibit properties approaching single-crystal properties by tailoring the microstructure texture to enhance crystallographic orientation of the constituent grains. The TGG process relies on the texturing effect of anisotropic grain growth of oriented single-crystal template particles dispersed in a dense, fine-grained matrix. In this presentation, we will discuss the synthesis crystal template particles dispersed in a dense, fine-grained matrix. In this presentation, we will discuss the synthesis of Sr2Nb2O7, template particle orientation, TGG of $Sr_2Nb_2O_7$, and the effect of heterovalent and isovalent doping on the grain growth, densification behavior and the electrical properties. Dielectric data will be presented which corroborates the texture development observed by scanning electron microscopy and x-ray diffraction. The effect of processing variables such as template particle size, concentration, sintering conditions, and dopant type and amount will also be presented.

MICROSTRUCTURE CONTROL OF AN OXIDE SUPERCONDUCTOR BY SOLID STATE CRYSTAL GROWTH

Masanobu Awano, Yasuharu Kodama, Yasumasa Takao, and Yoshitaka Kuwahara
National Industrial Research Institute of Nagoya
1-1, Hirate-cho, kita-ku
Nagoya 462, Japan

ABSTRACT

Highly oriented bulk bodies of HBCO superconductor was synthesized by the stress induced grain growth from the dense sintered body prepared by the spray drying and HIP sintering. Two dimensional selective grain growth of YBCO isotropic submicron grains under uniaxial pressure at elevated temperature below partial melting temperature was promoted by solid state reaction. Significant anisotropic grain growth vertical to x-axis and pressing direction occurred in optimum temperature and pressure condition at around 920 MPa. Microscopic and kinetic analysis indicated the mechanism of grain orientation not as the grain alignment but as the stress induced diffusion process resulting in the grain growth on the grains at the direction of c-axis parallel to compression direction. Superconducting transporting current drastically increased as the result of eliminated weak link at grain boundaries. Magnetic flux pinning centers introduced mainly in the superconducting matrix contributed to the enhancement of superconducting current.

MICROSTRUCTURAL EVOLUTION OF MELT TEXTURED $YBa_2Cu_3O_{7-X}$ DOPED WITH Pt OR CeO_2

Jeffrey D. Reding and Fatih Dogan
Department of Materials Science and Engineering
University of Washington, Roberts Hall, Box 352120
Seattle, WA 98195

ABSTRACT

The effect of Pt and CeO_2 doping on the Y_2BaCuO_5 (211) phase during melt texturing of $YBa_2Cu_3O_{7-x}$ (123) was observed by SEM and weight loss measurements. Both dopants greatly reduced the size of residual 211 inclusions after solidification, relative to the undoped 123. Pt is more effective than CeO_2 to reduce the loss of the liquid phase during the melt texturing process. It is proposed that capillary forces between the 211 particles and the liquid control the amount of the liquid phase in the semi-solid sample.

INTRODUCTION

Melt texturing techniques are widely used in processing of $YBa_2Cu_3O_{7-x}$ (123) superconductors to produce large oriented grains. This minimizes grain boundary weak links and therefore improves the superconducting properties of the material [1].

Fine non-superconducting inclusions, such as Y_2BaCuO_5 (211), greatly improve some superconducting properties, primarily the critical current density, J_c [2-3]. The flux pinning ability of these inclusions is improved as their size is reduced due to the short coherence length (4.5nm) of flux vortices in high T_c superconductors. Undoped melt textured 123 contains relatively coarse 211 inclusions, but small additions of Pt [4-17] or CeO_2 [15-22] significantly reduce the size of residual 211 in solidified 123. Both CeO_2 and Pt reduce the particle size of 211 and change its morphology to a long needle-like shape in the melt-textured microstructure.

It has been proposed that as Pt reacts with YBCO below the peritectic temperature, fine particles of $Ba_4CuPt_2O_9$, and free CuO are produced [8,9], which in turn leads to formation of excess 211 phase. These 211 particles act as nucleation sites for 211 formed during peritectic melting of 123 [10]. It has also been argued that Pt lowers the interfacial energy between the solid and liquid phases during melt texturing, thus reducing the critical nucleation size of 211 and slowing the coarsening of these particles [11-15]. Similar arguments are made for CeO_2 in understanding its role during melt processing of 123. Some researchers suggest that when CeO_2 reacts with 211, fine particles of Y_2O_3 are produced that act as nucleation sites for new 211 [18-19]. It has also been proposed that the interfacial energy between the solid and liquid phases during melt texturing can be reduced by the addition of CeO_2 [17].

An aspect that has not been adequately addressed, but one that yields some useful information in the interpretation of the liquid loss during melt texturing is the capillarity of the liquid phase in the semi-solid melt. An insight into the capillarity can be gained by measuring the amount of liquid lost by the sample during melt texturing and comparing it with the size of 211 particles. In this work, the relationship between the liquid loss and the role of CeO_2 or Pt additions towards the refinement of 211 particles is studied. The samples were prepared either by the melt textured growth method (MTG), or by quenching from the molten state. Capillary forces in the semi-solid melt and microstructural development during solidification are discussed.

EXPERIMENTAL PROCEDURE

$YBa_2Cu_3O_{7-x}$ powders, undoped and doped with 0.5 wt. % Pt or 1 wt. % CeO_2 were provided by Praxair Specialty Ceramics, Woodinville, WA.

The powders were pressed into 10mm diameter, 4g pellets by uniaxial pressing with subsequent isostatic compaction at 100 MPa. The pellets were then heated on a porous 211 substrate at 2°C/min to 1040°C, held for five hours, and either quenched to room temperature by opened furnace cooling, or cooled at 2°C/hr to 950°C for the MTG samples. Sample weights were measured before and after heat treatment to determine the amount of liquid lost for each pellet. The samples were mechanically polished, and etched with a 1% hydrochloric acid solution. Microstructural characterization of polished and etched samples was carried out by SEM and EDS analysis techniques.

RESULTS AND DISCUSSION

The differences in the role of Pt and CeO_2 doping in 123 become evident when considering liquid phase retention during melt texturing. The weight lost by the sample during melt processing is primarily due to loss of the liquid phase, and controlled by the capillary forces present in the partially molten state. Capillary forces can be affected by the wetting between the solid and liquid phases, viscosity of the liquid phase, or separation distance and particle size of the solid particles dispersed in the liquid. Undoped 123 samples show considerable weight-loss during melt processing due to loss of the liquid phase. The samples doped with CeO_2 show relatively less weight loss, while platinum doped samples exhibit no liquid phase depletion (fig. 1).

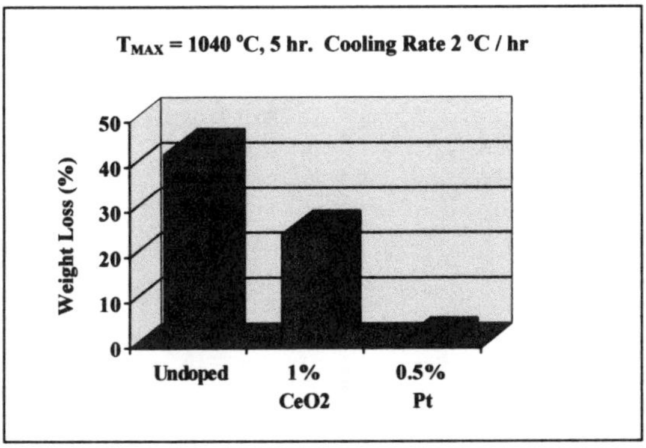

Figure 1. The average weight lost by 123 during melt texturing.

Figure 2 shows the microstructure of undoped melt-textured 123 with large, blocky 211 inclusions. The melt textured samples shown in figure 3 reveal that Pt and CeO_2 are effective in controlling the size and morphology of 211 particles dispersed in the 123 phase. The microstructures of the Pt and CeO_2 doped samples exhibit much smaller and needle-like 211 particles. In addition to 211 inclusions, the microstructure of the CeO_2 doped sample reveals cube shaped, relatively small $BaCeO_3$ inclusions which could act as more effective flux-pinning centers in $YBa_2Cu_3O_{7-x}$.

Figure 2- Melt textured microstructure of undoped $YBa_2Cu_3O_{7-x}$.

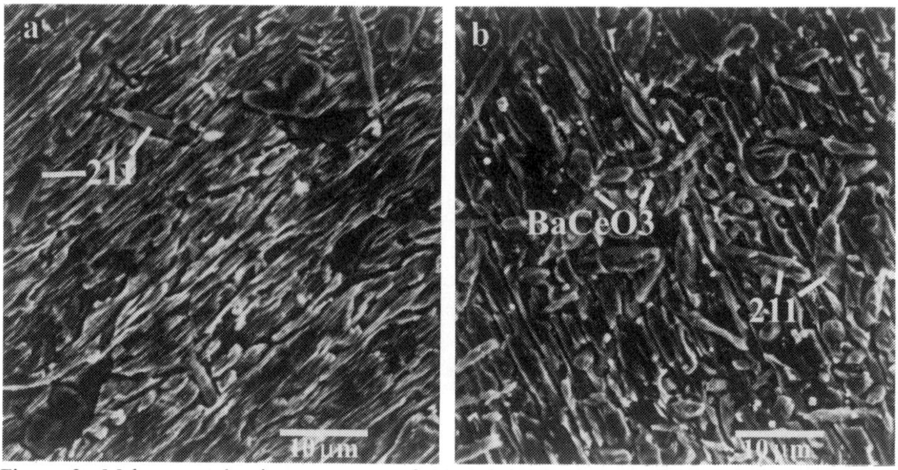

Figure 3- Melt textured microstructures of (a) 0.5 % Pt doped and (b) 1 % CeO_2 doped $YBa_2Cu_3O_{7-x}$. Pt and CeO_2 doped samples show small needle-like Y_2BaCuO_5 inclusions.

Some 211 inclusions that are present in the Pt doped sample (fig. 3a), are due to the reaction of Pt with the liquid phase to form $Ba_4CuPt_2O_9$ [9], so that the overall stoichiometry shifts towards the formation of the 211 phase in 123. Similarly, the reaction between CeO_2 and 123 leads to the formation of $BaCeO_3$ and 211 particles dispersed within the 123 phase. The final stoichiometry could be more precisely controlled by the addition of these reaction products, $BaCeO_3$ or $Ba_4CuPt_2O_9$, instead of CeO_2, PtO_2 or Pt [9, 14, 17].

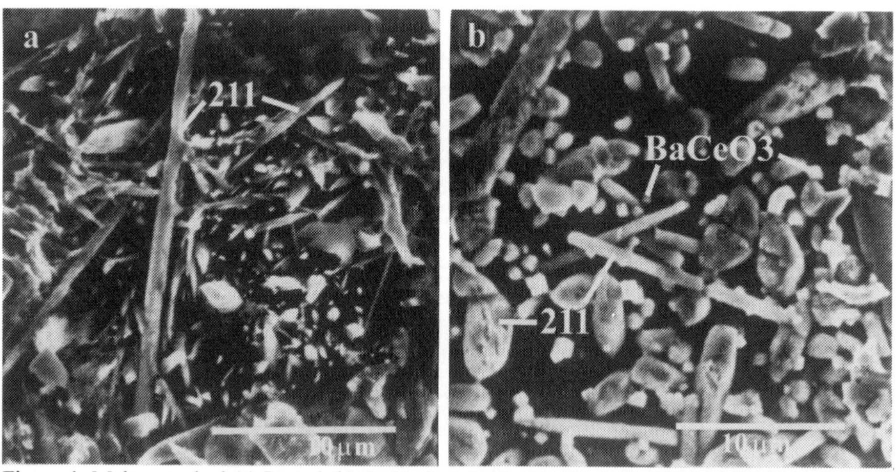

Figure 4- Melt quenched (a) Pt and (b) CeO_2 doped $YBa_2Cu_3O_{7-x}$. The Pt doped sample shows a smaller average particle size and a broader size distribution than the CeO_2 doped sample.

Some significant differences are observed between the quenched melt structures of the Pt and CeO_2 doped samples. The Pt doped melt quench sample shows a smaller average size, and a broader size distribution of 211 particles, as compared with the CeO_2 doped sample (fig. 4). Third phase $BaCeO_3$ is clearly shown in figures 3b and 4b as submicron cube-shaped particles. The Pt doped samples (figs. 3a and 4a) contain third phase particles, which are rich in Pt and Ba as confirmed by EDS analysis.

The narrow size distribution of 211 in the semi-solid phase of the CeO_2 doped sample suggests that much of the 211 nucleated within a short period of time. The relatively smaller 211 particle size (larger surface area) and broader size distribution in the quenched Pt doped 123 sample, than that of the undoped or CeO_2 doped 123, can be attributed to the lower interfacial energy between the solid and liquid phases. A reduction of interfacial energy would lower the capillary forces in the semi-solid sample. The smaller particle size of 211, on the other hand, would increase the capillary forces in the melt. The absence of the liquid loss in the Pt doped 123 sample suggests that reducing the size of 211 particles is more dominant than lowering of the interfacial energy with respect to stronger capillary forces. On the other hand, the liquid loss in the CeO_2 doped samples can be attributed to weaker capillary forces between relatively larger 211 particles in the melt (fig. 4b). When a certain amount of the liquid phase is depleted, the 211 particles may come to a close approach so that the remaining liquid phase is retained in the semi-solid melt [4]. Further studies are in progress to control the depletion of the liquid phase and to introduce fine $BaCeO_3$ particles as effective flux pinning centers in $YBa_2Cu_3O_{7-x}$.

SUMMARY

The microstructures of platinum and cerium oxide doped $YBa_2Cu_3O_{7-x}$ were compared after a melt quenching of the semi-solid phase at 1040°C, as well as a melt texturing process. It was found that Pt was more effective than CeO_2 addition with respect to reducing the size of 211 particles in the melt. The loss of the liquid phase in the Pt doped sample was completely stopped, which was attributed to strong capillary forces between fine 211 inclusions in the liquid. On the other hand, a loss of the liquid phase up to 25% occurred in the CeO_2 doped samples, which contained uniformly distributed fine $BaCeO_3$ particles.

ACKNOWLEDGMENTS

This work was supported by the Washington Technology Center and Praxair Specialty Ceramics Inc., Woodinville WA.

REFERENCES

[1] S. Jin, T.H. Tiefel, R.C. Sherwood, R.B. van Dover, M.E. Davis, G.W. Kammlott and R.A. Fastnacht, "Melt Textured Growth of Polycrystalline $YBa_2Cu_3O_{7-x}$ with High Transport J_c at 77K," Phys. Rev. B 37, pp. 7850-3 (1988).

[2] M. Murakami, H. Fujimoto, S. Gotoh, K. Yamaguchi and S. Tanaka, "Flux Pinning Due to Nonsuperconducting Particles in Melt Processed YBaCuO Superconductors," Physica C, 185-189, pp. 321-6 (1991).

[3] D.F. Lee, V. Selvamanickam and K. Salama, "Influences of Y_2BaCuO_5 Particle Size and Content on the Transport Critical Current Density of $YBa_2Cu_3O_x$ Superconductor," Physica C, 202, pp. 83-96 (1992).

[4] F. Dogan and J. Reding, "Effect of Y_2BaCuO_5 Particle Size on Undercooling and Solidification Rate of $YBa_2Cu_3O_{7-x}$," pp. 23-30, Ceramic Transactions, 84, Edited by U. Balachandran, W. Wong-Ng and A. Bhalla, The American Ceramic Society, Westerville, OH (1998).

[5] N. Ogawa, I. Hirabayashi and S. Tanaka, "Preparation of High-J_c YBCO Bulk Superconductor by the Platinum Doped Melt Growth Method," Physica C, 177, pp. 101-5 (1991).

[6] T. Izumi, Y. Nakamura, T.H. Sung and Y. Shiohara, "Reaction Mechanism of Y-system Superconductors in the MPMG method," J. Mater. Res. 7 [4], pp. 801-7 (1992).

[7] C. Varanasi and P.J. McGinn, "The Effect of $PtO_2 \bullet H_2O$ Additions on the Y_2BaCuO_5 Morphology in Melt Textured $YBa_2Cu_3O_{7-x}$," Physica C, 207, pp. 79-84 (1993).

[8] J. Langhorn, Y.J. Bi, M. Mathews and J.S. Abell, "Platinum Group Metals as Flux Pinning Additions to superconducting Y-Ba-Cu-O Thick Films," pp. 269-74 in High Temperature Superconductors: Synthesis, Processing, and Large Scale Applications, 1st ed. Edited by U. Balachandran, P.J. McGinn and J.S. Abell. The Minerals, Metals and Materials Society (1996).

[9] J.H. Park, H.W. Kim and J.T. Song, "$YBa_2Cu_3O_{7-x}$ High-T_c Superconductor Prepared by Melting Method," J. Mater. Sci., Mater. Electron., 4, pp. 77-82 (1993).

[10] C.J. Kim, K.B. Kim, I.H. Kuk and G.W. Hong, "Role of PtO_2 on the Refinement of Y_2BaCuO_5 Second Phase Particles in Melt-textured Y-Ba-Cu-O Oxides," Physica C, 281, pp. 244-52 (1997).

[11] Y. Nakamura, T. Izumi, Y. Shiohara and S. Tanaka, "Solidification Mechanism of $YBa_2Cu_3O_x$ Superconductors," pp. 675-80 in Conference on layered Superconductors: Fabrication, Properties, and Applications. Symposium, San Francisco CA, USA, 1st ed. Edited by D.T. Shaw, C.C. Tsuei, T.R. Schneider, and Y. Shiohara. Materials Research Society, Pittsburgh PA, USA (1992).

[12] T. Izumi, Y. Nakamura and Y. Shiohara, "Doping Effects on Coarsening of Y_2BaCuO_5 Phase in Liquid," J. Mater. Res., 8 [6], pp. 1240-46 (1993).

[13] C. Varanasi, M.A. Black and P.J. McGinn, "A Comparison of the Effects of PtO_2 and $BaSnO_3$ additions on the refinement of Y_2BaCuO_5 and Magnetization of Textured $YBa_2Cu_3O_{6+x}$," Supercond. Sci. and Technol., 7, [1] pp. 10-16 (1994).

[14] T. Meignan and P McGinn, "The Effect of Platinum Additions on Y_2BaCuO_5 Precipitation and Coarsening During Meit Texturing of $YBa_2Cu_3O_{7-\delta}$," Supercond. Sci. and Technol., 10, pp. 576-82 (1997).

[15] P. Diko, C. Wende, D. Litzkendorf, T. Klupsch and W. Gawalek, "The Influence of Starting $YBa_2Cu_3O_{7-x}$ Particle Size and Pt/Ce Addition on the Microstructure of $YBa_2Cu_3O_{7-x}$ -Y_2BaCuO_5 Melt Processed Bulks," Supercond. Sci. and Technol., 11, pp. 49-53 (1998).

[16] I. Monot, K. Verbist, M. Hervieu, P. Laffez, M.P. Delamare, J. Wang, G. Desgardin and G. Van Tendeloo. "Microstructure and Flux Pinning Properties of Melt Textured Grown Doped $YBa_2Cu_3O_{7-\delta}$," Physica C, 274, pp. 253-66 (1997).

[17] M.P. Delamare, I. Monot, J. Wang, J. Provost and G. Desgardin, "Influence of CeO_2, $BaCeO_3$, or PtO_2 Additions on the Microstructure and the Critical Current Density of Melt Processed YBCO Samples" Supercond. Sci. and Technol., 9 [7], pp. 534-42 (1996).

[18] N. Vilalta, F. Sandiumenge, S. Piñol and X. Obradors, "Precipitate Size Refinement by CeO_2 and Y_2BaCuO_5 Additions in Directionally Solidified $YBa_2Cu_3O_7$," J. Mater. Res., 12 [1], pp. 38-46 (1997).

[19] S. Piñol, F. Sandiumenge, B. Martínez, V. Gomis, J. Fontcuberta, X. Obradors, E. Snoeck and C. Roucau, "Enhanced Critical Currents by CeO_2 Additions in Directionally Solidified $YBa_2Cu_3O_7$," Appl. Phys. Lett., 65 [11], pp. 1448-50 (1994).

[20] C.J. Kim, K.B. Kim, G.W. Hong, D.Y. Won, B.H. Kim, C.T. Kim, H.C. Moon and D.S. Suhr, "Microstructure, Microhardness, and Superconductivity of CeO_2-added Y-Ba-Cu-O Superconductors," J. Mater. Res., 7 [9], pp. 2349-54 (1992).

[21] C.J. Kim, K.B. Kim, D.Y. Won, H.C. Moon, D.S. Suhr, S.H. Lai and P.J. McGinn, "Formation of $BaCeO_3$ and its Influence on Microstructure of Sintered/melt -textured Y-Ba-Cu-O Oxides with CeO_2 Addition," J. Mater. Res., 9 [8], pp. 1952-60 (1994).

[22] S. Piñol, F. Sandiumenge, B. Martínez, N. Vilalta, X. Granados, V. Gomis, F. Galante, J. Fontcuberta and X. Obradors, "Modified Growth Mechanism in Directionally Solidified $YBa_2Cu_3O_7$.," IEEE Transactions on Applied Superconductivity, 5 [2], pp. 1549-52 (1995).

ANISOTROPIC OSTWALD RIPENING IN β-Si_3N_4 WITH DIFFERENT LANTHANIDE ADDITIVES

Mikito Kitayama*, Kiyoshi Hirao**, Motohiro Toriyama** and Shuzo Kanzaki**
*Synergy Ceramics Laboratory, Fine Ceramics Research Association
**National Industrial Research Institute of Nagoya
1-1 Hirate-cho, Kita-ku, Nagoya 462 Japan

ABSTRACT

α-Si_3N_4 powder mixed with SiO_2 and Ln_2O_3 (Ln=La, Gd or Yb) was fired at 1750-1900°C. After acid rinsing, the morphologies of the resultant β-Si_3N_4 crystals were quantitatively analyzed. The width of the crystals doped with different lanthanide oxide increased in the order of La<Gd<Yb, while the aspect ratio showed the inverse relationship among them. The anisotropic Ostwald ripening model successfully simulated grain growth behavior of β-Si_3N_4, and predicted that there would be about two orders of magnitude difference in the interfacial reaction constants of the [100] facet between Ln = La and Yb.

INTRODUCTION

Since the high toughness of Si_3N_4-based ceramics has been attributed to its rod-like grain shape due to the anisotropic grain growth of β-Si_3N_4,[1] many works have focused on controlling the grain morphology of β-Si_3N_4 by changing the α-β ratio of the starting powder,[2] sintering conditions,[3,5] and the ratio and amount of additives.[4,5] Recently, it was reported that the radius of lanthanide (Ln) ions had a significant effect on grain growth behavior of β-Si_3N_4 in Ln-Si-Al-O-N glass during and after the α-β transformation.[6,7] Radii of lanthanide ions showed a good positive correlation with mean aspect ratios at the end of the phase transformation. If it is assumed to be due to the increase of viscosity through the increase of the field strength of lanthanide ions, the change of growth mechanism from interfacial reaction controlled to diffusion controlled could explain this result. However, it was not consistent with the observed phase transformation rates. A kind of catalytic effect of lanthanide ions on the prism and/or basal planes is proposed.[7]

In this work, only lanthanide oxide was used as a sintering additive to elucidate its effect on the grain growth behavior of Si_3N_4. The seed fabrication method was chosen to separate β-Si_3N_4 crystals from the lanthanide-silicate grain boundary phase for easier and more accurate observation.[8] Annealing at much higher temperatures than in the previous work[6,7] was done to further investigate grain growth behaviors of β-Si_3N_4 with different lanthanide additives after the phase transformation (Ostwald ripening). Grain growth simulations based on the anisotropic Ostwald ripening model[9] were performed for discussing the effect of lanthanide ions on the interfacial reaction constant.

EXPERIMENT

α-Si_3N_4 (E-5 Grade, Ube Industries Ltd., Japan) with 10mol% SiO_2 and 5 mol% Ln_2O_3 (Ln = La, Gd or Yb) (Hokko Chemicals Ltd., Japan) were planetary milled using methanol as a mixing medium. The powder mixture was dried, passed through a 60-mesh sieve, and fired in a BN crucible at 1750, 1800, 1850 or 1900°C for 2h under a nitrogen pressure of 0.9MPa.

The product, an aggregated porous body, could be easily screened through a 100-mesh sieve. The screened powders were then subjected to the following treatments: (1) a mixed solution of concentrated hydrofluoric acid (HF) and nitric acid (HNO_3) at 60°C to dissolve the residual glassy phase, (2) a concentrated sulfuric acid (H_2SO_4) at 160°C to eliminate lanthanide compounds possibly formed during the previous treatment.

X-ray diffraction analysis (XRD) was performed for phase identification. Morphologies of Si_3N_4 crystals were observed by scanning electron microscope (SEM). As β-Si_3N_4 had a hexagonal rod shape, its width and length were measured for each crystal.

RESULTS AND DISCUSSIONS

XRD identified that all Si_3N_4 powders obtained consisted only of the β-phase. No other phases were detected, which indicates that grain growth at 1750-1900°C is categorized as Ostwald ripening. Figures 1-3 (upper row) show width-length distributions of Si_3N_4 crystals fired at 1750-1900°C with La_2O_3, Gd_2O_3 and Yb_2O_3, respectively. At 1750°C, the difference in lanthanide additives did not make any differences in both distributions. However, it did cause significant differences at higher temperatures. La_2O_3 gave almost one-dimensional growth, while Yb_2O_3 gave an appreciable width growth. As a result, wide distributions were observed in length and width for La_2O_3 and Yb_2O_3 systems, respectively. Gd_2O_3 showed intermediate behavior between both additives. Aspect ratios of the crystals increased in the order of La>Gd>Yb. It was reported that radii of lanthanide ions show a good positive correlation with their mean aspect ratios at the end of the phase transformation.[6,7] In this work, it was found that lanthanide

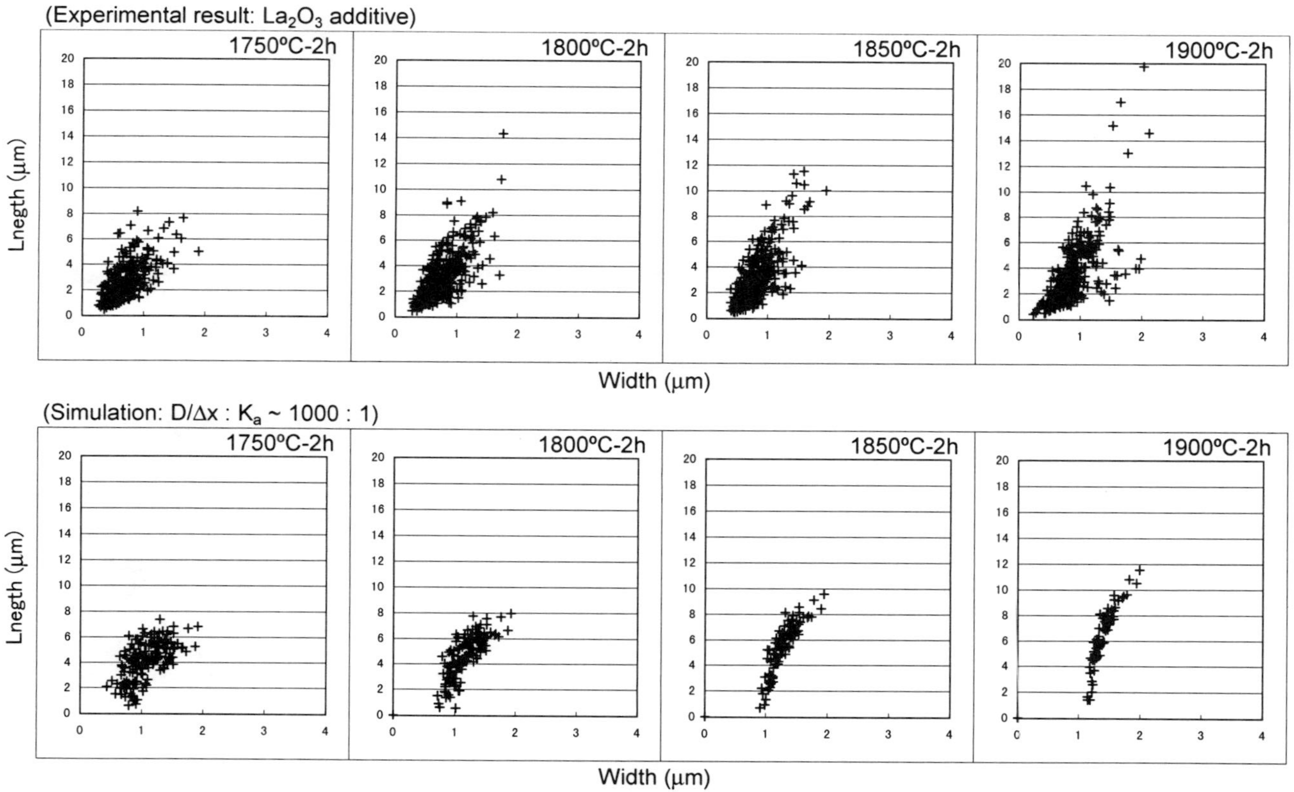

Fig. 1 Comparison of width-length distributions of β-Si_3N_4 crystals between experimental (La_2O_3) and simulation results.

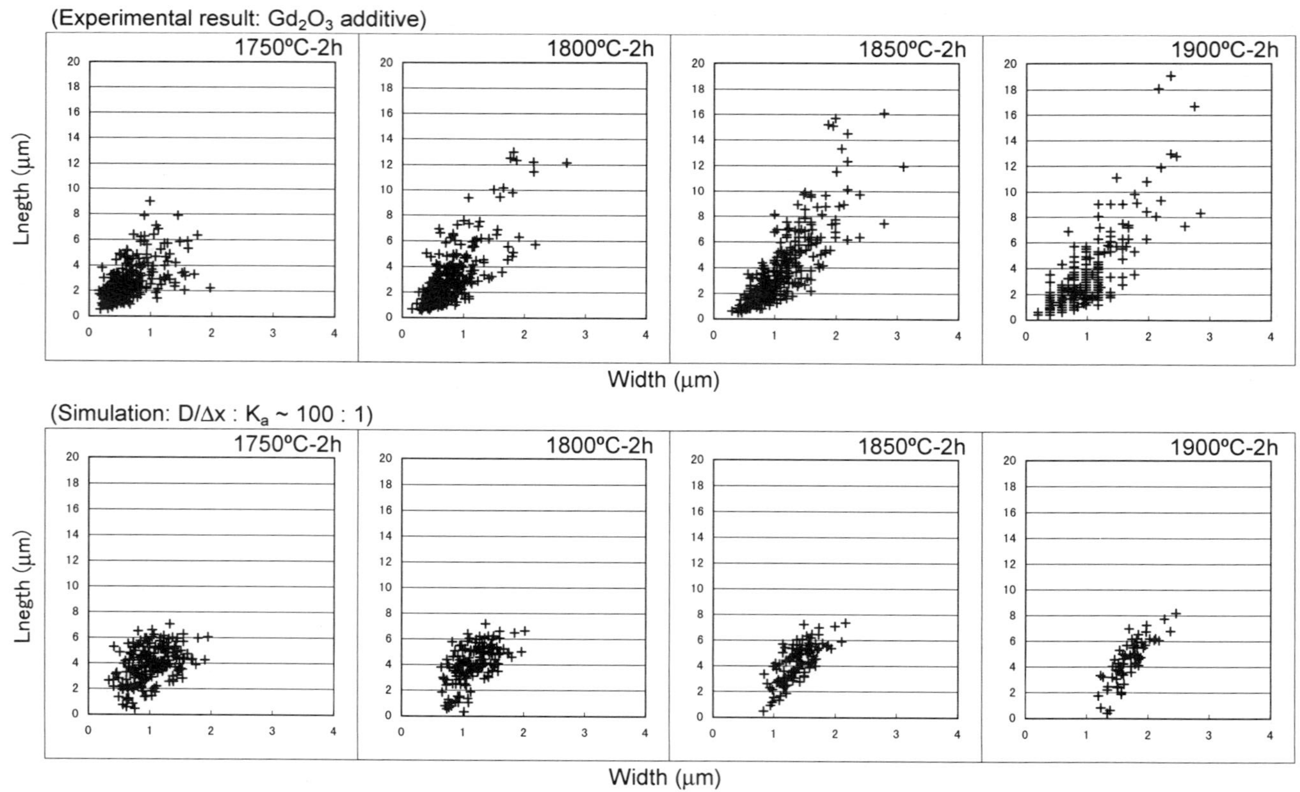

Fig. 2 Comparison of width-length distributions of β-Si_3N_4 crystals between experimental (Gd_2O_3) and simulation results

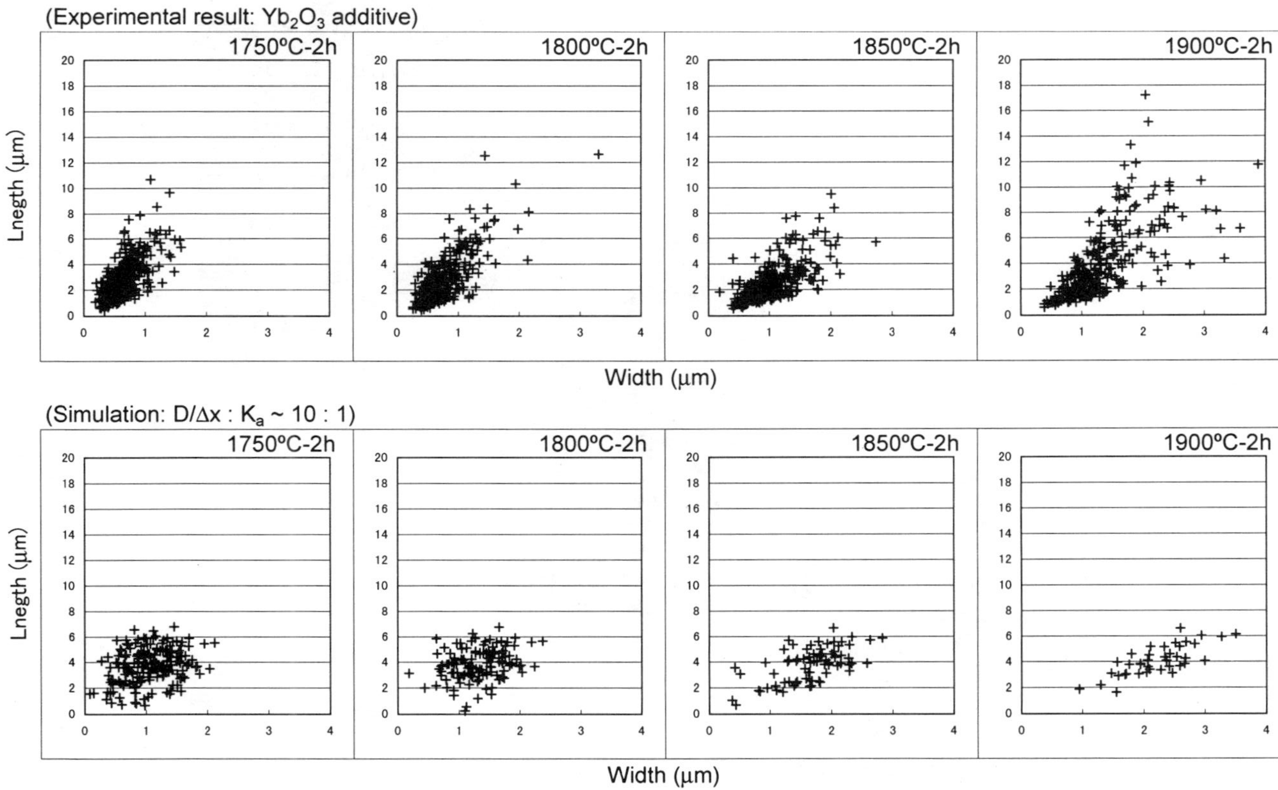

Fig. 3 Comparison of width-length distributions of β-Si_3N_4 crystals between experimental (Yb_2O_3) and simulation results.

ions caused a difference not at the end of the phase transformation but during Ostwald ripening. Figure 4 shows the morphologies of β-Si_3N_4 crystals fired at 1850°C with (a) La_2O_3, (b) Gd_2O_3 and (c) Yb_2O_3, which demonstrates the morphology difference among three additive systems. Fibrous grain shape of La system shows a remarkable contrast with rather isotropic shape of Yb system.

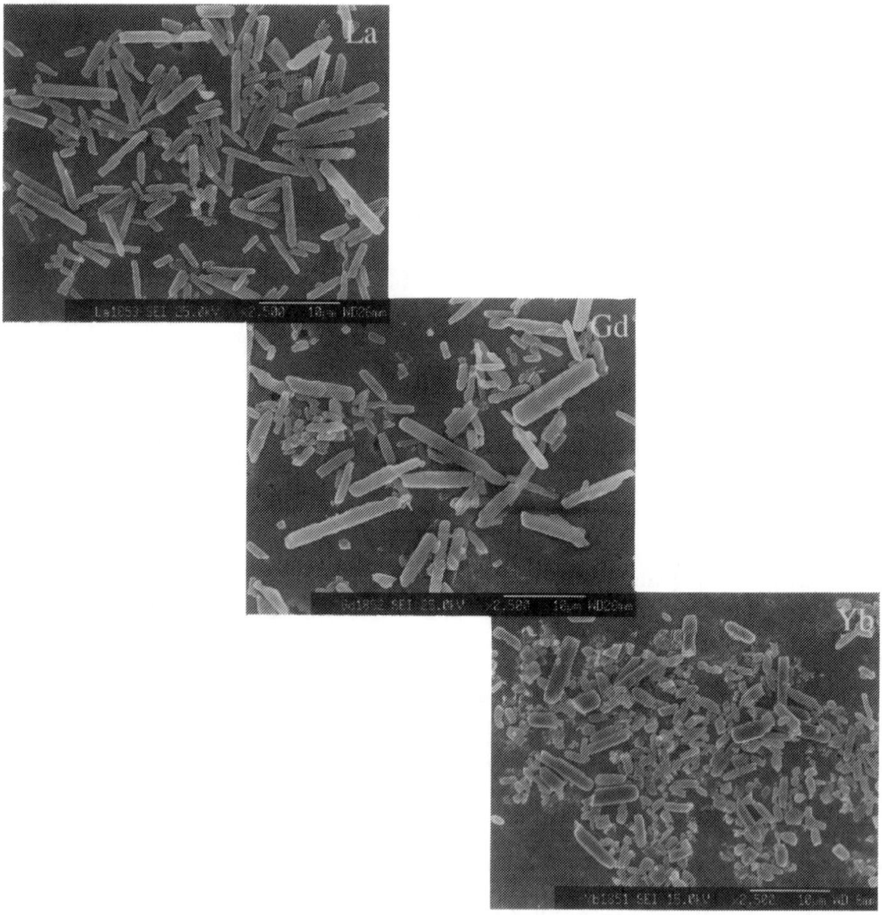

Fig. 4 Morphologies of β-Si_3N_4 crystals fired at 1850°C with (a) La_2O_3, (b) Gd_2O_3 and (c) Yb_2O_3

As already mentioned in the introduction, the difference in viscosity of glassy phase due to the difference of the ionic field strength could solely explain this tendency: interfacial reaction controlled and diffusion controlled kinetics for La and Yb systems, respectively. However, the observed grain morphology of Yb system is facetted as well as La system, which is a clear evidence of interfacial reaction controlled kinetics. Also, it contradicts with the observed α-β transformation rates: Yb_2O_3 gave a faster phase transformation rate than La_2O_3. Thus, a proposed catalytic effect of lanthanide ion on the prism and/or basal planes[7] would be expressed as that lanthanide ions alter the interfacial reaction rates at the (100) and/or (001) planes.

To further investigate this effect on grain growth of β-Si_3N_4, simulations based on the anisotropic Ostwald ripening model[9] were performed. It was reported that the lengths of β-Si_3N_4 crystals are independent of width, and are constant at the end of the phase transformation.[10] Based on this experimental fact, an initial particle size distribution of β-Si_3N_4 crystals was computer-generated assuming normal distribution. Mean grain width and length were set to be 0.8 and 4.0 μm, respectively. Since the tip shape of β-Si_3N_4 crystal was observed to be rounded, it was assumed that the growth in the [001] direction was diffusion controlled. The appearance of prismatic facets dictates interfacial reaction controlled kinetics in the [210] direction. Activation energies of grain growth in the [001] and [210] directions were taken from the literature[11], 686 and 772 (kJ/mol), respectively. The difference of ionic field strength of lanthanide ions were taken into account by assuming diffusion constants of the liquid phase as follows; $D(\text{La}) = 2\times D(\text{Gd}) = 4\times D(\text{Yb})$. The (100) interfacial reaction constants were set as $D/\Delta x{:}K_a = {\sim}1000{:}1$, ~100:1 and ~10:1 for La, Gd and Yb, respectively.

Figures 1-3 compare width-length distributions of β-crystals at 1750-1900°C obtained through simulations (lower row) with experimental results (upper row). Although there are some inconsistencies between them probably due to the existence of finely ground crystals and the effects of gas phase reaction at higher temperatures (1850°C<) where the decomposition of Si_3N_4 would be more active, the maximum widths agree quite well between them. This suggests that the [100] interfacial reaction rate in the liquid phase containing La would be about two orders of magnitude lower than in the liquid phase containing Yb. To rationalize this difference, the interface between β-Si_3N_4 crystal and glassy phase should be examined carefully.

CONCLUSION

The seed fabrication method was used for studying anisotropic Ostwald ripening of β-Si_3N_4 with different lanthanide (Ln) additives. The width of the crystals doped with different lanthanide oxide increased in the order of La<Gd<Yb, while the aspect ratio showed the inverse relationship among them. Excellent agreement between experimental results and simulations based on anisotropic Ostwald ripening model suggests that there would be about two orders

of magnitude difference in the interfacial reaction constants of the [100] facet between Ln = La and Yb.

ACKNOWLEDGEMENT

This work has been carried out as part of the Synergy Ceramics Project under the Industrial Science and Technology Frontier Program (ISTF) promoted by AIST, MITI, Japan. The authors are members of the Joint Research Consortium of Synergy Ceramics.

REFERENCE

[1]F. F. Lange, "Relation Between Strength, Fracture Energy, and Microstructure of Hot-Pressed Si_3N_4," *J. Am. Ceram. Soc.*, **56** [10] 518-22 (1973).

[2]G. Himsolt, H. Knoch, H. Huebner and F. W. Kleinlein, "Mechanical Properties of Hot-Pressed Silicon Nitride with Different Grain Structures," *J. Am. Ceram. Soc.*, **62** [1-2] 29-32 (1979).

[3]F. F. Lange, "Fracture Toughness of Si_3N_4 as a Function of the Initial α-Phase Content," *J. Am. Ceram. Soc.*, **62** [7-8] 428-30 (1979).

[4]G. Wötting and G. Ziegler, "Influence of Powder Properties and Processing Conditions on Microstructure and Mechanical Properties of Sintered Si_3N_4," *Ceram. Int.*, **10** [1] 18-22 (1984).

[5]G. Wötting, B. Kanka and G. Ziegler, "Microstructural Development, Microstructural Characterization and Relation to Mechanical Properties of Dense Silicon Nitride," pp. 83-96 in Non-Oxide Technical and Engineering Ceramics, Edited by S. Hampshire. Elsevier Applied Science (1986).

[6]M. Kanamaru, Ph.D. thesis, University of Stuttgart (1994).

[7]M. J. Hoffmann, pp. 59-72 in Tailoring of Mechanical Properties of Si_3N_4 Ceramics, Edited by M.J. Hoffmann and G. Petzow, Kluwer Academic Publishers, Netherlands (1994).

[8]K. Hirao, A. Tsuge, M. E. Brito and S. Kanzaki, "Preparation of rod-like β-Si_3N_4 Single Crystal Particles," *J. Ceram. Soc. Japan*, **101** [9] 1078-80 (1993).

[9]M. Kitayama, K. Hirao, M. Toriyama and S. Kanzaki, "Modeling and Simulation of Grain Growth in Si_3N_4–I. Anisotropic Ostwald Ripening," submitted to *Acta Mater.* (1998).

[10]M. Kramer, M. J. Hoffmann and G. Petzow, "Grain Growth Studies of Silicon Nitride Dispersed in an Oxynitride Glass," *J. Am. Ceram. Soc.*, **76** [11] 2778-84 (1993).

[11]K.-R. Lai and T.-Y. Tien, "Kinetics of β-Si_3N_4 Grain Growth in Si_3N_4 Ceramics Sintered under High Nitrogen Pressure," *J. Am. Ceram. Soc.*, **76** [1] 91-96 (1993).

PREPARATION, INTERFACE CHEMISTRY AND MECHANICAL PROPERTIES OF $SiC-Si_3N_4$ NANO/MICRO-COMPOSITES

P. Sajgalik, K. Rajan*, P. Warbichler,** F. Hofer,** and R. Riedel†
Institute of Inorganic Chemistry
Slovak Academy of Sciences
Dubravska cesta 9
SK-942 36 Bratislava, Slovakia

*RensselaerPolytechnic Institute
Materials Engienering Department
Troy, NY 12180-3590

**Forschunginstitut fur Elektronenmikroskopie und Feinstrukturforschung
TU Erzherzog-Johann-Universitat Graz
Steyrergasse 17
A-8010 Grax, Austria

†Fachbereich Materialwissenschaft
Fachgebiet Disperse Feststoffe
TH Darmstadt
Petersenstrasse 23
D-64287 Darmstadt, Germany

ABSTRACT

$SiC-Si_3N_4$ nano/micro-composites were prepared from crystalline Si_3N_4 powder and amorphous Si-N-C powder using forced homogeneous nucleation. Three classes of interfaces were identified in the densified composite:

SiC nano-inclusion – Si_3N_4 host micrograin
SiC micrograin – Si_3N_4micrograin
Si_3N_4micrograin – Si_3N_4micrograin.

Based on the chemistry of these interfaces and the thermal expansion coefficients of SiC and $Si_3N_4$the observed interfaces were divided into the two principal groups: the weak and strong interfaces. Mechanical properties of the composites prepared in present work are quite promising; 4-point bending strength 1.2 GPa, fracture toughness 6.9 $MPa.m^{1/2}$ and Weibull modulus 19. the role of the interface strength with respect to its chemistry and mechanical properties of composite will be discussed.

PROCESSING AND MECHANICAL PROPERTIES OF HIGHLY ANISOTROPIC Si_3N_4

Hisayuki Imamura, Kiyoshi Hirao,* Manuel Brito,* Motohiro Toriyama,* and Shuzo Kanzaki*
Fine Ceramics Research Association
Synergy Ceramics Laboratory
Nagoya, Japan

*National Industrial Research Institute of Nagoya
Nagoya, Japan

ABSTRACT

Silicon nitride with anisotropic microstructure was fabricated by tape-casting of raw powder slurries with 0–10 vol.% seed particles of β-Si_3N_4 single crystal, followed by gas pressure sintering of stacked sheets. The effect of seeding on microstructure development and mechanical properties were investigated for two kinds of seed particles. The seeding of an appropriate amount of β-Si_3N_4 particles with a small size allowed the development of optimal microstructures, where elongated grains grown from seed particles were fine an highly dispersed. In both cases, fracture toughness (K_{Ic}) and bending strength measured in the direction perpendicular to the grain alignment increased with an increase in seeds amount. this behavior was associated with the fact that both orientation degree and area fraction of elongated grains increased with increasing amount of seeds. In particular, the specimen with 5 vol.% small size seeds, in which elongated grains occupy a cross-sectional area fraction of about 45%, exhibited a dominant orientation parallel to the casting direction. The specimen showed a high K_{Ic} of 12.5 $MPa.m^{1/2}$ and a high bending strength of 1400 MPa. The improved fracture toughness can be associated with an increase in the proportion of elongated grains which are effective on bridging cracks. The increased bending strength of the specimen can be explained by the improved toughness and the decrease in flaw size owing to additions of small size seed particles.

MULTIPHASE COMPOSITES OF SILICON NITRIDE WITH SYNERGIC EFFECT

J. Hojo, N. Oyama, K. Kishi,* and S. Umebayashi
Fac. Engineering
Kyushu University
Fukuoka, 812 Japan

*KNIRI
Tosu, 841 Japan

ABSTRACT

Multiphase composites of Si_3N_4-BN-SiC system were produced from the powder mixture of Si_3N_4 with nano-size BN particles and micro-size SiC particles. The composites were densified by hot-pressing using Al_2O_3 and Y_2O_3 as sintering aids. The sintered bodies had fine-grained microstructures, because the growth of Si_3N_4 grains was retarded by BN and SiC inclusions. The thermal shock resistance of Si_3N_4 was improved by dispersion of the fine h-BN particles which absorbed the thermal stress. However, the fracture toughness was lowered. The fracture toughness was improved by addition of coarse SiC particles to Si_3N_4-BN system because of the crack deflection by SiC inclusions. The Si_3N_4-BN-SiC composite had excellent thermal shock resistance and retained high fracture toughness similar to those of monolithic Si_3N_4 and Si_3N-SiC composite. In this multiphase composite, BN and SiC worked cooperatively to improve the mechanical properties, meaning the synergic effect.

CONTROLLING GRAIN MORPHOLOGY FOR Si_3N_4 AND SiC MICROSTRUCTURES

Craig D. Hillman and William J. Clegg
Materials and Metallurgy Department
Cambridge University
Cambridge, England CB4 3ES

ABSTRACT

It has been determined that the addition of rare-earth oxides (CeO_2, Y_2O_3, Ln_2O_3) to alumina-silica ceramics results in the formation of a "silicon-carbide" type microstructure, composed of platelet-like grains within an amorphous phase matrix. The development of this microstructure could lead to a high-toughness, high-strength oxide ceramic. It is hypothesized that this microstructure arises due to the growth of a ceria-aluminosilicate phase with a magnetoplumbite structure on certain crystallographic surfaces. Various analytical tools (SEM, XRD, EDX) were used to observe the effect of composition and heat treatment on microstructure. Experiments on the change in wettability were also carried out. Further heat treatments were carried out to crystallize the glassy phase and to determine its effect on the fracture toughness and creep strength of these materials. The properties of these materials were then compared to stoichiometric mullite.

DEVELOPMENT OF ELONGATED Ca α-SiAlON GRAINS BY PRESSURELESS SINTERING

Chris Wood, Hong Zhao and Yi-Bing Cheng
Department of Materials Engineering
Monash University, Melbourne, 3168
Australia

ABSTRACT

The work presented here concerns the development of high m-value Ca-stabilised α-SiAlONs that develop a predominantly elongated α-SiAlON microstructure. The materials are pressureless sintered and utilise relatively cheap starting powders, thus presenting the possibility of fabricating low-cost, high performance in situ toughened Ca α-SiAlON materials. Some toughening mechanisms have been observed in the materials.

INTRODUCTION

It is well established that the elongated grain morphology of β-Si_3N_4/SiAlON ceramics results in improved fracture toughness through microstructural toughening mechanisms such as crack deflection, crack bridging and grain pull-out.[1] α-SiAlON (α') commonly occurs in an equiaxed grain morphology, hence it has inferior fracture toughness to β-SiAlON (β') which often appears in an elongated grain morphology [2, 3]. α-SiAlON, however, has a significantly higher hardness than β-SiAlON, making it a more desirable material for high wear applications [4]. To date, β-SiAlONs have been more extensively developed than α-SiAlONs due to their ease of processing via pressureless sintering and their good combination of properties.

α-SiAlONs are solid solutions of the α-Si_3N_4 crystal structure $Me_xSi_{12-(m+n)}Al_{(m+n)}O_nN_{(16-n)}$ (Eqn. 1), where the substitution of m (Al-N) and n (Al-O) bonds for (m+n) (Si-N) bonds occurs. The charge imbalance created by replacing m (Si-N) bonds with m (Al-N) bonds is compensated for by the incorporation of m/v Me cations into the interstices of the α-SiAlON unit cell. If the valency of Me is v, charge balance requires that $x = m/v$. The cations Me that may be used include most rare-earths, Y, Ca, Li and Mg. α-SiAlONs stabilised with rare earths tend to exhibit an equiaxed α-SiAlON grain morphology in their microstructures. Recent progress has seen the development of elongated α-SiAlON grains in hot-pressed α-SiAlON [5] and in rare-earth SiAlONs that utilise β-Si_3N_4 starting powders [6], however both these are expensive processing routes.

The use of calcium as the stabilising cation rather than rare-earth oxides has many advantages. Calcium carbonate is a cheap and abundant mineral and thus Ca α' is significantly cheaper to produce. Ca α' is more thermally stable than its rare-earth counterparts which may undergo an α' to β' decomposition at temperatures exceeding 1000°C [4, 7]. Ca is also reported to have the highest solubility per unit cell in the α' lattice [2] which can assist in controlling quantities of residual glass in the sintered product. Furthermore, the materials prepared in this work utilise α-Si_3N_4 as a starting powder rather than the more expensive β-Si_3N_4 used by Chen and Rosenflanz [6].

This paper presents recent work on the Ca α-SiAlON system, and additionally compares the Ca-stabilised material with four rare-earth stabilised α-SiAlONs that were processed under identical conditions.

EXPERIMENTAL

Five compositions, each utilising a different stabilising cation, were examined in this work. Stabilising cations used were Ca^{2+}, Sm^{3+}, Nd^{3+}, Y^{3+} and Yb^{3+}. The m-, n- and x-values of the five compositions are listed in Table I. Samples are designated by their m and n values in Equation 1 according to their positions on the α' plane, where the name CA3618 refers to a design composition m=3.6 and n=1.8, and so on. Corrections were made to account for the presence of surface oxides on the starting powders. All these samples are designed with a higher m-value than the reported α' composition in their respective SiAlON systems. This positions the designed compositions outside the α' phase region and towards the AlN corner of the phase diagram.

Table I. Five compositions studied in this work

Composition	m-value	n-value	x-value
CA3618	3.6	1.8	1.80
SM2814	2.8	1.4	0.93
ND2613	2.6	1.3	0.87
Y3015	3.0	1.5	1.00
YB4422	4.4	2.2	1.47

Powders of Si_3N_4 (HCST LC10), AlN (HCST Grade B) and either $CaCO_3$ (Ajax), Sm_2O_3 (99.9% pure, Heraeus), Nd_2O_3 (99.9%, Heraeus), Y_2O_3 (99.99%, Commercial Minerals) or Yb_2O_3 (99.99%, Commercial Minerals) were ball milled in isopropanol for 24 hours using Si_3N_4 milling media. Subsequent to milling the slurries were dried and hand mixed. Approximately 9g of powder was then uniaxially pressed into 25 mm diameter pellets, followed by cold isostatic pressing at 200 MPa. Prior to sintering, the Ca-containing specimen was calcined in a graphite resistance furnace at 900°C for 1 hour under vacuum to decompose $CaCO_3$ to CaO. For sintering, pellets were packed in a 50 wt% Si_3N_4 : 50 wt% BN powder bed inside a graphite crucible. The pellets were then pressureless sintered in a graphite resistance furnace for 4 hours at 1800°C in a nitrogen atmosphere at 0.1 MPa pressure. Furnace heating rates were 25°C/min. to 1000°C and 10°C/min. thereafter. Cooling took place in the furnace by switching off power after sintering and allowing the samples to furnace cool.

Bulk densities were determined by the Archimedes method (for the Ca sample) after the sample was boiled in water for 1 hour, and by the mercury immersion method (for all the other samples). Phase identification was performed on a Rigaku-Geigerflex XRD using Ni-filtered CuKα radiation. Fracture resistance measurements for R-curve analysis were performed using the indentation method described by Kim et al. [8] Microstructural observations of both polished and fracture surfaces were performed on a JEOL JSM 840-A scanning electron microscope (SEM). Polished Ca α-SiAlON surfaces were etched by immersion in molten NaOH for 10 seconds and then carbon coated prior to SEM observation. Fracture surfaces were gold coated. Energy dispersive x-ray spectroscopy (EDXS) was utilised on the SEM to chemically analyse specific microstructural features in the etched sample.

RESULTS AND DISCUSSION

1. Crystalline Phases

The sintered densities of the five samples are contained in Table II. XRD analysis showed that α-SiAlON was the majority phase in the Ca, Sm, Nd and Y containing samples and a major phase in the Yb-stabilised sample. XRD spectra are shown in Figure 1.

Table II. Density and phase analysis of five sintered compositions

Sample	Density (gcm^{-3})	α'	AlN*	$2H^{\delta}$	others
CA3618	3.20	vs	w	-	-
SM2814	3.52	vs	-	vw	M' (ms)
ND2613	3.24	vs	-	vw	JEM (ms), M' (m)
Y3015	2.96	vs	-	vw	M' (s)
YB4422	3.34	s	-	vw	YAM (vs)

Note: M' = $Me_2Si_2AlO_4N_3$, JEM = $NdAl(Si_{6-z}Al_z)(N_{10-z}O_z)$, YAM = $Yb_4Si_2O_7N_2$, $2H^{\delta}$ = $Si_{2.4}Al_{8.6}O_{0.6}N_{11.4}$

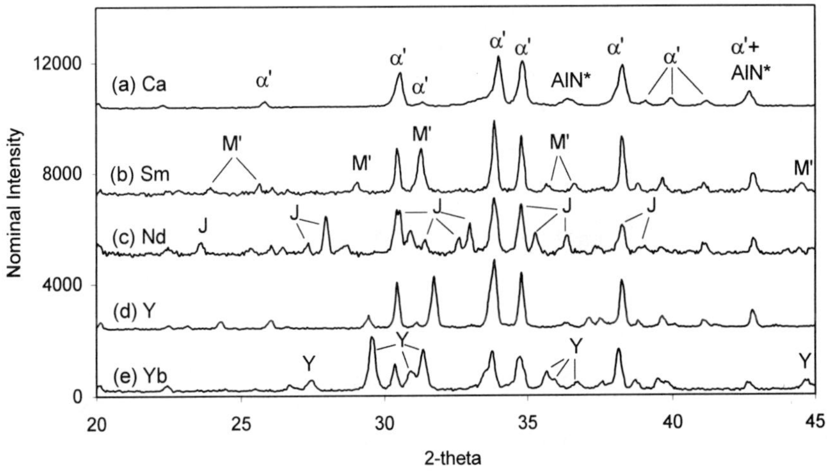

Figure 1 - XRD spectra of different SiAlON compositions fired at 1800°C for 4 hours. (a) Sample CA3618, (b) SM2814, (c) ND2613, (d) Y3015 and (e) YB4422. α' = α-SiAlON, M' = melilite-type phases, J = JEM, Y = YAM. $2H^{\delta}$ polytypoid peaks are obscured due to peak overlaps.

The high m-values of the compositions place them in equilibrium with a number of AlN-rich phases, hence AlN was observed in the Ca-containing composition and the AlN polytypoid 2H[δ] was found to be a constituent in all of the other samples examined in this work. The Al-containing nitrogen melilite, M' ($Me_2Si_2AlO_4N_3$) [9], was the grain boundary phase in the Sm, Nd and Y containing samples, but it was not stable in the Yb SiAlON system and the YAM phase ($Yb_4Si_2O_7N_2$)was found in the Yb sample. All the compositions lie well outside their respective β-SiAlON phase fields, hence there was no expectation of any β-SiAlON being found in the samples. The AlN phase identified in the sintered sample CA3618 had a very similar XRD pattern to the AlN used as the starting powder. Subsequent microstructural observation of the material led to the conclusion that this AlN was not unreacted starting powder, but was a new precipitated AlN-like phase designated AlN*. The exact nature of this phase is the subject of current work.

2. Grain Morphology

The microstructure of the Ca stabilised sample, CA3618, featured two distinct grain morphologies (Fig. 2). EDXS analysis found the two phases to be Ca α-SiAlON and AlN*.

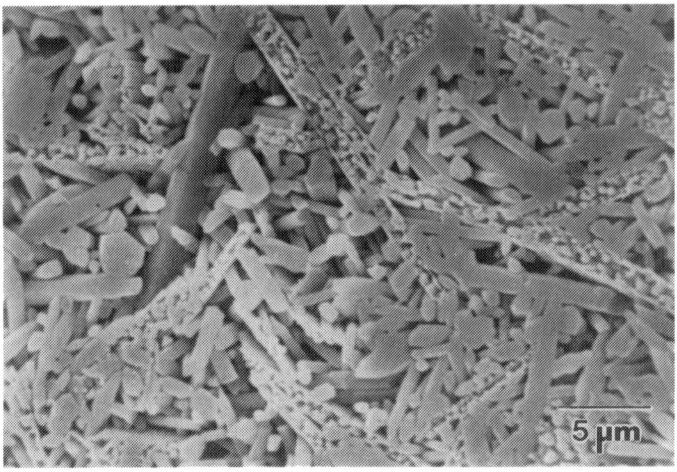

Figure 2 - SEM micrograph of Sample CA3618. The elongated solid grains are α-SiAlON and the elongated grains consisting of small dots are AlN*.

These two phases were contained in a glassy matrix, which was etched away for SEM observation of the sample. Many α-SiAlON grains showed an elongated morphology and a solid appearance. The average grain length observed was ~10μm, with some α-SiAlON grains exceeding 20μm in length. Aspect ratios of the α-SiAlON grains averaged ~8 and exceeded 12 for some longer grains. Anisotropic grain growth in the bulk material can be restricted due to grains impinging upon one another. However, observation of grain growth into pores in the sample, where there is little restriction to growth, showed the α-SiAlON grains grow freely into highly elongated grains (Fig. 3). Hence, the elongated morphology of Ca α-SiAlON observed here appears to be its intrinsic morphology.

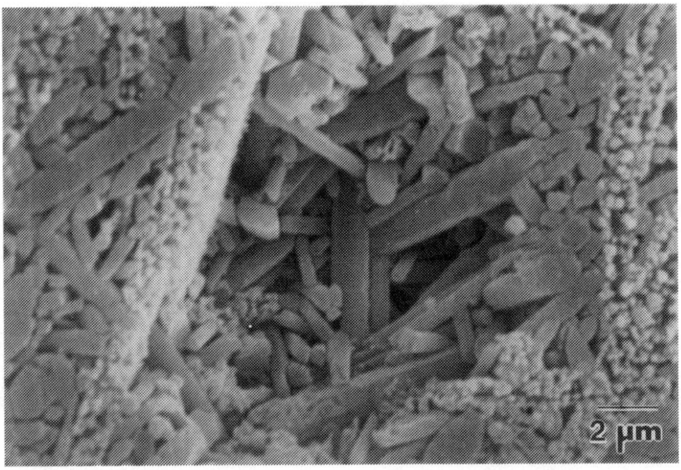

Figure 3 - The free growth of α' grains in pores reflects the intrinsic nature of the elongated morphology for Ca α-SiAlON grains.

The AlN* phase featured a distinctive appearance, characterised by grains that appeared to be comprised of small sub-micron particles or dots. The AlN* grains were present as elongated rods or plates displaying extreme aspect ratios. Grain lengths ranged from ~15µm to over 35 µm. Aspect ratios ranged from ~10 to over 40.

Backscattered SEM micrographs (Figs. 4(a) and (b)) of the rare-earth and yttrium stabilised materials revealed distinct morphologies from the Ca α-SiAlON material. In the Sm and Nd samples, equiaxed α-SiAlON grains (mid-grey) and acicular $2H^{\delta}$ grains (dark) were observed, surrounded by a rare-earth-rich matrix (light). The Yb sample had similar grain morphologies, the differences being some α-SiAlON grains taking on an elongated morphology and a higher proportion of AlN-polytypoid phase, but the majority of the α-SiAlON in the Yb sample was still equiaxed. From the XRD result, the Y-containing sample had a three phase microstructure of α-SiAlON, $2H^{\delta}$ and Y-rich melilite. However, the α-SiAlON phase was difficult to distinguish from the polytypoid phase due to low contrast differences between the two phases. The melilite was distributed inhomogeneously throughout the sample.

Although all the compositions studied in this work were selected from similar phase regions and prepared under identical conditions, the Ca α-SiAlON grains are predominantly elongated and the α-SiAlON phases in other systems have taken a more commonly observed equiaxed form. This comparison clearly suggests some uniqueness in the microstructural development for the Ca α-SiAlON phase. The Ca-SiAlON system has a relatively low temperature eutectic that promotes the early formation of a liquid phase. This is beneficial during sintering as it assists in densifying the material at lower temperatures. Work on this system also appears to indicate that the presence of the liquid phase promotes the anisotropic grain growth of Ca α-SiAlON. A drawback to this, however, is the presence of a residual grain boundary glass which may reduce the performance of the material at elevated temperatures. Engineering of this grain boundary phase would be required in order to optimise the material's high temperature properties.

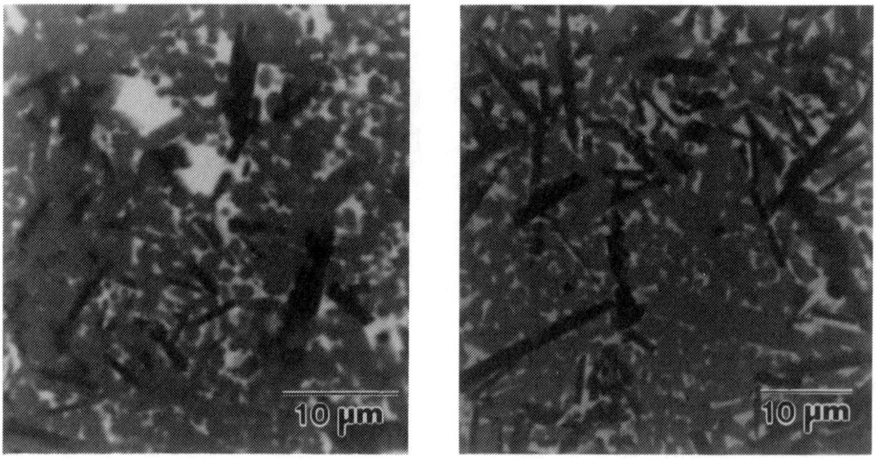

(a) (b)

Figure 4 – Backscattered SEM micrographs of Sm- (a) and Yb-stabilised (b) α-SiAlONs. Dark grains are $2H^{\delta}$ polytypoid, mid-grey grains are α-SiAlON and light phases are rare-earth rich grain boundary phases. α-SiAlON grains are seen mostly with an equiaxed morphology, though in (b) some elongated α' grains are observed.

3. Mechanisms Of Toughening

The toughening effect of propagating cracks interacting with elongated grains is well established in the β-Si_3N_4/-SiAlON system. Therefore, the elongated α-SiAlON grain morphology observed in CA3618 presents the potential for in-situ toughening. SEM micrographs taken from fracture surfaces of CA3618 (Fig. 5) revealed that fracture toughness enhancing mechanisms were present in the material. Examples of elongated grain debonding and grain pullout were observed throughout the sample.

Measurement of the fracture toughness of CA3618 under different indentation loads showed the material exhibited R-curve behaviour, whereby resistance to crack propagation increases with crack length (Fig. 6). R-curve behaviour is characteristic of in-situ toughened whisker-containing ceramics [8]. Crack deflections and bridging were visible in crack paths. Examination of many cracks showed this to be common throughout the sample and may account for the improved fracture toughness and R-curve behaviour observed in the material.

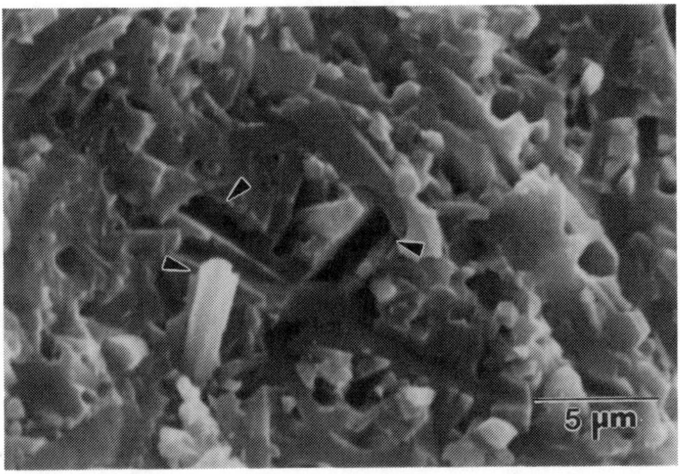

Figure 5 - Fracture surface of Sample CA3618. Sites of grain debonding and pullout are clearly visible

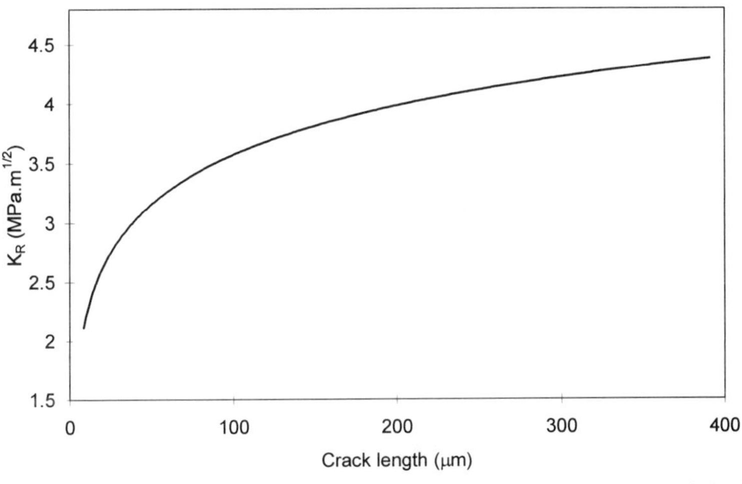

Figure 6 – Fracture resistance curve for Sample CA3618 showing R-curve behaviour.

The role of the residual grain boundary glass is critical to in-situ toughening as the strength of the interfacial bond between a grain and the matrix will affect how a propagating crack front will interact with a grain in its path. The bond must not be so strong as to prevent a grain from debonding from the matrix, but conversely, the interfacial bonding must be strong enough to not overly compromise the materials strength. Clearly, grain boundary phase engineering plays an important role in determining the overall properties of the material.

CONCLUSION

A SiAlON material that features an elongated α-SiAlON grain morphology has been developed in the Ca-SiAlON system. This is in distinct contrast to the more commonly observed equiaxed α-SiAlON morphology in the rare-earth and yttrium SiAlON systems. Microstructural toughening mechanisms such as crack deflection, grain debonding and pullout have been observed in the Ca α-SiAlON material, and the material exhibits R-curve behaviour. The material utilises relatively low-cost starting powders and pressureless sintering, hence with further work a class of low-cost, high-performance ceramics may be developed for broad structural applications.

ACKNOWLEDGMENT

This work is supported by the Australian Research Council.

REFERENCES

[1]M.J. Hoffmann, pp. 59-72 in "Tailoring of Mechanical Properties of Si_3N_4 Ceramics", ed. by M.J. Hoffmann and G. Petzow, Luwer Acad. Publ., 1994

[2]G.Z. Cao and R. Metselaar, "α-SiAlON Ceramics: A Review", Chem. Mater., Vol. 3, No. 2, 242-52 (1991)

[3]T. Ekström and M. Nygren , "SiAlON Ceramics", J.Am. Ceram. Soc., 75, [2] 259-76 (1992)

[4]T. Ekström, "α-SiAlON and α-β SiAlON Composites; Recent Research", pp. 1-21 in Engineering Ceramics '96', Slovakia, 1996

[5]H.Wang, Y.-B. Cheng, B.C. Muddle, L. Gao & T.S. Yen, "Preferred Orientation in Hot-Pressed Ca α-SiAlON Ceramics", J. Mater. Sci. Lett., 15 [16] 1447-9 (1996)

[6]I.-W. Chen & A. Rosenflanz, "A Tough SiAlON Ceramic based on α-Si_3N_4 with a Whisker-like Structure", Nature, Vol. 389, pp. 701-704 (1997)

[7]C.L. Hewett, Y.-B. Cheng, B.C. Muddle and M.B. Trigg, "Thermal Stability of Calcium α-SiAlON Ceramics", J. Euro. Ceram. Soc., in press

[8]Y.W. Kim, M. Mitomo and N. Hirosaki, "R-curve Behaviour of Sintered Silicon Nitride", J. Mater. Sci., 30, 4043 (1995)

[9]Y.-B. Cheng and D.P. Thompson, "Aluminium-Containing Nitrogen Melilite Phases", J. Am. Ceram. Soc., 77, 143-8 (1994)

EFFECT OF ZrO_2 INCLUSIONS ON THE DENSIFICATION OF Al_2O_3 AND ZnO MATRIX COMPOSITES

Yohsuke Nakada, Makoto Itoh and Toshio Kimura
Faculty of Science and Technology, Keio University
3-14-1 Hiyoshi, Kohoku-ku, Yokohama 223-8522 Japan

ABSTRACT

Rigid ZrO_2 particles hinder densification of Al_2O_3- and ZnO-matrix composites. The degree of hindrance is dependent on the characteristics of the matrix powder (chemistry and particle size). The particle size of matrix influences the degree of hindrance in densification of the Al_2O_3-matrix composites at the later stage of sintering. The ZrO_2 inclusions give a larger degree of hindrance to the ZnO-matrix composites than to the Al_2O_3-matrix composites.

INTRODUCTION

Ceramic-matrix composites have been developed to increase fracture toughness. Although ceramics with large fracture toughness can be obtained, the densification of matrix is hindered by the presence of second phase particles (inclusions). Several mechanisms have been proposed to account for the hindrance in densification by rigid inclusions,[1-5] but the sintering behavior of ceramic-ceramic composites is not fully understood. Previous studies in the literature indicate that an increase in the volume fraction and a decrease in particle size of inclusion increase the degree of hindrance.[6]

This work focused on the effects of characteristics of matrix material (chemistry and particle size) on the degree of hindrance. The sintering behavior of Al_2O_3 and ZnO containing the same ZrO_2 inclusion was examined. The degree of hindrance was found to be different for the ZnO- and Al_2O_3-matrix composites. The densification behavior of composites made up of three Al_2O_3 powders with different particle sizes was also compared.

METHODS AND MATERIALS

As-received Al_2O_3 and ZnO powders were used without further treatment. Three kinds of Al_2O_3, which will be called A2, A4, and A6, were used as a matrix. The trade names of A2, A4, and A6 are TM-D (Taimei Kagaku Kogyo, Nagano, Japan), AKP-30, and AKP-20 (both Sumitomo Chemical, Osaka, Japan), respectively. The average particle sizes were 0.25, 0.40, and 0.60 μm for A2, A4, and A6, respectively, as determined by the centrifugal sedimentation method (CAPA-500, Horiba, Tokyo,

Japan). ZnO was obtained from Seido Chemical, Osaka, Japan. The average particle size of ZnO was 0.61 μm.

The inclusion particles were prepared from a ZrO_2 powder (Daiich Kigenso Kagaku Kogyo, Tokyo, Japan); it was pressed, heated at 1500°C for 4h, and then ground with a ball mill. The fraction with particle sizes between 3 and 6 μm was classified by sedimentation using water as a medium. The average size of inclusion particles was 4.0 μm.

The matrix and inclusion powders were mixed in a ball mill for 24h, using ethanol as a medium. The inclusion contents were 10, 20, and 30 vol% based on the total solid content. The densities used for calculating oxide volumes were 3.99, 5.68, and 5.83 g/cm^3 for Al_2O_3, ZnO, and ZrO_2, respectively. The mixed powders were pressed by cold isostatic pressing. Applied pressure was adjusted to obtain compacts with the matrix relative densities of 50±1% for the A2- and A4-matrix composites and 55±1% for the A6- and ZnO-matrix composites. The density of the compacts was calculated from the weight and size of the compacts. The matrix density ρ_m was calculated from the measured density d_c using the following relation

$$\rho_m = \frac{d_c(1-F)}{F(d_{i,th} - d_c) + d_{m,th}(1-F)} \tag{1}$$

where $d_{m,th}$ and $d_{i,th}$ are the theoretical densities of matrix and inclusion, respectively, and F is the volume fraction of ZrO_2 based on the total solid content.

Densification behavior of the composites as well as pure Al_2O_3 and ZnO was studied by a constant-rate heating (5°C/min) experiment. The linear shrinkage was measured with a thermomechanical analyzer (TMA320, Seiko Instruments, Tokyo, Japan). The matrix relative density was calculated from the composite density obtained from the linear shrinkage data assuming isotropic shrinkage. The microstructure of the sintered compacts was observed with a scanning electron microscope (SEM; JSM-5200, JEOL, Tokyo, Japan). The samples were ground, polished, and sputtered with gold, prior to the microstructure observation.

RESULTS AND DISCUSSION

Densification Behavior

Figure 1 shows the effect of inclusion content on the densification behavior of the A2-matrix composites. Similar results were obtained for the A4- and A6-composites. An increase in the inclusion content did not change the starting temperature of densification, but reduced the slope of matrix density vs. temperature curves.

The densification behavior was compared by using a densification rate. The densification rate $\dot{\varepsilon}_m$ is defined by the following equation,

$$\dot{\varepsilon}_m = \frac{\alpha}{\rho_m} \frac{d\rho_m}{dT} \tag{2}$$

where ρ_m is the matrix density, α is the heating rate and T is temperature. Three kinds of

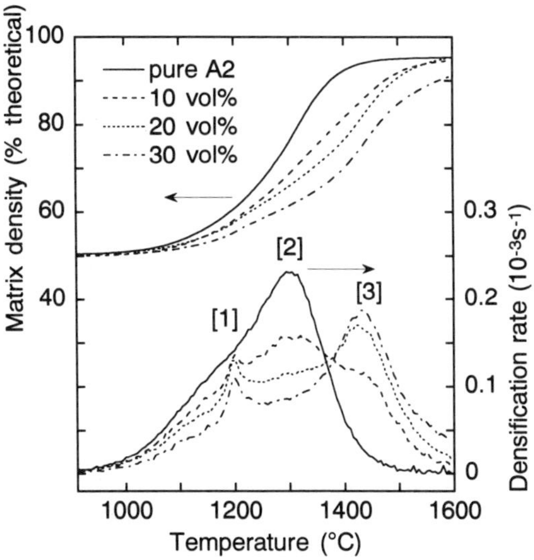

Fig. 1 Densification behavior of A2-matrix composites.

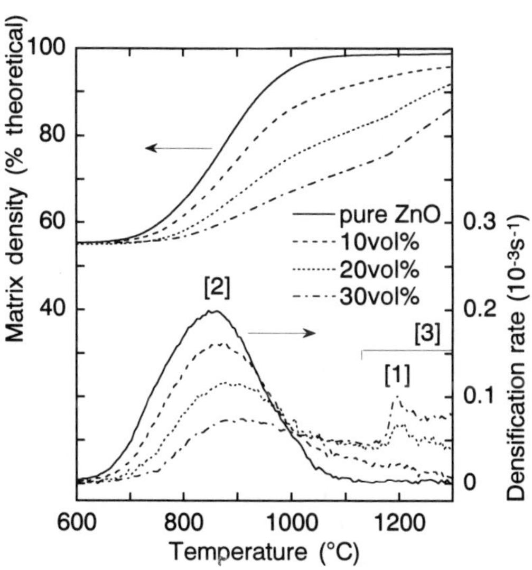

Fig. 2 Densification behavior of ZnO-matrix composites.

peaks were observed in Fig. 1. The first peak (indicated by [1]) was observed at about 1200°C and the peak height increased as the inclusion content increased. The heights of the second peak at about 1300°C (indicated by [2]) and the third peak at about 1450°C (indicated by [3]) decreased and increased, respectively, as the inclusion content increased.

The first peak is associated with the phase transformation of ZrO_2 from the monoclinic to tetragonal structures with a volumetric decrease. The second peak is associated with the constrained densification process of Al_2O_3 matrix. Figure 1 indicates that the inclusion particles reduce the densification rate at this temperature range. The third peak is associated with the densification process of Al_2O_3 matrix which has been subjected to the hindrance effect. The presence of the third peak indicates that inclusion particles shift the temperature range, at which the matrix densifies extensively.

Figure 2 shows the effect of inclusion content on the densification behavior of the ZnO-matrix composites. As in the Al_2O_3-matrix cases, the increase in inclusion content increased the degree of hindrance in densification, and there were three peaks in the temperature dependence of the densification rate. The peaks

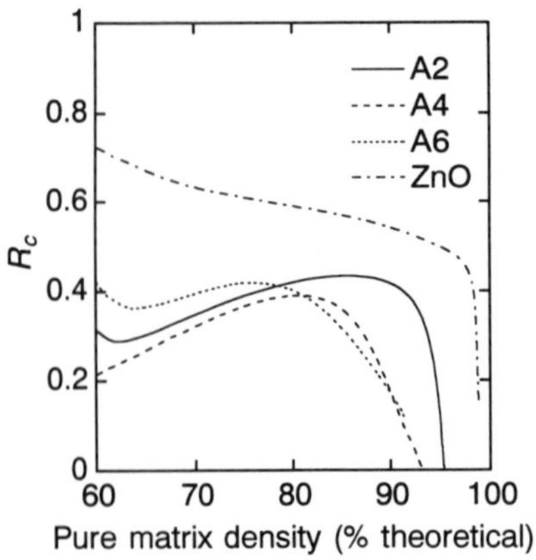

Fig. 3 Degree of hindrance in densification of composites containing 20 vol% ZrO_2 as a function of the relative density of pure matrix.

at about 900°C, above 1100°C, and about 1200°C correspond to peak [2], [3], and [1] for the Al_2O_3-matrix composites, respectively.

Figures 1 and 2 indicates that the heights of the peak [2] and the peak [3] decreased and increased, respectively, as the inclusion content increased. The maximum temperature of peak [2] was not changed for the Al_2O_3-matrix composites, but it shifted to high temperatures for the ZnO-matrix composites.

Effects of Characteristics of Matrix Powders

The degree of hindrance in densification of the composites was compared with respect to the density of pure matrices, which were heated under the same conditions for the composites. The degree of hindrance is defined as R_c,

$$R_c = 1 - \frac{\rho_m - \rho_{m0}}{\rho_u - \rho_{u0}} \qquad (3)$$

where ρ_u and ρ_m are the sintered matrix densities of pure matrix and composite, respectively, heated under the same conditions, and ρ_{u0} and ρ_{m0} are the green matrix densities of pure matrix and composite, respectively. Figure 3 shows the degree of hindrance in densification of composites containing 20 vol% ZrO_2 as a function of the relative density of pure matrix. When the inclusions give no effect on the densification of matrix, $R_c = 0$ is expected. When the inclusions fully hinder the densification (no shrinkage in the composites), $R_c = 1$ is expected. The observed values lay between 0 and 1, indicating that the inclusions partly hindered the densification of matrix.

Figure 3 indicates that the degree of hindrance in densification was dependent on the matrix powders. For the Al_2O_3-matrix composites, the degree of hindrance was almost the same up to the relative density of pure matrix of 80% irrespective of the kind of Al_2O_3 powder. Above 80%, A2 was subject to the largest hindrance. The comparison of the degree of hindrance of the ZnO-matrix composites with that of Al_2O_3-matrix composites indicates that the extensive hindrance occurred from the early stage in the ZnO-matrix composites. This result indicates that the same ZrO_2 inclusions give the different degree of hindrance depending on the chemistry of matrix.

Microstructures

Figure 4 shows the microstructures of pure A6 and ZnO and composites with the

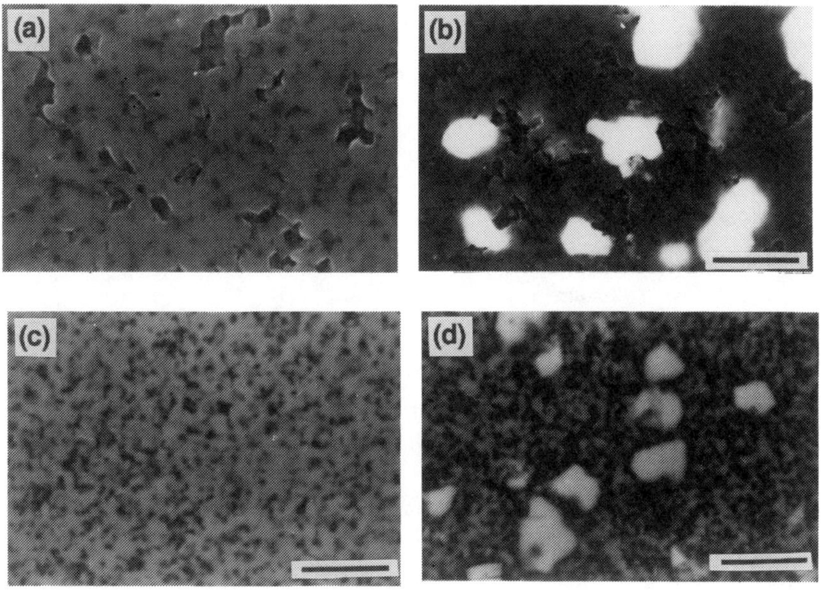

Fig. 4 Microstructure of (a) pure A6, (b) A6/20 vol% ZrO_2, (c) pure ZnO, and (d) ZnO/20 vol% ZrO_2 heated under the same conditions as those under which pure Al_2O_3 and ZnO reached the relative density of 90% (bar = 5 μm).

ZrO_2 content of 20 vol% heated under the same conditions as those under which the relative density of pure A6 and ZnO reached about 90%. The A6 powder has the average particle size almost equal to that of the ZnO powder. The backscattered electron image of SEM was used for the microstructural observation of the sintered compacts. The white large particles are ZrO_2 in the Al_2O_3- and ZnO-matrix composites. The microstructure of matrix in the ZnO-matrix composite was different from that of pure ZnO and more porous than that of matrix in the A6-matrix composite.

Model for the sintering behavior

Figure 5 illustrates a model of composite. It is assumed that the matrix and inclusion particles are spheres with an equal size. The cube enclosed by solid lines in Fig. 5(a) is a unit cell which composes the composite with the matrix density of 49% and the inclusion content of 12.5%. Stresses generate due to the differential shrinkage between matrix/matrix (m/m) and matrix/inclusion (m/i) pairs. The mechanism of stress generation is based on the principle analogous to the previous analyses.[2,3] But the assumptions that the matrix is homogeneous and continuous are not used in the present model. Tensile and compressive stresses generate at the points of contacts between matrix particles and between matrix and inclusion particles, respectively. These stresses transmit through the points of contact between the particles.

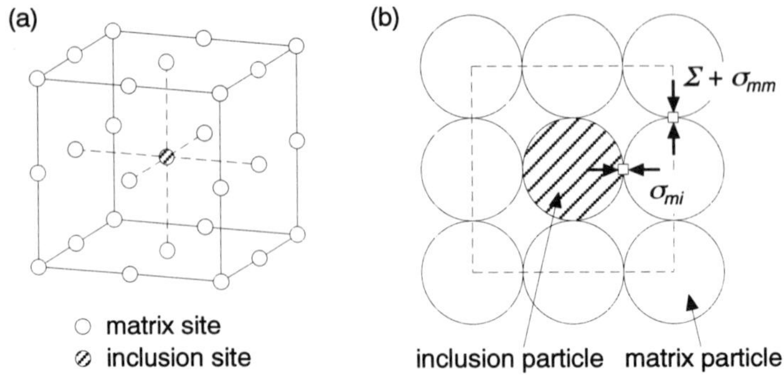

Fig. 5 Schematic model of (a) unit cell of composite and (b) stresses which generate in composite.

The previous models demonstrate that the stresses are relaxed by the diffusional creep for a viscoelastic or viscous matrix. In that case, it is reported that the stresses are too small to explain the reduction of densification rate of matrix.[3] In the present model, when the particle size of inclusion is much larger than that of matrix, it is expected that stress relaxation by rearrangement of matrix particles reduces the hindrance effect. On the other hand, the stress relaxation by rearrangement is impossible for the composite with the comparable particle sizes as shown in Fig. 5. Therefore, sintering stress which acts at the neck between matrix particles is offset by the tensile backstress, suppressing the shrinkage. This model is applicable to the densification process at peak [2] in the densification rate. The degree of hindrance depends on the particle size of inclusions; a decrease in particle size increases the degree of hindrance.[6] This model could explain the effect of inclusion size.

Peaks [2] and [3] partially overlapped for the Al_2O_3-matrix composites, whereas each peak was almost isolated for the ZnO-matrix composites. The densification of composites at peak [3] is related to the sintering of a *m*/*i* pair, as shown by the following experiment. Powder compacts of the mixture of Al_2O_3 with the particle size of 0.5 µm (Sumi-corundum AA-0.5, Sumitomo Chemical) and ZrO_2 with the particle size of 1 µm were prepared and heated at 1600°C. Figure 6 shows the micrographs of fracture surfaces of sintered compacts. The characteristics were formation of necks between Al_2O_3 and ZrO_2 and the sintering of ZrO_2.

Figure 7 illustrates the model of the sintering behavior of an Al_2O_3/ZrO_2 composite. The two-sphere model is applied for a *m*/*m* or *m*/*i* pair. In the pure matrix composed of *m*/*m* pairs, the center-to-center distance becomes closer due to the sintering stress Σ (process A). On the other hand, the *m*/*m* and *m*/*i* pairs coexist in the composite. The sintering stress which acts at the neck between *m*/*m* pair particles is partly offset by backstress σ_{mm} (process A'), as shown in Fig. 5(b). For the *m*/*i* pair, the mass transport to the neck from the matrix particle at lower temperature (process B) and that from the

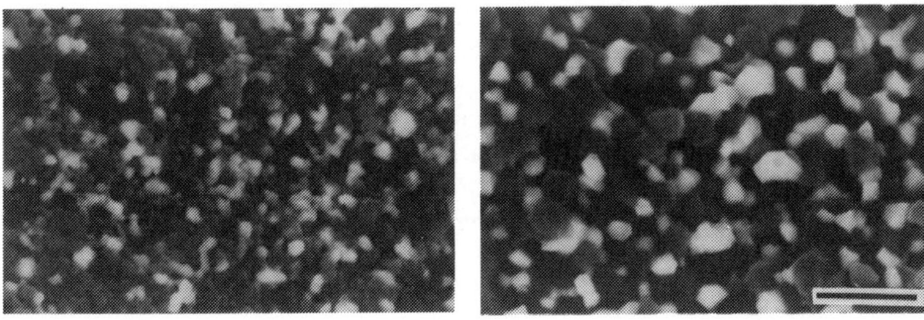

Fig. 6 Scanning electron micrographs of fracture surfaces of composites containing 20 vol% ZrO_2 heated at 1600°C for (a) 10 min and (b) 1000 min (bar = 5 μm).

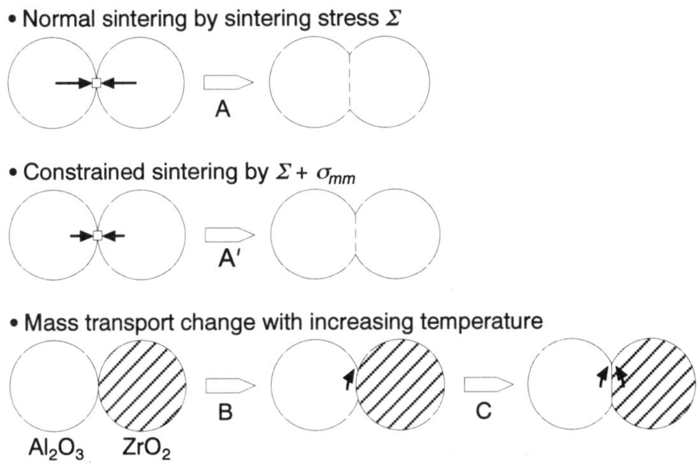

Fig. 7 Schematic model for the sintering behavior of Al_2O_3/ZrO_2 composite.

matrix and inclusion particles at higher temperature (process C) contribute the approach of particle centers which reduces σ_{mm}. Thus, the dominant densification process of composites may change from processes A' and B at peak [2] in Fig. 1 to processes A' and C at peak [3] with an increasing temperature.

For the Al_2O_3-matrix composites, the A2 matrix was subjected to the largest hindrance in densification (Fig. 3). Peak [2] appeared at the lower temperature range for the composites with the smaller matrix particle size, thus, increasing the difference in temperature ranges of peaks [2] and [3]. Because the degree of hindrance is defined with respect to the densification of pure matrix, the large difference in temperature ranges caused the large hindrance for the A2-matrix composites. The densification of the ZnO matrix suffered the hindrance effect larger than that of the Al_2O_3 matrix. For the ZnO-matrix composites, the difference in temperature ranges was still larger than that of the Al_2O_3-matrix composites. The microstructural observation supports that the dominant densification process of composites changes with an increasing temperature; the *m*/*i* pair particles sintered well in the A6-matrix composites heated to 1550°C (Fig. 4(b)), whereas they had poor contact in the ZnO-matrix composites heated to 900°C (Fig. 4(d)).

CONCLUSIONS

The Al_2O_3- and ZnO-matrix composites containing ZrO_2 inclusions sinter through the two kinds of densification processes. In the densification process at lower temperatures, the differential sinterability between the *m*/*m* and *m*/*i* pairs causes the hindrance in densification of the matrix. In the densification process at higher temperatures, the sintering of both *m*/*m* and *m*/*i* pairs lead to the further densification of the matrix which has been subjected to the hindrance effect. The degree of hindrance is dependent on the characteristics of the matrix powder. For the ZnO-matrix composites, the difference in temperature ranges of the densification processes is larger than that of the Al_2O_3-matrix composites. Therefore, the densification of the ZnO matrix is subjected to the hindrance effect larger than that of the Al_2O_3 matrix.

ACKNOWLEDGMENT

We would like to thank Sumitomo Chemical Co., Ltd. for the supply of Al_2O_3 powder (AKP-30, AKP-20 and Sumi-corundum series).

REFERENCES

[1]C.-L. Fan and M. N. Rahaman, "Factors Controlling the Sintering of Ceramic Particulate Composites: I, Conventional Processing," J. Am. Ceram. Soc., 75 [8] 2056-65 (1992).

[2]R. Raj and R. K. Bordia, "Sintering Behavior of Bi-Modal Powder Compacts," Acta Metall., 32 [7] 1003-19 (1984).

[3]G. W. Scherer, "Sintering with Rigid Inclusions," J. Am. Ceram. Soc., 70 [10] 719-25 (1987).

[4]F. F. Lange, "Constrained Network Model for Predicting Densification Behavior of Composite Powders," J. Mater. Res., 2[1] 59-65 (1987).

[5]O. Sudre, G. Bao, B. Fan, F. F. Lange and A. G. Evans, "Effect of Inclusions on Densification: II, Numerical Model," J. Am. Ceram. Soc., 75[3] 525-31 (1992).

[6]M. W. Weiser and L. C. De Jonghe, "Inclusion Size and Sintering of Composite Powders," J. Am. Ceram. Soc., 71 [3] C-125-C-127 (1988).

MICROSTRUCTURE AND IONIC CONDUCTIVITY OF CaO-MODIFIED CeO_2-Y_2O_3 SOLID SOLUTIONS PROCESSED BY A SOLUTION PRECIPITATION ROUTE

C. Moure, M. Villegas, J. Tartaj, P. Duran

Instituto de Ceramica y Vidrio, CSIC, Electroceramics Department, 28500, Arganda del Rey, Madrid, Spain.

ABSTRACT

The present work is focused on the preparation, by both solid state reaction and solution coprecipitation of ceria-yttria solid solutions, modified with small amounts of CaO to improve the grain boundary conductivity. This second method led to a strong decrease of the sintering temperature, >300°C. SEM of these sintered samples showed a homogeneous submicron microstructure. The ionic conductivity, measured by Complex Impedance Spectroscopy, showed a drastic decrease of the grain boundary resistivity, which attains values equivalent to that of grain bulk, i.e. an order of magnitude lower than those of undoped solid solutions of the same composition.

INTRODUCTION

Solid electrolytes based on fluorite-type CeO_2 solid solutions are excellent oxygen ion conductors, at intermediate temperatures (IT electrolytes)(1-4). However, the ceramic samples have a high-resistivity grain boundary which is an order of magnitude higher than the bulk conductivity and is thus detrimental for use as solid electrolyte for Solid Oxide Fuel Cells (SOFC's) devices (5,6). Modifying the grain boundary nature could make possible to reduce that high resistivity.

The main goal of this work is to establish the effect of small additions of CaO on the microstructure and electrical behavior of ceramic electrolytes belonging to the system CeO_2-Y_2O_3, which have good economical perspectives against the more conductive, but much more expensive CeO_2-Gd_2O_3-based electrolytes.

EXPERIMENTAL PROCEDURES

Solid electrolytes based on a solid solution of 80 mol% CeO_2-20 mol% $YO_{1.5}$, modified with 2 at% Ca^{2+} have been prepared by solid state reaction and by solution-precipitation procedures. Solid state reaction was accomplished by heating an intimate mixture of crystalline cerium and yttrium oxalates and calcium carbonate, up to 800°C for 1 h. The resulting powders were attrition milled, and XRD characterised with CuKα radiation. Mixed aqueous solutions of cerium and yttrium nitrates were added into a vigorously stirred aqueous calcium carbonate suspension. Ammonium hydroxide was then added to precipitate the corresponding hydroxides. The precipitate was thoroughly washed with a mixture of water and isopropyl alcohol, and slowly oven dried at 60°C. The dried powder was X-ray amorphous. This powder was heat-treated at 500°C for 1h. Homogeneous, nanosized, fully reacted powders were obtained. The particle size was measured by Coulter and by XRD peak width. Specific surface areas was established by BET method

Isopressed bars were prepared at 200 MPa. Pellets were cut out and sintering was carried out on them. The solid state reacted samples (SSR) were sintered between 1450 and 1600°C, whereas the chemically processed samples (CP) were sintered between 1300 and 1400°C. Apparent density was measured by water displacement. Microstructure was established by means of Scanning Electron Microscopy (DMS 950 Model of Carl Zeiss), on polished and thermally etched sample surfaces.

Discs were sliced from sintered pellets and electroded with Ag paste. The silver electrodes have proved to be a good alternative to the more expensive, catalytic Pt electrodes for use at the intermediate temperature ranges(4). The painted electrodes were sintered at 800°C. Complex Impedance Spectroscopy was performed by using an Impedance Meter model HP 4192A, from 10 to 10^7 Hz. Nyquist diagrams were obtained for temperatures between 200 and 700°C.

RESULTS

Complete solid solution formation of the fluorite type phase was attained at 800°C, 1h, for the solid state reaction procedure, and at 500°C, 1h, from the coprecipitated, amorphous product. The particle sizes of milled powders were 400 nm for the SSR powder, and 20 nm for the CP, respectively.

Figure 1 shows the densification curves as a function of A) temperature for the SSR powder, and B) the time for the CP powder. Owing to the scarce differences appreciated in the apparent density of samples sintered at different temperatures in the range 1250-1350°C, we have chosen the last curve. Relative density values are shown, taking D_{th} = 6.7 g/cm^3 as calculated from XRD data. The different reactivity degrees are reflected by the strong difference in the temperatures for the maximum density, 1600 and 1300°C, respectively. Besides that, the values corresponding to the maximum density are lower for the SSR powders, as it also seen from the curves. Samples with apparent density >99.5 D_{th} have been obtained by sintering the chemically synthesized powders at temperatures as low as 1300°C. The samples were translucent for thickness lower than 500 μm. These results are similar to those reported by van Herle et al(7), who used an oxalate precipitation way and a temperature of 700°C for fluorite synthesis. The present method is somewhat easier and leads to the solid solution at temperatures as low as 500°C. Higher density (>99%D_{th}) is attained because of the higher powder reactivity caused by its smaller particle size.

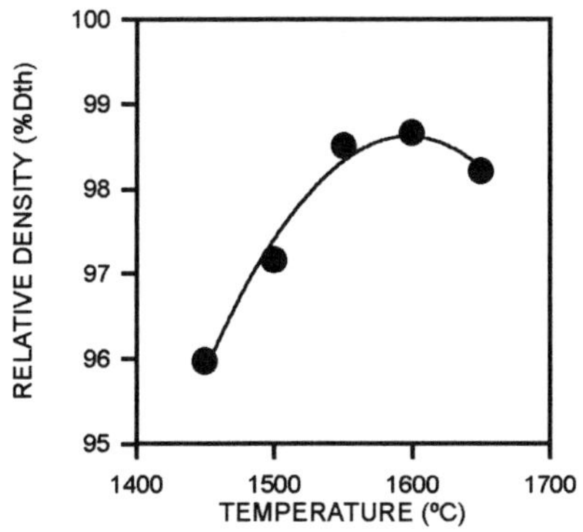

Figure 1 A), relative density vs temperature for SSR samples sintered at 4 h constant time

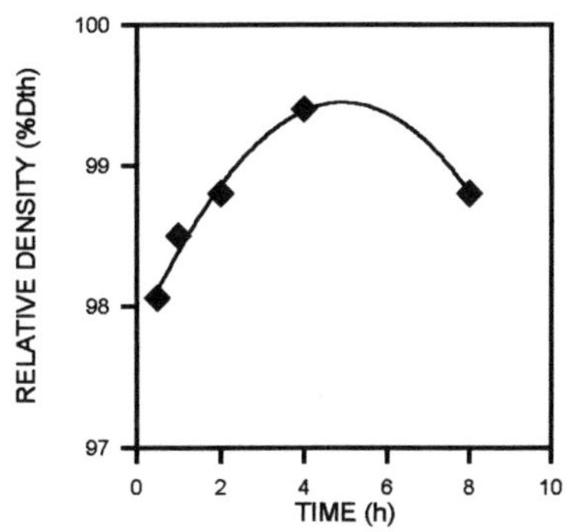

Figure 1 B), relative density vs time for CD samples sintered at 1300°C

Figure 2 shows the SEM micrographs of the polished surfaces of sintered: A) CP samples and B) SSR samples. As a first statement it can be seen that the CP sample has a smaller grain size and a higher homogeneity than that corresponding to the SSR sample. As a second feature it is possible to see that the CaO is located preferentially on the grain boundaries and on some triple points as small, almost transparent crystallites. Nevertheless, it is not possible to establish if this microstructure has appeared as a consequence of thermal etching, necessary for revealing the grain boundaries, or whether is actually as sintered microstructure. A set of polished samples was chemically etched, but the results were inconclusive. Current studies using HRTEM will allow establish with a higher accuracy what is the exact location of the Ca^{2+} cations.

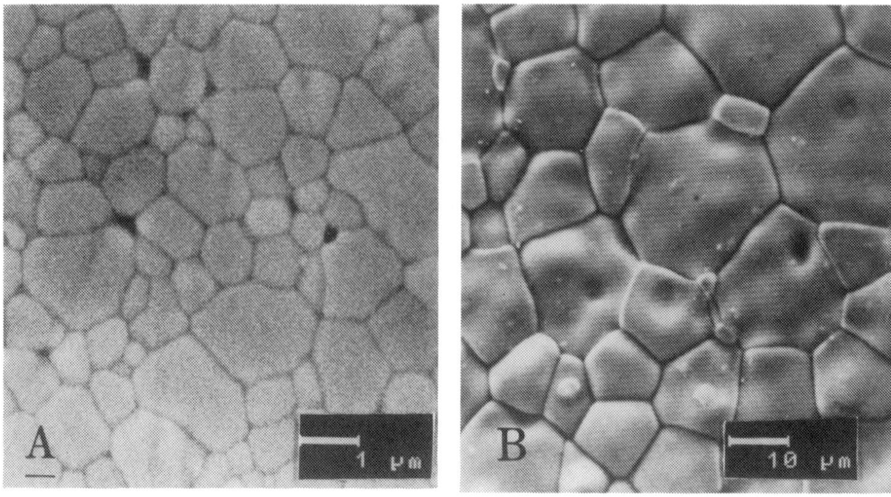

Figure 2. SEM micrographs corresponding to A) CP sample sintered at 1300°C, and B) SSR sample sintered at 1600°C

Figure 3 shows the impedance curves of ceria-yttria compositions with and without CaO. The most striking effect is the strong decrease of the Grain

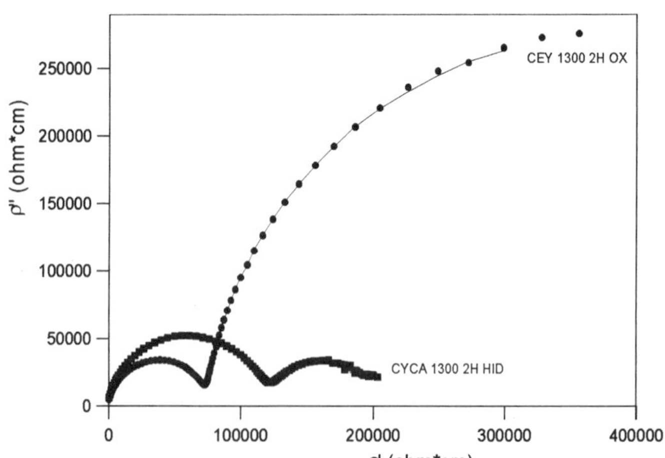

Figure 3. Impedance curves corresponding to a Yttria-Ceria solid solution with and without CaO doping

Boundary (GB) resistivity when CaO is present. It can be seen that the grain boundary resistivity which is represented by the second arc taken from the co-ordinate origin decreases an order of magnitude and attains a value which is comparable to that of the bulk resistivity. On the other hand, a decrease in the bulk conductivity is also visible. The reason is not clear, taking into account that the CaO incorporation induces a higher oxygen vacancy concentration.

The explanation could be the location of the CaO on the grain boundaries, such as has been observed in the SEM analysis (Figure 2) This displacement supposes a slight increase of the ratio Ce/aliovalent cation in the bulk grain. Nevertheless, is necessary to make further studies to establish the model explaining that decrease.

Figure 4 depicts the impedance curves corresponding to two samples of SSR powders, sintered at 1550 and 1600°C against one CP sample sintered at

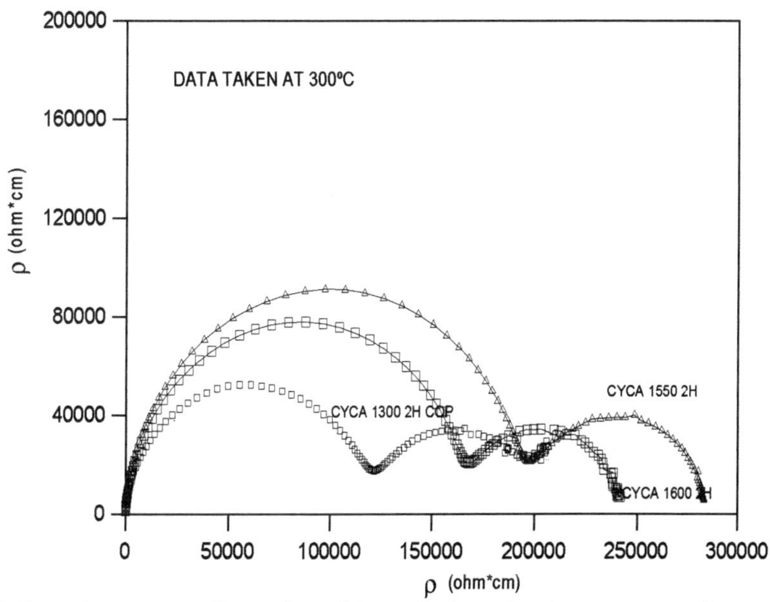

Figure 4. Impedance arcs of samples with equivalent density corresponding to different starting powders

1300°C. It can be seen that the bulk conductivity is affected by the preparation method. Bulk conductivity is higher for the sample with higher density and more homogeneous microstructure, i.e. for the CP sample. On the other hand, the GB conductivity is similar for all the samples. The higher the sintering temperature, the higher the bulk conductivity, in close agreement with the sintering temperature increasing. As compared with the conductivity of the more conducting Gd-based Ceria electrolytes, we have observed that at 300°C there is a strong difference between the conductivity of both systems. However, because of the higher activation energy of the yttria ceramics, the conductivity at 700°C is more similar for both samples (8). It supposes an advantage for yttria-based solid electrolytes when one compares them from an economic point of view.

Figure 5 represents the evolution of the complex impedance diagrams corresponding to SSR samples sintered at several temperatures and constant 4h time. It is possible to appreciate that both bulk and grain boundary conductivity mechanisms increase with temperature, as a result of the higher sample density

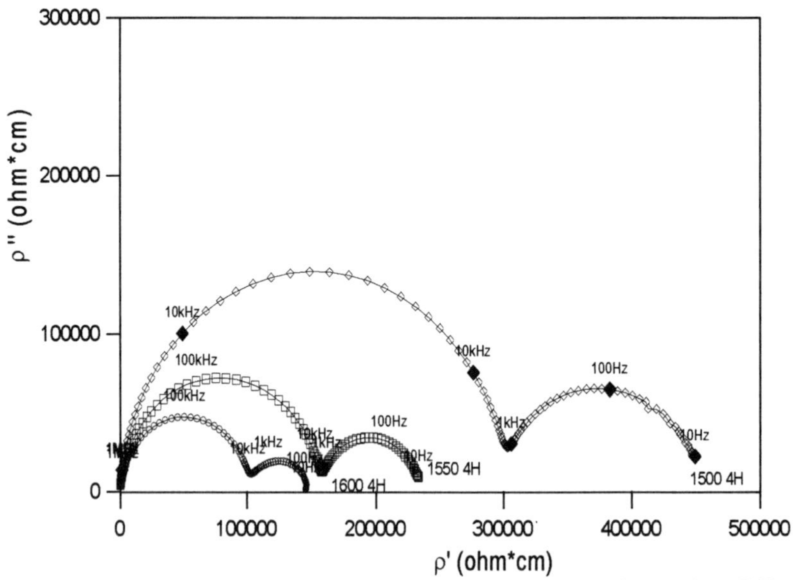

Figure 5. Impedance curves corresponding to SSR samples sintered at different temperatures

Figure 6 shows the Arrhenius plots corresponding to bulk and GB conductivity of two CP samples sintered at the same temperature and different times. It can be seen that the GB conductivity is in all cases higher than the Bulk conductivity in the intermediate and high temperature range. It can be seen in the figure that the bulk conductivity remains almost constant, i.e. is independent of the grain size, whereas the GB conductivity increases when the grain boundary density decreases, as corresponding to the nature of both class of conductivity.

The activation energy values calculated from those fits ranging about the following ones:

Ea(B) =1.05 eV
Ea(GB) =1.11 eV

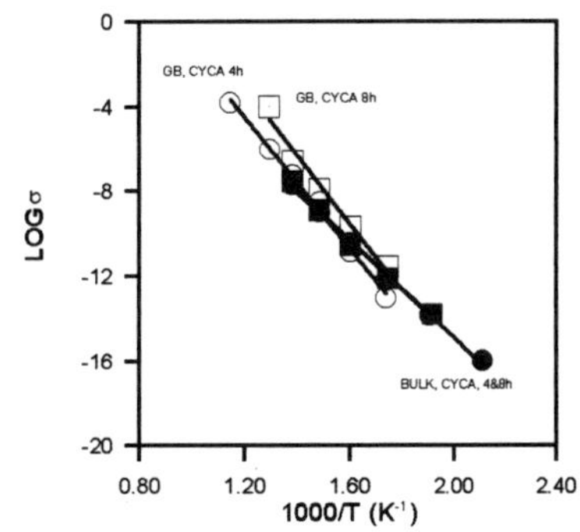

Figure 6. Arrhenius plots of the bulk and GB conductivity corresponding to two CP samples sintered at different times

The effect of the presence of the Ca cations on the GB conductivity is not yet established. It is well known that the alumina inclusions in ceramics based on Y-FSZ and Y-TPZ causes a similar enhancement on the GB conductivity, attributed to a scavenge effect which cleans the grain boundary of the existing silica. This impurity seems to be the cause of the high resistivity measured on the grain boundary region both in the zirconia and ceria based ceramics.

In the present case, a different mechanism could be responsible for the conductivity enhancement. The CaO seems to be located near the grain boundary. In this region the presence of the divalent cations could cause an increment in the oxygen vacancy concentration, and therefore an improvement of the transport mechanism through the grain boundaries. On the other hand is also possible the formation of some new compound, such as a calcium silicate which migrate out of the grain boundary region, and could provoke a cleaning effect

CONCLUSIONS

1° Processing of ceramic materials belonging to the CeO_2-Y_2O_3 system by coprecipitation of hydroxides reduces the sintering temperatures by as much as 300°C, and yields a more uniform microstructure.

2° Doping with CaO, which seems to locate in the GB, reduces the GB resistivity, by an order of magnitude relative to undoped ceramic materials.

3° For the intermediate temperatures, (600-800°C), the conductivity of the ceramics belonging to the CeO_2-Y_2O_3 system is comparable to that of the CeO_2-Gd_2O_3 .

ADKNOWLEGMENTS

This work is supported by the Spanish CICYT, Project MAT 94-871

REFERENCES

1.- T. Kudo and H. Obayashi, "Oxygen ion conduction of the fluorite-type $Ce_{1-x}Ln_xO_{1-x/2}$ (Ln= Lanthanide)", J. Electrochemical Soc. , 122, 142-147, 1975.

2.- C. Pascual, J.R. Jurado, G.F. Arroyo, L. Del Olmo, C. Moure, P. Duran, "Electrical conductivity of solid solutions in the system CeO_2-CaO and CeO_2-Ln_2O^3 (Ln=La, Nd, Sm, Gd, Er, Y)", Sci. of Ceramics, 12, 729-734, 1982.

3.- K. Eguchi, T. Setoguchi, T. Inoue, and H. Arai, "Electrical properties of ceria-based oxides and their application to solid oxide fuel cells", Solid State Ionics, 52, 165-172, 1992.

4.- C. Moure, J.R. Jurado, P. Duran, "Ionic conductivity and ionic domain of CeO_2-Gd_2O_3 based ceramic materials prepared from nanoscaled particles obtained by coprecipitation procedures", Electroceramics V, Vol 2, 139-142, 1996

5.- R. Gerhardt and A.S. Nowick, "Grain-Boundary effect in Ceria doped with trivalent cations: I Electrical measurement", J. Amer. Ceram. Soc, 69, 641-646, 1986.

6.- R. Gerhardt , A.S. Nowick and M.E. Mochel, "Grain-Boundary effect in Ceria doped with trivalent cations: II Microstructure and Microanalysis" , J. Amer. Ceram. Soc, 69, 641-, 1986.

7.- J. Van Herle, T. Horita, T. Kawada, N. Tasaki, H. Yokokawa and M dikiya, "Fabrication and sintering of fine Yttria-doped Ceria powder", J. Amer. Ceram. Soc., 80, 933-940, 1997

8.- J.R. Jurado, B. Ghargabe, M.T. Colomer, C. Moure, P. Duran, "Electrochemical properties of ceramics based on CeO_2-Gd_2O_3 system", Electroceramics V, Vol. 2, 215-218, 1996

High Temperature Hardness and Toughness of MgO Partially Stabilized Zirconia and Y_2O_3 Tetragonal Zirconia Polycrystal

Seiichi MUNEKI, Fujio ABE
Frontier Research Center for Structural Materials in NRIM
1- 2- 1 Sengen, Tsukuba, Ibaraki, 305, Japan
and
Hirosada IRIE
Mechanical Mechanism Division in NRIM

ABSTRACT

In order to be clear the relationship between hardness and fracture toughness of zirconia compositions and metallic materials with high specific strength, hardness and fracture toughness of them at room and high temperatures have been investigated. To evaluate fracture toughness of the structural materials it is very effective to investigate the relationship at room temperature. Hardness of zirconia compositions decreased with increasing temperature, but fracture toughness did not increase due to an occurrence of long cracking at the end of indentation. It is an important to investigate fracture toughness at high temperatures at the time to design structural materials for applications.

1. INTRODUCTION

Structural materials with high strength- to- density ratio have been used in aerospace industry, ocean and atomic energy development fields. For these applications, several kinds of materials have been developed in steels, non- ferrous alloys, and ceramics. At that time it is an important to compare strength and fracture toughness of them.

This study was discussed on fracture toughness of ultra high strength maraging steels, beta titanium alloys and partially stabilized zirconia compositions. In that case it is very difficult to estimate the normal strength of zirconia compositions due to an occurrence of catastrophic failure during tensile tests. Therefore hardness was used instead of their strength in this study.

2. EXPERIMENTAL

2.1 Materials

Seven beta titanium alloys used in this study were double melted by a plasma beam furnace for approximately 5.5kg in each alloy[1]. They are Ti- 10V- 2Fe- 3Al, Ti- 10V- 2Fe- 5Zr- 3Al, Ti- 8Mo- 2Fe- 3Al, Ti- 15V- 3Cr- 3Sn- 3Al, Ti- 3Al- 8V- 6Cr- 4Mo- 4Zr, Ti- 10V- 5Mo- 2Fe- 3Al and Ti- 8V- 8Mo- 2Fe- 3Al. They were homogenized at 1273K for 18ks in vacuum, followed by cooling to room temperature. After heating at 1273K for 3.6ks in air, these ingots were pressed and rolled into 30mm square bars. After reheating at 1273K for 1.8ks, these bars were again hot rolled into 11mm square bars with 86% reduction between 1173K and 1073K.

Twelve maraging steels were melted by a vacuum induction furnace for 7kg[2][3]. They are Fe- 16~17Ni- 15Co- 4~6Mo- 1~3.5Ti- 0~1.5Al and Fe- 10Ni- 18Co- 12~13Mo- 0.2Ti. They were homogenized at 1473K for 86.4ks in hydrogen. After heating at 1523K for 3.6ks in air, these ingots were pressed and rolled into 23mm square bars. After reheating at 1473K for 3.6ks, these bars were thermo- mechanical heat treated between 1473K and 1073K.

Two zirconium alloys with the composition of 9.7mol% MgO PSZ and a 2mol% Y_2O_3 TZP powders by TOSO cooperation in Japan were used in this study[4]. As the grain size of metallic materials was affected on their fracture toughness, samples fabricated were processed in both a melting and a sintered methods in order to control their grain sizes.

2.2 Measurements of fracture toughness

Measurement of plane strain fracture toughness (K_{Ic}) of beta titanium alloys and ultra high strength maraging steels was carried out using V- notched Charpy type specimens with a 0.1mm notch- root radius and fatigue pre- cracking under as solution treated condition in accordance with ASTM E399[5].
The K_{Ic} test was carried out after a given aging treatment. While the value of K_{Ic} in zirconia compositions was estimated by the indentation method and through the Niihara's equation[6].

2.3 High temperature hardness testing

High temperature hardness test was carried out by vickers hardness testing machine attached with chamber which was possible to heat and cool between room temperature and 1373K by program controller. Hardness was measured and the indentation was in-situ observed every 50K between 473K and 1173K during heating and cooling treatments. Cracking at the end of the indentation was in-situ recorded by video printer. After cooling the indentations were observed again by optical microscope. The value of K_{Ic} at high temperature in zirconia compositions was evaluated from the crack length at the end of indentation.

3. RESULTS and DISCUSSION

Figure 1 shows the relationship between tensile strength and fracture toughness of beta titanium alloys. K_{Ic} values are reduced from approximately 100 to 20MPa$\sqrt{}$m with increasing tensile strength from 1000 to 1700MPa.

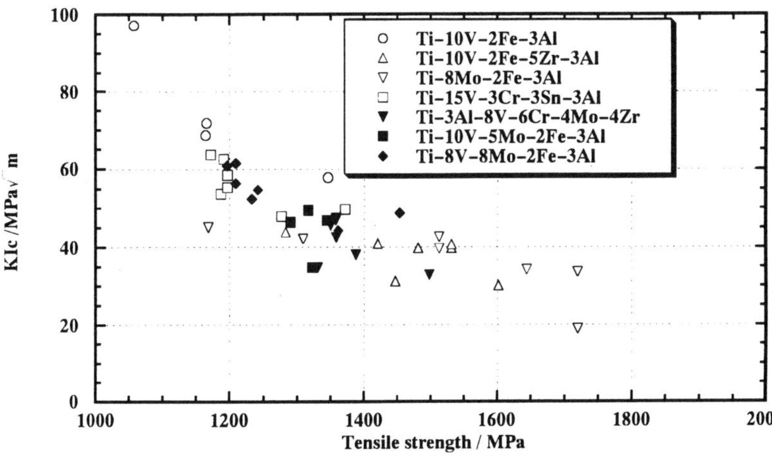

Fig.1 Relationship between tensile strength and fracture toughness of beta titanium alloys.

Figure 2 shows the similar relationship of ultra high strength maraging steels. K_{Ic} values are monotonously reduced from 40 to 15MPa$\sqrt{}$m with increasing tensile strength from 2500 to 3200MPa.

Figure 3 shows the relationship between hardness and fracture toughness of 9.7mol% MgO partially stabilized zirconia conposition. Grain size of samples fabricated by a sintered process indicated approximately 50μm. On the other hand, the size of arc melted samples indicated about 300μm. K_{Ic} of zirconia compositions monotonously decreased from 7 to 3 MPa$\sqrt{}$m with increasing hardness from 9 to 14GPa independent of their grain sizes.

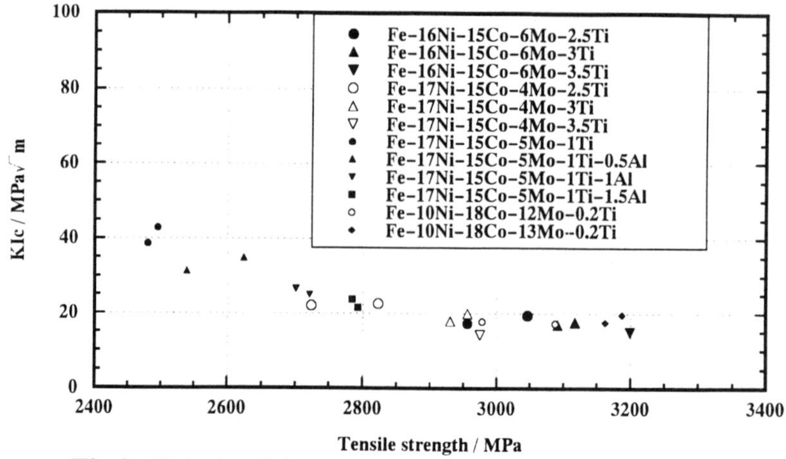

Fig.2 Relationship between tensile strength and fracture toughness of maraging steels.

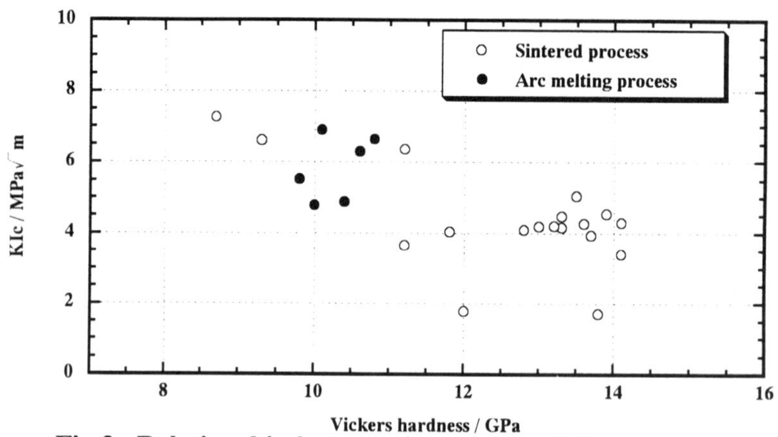

Fig.3 Relationship between hardness and fracture toughness of 9.7mol% MgO PSZ composition.

Figure 4 shows hardness and fracture toughness of three kinds of materials. As clearly shown in this figure, 9.7mol% MgO PSZ composition showed a reasonable characteristic. Namely, plots of MgO PSZ composition located at the area of higher hardness level on the extended line of metallic materials.

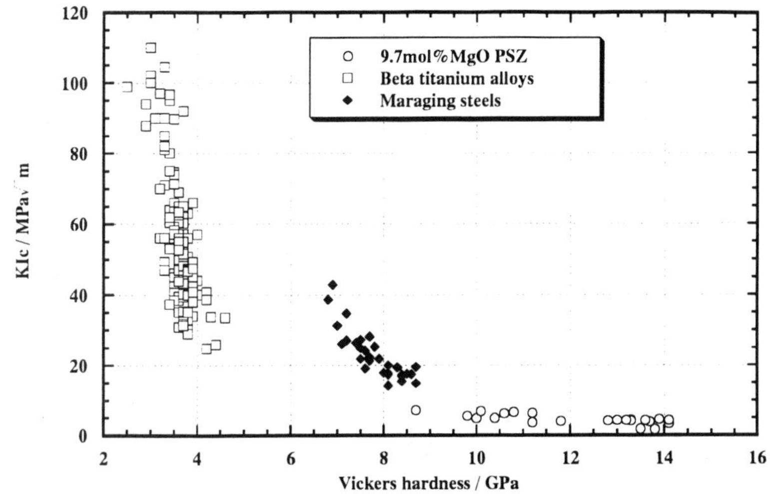

Fig.4 Hardness and fracture toughness of three kinds of materials at room temperature.

At the same time beta titanium alloys indicated less balance than that of ultra high strength maraging steels as compared to open and solid square symbols into the figure.

Next we discussed about high temperature characteristics. Figure 5 shows changes of hardness at high temperatures of 9.7mol%MgO PSZ composition.

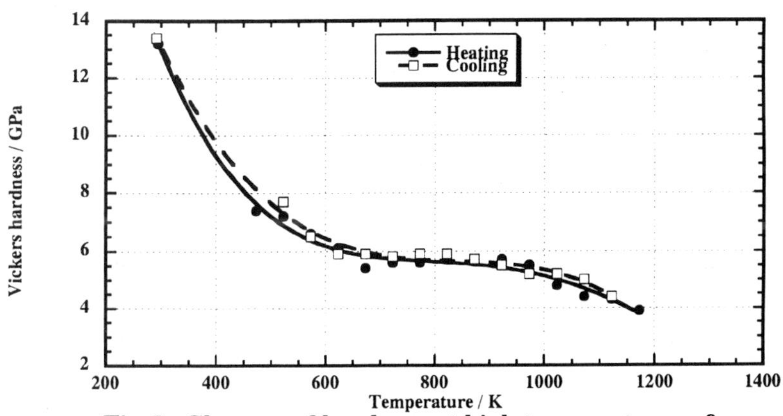

Fig.5 Changes of hardness at high temperatures of 9.7mol%MgO PSZ composition.

Hardness at room temperature drastically decreased to 8GPa at 473K with increasing the temperature. After that hardness gradually decreased from 473K to 673K, then it indicated the region of unchanged values until 923K, and then it gradually decreased to 4GPa at 1173K with increasing the temperature. When the specimen was cooled from 1173K to room temperature , plots of hardness increased in reverse on the same trace on heating.

Figure 6 shows an optical micrograph of indentation cracks on heating in 9.7mol% MgO PSZ composition. Although hardness indicated the lowest value at 1173K, long indentation cracks were observed.

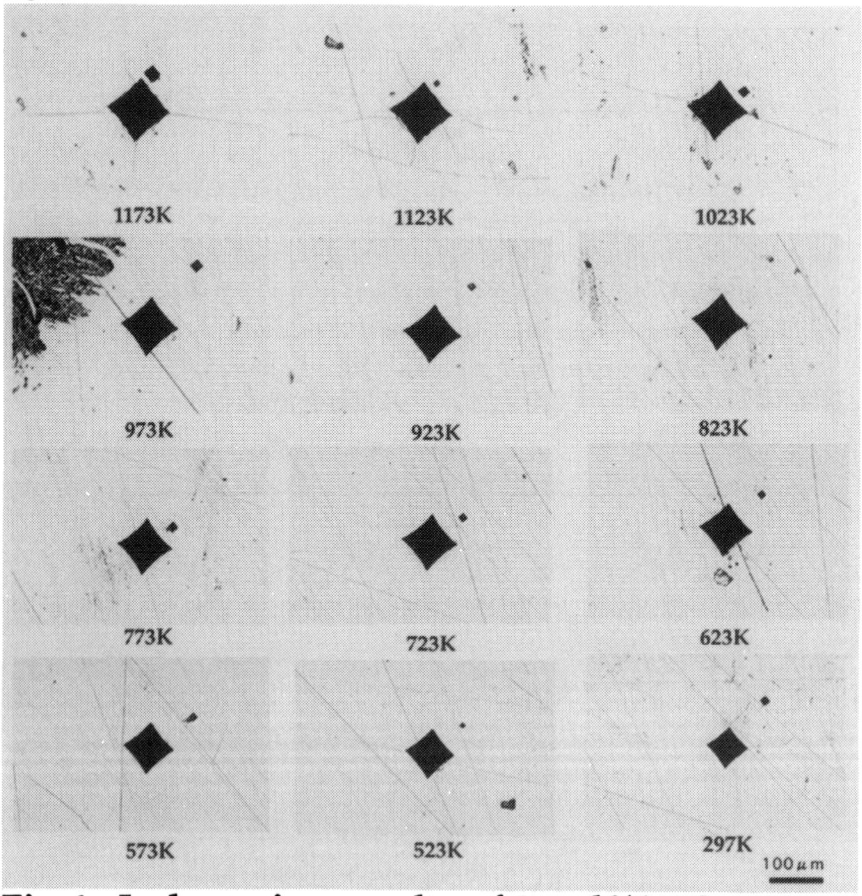

Fig.6 Indentation cracks of 2mol%Y2O3 TZP composition on cooling.

Figure 7 shows the relationship between hardness and fracture toughness in the test temperatures between 523K and 1173K of 9.7mol% MgO PSZ composition. Hardness decreased from 10 to 4GPa, but fracture toughness slightly decreased, too. It is considered to be accelerated the sensitivity of cracking at high temperatures. From these results, it is very attention to select and design of materials for structural applications at high temperatures as shown in figure 8.

4. CONCLUSION

In order to be clear the relationship between hardness and fracture toughness of zirconia compositions and metallic materials with high specific strength, hardness and fracture toughness of them at room and high temperatures have been investigated. To evaluate fracture toughness of the structural materials it is very effective to investigate the relationship at room temperature. Hardness of zirconia compositions decreased with increasing temperature, but fracture toughness did not increase due to an occurrence of long cracking at the end of indentation. It is an important to investigate fracture toughness at high temperatures at the time to design structural materials for applications.

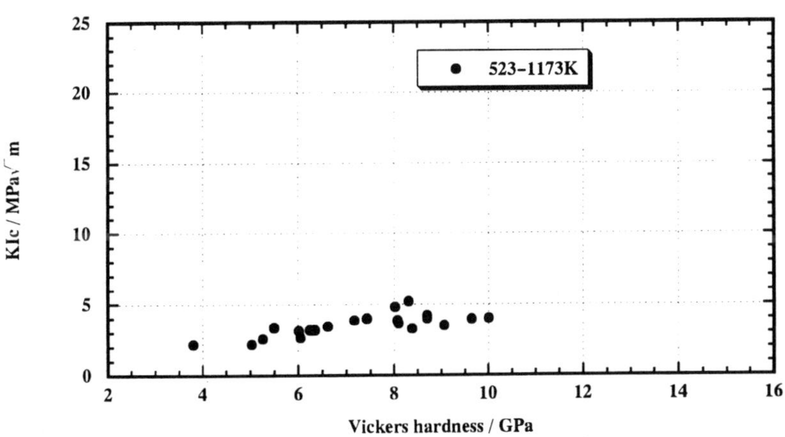

Fig.7 Relationship between fracture toughness and hardness at high temperatures of 9.7mol% MgO PSZ alloy.

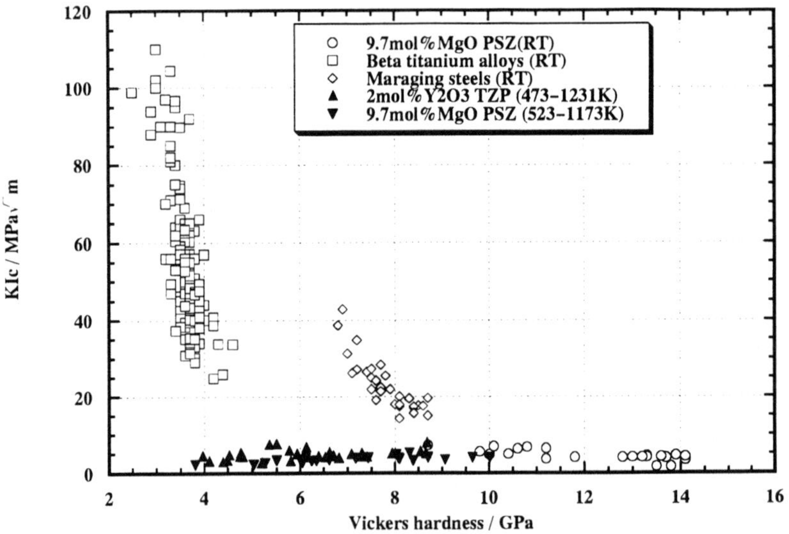

Fig.8 Hardness and fracture toughness balance at room temperature (RT) and high temperatures of structural materials.

REFERENCES

1) S.Muneki, Y.Kawabe, T.Kainuma and J.Takahashi : Proc.2nd Japan International SAMPE Symposium(1991), pp.418- 424
2) S.Muneki, Y.Kawabe, K.Nakazawa and H.Yaji : Trans.ISIJ, 20(1980), pp.309- 317
3) S.Muneki and Y.Kawabe : Tetsu- to- Hagane, 66(1980), pp.1177- 1186
4) S.Muneki, F.Abe and H.Irie : Proc.13h Korea- Japan Seminar on New Ceramics, (1996), pp.216- 220
5) ASTM Annual Book of ASTM Standards, 03- 01(1983), p.518
6) K.Niihara : J.Mater.Sci., 221(1983)

SINTERING MODEL FOR MIXED OXIDE DERIVED LEAD ZIRCONATE TITANATE (PZT) CERAMICS

Marianne Hammer and Michael J. Hoffmann
University of Karlsruhe
Institute of Ceramics in Mechanical Engineering
Haid-und-Neu-Str. 7
D-76131 Karlsruhe, Germany

ABSTRACT

The properties of PZT ceramics are determined by the microstructure and chemical homogeneity of Zr, Ti and dopants within the grains as well as the presence of secondary grain boundary phases. Stoichiometric and 3 mole-% PbO excess containing PZT compositions (53/47) were prepared by the mixed oxide process and densified by pressureless sintering in oxygen atmosphere. The influence of the PbO content and different La-concentrations on the densification behaviour was analyzed by dilatometric measurements. Quantitative image analysis showed a different relative density and grain size dependence for samples containing >0.5 mole-% additives compared to the samples with <0.5 mole-% La. On the basis of a model experiment and by using different analytical methods (microprobe analysis, HRTEM, STEM and Auger spectroscopy) three types of inhomogeneities could be detected in conventional prepared PZT ceramics: the existence of a Ti- and La-enrichment in the core of PZT grains, the presence of PbO-rich secondary phases in triple junctions as well as in grain boundary films. The results of the microstructural characterization and the analysis of the densification behaviour were finally combined to deduce a sintering model based on a Pb-vacancy concentration gradient within the PZT grains.

KEYWORD AND AUTHOR INDEX